The Biology of Glycoproteins

The Biology of Glycoproteins

Edited by

Raymond J. Ivatt

M. D. Anderson Hospital and Tumor Institute
University of Texas System Cancer Center
Houston, Texas

Plenum Press • New York and London

Library of Congress Cataloging in Publication Data

Main entry under title:

The Biology of glycoproteins.

Includes bibliographical references.
1. Glycoproteins—Physiological effect. 2. Cell interaction. I. Ivatt, Raymond J.
QP552.G59B57   1984   574.19′245   84-8321
ISBN 0-306-41596-8

Contributors

Clayton A. Buck • The Wistar Institute of Anatomy and Biology, Philadelphia, Pennsylvania

Caroline H. Damsky • The Wistar Institute of Anatomy and Biology, Philadelphia, Pennsylvania

Michiko N. Fukuda • Cancer Research Center, La Jolla Cancer Research Foundation, La Jolla, California

Minoru Fukuda • Cancer Research Center, La Jolla Cancer Research Foundation, La Jolla, California

Ellen J. Henderson • Department of Biology, Georgetown University, Washington, D.C.

A. Tyl Hewitt • Wynn Center for the Study of Retinal Degenerations, The Wilmer Ophthalmological Institute, The Johns Hopkins Hospital, Baltimore, Maryland

Raymond J. Ivatt • Department of Tumor Biology, University of Texas, M. D. Anderson Hospital and Tumor Institute at Houston, Houston, Texas

Paul T. Kelly • Division of Biology, Kansas State University, Manhattan, Kansas

Karen A. Knudsen • The Wistar Institute of Anatomy and Biology, Philadelphia, Pennsylvania

George R. Martin • Laboratory of Developmental Biology and Anomalies, National Institute of Dental Research, National Institutes of Health, Bethesda, Maryland

Christopher L. Reading • The Department of Tumor Biology and the Bone Marrow Transplantation Center, The University of Texas, M. D. Anderson Hospital and Tumor Institute at Houston, Houston, Texas

Preface

This book is in many ways a sequel to *The Biochemistry of Glycoproteins and Proteoglycans*. The enormous recent progress in understanding the biological roles of glycoproteins has prompted the present volume. The reasons for studying glycoproteins have multiplied, and in the present volume the roles played by glycoproteins are explored in a variety of biological situations. The first two chapters describe molecules involved in cell–substratum and cell–cell interactions in a broad sense, and also focus on recent progress in identifying specific attachment molecules. Our understanding of how normal processes, such as cellular differentiation and tissue organization, are regulated is dependent on understanding how cells interact with the extracellular matrix. When these processes go awry the consequences can be tragic, for example, when manifest as birth defects and cancer. Our ability to devise appropriate therapies is in many cases limited by our understanding of such cell–matrix interactions. The third chapter explores the roles by glycoproteins during early mammalian development. The carbohydrate portions clearly play very important roles in presenting information during early embryogenesis, and an unusual tumor stem cell, the embryonal carcinoma, looks very promising in providing an experimental system for understanding how the expression of these complex carbohydrate determinants is regulated. The next three chapters explore the biology of glycoproteins in distinct situations: in the immune system, in the nervous system, and during erythropoiesis. Each chapter presents a wealth of information regarding the programmed expression of glycoproteins and emphasizes their functional involvement in various biological processes. The last chapter describes the life cycle of an unusual organism—the cellular slime mold, which can exist both as a unicellular vegetative ameba and as a multicellular sporulating organism—and summarizes the roles played by cell surface glycoproteins in regulating the complex cellular interactions that occur during its life cycle.

I have enjoyed interacting with these authors and have learned an enormous amount from each of them. I thank them for the conscientious efforts they have made to make their chapters concise, comprehensive, and yet still very readable.

I would like to take this opportunity to acknowledge the debt I owe to the following scientists from whom I have learned so much: Michael Rosemeyer, Charles Gilvarg, Phillips Robbins, Richard Hynes, and Garth Nicolson.

Raymond J. Ivatt

Contents

Chapter 3

Role of Glycoproteins during Early Mammalian Embryogenesis

Raymond J. Ivatt

Chapter 4

Cell Surface Glycoproteins and Carbohydrate Antigens in Development and Differentiation of Human Erythroid Cells

Minoru Fukuda and Michiko N. Fukuda

Chapter 5

Carbohydrate Structure, Biological Recognition, and Immune Function

Christopher L. Reading

Chapter 6

Nervous System Glycoproteins: Molecular Properties and Possible Functions

Paul T. Kelly

Chapter 7

The Role of Glycoproteins in the Life Cycle of the Cellular Slime Mold Dictyostelium discoideum

Ellen J. Henderson

1

Integral Membrane Glycoproteins in Cell–Cell and Cell–Substratum Adhesion

Caroline H. Damsky, Karen A. Knudsen, and Clayton A. Buck

1. INTRODUCTION

The focus of this chapter is on the integral membrane glycoproteins of the animal cell involved in cell–cell and cell–substratum adhesion. The approach to the subject, both intellectually and experimentally, reflects not only the prejudices of the authors but also the ideas and concepts gleaned from the literature and from conversations with colleagues. For the free interchange of ideas, we are grateful; for misinterpretations or oversights, we apologize. We have focused attention on that body of experimental data that has used both biochemical and biological approaches in an attempt to understand the adhesion process. Immunology, with its potential for specificity, has played a particularly crucial role in the discovery of adhesion-related glycoproteins and, with the increased use of monoclonal antibody technology, will play an increasingly important role in exploring this area. It is a tool that must be used with caution, however, as will be noted throughout the chapter.

Caroline H. Damsky, Karen A. Knudsen, and Clayton A. Buck • The Wistar Institute of Anatomy and Biology, Philadelphia, Pennsylvania 19104.

1

From our point of view, surface membrane glycoproteins may be viewed as decoders of extracellular information. As such, when a cell interacts with a particular substratum (fibronectin, laminin, the collagens, proteoglycans, etc.) or with ligands on the surface of another cell, the information as to whether to form an adhesive complex or not must rest in the molecular architecture of these glycoproteins and be transmitted to elements of the cytoskeleton and to other membrane proteins. If the correct combination of molecular signals is present, the cell will begin to organize an adhesive complex. Ultrastructural studies and a careful examination of the adhesion process in culture show that there are several types of adhesive interactions, and that adhesion is a stepwise process. According to this idea, the cell initially forms loose attachments to its substratum or to a neighboring cell. This attachment triggers a series of molecular events in which more glycoproteins are recruited into the attachment areas, resulting in a stabilization of adhesion and perhaps eventually in the formation of adhesive plaques or well-defined junctional complexes. This is a complicated series of events that could be disrupted (and perhaps controlled) at any point. Thus, it will be difficult both to prove whether a particular glycoprotein is involved directly or indirectly with the adhesion process, and to establish precisely what role the glycoprotein plays in the series of events leading to cellular adhesion. Such proof will require carefully correlated biochemical, immunological, biological, and ultrastructural studies. While no molecule has as yet been subjected to the rigors of this combined approach, there are several candidates that must be considered as major participants in the adhesion process. We will begin by examining recent developments in the study of cell–cell adhesion and then move on to the subject of cell–substratum adhesion.

2. CELL–CELL ADHESION

The process whereby cells recognize and adhere to each other is obviously one of the most crucial and complicated events in biology. When this process goes well as in normal embryonic development, the result is absolutely fascinating—the organization of cells into fully functional tissues and organs and the establishment of the intricate circuitry of the brain and peripheral nervous system. But when this process fails, the results are disastrous: terribly malformed individuals on one extreme or rampant metastasis on the other. The information for establishing and maintaining the correct contacts must lie in the expression and organization of molecules on the cell surface. Therefore, an understanding of the molecular basis of this drama requires an identification of the participating characters and a knowledge of how they move, respond, and interact with each other. As will be seen below, the cast of molecular characters thought to be involved is large, diverse, and will undoubtedly continue to grow. In some cases, our biochemical casting will be imperfect and certain characters will be lost by the wayside due more than likely to the uncritical eye or excessive enthusiasm of their biochemist casting agents, whereas other molecules will be placed on the scene as new approaches and new systems are developed for further investigation. The purpose of this section is to enumerate the various glycoproteins implicated in the adhesion process and to see if any meaningful generalizations as to their various roles can be made. The focus will be on those glycoproteins that have been biochemically identified on the basis of some biologically functional parameters either as related to *in vitro* cellular behavior or to some *in vivo* tissue organizational phenomenon. We will deal with those molecules more likely involved in the early and perhaps more transitional events of intercellular adhesion rather than those involved in the later organization of more permanent junctional complexes such as desmosomes, gap junctions, and tight junctions.

Two general approaches have been taken in identifying the molecules involved in cell–cell adhesion. One has been to establish *in vitro* systems of cell–cell adhesion involving dispersed embryonic tissue or even cultured cell lines, and to identify factors that either promote or inhibit the particular cell–cell interaction being studied (see reviews by Moscona, 1974; Lilien *et al.*, 1979; Roth, 1982; Frazier and Glaser, 1979; Steinberg, 1978). A second approach, patterned after that used by Gerisch's group (Gerisch, 1977) to identify surface molecules involved in aggregation of individual slime mold amebae (see Henderson, this volume), consists of preparing a broad-spectrum antiserum capable of disrupting some form of cell–cell interaction, and then attempting to identify the antigen(s) involved. The work utilizing the latter approach will be dealt with in more detail here as it has been responsible for the latest surge of interest in the molecular basis of adhesion events.

The use of antibodies to dissect adhesion events has been fruitful, but the results must be interpreted with caution. The antiserum produced is only as good as the antigen injected, and the most immunogenic constituent of any injected material may not be quantitatively the major component. The final absolute identification of the cell surface molecule with which a particular antiserum is reacting to produce the desired biological effect requires careful, precise biochemical analysis and the eventual production of monospecific antibodies. The isolation of hybridomas producing monoclonal antibodies has proven invaluable in this regard. However, even these reagents must be used with extreme care and a wary eye for the existence of cross-reacting sequences involving either oligosaccharides or polypeptides on different proteins.

With these reservations in mind, let us examine the validity of using antibodies in a search for cell surface molecules involved in cell–cell adhesion. This approach makes different assumptions from those used previously to study cell–cell interactions by the phenomenological and biochemical dissection of *in vitro* aggregating systems (as discussed by Moscona, 1974; Lilien *et al.*, 1979; Steinberg, 1978) and hence may reveal molecules involved in other aspects of intercellular adhesion. It must be kept in mind that the principal antigen recognized by any antibody may not be the ligand directly involved in holding two cells together, but may instead represent another cell surface molecule somehow required to maintain the structure or organization of molecules at the site of adhesion. Such molecules, although not involved directly at the site of attachment, would have to be present and functioning for the adhesion process to work. Antibodies may well react with such molecules and thereby perturb cellular behavior. Obviously, in the grand lottery of antibody production (especially when complex immunogens are used), reagents will be prepared that interfere with the biology of adhesion in different ways leading to the discovery of many different antigens each involved in an important aspect of the adhesion process. As will be seen below, antibodies have already presented the biochemist and biologist with an interesting group of molecules whose precise roles in the adhesion process have yet to be determined.

2.1. Adhesion Glycoproteins Discovered Using Antibodies to Perturb the Adhesion Process

2.1.1. Cell Adhesion Molecules from Developing Nervous Tissue: N-CAM

2.1.1a. Early Studies on N-CAM. Probably the most thoroughly studied cell–cell adhesion glycoprotein to be discovered using antibodies is N-CAM (nerve-cell adhesion molecule) (see review by Edelman, 1983). This work extends over several years and will be discussed in some detail. In 1976, Rutishauser *et al.* published the results of preliminary studies in which antibodies were prepared whose Fab' fragments could inhibit chick neural retinal cell or chick brain cell aggregation. Identification of the antigen with which the antibody reacted was complicated by the presence of proteolytic fragments in the cellular extracts (Edelman, 1983). Subsequently, a new broad-spectrum antiserum (anti-R10) was produced against 10-day neural retinal cells that had recovered from trypsinization prior to their injection into rabbits. Anti-R10 Fab interferred with the aggregation of chick neural retinal or brain cells (Brackenbury *et al.*, 1977). As nonspecific controls, it was shown that neither antibodies against cell surface carbohydrates nor succinyl Con A blocked intercellular aggregation. Material shed by neural retinal cells into culture medium was isolated and fractionated according to its ability to neutralize anti-R10 inhibition of neural retinal aggregation. Active fractions contained proteins of 140, 120, and 65K that were immunoprecipitated by anti-R10 serum. Because of the limited availability of material, the "active" gel slices were pooled and injected into rabbits. The resulting antiserum, later referred to as anti-N-CAM, blocked neural retinal aggregation and immunoprecipitated polypeptides of 140, 120, and 65K from unfractionated tissue culture supernatants, suggesting its specificity for these polypeptides. A 140K polypeptide was the principal component immunoprecipitated from detergent extracts of neural retinal cells using this antiserum. As was pointed out by these investigators, it is possible that CAM was precipitated by antibodies unrelated to neutralizing activity of the antiserum. It was also possible that the components that neutralized anti-N-CAM's ability to block retinal cell aggregation were associated with a different polypeptide poorly iodinated by the chloramine-T procedure.

2.1.1b. Monoclonal Antibodies Directed against N-CAM. The above-mentioned problems illustrate the difficulties in identifying the antigens relevant to the biological activity of a complex antiserum. Clear proof that N-CAM was involved in neural aggregation required that the molecule be purified and shown to block antibody inhibition of aggregation and that monoclonal antibodies be produced against N-CAM and used to interfere with neural retinal aggregation. This is precisely what this same group has accomplished (Hoffman *et al.*, 1982). Hybridomas producing monoclonal antibodies that react with N-CAM were obtained following inoculation of mice with partially purified N-CAM from chick neural tissue. Of 1200 hybrids tested, nine were found to bind to N-CAM. Six of these produced antibodies that would block or partially block brain or neural retinal cell aggregation. A combination of these monoclonal antibodies was found to be more effective in blocking aggregation than any single one. This might be expected considering the specificity of monoclonal antibodies. It would be possible to obtain a series of antibodies, some of which would react with the site on N-CAM directly involved in cell–cell adhesion and others that would not. A mixture of such monoclonal antibodies would be expected to have a biological effect more nearly similar to that of polyclonal antiserum.

2.1.1c. Biochemical Characterization of N-CAM. Immunoabsorbent columns using monoclonal antibodies were prepared and used to remove N-CAM from detergent extracts of membranes from neural retinal cells. The N-CAM isolated in this manner migrated on SDS–polyacrylamide gel electrophoresis (SDS–PAGE) as a broad, continuous band of apparent M_r between 200 and 250K. The addition of reducing agents did not produce smaller peptides, suggesting no interchain disulfide bonding. Boiling the polypeptide and/or digestion with neuraminidase resulted in a polypeptide migrating on reduced SDS–PAGE as a closely spaced doublet of 140K. Analysis of proteolytic peptides of bands cut from this 140K region of gels or sections cut from various places along the 200–250K region of gels, showed that this heterogeneity was not due to differences in the polypeptide composition of the molecule and further strengthened the argument that the heterogeneity was due to the sialic acid content of the N-CAM. Further evidence for this was found in the observation that heterogeneity seen on isoelectric focusing was lost after neuraminidase treatment.

Analysis of the N-CAM showed that phosphate was present but no lipid was found. Analysis of the carbohydrate content of the molecule showed that there was 13 moles of sialic acid to every 1.4 moles of galactose per 100 moles of amino acid. The sialic acid was apparently released as a polymer following prolonged boiling. Gel filtration and ultracentrifuge analysis showed a strong tendency of N-CAM to self-aggregate. This could account for the previous difficulties in obtaining homogeneous material. N-CAM isolated from neural retina behaved differently on SDS–PAGE analysis from that obtained from brain cells. The difference seems to be due to the sialic acid content of the molecules as digestion with neuraminidase resulted in identical polypeptides migrating at 140K.

A comparison of N-CAM from adult and embryonic nervous tissue was made by Rothbard *et al.* (1982). The two N-CAM preparations cross-reacted serologically, and each was able to neutralize the adhesion-blocking activity of anti-N-CAM. They each gave rise to similar peptides on proteolysis, and after removal of sialic acid, they both migrated as polypeptides of 140K on SDS–PAGE. Prior to removal of sialic acid, however, the embryonic form of N-CAM appeared to have a higher molecular weight than the adult form. There was about one-third the amount of sialic acid present on the adult form of N-CAM as was present on the embryonic form. In both cases the sialic acid was present as an as yet undefined polymer.

2.1.1d. Role of N-CAM in Cell–Cell Adhesion. The role of N-CAM in histogenesis was explored in several ways (Rutishauser *et al.*, 1978a,b). Staining tissue or cells with either horseradish peroxidase-conjugated antibodies or fluoresceinated antibodies demonstrated that anti-CAM was found on retinal optic nerve, spinal cord, and to some extent on muscle and liver cells in developing embryos. No staining was detected on fibroblasts. Anti-N-CAM Fab was found to block neurite fasciculation of cultured chick dorsal root ganglia (Fig. 1). This effect was blocked by purified N-CAM. Interestingly, the halo of neurites extending from the ganglia were the same size in the presence and absence of the antibody,

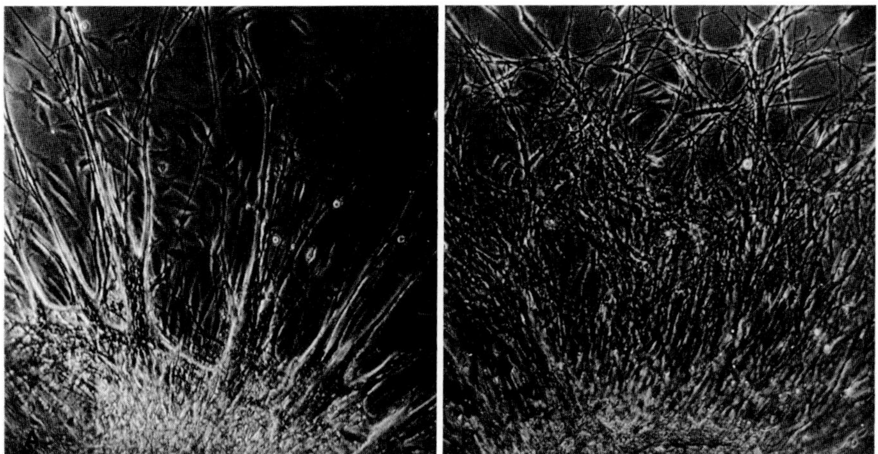

Figure 1. Representative outgrowth of neurites from thoracic ganglia (G) cultured for 24 hr in medium containing (A) 1 mg/ml Fab' from unimmunized rabbits and (B) 1 mg/ml anti-CAM Fab'. Ganglia cultured in medium containing 1–5 mg/ml antifibroblast Fab' were identical to those in (A). The presence of anti-CAM Fab' has resulted in a tangled outgrowth of fine processes rather than the thick and relatively straight fascicles formed in cultures without this antibody. (From Rutishauser *et al.*, 1978b, with permission.)

suggesting that only interactions between extending neurites were being blocked by the antibodies. They had no effect on the interaction of the growth cone with the substratum (Rutishauser *et al.*, 1978b). A similar effect has been reported for antisera prepared against the rat nervous system-specific protein D2 (Jørgensen *et al.*, 1980).

Histogenesis of cultured chick retina was also disrupted by anti-N-CAM Fab (Buskirk *et al.*, 1980). In this case, the layer of ganglionic cells was less well organized when compared to control tissue exposed to preimmune sera. Ganglionic cells were found scattered throughout the inner plexiform layer. The organization of cells within the nuclear layer was also disrupted. These observations are all consistent with N-CAM being a molecule involved in cell–cell adhesion.

The presence of N-CAM on muscle cells suggested that it might play a role in the neuromuscular interactions leading eventually to the formation of myoneural junctions (Grumet *et al.*, 1982; Rutishauser *et al.*, 1983). This idea is supported by the observation that myoblasts could bind to monolayers of neural retinal cells. This binding was blocked by anti-N-CAM (Grumet *et al.*, 1982). Similarly, lipid vesicles containing purified N-CAM were found to bind to muscle cells and this binding was also blocked by anti-N-CAM. No interaction was seen between muscle cells or vesicles and fibroblasts, a cell that does not carry N-CAM. A role for

N-CAM in initial nerve–muscle interactions was further supported by the observation that spinal cord neurites bound to myotubes in culture and that this binding could be blocked by anti-N-CAM (Rutishauser *et al.*, 1983). The effect of anti-N-CAM on neurite–myotube interactions could be neutralized by purified N-CAM, further demonstrating that N-CAM was the molecule being affected by the antiserum.

Preliminary comparisons of N-CAM from myotubes and nerve cells show that both are highly sialylated glycopeptides whose behavior on SDS–PAGE is greatly affected by neuraminidase digestion. Both N-CAM preparations yielded proteins that migrated in the 140K region of SDS–PAGE gels following neuraminidase digestion. However, an additional 170K polypeptide was seen in preparations from nerve cells, suggesting a difference in the susceptibility of the oligosaccharide in the two N-CAMs to neuraminidase. A more detailed comparison of the polypeptides of the two N-CAMs will be required to confirm this.

These observations are particularly interesting when considered along with the data of Thiery *et al.* (1982) and Edelman *et al.* (1983) showing the presence of N-CAM in nonneural tissue during early development. What was initially thought to be a molecule involved in early adhesion events between nerve cells may turn out to be a molecule that plays a broader role in early cellular interactions, as expression is not limited to neuronal cells and cells of ectodermal origin. Thus, N-CAM may function at any time during development when it is necessary to innervate tissue.

N-CAM has also been detected in rat neural tissue (Jørgensen *et al.*, 1980) and in human and mouse neural tissue (McClain and Edelman, 1982). The identification and purification of N-CAM from nonavian cells was greatly facilitated by the fact that antibodies could be prepared against purified chick N-CAM that would cross-react with N-CAM from other species (Edelman, 1983). Out of this work has come the observation that the change in sialic acid content from the embryonic to the adult form of N-CAM did not occur in the "staggerer" mouse mutant (Edelman and Chuong, 1982). It has been suggested that the embryonic to adult conversion would lead to an increase in the free energy of binding of N-CAM to its ligand on another cell and thus might serve to regulate intercellular interactions requiring N-CAM. Edelman (1983) has speculated that the failure of this embryonic to adult conversion might result in incorrect neuronal connections causing the neurological disorder seen in staggerer mice.

Although work with N-CAM has progressed rapidly, there is much we do not know. For example, we do not know how it is oriented within the membrane. The genetic control of both its appearance and modification, and precisely at what point in cell–cell interactions N-CAM begins to function, is not known. Its relationship to other cell adhesion molecules such as cognins (Hausman and Moscona, 1976), ligatin (Jakoi and Marchase, 1979), and the many factors shed into tissue culture medium by neural retinal cells (Lilien *et al.*, 1979) is not clear.

2.1.2. Cell Adhesion Molecules from Liver Cells: L-CAM, LAM, Cell-CAM 105

An immunological approach similar to that used to detect N-CAM has been used by two groups to isolate a molecule from chick liver cells that is involved in liver cell adhesion (Bertolotti *et al.*, 1980; Nielsen *et al.*, 1981). Initially, broad-spectrum antisera were prepared against chick embryo liver cells and used to inhibit embryonic liver cell aggregation. A factor capable of blocking the activity of these antibodies was isolated and partially purified from chick liver cells. This factor was first identified as a glycopeptide of about 65 to 68K. The molecule was designated L-CAM by Bertolotti *et al.* (1980) and LAM (liver adhesion molecule) by Nielsen *et al.* (1981). Antibodies prepared against purified L-CAM immunoprecipitated primarily a 68K polypeptide from various cell extracts. However, other polypeptides of 90, 135, and 160 K were also seen. The antiserum did not cross-react with N-CAM.

Partially purified L-CAM was used to generate monoclonal antibodies (Gallin *et al.*, 1983). More careful studies of L-CAM using these antibodies in addition to monospecific polyclonal antibodies revealed the cell surface-associated form of the molecule to have an M_r of 124K. L-CAM is clearly distinct from N-CAM. Purified L-CAM does not interact with anti-N-CAM, nor will anti-L-CAM neutralize N-CAM activity. The aggregation of liver cells involving L-CAM is calcium-dependent, whereas N-CAM-mediated aggregation is calcium-independent (Brackenbury *et al.*, 1981). The tissue distribution of L-CAM, in the adult, does not appear to overlap with that of N-CAM. However, during early development, L-CAM and N-CAM colocalize in some areas (Edelman *et al.*, 1983). Finally, the sialic acid content of the two molecules is distinctly different. These observations emphasize the fact that more than one cellular adhesion molecule exists and that all tissue and cell types may not use precisely the same process for adhering to one another.

Liver cell adhesion has also been studied extensively in rats. Ocklind and Öbrink (1982) identified a glycoprotein of 105K that was capable of neutralizing the effect of antihepatocyte antibodies on hepatocyte aggregation. This molecule was designated cell-CAM 105. Ocklind *et al.* (1983) have shown that its distribution in the adult rat is restricted to cell surfaces adjacent to the bile canaliculi in liver and to the simple epithelia of the

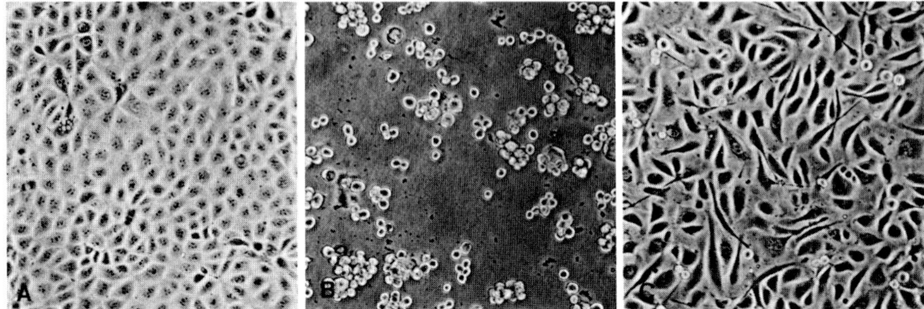

Figure 2. Effects of two different broad-spectrum antisera on the adhesion properties of an MMTE cell line. (A) Control monolayer of MMTE cells. (B) MMTE treated with a 1:256 dilution of anti-SFM I made against material shed into SFM by MMTE cells. Cells are detached from the substrate. Although it cannot be determined from this figure, cell–cell interactions are also affected (see Fig. 3). (C) MMTE treated with a 1:64 dilution of anti-SFM II made against material shed into SFM by the mammary carcinoma cell line MCF-7. Cells remain attached to the substratum but have severed their interactions with one another. (From Damsky *et al.*, 1981, with permission.)

intestinal tract, kidney tubules, and certain glandular tissue. Further studies will be required to show whether or not cell-CAM 105 is the "rat" equivalent of L-CAM.

2.1.3. Cell Adhesion Molecules from Cultured Epithelial Cells: Cell-CAM 120/80

2.1.3a. Effect of Antibodies on Cells Growing as Monolayers. CAMs discussed above have all been discovered studying systems in which Fab' fragments of antibodies were used to prevent the aggregation of cells in suspension. In these instances, preformed tissue must be disrupted, the cells allowed to recover from any injuries to the membrane that may have occurred during the disruption process, and then the cells allowed to aggregate. An alternative to this approach that does not require the disruption of preformed cellular interactions prior to antibody addition is to use antibodies themselves to disrupt established cellular interaction. This can be done using certain epithelial cell lines known to form intercellular adhesion complexes when grown as monolayers in culture. In fact, it is possible, with the right antibodies, to show that cell–cell and cell–substratum adhesion are completely separate events requiring different cell surface molecules. This is illustrated in Fig. 2, which shows the effect of two different antibodies on the behavior of cultured epithelial cells (Damsky *et al.*, 1981). The antisera used here were broad-spectrum antisera capable of reacting with numerous cellular antigens. One antiserum, anti-serum-free medium (SFM) I, disrupts cell–substratum adhesion (Fig. 2B). It was prepared against material shed into SFM by mouse mammary tumor epithelial cells (MMTE). The other antiserum, anti-SFM II (Fig. 2C), disrupts cell–cell adhesion, and was prepared against SFM from a human mammary tumor cell line MCF-7. The "target" cell line used for both antisera was the MMTE cell line. These antisera or any others may interfere with intercellular adhesion (1) by reacting with a single cell surface molecule and interfering with its function; (2) by causing a general rearrangement of proteins within the lipid bilayer and thus indirectly and nonspecifically perturbing cellular interactions; (3) by generally coating the cell surface with immunoglobulin in such a way as to prevent cellular interactions; (4) by somehow disrupting cytoskeletal organization and thereby indirectly affecting cellular behavior and/or morphology; and (5) by being cytotoxic. The effect of anti-SFM II on the epithelial cells (Fig. 2C) is specific for the disruption of cell–cell interactions. The cells remain firmly attached to the substratum, thus leaving the surface molecules involved in cell substratum adhesion properly organized and functional. It is possible, however, to perturb cell–substratum interactions with anti-SFM I (Fig. 2B). Thus, it is unlikely that the effect of anti-SFM II is due to coating of the cells with antibodies and thus disrupting all cellular adhesion interactions in a nonspecific manner. Anti-SFM II is not cytotoxic, for the cells continue to replicate at normal rates in the presence of the antibody and the effect of the antiserum is reversible (Damsky *et al.*, 1981). The antiserum did not function by producing a nonspecific aggregation of cell surface molecules, as Fab fragments produced the same loss of cell–cell interactions. Finally, a comparison of the effects of anti-SFM I and anti-SFM II on these cells demonstrates dramatically that cell–cell and cell–substratum adhesion are two separate and distinct processes utilizing different cell surface antigens. Not only is this suggested by the phenomenology demonstrated in Fig. 2, but, as will be discussed further later, the antigens involved in each process can be biochemically separated and shown to have distinctly different characteristics (Fig. 3).

2.1.3b. Identification of the Cell–Cell Adhesion Antigen from Mammary Tumor Epithelia. Two important questions arise out of the preceding observations. First, what is the antigen or antigens with which anti-SFM II reacts to disrupt cell–cell interactions? Second, what is the physiological relevance of this observation? Initially, attempts were made to isolate the relevant antigens from detergent extracts of cells (Damsky *et al.*, 1981). This proved impossible for reasons not yet understood. More recently, attempts to isolate the antigen shed into serum-free tissue culture medium have been successful, although this form of the antigen is most likely a proteolytic fragment of the actual cell surface molecule. As in

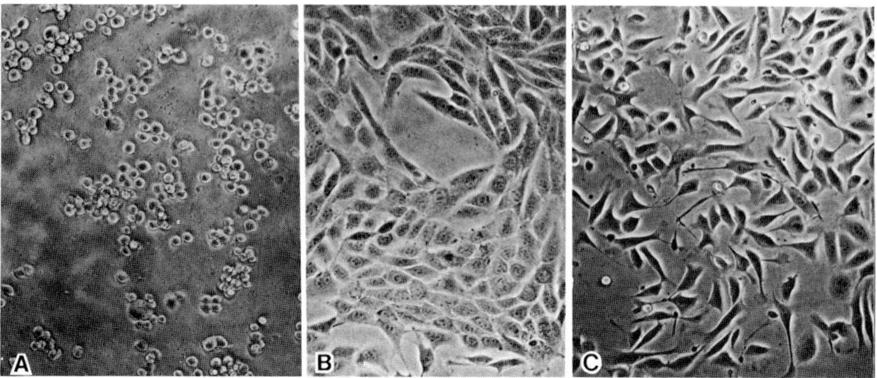

Figure 3. Differential inhibition of effects of anti-SFM I on cell–cell and cell–substratum adhesion. (A) MMTE cells treated with a 1:256 dilution of anti-SFM I. (B) MMTE cells treated with a mixture containing 1:256 dilution of anti-SFM I plus about 40 μg of an NP-40 extract of MMTE cells. These cells resemble a control monolayer (see Fig. 1A). (C) MMTE cells treated with a mixture of a 1:256 dilution of anti-SFM I plus about 1 μg of a fraction purified from an NP-40 extract of MMTE cells using ion-exchange and lectin affinity chromatography. Cell–substratum adhesion has been restored, but the MMTE cells have little cell–cell interaction. This result suggests that anti-SFM I has antibodies that affect both cell–cell and cell–substratum adhesion, that of an unfractionated NP-40 extract of MMTE cells contains antigens relevant to both adhesion systems, and that a subset of antigens involved only with cell–substratum adhesion can be purified from the NP-40 extract [see Damsky *et al.* (1981) for purification protocol and description of assays].

other instances, it was possible to detect the antigen by its ability to block anti-SFM II disruption of cell–cell interactions. This made it possible to follow the antigens throughout the various biochemical fractionation procedures. Anti-SFM II blocking activity was isolated from MCF-7-conditioned SFM by a combination of gel filtration, differential precipitation, ion-exchange, and lectin affinity chromatography. The final purification step involved antibody affinity chromatography using immobilized anti-SFM II immunoglobulin that had been exhaustively adsorbed to human fibroblasts. This absorption removed antibodies that reacted with other antigens on the surface of human cells. The material eluted from this antibody affinity column was analyzed by SDS–PAGE under nonreducing conditions. When the gels were sliced and the glycopeptides eluted, only material associated with a band at about 80K (gp80) was able to block the effects of anti-SFM II on MMTE cells (Fig. 4). As further proof of the relevance of this fraction to cell–cell adhesion, antiserum produced in rabbits against the gel-purified gp80 (anti-gp80) was able to disrupt cell–cell adhesion of MMTE cells in culture.

The specificity of anti-gp80 was examined by immunoblotting. By this procedure, it was shown that the antibody recognized a single gly-

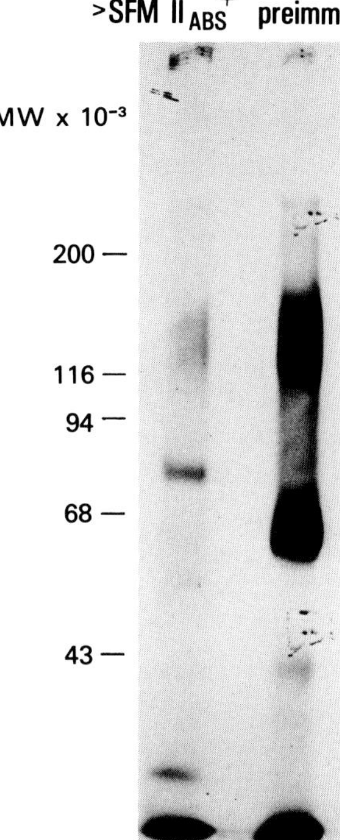

Figure 4. Analysis of SDS–PAGE, under nonreducing conditions, of material partially purified from MCF-conditioned SFM and eluted from either immobilized (A) anti-SFM II or (B) preimmune IgG. When such gels are sliced and the slice eluates tested for anti-SFM II blocking activity, only material present in the 80K region of the gel in lane A is able to block disruption of the cell–cell interaction of MMTE caused by anti-SFM II.

coprotein of 80K in MCF-7-conditioned medium. Immunoblotting experiments in which anti-gp80 were reacted with unfractionated Nonidet-P40 (NP-40) extracts of either MCF-7 cells or another human epithelial cell line (JAR choriocarcinoma) showed that the antiserum reacted only with a 120K constituent. In control experiments using NP-40 extracts of fibroblasts, anti-gp80 failed to react with anything in the NP-40 extracts. Immunofluorescence studies revealed that anti-gp80 stained the periphery of cultured epithelial cells at sites of cell–cell contact, but did not react with anything on the surface of control fibroblast cells. Thus, the gp80 purified from conditioned medium appears to be a fragment of a 120K molecule found on the surface of epithelial cells that plays some role in the adhesion of these cells to one another. This molecule has been designated cell-CAM 120/80 (Damsky *et al.*, 1983a).

2.1.3c. In Vivo Relevance of the Mammary Tumor Cell–Cell Adhesion Antigen Cell-CAM 120/80. Preimplantation mouse embryos were used to study the effect of anti-SFM II on early development. Initially, the effect of anti-SFM II on compaction was examined. Compaction is an early morphological event in mammalian embryonic development oc-

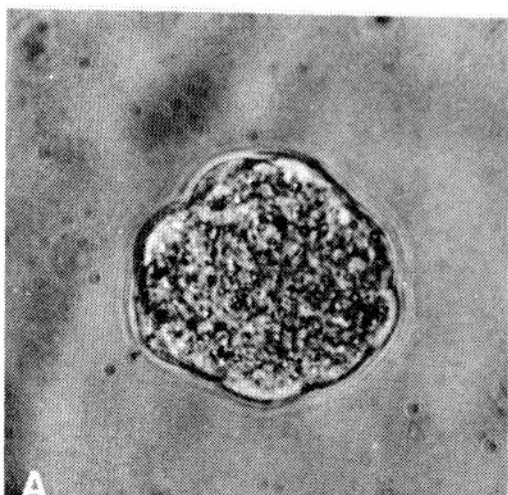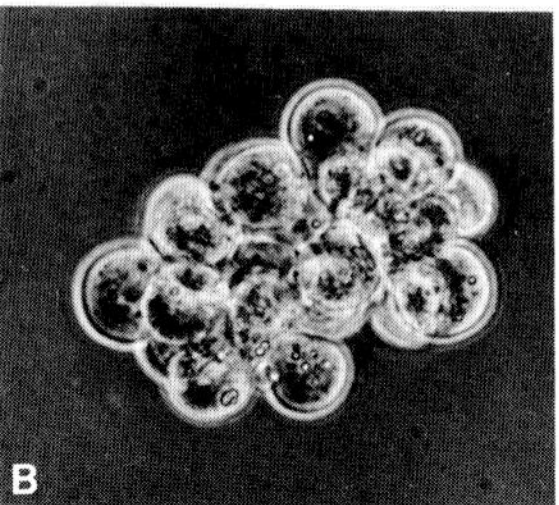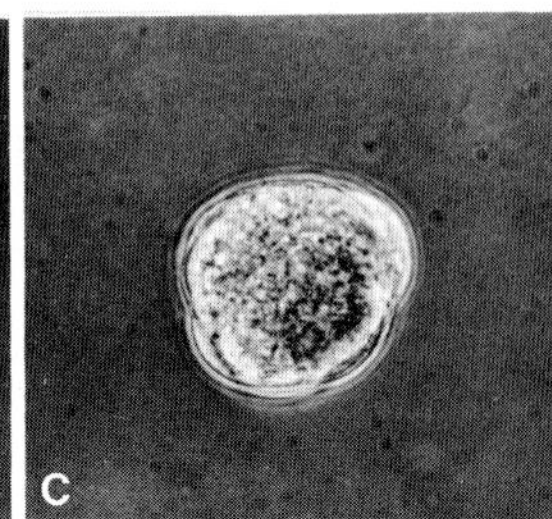

Figure 5. Effects of anti-SFM II on precompaction mouse embryos. (A) Embryo that has been cultured in normal medium for 24 hr following removal from mouse. Compaction has proceeded normally. (B) Embryo that has been cultured 24 hr in a 1:50 dilution of anti-SFM II. Compaction has been prevented. Cell division, however, is not affected. If embryos tested in this way are returned to normal medium, they reverse within 24 hr and form normal blastocysts at the same time as embryos cultured in control medium for the entire 48 hr (not shown). (C) Embryo that has been incubated with a mixture containing a 1:50 dilution of anti-SFM II and about 0.1 μg of purified gp80 eluted from SDS–polyacrylamide gel slices. Compaction has proceeded normally. Embryos treated in this way are indistinguishable from controls. Similar results have been reported for gp85 isolated from teratocarcinoma stem cells by Hyafil *et al.* (1981).

curring at the eight-cell stage. At this time, individual blastomeres of the eight-cell embryo become less distinct, forming a compact ball (Fig. 5A). During the transition from 8 to 32 cells, specialized junctions appear (Ducibella *et al.*, 1975; Ducibella and Anderson, 1975) in preparation for blastocoel formation and prior to the separation of embryonic layers into trophoblast and inner cell mass. When either anti-SFM II or anti-gp80 was added to embryos before or during compaction (Fig. 5B), compaction was completely disrupted (Damsky *et al.*, 1983a). These antisera were not toxic to embryonic cells as they continued to replicate in the presence of the antiserum, and upon removal of the antisera, the cells recompacted and continued normal *in vitro* development. Antibody-induced decompaction did not require the use of Fab fragments. Decompaction was not due to nonspecific reorganization of molecules on the surface, as other antisera that bind to the morula cells had no effect on compaction (Solter and Knowles, 1975). The decompaction could be blocked by the gp80 fragment isolated from SDS–PAGE gels (Fig. 5C). Thus, cell-CAM 120/80 is clearly involved in the process of compaction.

Cell-CAM 120/80 was also found on the surface of normal adult epithelial tissue as demonstrated by the ability of human foreskin epidermis, adult mouse kidney and epithelial-like cultured cell lines to adsorb the cell–cell disrupting activity as well as the decompaction activity from anti-SFM II and anti-gp80. Mouse fibroblasts, human foreskin dermis and parietal yolk sac (PYS) cells, an endodermal derivative of F9 embryonal carcinoma (EC) cells, were not able to absorb decompacting activity from anti-SFM II. In addition, a survey of frozen tissue sections using fluorescent antibody staining showed that anti-gp80 reacted with many epithelia, including stratified epithelium of skin, intestinal epithelium, kidney-collecting ducts, bile ducts, and mammary gland ducts (Damsky *et al.*, 1983c). The antibody did not stain connective tissue or muscle. Thus, the antigen recognized by anti-gp80, cell-CAM 120/80, is found primarily on epithelial cells and is expressed not only in embryonic tissue during early development, but also in adult, fully differentiated tissue.

2.1.4. Cell Adhesion Molecules from Embryonal Carcinoma Cells: Uvomorulin and Related Glycopeptides

The studies reported above suggest that cell-CAM 120/80 is similar to the cell–cell adhesion antigen uvomorulin identified earlier by other laboratories using broad-spectrum antisera against EC cells. Studies leading to the identification of uvomorulin were begun using antisera prepared against F9 EC cells (Kemler *et al.*, 1977; Dunia *et al.*, 1979; Ducibella, 1980). Fab fragments of anti-F9 serum prevented compaction of eight-cell-stage mouse embryos and induced decompaction of embryos prior to blastocoel formation. The antiserum was nontoxic, as embryonic cells continued to divide in culture in the presence of the antiserum. Decompaction appeared to be due to the interaction of the anti-F9 Fab with specific antigens on the cell surface, as other antisera prepared against mouse brain, lymphocytes, or embryonic liver reacted with the embryos, but did not induce decompaction (Kemler *et al.*, 1977). By adsorption, it was shown that the antigen was not expressed on the surface of PYS cells, lymphocytes, splenocytes, or various fibroblastic cell lines. The effect of anti-F9 Fab on compaction was reversible. Upon removal of the antiserum, the cells recompacted and normal development occurred. Blastocysts derived from such "reversed" cells were put into pseudopregnant mice and normal full-term litters were obtained. Thus, anti-F9 Fab seemed to interact with a cell surface antigen important to early organizational events in the mouse embryo. Interestingly, this antiserum also decreased the cell–cell interactions of EC cells in culture without affecting their cell–substratum interactions (Nicolas *et al.*, 1981). This property was used subsequently to follow the antigenic activity during purification (Hyafil *et al.*, 1980). In attempts to identify the antigen with which anti-F9 reacted, Hyafil *et al.* (1980) isolated a tryptic fragment from PCCAAzaR1 EC cells that blocked the effect of anti-F9 Fab on both decompaction and *in vitro* organization of PCCAAzaR1 cells. This fragment was an 84K glycopeptide as determined by SDS–PAGE. That it was derived from a cell surface glycoprotein was demonstrated by several lines of evidence (Hyafil *et al.*,

1980, 1981). First, trypsinates containing this glycopeptide blocked antibody-induced decompaction; second, this was the principal glycopeptide found in anti-F9 immunoprecipitates of F9 membrane trypsinates; third, cells that did not adsorb antibodies capable of inducing decompaction also had no effect on the ability of this antiserum to immunoprecipitate the 84K glycopeptide; fourth, only regions cut from nonreduced SDS–PAGE gels containing the 84K glycopeptide were capable of blocking antibody-induced decompaction; and finally, hybridomas were produced whose antibodies specifically reacted with this glycopeptide and were capable of blocking anti-F9-induced decompaction of mouse embryos (Hyafil *et al.*, 1981). Thus, this glycopeptide appears to be from a cellular glycoprotein that is involved in the process of compaction.

Takeichi *et al.* (1981) and Yoshida and Takeichi (1982) have prepared antiserum against F9 teratocarcinoma cells that disrupts the interaction of EC cells *in vitro* in much the same manner as anti-gp80 and antiuvomorulin. They isolated a 34K tryptic peptide from F9 teratocarcinoma cells that could inhibit the effect of their antiserum on the interactions of EC cells in culture. It is likely that this peptide, uvomorulin, and cell-CAM 120/80 are all from the same molecule, as all three peptides blocked antibody-induced changes in cell–cell adhesion of cultured cells, disrupted compaction, and appear to be from the class of Ca^{2+}-protected trypsin-sensitive molecules discussed below.

2.2. Calcium and Cellular Aggregation

2.2.1. Calcium-Dependent and Calcium-Independent Adhesion Sites

Adhesion molecules can be divided into two groups defined by the calcium ion requirement of the reaction they mediate in a particular *in vitro* assay. If cellular aggregation requires Ca^{2+}, the reaction is said to be Ca^{2+}-dependent and the receptor is protected by the presence of Ca^{2+} from digestion with trypsin (0.01%). If the *in vitro* cellular aggregation does not require Ca^{2+}, the reaction is considered Ca^{2+}-independent. The membrane constituents involved in the Ca^{2+}-independent reaction are preserved on the cell surface when digested with low levels of trypsin (0.0001%) in the absence of Ca^{2+}, whereas exposure to trypsin under these conditions inactivates the Ca^{2+}-dependent receptors. These two classes of receptors were first demonstrated by Takeichi (1977), who showed that hamster V79 fibroblasts and BHK cells (Urushihara *et al.*, 1977) digested with trypsin in the presence of Ca^{2+} would aggregate with one another but only if Ca^{2+} was included in the reaction. V79 fibroblasts digested with low levels of trypsin in the absence of Ca^{2+} lose their Ca^{2+}-

dependent aggregation properties and instead exhibit only Ca^{2+}-independent aggregation (Urushihara *et al.*, 1979).

2.2.2. Calcium-Dependent and Calcium-Independent Adhesion Glycoproteins on the Surface of Fibroblasts

By comparing SDS–PAGE profiles of V79 cells trypsinized in the presence and absence of Ca^{2+}, Takeichi *et al.* (1981) concluded that a protein of 150K was responsible for Ca^{2+}-dependent aggregation of fibroblasts. This has not been confirmed either by isolating the polypeptide and using it to block Ca^{2+}-dependent aggregation or by preparing antibodies to the polypeptide and testing them for their ability to inhibit Ca^{2+}-dependent aggregation. Hence, the designation of this glycopeptide as the Ca^{2+}-dependent aggregation receptor must be considered tentative.

Urushihara and Takeichi (1980) have also attempted to identify the receptor involved in Ca^{2+}-independent aggregation of V79 fibroblasts using an antiserum prepared against the factors released from cells by vigorous trypsinization of cells previously digested with low levels of trypsin in the absence of Ca^{2+} (conditions that destroy Ca^{2+}-dependent receptors). The Fab fragments of this antiserum, referred to as anti-TRF (trypsin-released factor), blocked Ca^{2+}-independent aggregation of V79 cells. This antiserum precipitated several radioactive proteins from V79 fibroblasts digested with low levels of trypsin in the absence of Ca^{2+}. The major unique protein in the immunoprecipitate was a polypeptide with an M_r of 125K. This protein was lost from these same cells on further digestion with trypsin. This polypeptide could be surface-labeled with ^{125}I and metabolically labeled with radioactive methionine or fucose. A 120K protein with similar properties was found in immunoprecipitates of extracts from cells trypsinized under conditions that render the Ca^{2+}-independent aggregation inoperative. It was then tentatively suggested that this peptide represented a cleaved portion of the 125K protein. These authors concluded that the 125K glycoprotein was the most likely candidate for the membrane molecule involved in Ca^{2+}-independent adhesion, but again, no direct evidence has been presented.

The precise *in vivo* role, if any, of fibroblast glycoproteins involved in Ca^{2+}-independent adhesion has not been determined. It is interesting to note that anti-TRF has no effect on cell morphology *in vitro* and therefore these receptors do not appear to play a role in the adhesion of cultured cells to a substratum. Urushihara and Takeichi (1980) did, however, observe that the most effective inhibition of Ca^{2+}-independent aggregation by anti-TRF occurred after treatment of the cells with hyaluronidase, which suggests that these Ca^{2+}-independent receptors might act as mediators in the interaction of cells with glycosaminoglycans (GAGs). Be-

70K glycoprotein not observed in extracts of cells treated with trypsin in the absence of Ca^{2+}. Thus, a 130K glycoprotein was implicated as the surface molecule protected from trypsin by the presence of Ca^{2+} and perhaps involved in Ca^{2+}-dependent aggregation. To substantiate this, these workers used gel filtration and ion-exchange chromatography to fractionate tissue culture medium in which neural retinal cells had been grown. The fraction of interest was followed throughout the purification procedure by its ability to block the effect of anti-CaT on Ca^{2+}-dependent neural retinal cell aggregation. This resulted in the discovery of an active peptide of 85K that blocked the immunoprecipitation of the 130 and 70K glycoproteins from cell extracts. These authors, therefore, concluded that the 130K glycoprotein was the Ca^{2+}-protected, trypsin-sensitive receptor involved in chick neural retinal cell adhesion. Based on its presence on Ca^{2+}-dependent aggregating cells, its absence from nonaggregating cells, its immunoprecipitation by aggregation-inhibiting antibodies, and the ability of a blocking competitor to inhibit its immunoprecipitation, they speculate that the 85K fragment found in the medium and the 70K component found in the immunoprecipitate might represent proteolytic fragments of the 130K receptor.

Other proteins, described earlier, may also be involved in Ca^{2+}-dependent cell–cell interactions. The 84K tryptic glycopeptide of uvomorulin, the glycoprotein involved in compaction in preimplantation mouse embryos, undergoes a Ca^{2+}-dependent conformational transition such that it becomes sensitive to trypsin in the absence of Ca^{2+}. After incubation with EGTA, this glycopeptide is no longer precipitable by the monoclonal antibody DE1, which reacts with uvomorulin (Hyafil *et al.*, 1981). This transition is reversible, as the glycopeptide will react with DE1 on readdition of Ca^{2+}. The fact that eight-cell embryos are reversibly decompacted in Ca^{2+}-free medium (Ducibella and Anderson, 1975), together with the Ca^{2+} sensitivity of the conformation of the 84K glycopeptide, suggests that Ca^{2+}-protected antigens play an important role in the cell–cell adhesion events leading to embryo compaction. Evidence has also been presented that the compaction-related molecule, cell-CAM 120/80, discussed earlier also requires Ca^{2+} to maintain an active configuration (Damsky *et al.*, 1983a). Thus, Ca^{2+}-protected molecules appear to be important to cellular adhesion from the earliest times of development.

In summary, the major approach used in studies of Ca^{2+}-dependent and -independent cell aggregation has been to take cells either from cell culture or from embryonic tissue, digest them with trypsin in the presence or absence of Ca^{2+}, and then determine the Ca^{2+} dependency of their ability to form homotypic aggregates. Most investigators feel that the Ca^{2+}-dependent aggregation is the more physiologically significant of the two. This is due to the fact that Ca^{2+}-dependent aggregation expresses some degree of cell-type specificity and the fact that its expression in cells derived from embryonic neural retina correlates with the time of maximum morphogenesis. However, there is not universal agreement on this issue. N-CAM, the glycoprotein discussed in detail in the previous section and postulated to be responsible for aggregation in chick neural tissue, participates in Ca^{2+}-*independent* cell aggregation.

2.3. Summing Up Data on Cell–Cell Adhesion

Molecular Diversity

Table 1 has been assembled in an attempt to arrange a conceptually useful tableau of the molecular characters involved in intercellular adhesion and to summarize much of the preceding information. The apparent diversity is great. It is possible, however, that some of the molecules being examined in various laboratories will turn out to be similar or identical. At this stage, however, we are presented with molecules ranging in subunit size from 10K to 250K. A few have been purified to homogeneity and partially characterized. Others have been identified only indirectly by immunoprecipitation or comparative SDS–PAGE. This makes it difficult to make comparisons. Many molecules have been identified by their ability to enhance cellular aggregation (cognins) or their ability to block cellular aggregation (ligatin). Others have been discovered on the basis of their ability to block antibody-induced disruption of intercellular adhesion (uvomorulin gp84, cell-CAM 120/80, N-CAM, L-CAM, LAM, cell-CAM 105, D2, 34K peptide). Most are glycosylated proteins that can be released from the membrane with nonionic detergents, and hence are probably integral membrane glycoproteins or perhaps peptides arising from integral membrane proteins.

In addition to the adhesion molecules shown in Table 1, the molecules participating in morphologically distinguishable junctional complexes must also be characterized in order for us to completely understand the adhesion strategy of developing and mature tissue systems (for reviews, see Staehelin, 1974; Hull and Staehelin, 1979). To this end, antibodies are now being used to identify families of membrane glycoproteins present in desmosome complexes (Gorbsky and Steinberg, 1981; Cohen *et al.*, 1983).

The adhesion process is complicated, probably occurring in a stepwise manner as discussed earlier. Therefore, it undoubtedly involves a large number of molecules each acting at different points in the process. Thus, the molecule detected by an investigator would depend very much on the assay system used. For example, the aggregation assays used to follow N-CAM activity are of brief duration monitoring intracellular in-

Table 1. Integral Membrane Glycoproteins Involved in Cell–Cell Adhesion

Surface membrane glycoprotein	M_r ($\times 10^{-3}$)	Source	Putative function	How identified	Associated antibody (see text)	Additional characteristics	References
—	150	V79 fibroblasts	Ca^{2+}-dependent aggregation of fibroblasts	Comparative SDS–PAGE of V79 fibroblasts digested with trypsin in presence and absence of Ca^{2+}	—		Takeichi (1977)
—	125	V79 fibroblasts	Ca^{2+}-independent aggregation of fibroblasts	Immunoprecipitation of lysates from cells trypsinized in presence of EGTA using anti-TRF	Anti-TRF		Urushihara and Takeichi (1980)
Uvomorulin	84 (trypsin fragment)	EC cells	Ca^{2+}-dependent aggregation of EC cells; involved in compaction of early mouse embryos	Isolated from tryptic digests of EC cells digested in presence of Ca^{2+}	Anti-F9	Blocks anti-F9-induced decompaction of eight-cell mouse embryos	Kemler *et al.* (1977), Hyafil *et al.* (1980, 1981)
				Purified by ability to block anti-F9-induced disruption of cell–cell adhesion in EC cells	DE-1	Monoclonal antibody DE-1 against gp84 interacts with gp84 in Ca^{2+}-dependent manner and inhibits anti-F9-induced disruption of EC cell monolayer	

Cell-CAM 120/80	80	MCF-7 human mammary tumor cell-conditioned medium	Ca^{2+}-dependent cell–cell adhesion molecule in early embryonic and mature epithelial tissue	Blocks cell–cell disruption activity of anti-SFM II. This activity used to purify molecule from serum-free tissue culture medium	Anti-SFM II, anti-gp80	Fragment of larger cell surface molecule	Damsky et al. (1981, 1983a)
	120	MCF-7 NP-40 extract	Cell surface form of 80K molecule above	Anti-gp80 recognizing 120K protein in immunoblot of NP-40 extracts of epithelial cells		Antiserum to gel-purified gp80 and anti-SFM II disrupt compaction of eight-cell mouse embryos. Disruption blocked by gp80; activity of gp80 lost after trypsinization in absence of Ca^{2+}. Present on mature epithelia in addition to tumor and embryonic cells	
—	124	F9 teratocarcinoma stem cells	Ca^{2+}-dependent aggregation of teratocarcinoma cells	Differential immunoprecipitation of extracts of F9 cells trypsinized in presence and absence of Ca^{2+} using antibody to F9 cells trypsinized in presence of Ca^{2+} (anti-TCF9)	Anti-TCF9	124K component likely to be native molecular weight. Probably related to gp80 (uvomorulin) and cell-CAM 120/80 described above	Takeichi et al. (1981), Yoshida and Takeichi (1982), Ogou et al. (1983)

(Continued)

Table 1. (Continued)

Surface membrane glycoprotein	M_r ($\times 10^{-3}$)	Source	Putative function	How identified	Associated antibody (see text)	Additional characteristics	References
—	34	F9 teratocarcinoma cells	Ca^{2+}-dependent aggregation of teratocarcinoma cells	Released by retrypsinization in presence of EDTA of cells previously digested with trypsin + Ca^{2+}	Anti-TCF9	Blocks anti-TCF9 effects on Ca^{2+}-dependent F9 aggregation. Thought to be a fragment of gp124 (above). 34K fragment contains no carbohydrate	Yoshida and Takeichi (1982)
—	130	Chick neural retina, avian and mammalian neural tissue	Ca^{2+}-dependent adhesion of chick neural retina	Comparative two-dimensional gel electrophoresis and comparative immunoprecipitation using anti-CaT serum of neural retinal cells trypsinized in presence and absence of Ca^{2+}	Anti-CaT	130K likely to be native mature molecular weight	Grunwald *et al.* (1982)
—	70						Cook and Lilien (1982)
—	80	Chick neural retina-conditioned medium	Ca^{2+}-dependent adhesion in chick neural retina	Isolated from conditioned medium based on ability to block disruption of cell–cell adhesion by anti-CaT (see above)		Probably a fragment of gp130 above	Cook and Lilien (1982)

N-CAM	200–250 before removal of sialic acid. Doublet of ~140 after removal of sialic acid	Chick neural retina, avian and mammalian nervous tissue	Tissue-specific interreactions in neural retinal and brain cells	Anti-R10 blocking activity isolated from neural retina-conditioned medium used to produce polyclonal anti-N-CAM	Anti-R10, anti-N-CAM, anti-N-CAM monoclonal antibody	Involved in neurite fasciculation and neural cell or neural–muscle cell interaction; involved in Ca^{2+}-independent adhesion of neural cells	Brackenbury *et al.* (1977, 1981), Thiery *et al.* (1977), Grumet *et al.* (1982), Rustishauser *et al.* (1982, 1983)
				Immunoprecipitation of 140K glycoprotein from NP-40 extracts of neural retina by anti-N-CAM Monoclonal antibody that inhibits neural cell aggregation. Effect blocked by purified N-CAM		Mobility on gels depends on extent of boiling and treatment with neuraminidase (200–250K vs. 140K). Contains large amount of sialic acid in a heat-labile association with polypeptide. Found transiently on several other types of tissues. Cross-reacting material found in humans	Hoffman *et al.* (1982), Rothbard *et al.* (1982), Thiery *et al.* (1982), McClain and Edelman (1982)

(Continued)

Table 1. (Continued)

Surface membrane glycoprotein	M_r ($\times 10^{-3}$)	Source	Putative function	How identified	Associated antibody (see text)	Additional characteristics	References
D_2	139	Rat neuron tissue	Neurite fasciculation	Immunoprecipitate of detergent extract of rat neuronal tissue by anti-N-CAM; production of anti-rat material by injection of immunoprecipitate into rabbits	Anti-D_2	Presumed to be a mammalian homolog to N-CAM of chicks	Jørgensen *et al.* (1980)
L-CAM	124	Embryonic chick liver	Cell–cell adhesion in embryonic liver	Extraction of liver cells with aqueous solution + EDTA. Partial purification based on ability to block activity of anti-L-CAM Monoclonal antibody recognizes 124K form in whole cell extracts. 92 and 81K forms generated by trypsin + Ca^{2+}	Anti-L-CAM	Involved in Ca^{2+}-dependent aggregation in liver Immunologically and biochemically distinct from N-CAM 120K form easily converted to 92 and 81K	Bertolotti *et al.* (1980), Brackenbery *et al.* (1981) Gallin *et al.* (1983)
Cell-CAM 105	105	Adult rat liver	Cell–cell adhesion in liver	Inhibits disruption of liver cell–cell aggregation caused by polyspecific antiserum to liver membranes	Anti-cell-CAM 105	Present on hepatocytes at bile front junctional complexes. Also present on apical cell–cell borders of some but not all epithelia	Ocklind and Öbrink (1982), Ocklind *et al.* (1983)

LAM	68	Chick embryo liver	Cell–cell adhesion in liver	Neutralized antiserum used to block embryonic chick liver cell aggregation in aggregation collection assay	Binds to lens culinaris column, is protease-sensitive. Blocking activity not sensitive to neuraminidase or peroxidase oxidation	Nielsen et $al.$ (1981)
Cognins	50	Chick neural retina and other tissues	Enhances tissue-specific cell reaggregation			Hausman and Moscona (1976)
Ligatin	10	Chick neural retina and other tissues			Released by Ca^{2+} treatment as large filaments that reversibly assemble and disassemble with Ca^{2+} and EGTA. Can block neural retina cell aggregation in $vitro$. Phosphorylated. Binds glycoprotein carrying oligosaccharides with phosphodiester-linked terminal mannose	Jakoi and Marchase (1979), Marchase et $al.$ (1982)

teractions over a period of 30 min, and involving the inhibition of aggregation by a specific antiserum; whereas the assays for cognins are of longer duration monitoring the enhanced aggregation of cells over a period of 24 hr. It is therefore possible that the initial interaction of neuronal cells with one another involves N-CAM and must occur prior to the events in which cognins might be involved. Thus, a developmental scheme could be envisioned in which N-CAM or N-CAM-like molecules serve as signals to arrest cells in their proper location as they migrate systematically throughout the developing embryo, and other molecules function later as the cells specialize as a result of their interactions in different parts of the embryo, and tissue organization commences.

Most investigators agree that the molecules involved in cell–cell interactions can be divided into two general classes on the basis of the Ca^{2+} dependence of the aggregation process. Both Ca^{2+}-dependent and -independent receptors are present on the same cell. There is a difference of opinion as to which are involved in tissue-specific aggregation. N-CAM, for example, has been shown to be important for cell–cell interaction both *in vitro* and *in vivo*, and is a Ca^{2+}-independent receptor. In contrast, Grunwald *et al.* (1980) found that in the neural retina, Ca^{2+}-dependent receptors are maximally expressed during the height of morphogenetic movement. Consistent with this is also the fact that cellular specificity is exhibited in Ca^{2+}-dependent aggregation and not in Ca^{2+}-independent aggregation when several different cell types are compared (Takeichi *et al.*, 1979, 1981). Takeichi *et al.* (1981) proposed that Ca^{2+}-dependent aggregation receptors be further classified on the basis of whether they were expressed during early development or present on mature, fully developed tissue. Thus, they suggested that the glycoprotein involved in compaction [uvomorulin (Hyafil *et al.*, 1981), cell-CAM 120/80 (Damsky *et al.*, 1983a), 34K tryptic peptide (Yoshida and Takeichi, 1982; the 124Kd parent molecule of this peptide, Ogou *et al.*, 1983)] represents the class of Ca^{2+}-dependent receptors characteristic of early nondifferentiated tissue, whereas the Ca^{2+}-dependent glycoproteins found on fibroblasts [150K glycoprotein (Takeichi, 1977)] and neural retinal cells [130K glycoprotein (Cook and Lilien, 1982; Grunwald *et al.*, 1982)] represent receptors characteristic of fully differentiated tissue. This is probably not the best classification of these molecules, as Damsky *et al.* (1983a) have demonstrated that the Ca^{2+}-dependent receptor cell-CAM 120/80 involved in mouse embryo compaction is also found in adult tissue.

We do not have conclusive evidence that any of these molecules are directly involved in intercellular adhesion. It may be that they serve as keystones around which other molecules within the membrane organize to form an adhesion complex. They are, however, specifically localized at sites of intercellular adhesion, and in one case (N-CAM), liposomes containing the adhesion molecule adhere only to appropriate cells. The ligands on adjacent cells to which these glycoproteins bind are not known. It may be that the molecules themselves serve as both ligands and receptors. While the molecular details required for understanding how these molecules function in the adhesion process must still be worked out, the biochemical and immunological technology required for doing this now exists. In the near future, we will not only understand these events as they occur at the molecular level in the membrane itself, but we will be able to isolate the genes coding for these glycoproteins. This will make possible detailed studies of the genetic basis of the regulation of their synthesis in developing and transformed systems.

3. CELL–SUBSTRATUM ADHESION

3.1. Introduction

Cell–substratum adhesion, particularly that of fibroblasts *in vitro*, has been studied intensively over the past decade. Much of this research has focused on adhesion as a dynamic, stepwise process. Such studies have suggested that there is an initial attachment of cells to a matrix-associated ligand via ligand-specific cell surface receptors. This is followed by a recruitment of receptors to the ventral surface of the cell and a reorganization of subsurface cytoskeletal elements. These events are thought to be necessary for cell spreading and the subsequent development of several kinds of morphologically distinguishable cell–substratum adhesion sites. The work leading to the development of this model has been discussed and reviewed recently by Grinnell (1978, 1980, 1982), Hughes *et al.* (1979), and Rees *et al.* (1981). A second general approach to the study of cell–substratum adhesion has been directed toward the identification of particular molecules present at sites of cell–substratum adhesion, and the dissection of the morphological relationships among them. These efforts include the study of both adhesion-promoting matrix-associated glycoproteins, such as fibronectin, laminin, collagens, and proteoglycans, and of subsurface microfilament-associated proteins, such as vinculin and α-actinin and more recently the src protein, which are concentrated at sites of cell–substratum adhesion. There have been several recent reviews on the biological properties and structure of fibronectin (Yamada and Olden, 1978; Hynes, 1981a,b; Ruoslahti *et al.*, 1981; Yamada, 1981, 1983; Kleinman *et al.*, 1981; Hynes and Yamada, 1982), the structure and function of laminin and other matrix glycoproteins (Kleinman *et al.*, 1981; Yamada, 1981, 1983; Hay, 1981; Hewitt and Martin, this volume), as well as proteoglycans (Culp, 1978; Rollins *et al.*, 1982; Hascall and Hascall, 1981). The properties of cytoskeleton-associated molecules and the transmembrane relationships between matrix and cytoskeletal components at sites of adhesion have also been reviewed recently (Hynes, 1981a,b; Geiger, 1981).

There are two generalizations emerging from the above studies that have particular significance for our discussion of integral membrane glycoproteins. The first is that many cell types grown in culture appear to express surface membrane receptors for more than one adhesion ligand. The second is that cell–substratum adhesion is a transmembrane process involving the coordinated behavior of both matrix- and cytoskeleton-associated molecules, none of which have been reported to be integral membrane components. Both of these generalizations support the hypothesis that cells must have at least one, and more likely several, integral membrane components that span the membrane and mediate or coordinate the interaction of the matrix ligands that provide the adhesion signals and the cytoplasmic effectors of these signals.

This section will focus primarily on experiments designed to identify and purify integral membrane glycoproteins that may be involved in the transmembrane control of cell–substratum adhesion. Just as in the section on cell–cell adhesion, studies using the antibody approach will be emphasized, but not to the exclusion of other approaches. Before turning full attention to adhesion-related integral membrane glycoproteins, however, we would like to review briefly the evidence supporting the two generalizations introduced at the beginning of the section. This should provide some necessary background for our discussion of integral membrane glycoproteins.

3.1.1. Cells Have Cell Surface Receptors for More Than One Adhesion Ligand

Fibrogenic cells *in vivo* are embedded in an interstitial matrix consisting of type I and III collagen and fibronectin along with proteoglycans and GAGs. Normal fibroblasts *in vitro*, when presented with an appropriate surface (Grinnell, 1983), will secrete adhesion molecules in the course of attachment and spreading and with time will produce an elaborate pericellular matrix containing fibronectin, collagen, proteoglycans, GAGs, and smaller amounts of several other glycoproteins (Hedman *et al.*, 1979, 1982; Carter and Hakomori, 1981; Carter, 1982; Lehto *et al.*, 1981, 1983). Fibroblasts *in vitro* can attach to fibronectin-coated substrates and to denatured collagen using their own or exogenously added fibronectin as an intermediary ligand (Grinnell and Minter, 1978). Fibroblasts can adhere to native collagen in the absence of fibronection (Grinnell and Minter, 1978; Schor and Court, 1979) and in the presence of antifibronectin (Grinnell and Minter, 1978), suggesting that they have receptors for native collagen as well as for fibronectin. Recent results suggest that fibroblasts can also adhere to laminin coated substrates (Couchman *et al.*, 1983).

Epithelial and endothelial cells *in vivo* adhere to a thin sheet of extracellular matrix, the basal lamina, which is differentiated into a lamina lucida directly under the epi/endothelium and a thicker lamina densa (see Hewitt and Martin, this volume). The large glycoprotein laminin, which is synthesized by epithelial cells, has been localized to the lamina lucida (Madri *et al.*, 1980; Foidart *et al.*, 1980), while basement membrane collagen (type IV), also produced by epithelial cells, is in the lamina densa. Proteoglycans, in particular heparan sulfate proteoglycan, are also present in basal laminae. Laminin is considered to be the principal attachment protein for epithelial cells. However, fibronectin has also been localized in the basal lamina of adult kidney (Courtoy *et al.*, 1980) and is present in embryonic basement membranes (Mayer *et al.*, 1981; Newgreen and Thiery, 1980) and at sites of epithelial cell migration during wound healing. Endothelial cells secrete fibronectin (Birdwell *et al.*, 1978) as well as laminin and incorporate both into their basal lamina. Whether epithelial cells use both fibronectin and laminin as adhesion ligands *in vivo* is not known. Corneal epithelial cell sheets stripped of their basement membrane can bind either soluble laminin or collagen type IV directly to their exposed basolateral surface. Following such binding, the blebbing basal surfaces of these epithelia become smooth and revert to the morphology of the intact corneal epithelium (Sugrue and Hay, 1982). Hepatocytes and hepatoma cells have been reported to adhere *in vitro* directly to collagen (Rubin *et al.*, 1981), laminin (Jöhansson *et al.*, 1981), or fibronectin (Höök *et al.*, 1977; Jöhansson *et al.*, 1981). The hepatocytes will adhere and spread onto laminin-coated substrates in the presence of both antifibronectin and cycloheximide and onto fibronectin-coated substrates in the presence of cycloheximide and antilaminin. The spreading onto either ligand can only be reversed by the homologous antibody. Whether the same cell surface molecules bind to each of the extracellular matrix components, or whether there are unique membrane molecules required for each substratum component is not resolved unequivocally in these experiments. Adhesion variants of Chinese hamster ovary (CHO) cells, which are unable to attach to fibronectin-coated substrates, will attach to a substrate formed from material left behind by cell monolayers following their removal with EGTA [substrate-attached material (SAM)]. The identity of the ligand in SAM to which cells are able to attach is unknown, but attachment to SAM by either wild-type or variant cells takes place in the presence of antifibronectin and is much more sensitive to pretreatment of cells with low levels of trypsin than its attachment to fibronectin (Harper and Juliano, 1980, 1981a,b). These data suggest that CHO cells have distinct receptors on their surfaces for at least two different adhesion ligands.

Other glycoproteins besides laminin, collagen, and fibronectin that may play a role in cell–substratum adhesion are present in extracellular matrices of epithelial and fibroblastic cells *in vivo* and *in vitro*. Some of these have tissue specificity (see Hewitt and Martin, this volume). For example, the basal lamina of stratified squamous epithelia contains the bullous pemphigoid antigen (BPA) (Diaz and Marcello, 1978; Stanley *et al.*, 1981b). This is a disulfide-linked glycoprotein with chains of 220K distinct from fibronectin. It is clearly involved in adhesion, as patients with antibodies to this protein have peeling and blistering of their skin at the dermal–epidermal border. In the new outgrowth of skin epithelium at a wound site, BPA localizes to the leading edge of the advancing (and proliferating) epithelial sheet and appears before laminin and type IV collagen, which are organized into the new basal lamina behind the advancing epithelial tip (Stanley *et al.*, 1981a). Nothing is known about the cell surface receptor for this molecule. Finally, analysis of sera used for tissue culture has shown that there is at least one glycoprotein of 70K (Knox and Griffiths, 1979; Hayman *et al.*, 1982) distinct from fibronectin that will promote the attachment and spreading of cultured cells.

The studies summarized above demonstrate that there are many potential adhesion ligands available for cells to use. There is experimental evidence that a particular cell type can express receptors for more than one adhesion ligand *in vitro*, although the *in vivo* relevance of these multiple ligand–receptor systems has not been shown. How cells decide which ligand(s) to use at particular times during attachment, spreading, migration, or differentiation remains a fascinating but unanswered question.

3.1.2. Adhesion Is a Transmembrane Process

Fibroblasts adhere to a substratum *in vitro* (produced with contributions from the cells themselves and the medium) via specialized regions of their ventral surfaces. Two types of specializations, close contacts and focal contacts, have been distinguished by both interference reflection microscopy (IRM) (Curtis, 1964; Izzard and Lochner, 1976; Badley *et al.*, 1978; Heath and Dunn, 1978) and electron microscopy (Abercrombie *et al.*, 1971; Revel and Wolken, 1973; Badley *et al.*, 1980; Heath and Dunn, 1978; Chen and Singer, 1980, 1982). Close contacts are defined as regions of the ventral cell surface that come to within 30–50 nm of the substratum. These are relatively broad areas of membrane that appear gray by IRM. They appear to be the principal type of adhesion structures in highly motile cells as well as in many transformed fibroblast lines. Subcellular accumulations of microfilaments visible by electron microscopy appear on the underside of the surface membrane in these regions (Fig. 6). The filaments have both lateral and end-on associations with the surface membranes. Focal contacts (adhesion plaques) are relatively discrete sites of very close (< 20 nm) association of the cell surface with the substratum. These appear as arrays of elongated black streaks when well-spread cells are observed by IRM. These are particularly prominent near the cell periphery. Focal contacts are sites of end-on terminations of highly organized subsurface microfilament bundles (MFB) (Heath and Dunn, 1978; Singer, 1979; Chen and Singer, 1982) (Fig. 6). Correlated IRM and immunofluorescence have shown that the microfilament-associated proteins α-actinin (Wehland *et al.*, 1979; Geiger *et al.*, 1980) and vinculin (Geiger, 1979; Burridge and Feramisco, 1980) as well as the src gene product (Rorschneider *et al.*, 1981) are concentrated where MFB terminate at sites of focal contacts (reviewed in Hynes, 1981a,b; Heath, 1982). Using electron microscopy, Chen and Singer (1982) have recently identified additional sites of cell–substratum interaction called extracellular matrix (ECM) contacts. These occur in regions where the cell surface comes in close contact with substantial aggregations of extracellular material (Fig. 6). Vinculin is associated with some, but not all, of these sites.

The relationship of ECM molecules, in particular fibronectin, to the submembranous cytoskeleton at these adhesion specializations has been closely studied. It is agreed that both fibronectin and the submembranous actin-containing MFB that terminate at focal contacts become arranged in very similar fibrillar patterns in well-spread, growth-arrested, non-transformed fibroblasts and that the organization of these two orderly arrays can be manipulated in a coordinate fashion by treating cells with antifibronectin, cytochalasin B, transforming agents, and exogenous fibronectin (reviewed by Hynes, 1981a,b; Hynes *et al.*, 1981; Geiger, 1981; Birchmeier *et al.*, 1981). Some studies suggest that in growth-arrested cultured fibroblasts, the fibrillar fibronectin pattern is colinear (not necessarily coextensive) with the termini of MFB at focal adhesions as identified by IRM and immunofluorescence using antivinculin (Hynes and Destree, 1978; Singer, 1979, 1982). This is disputed by other investigators (Chen and Singer, 1980, 1982; Birchmeier *et al.*, 1980; Fox *et al.*, 1981) whose studies suggest that fibronectin is distributed adjacent to the focal adhesion sites in areas of close contacts. The studies of Chen and Singer (1982) using the frozen thin-section technique (Tokuyasu and Singer, 1976) and double-label immunoelectron microscopy provide excellent resolution of cell–substratum contact sites. Their results indicate that α-actinin is present near all types of contact sites, whereas vinculin and fibronectin have a complementary distribution; vinculin is present at focal contacts

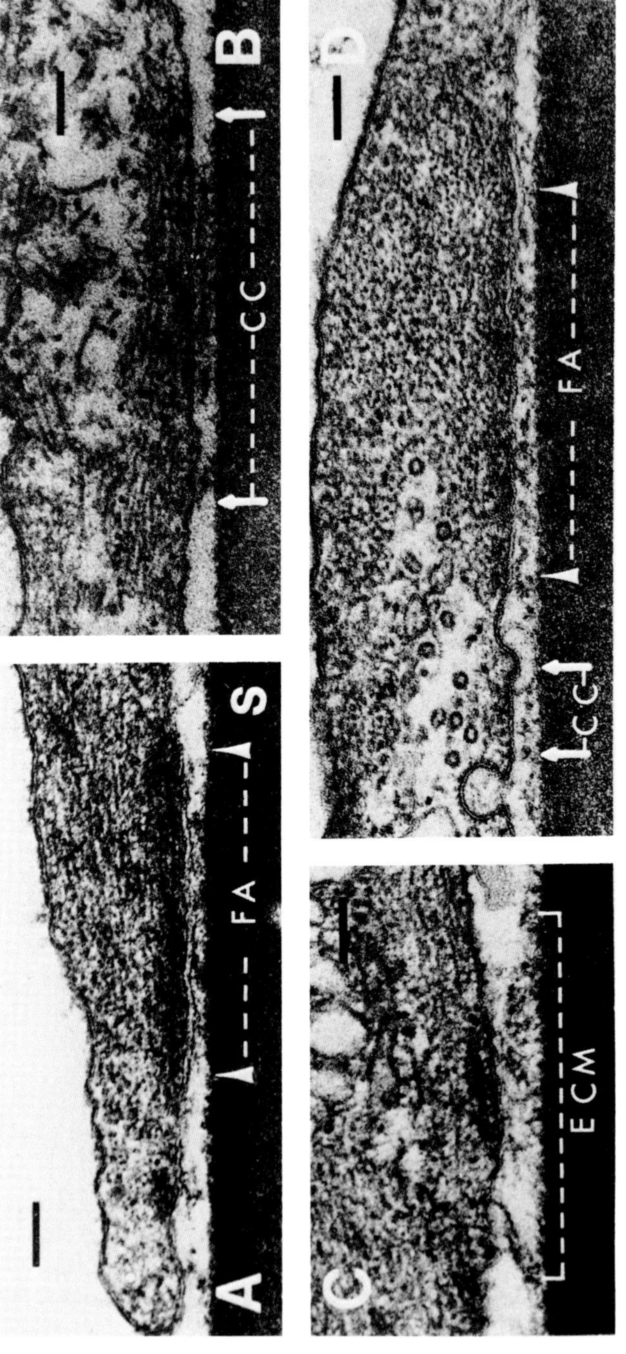

Figure 6. Transmission electron micrographs of regions where the ventral surface of a fibroblast approaches the substratum. Such areas are presumed to be sites of adhesion of these cells to the substratum. Based on the distance of the cell surface from the substrate, attachment sites have been classified (originally using IRM) as either focal adhesions (FA; A, D) or close contacts (CC; B, D). More recently, a third category, extracellular matrix contact (ECM; C), has been proposed (see text for further description). (From Chen and Singer, 1982, with permission.)

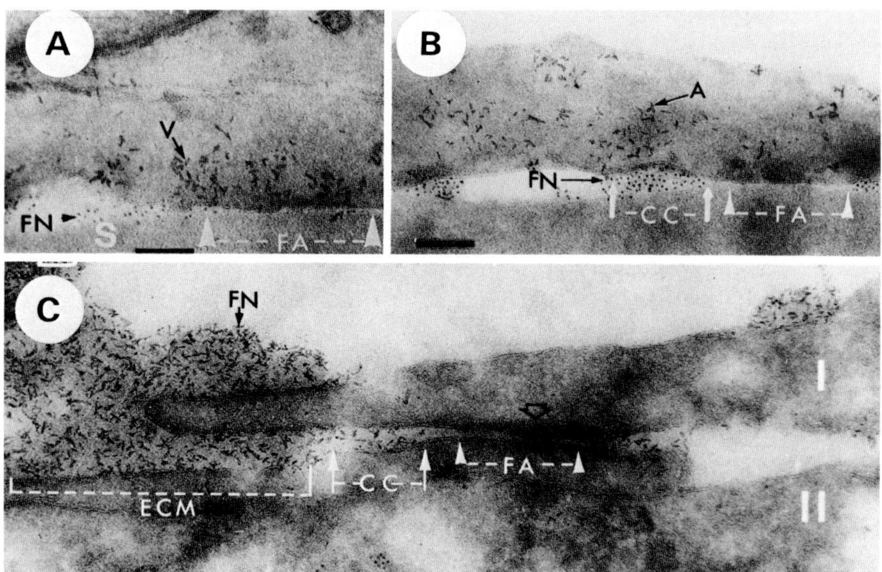

Figure 7. Immunoelectron microscopy of cell–substratum adhesion sites using antibodies to fibronectin (A–C) and either vinculin (A) or α-actinin (B). Double-immunoelectron microscopy provides excellent resolution of the distribution of molecules at sites of cell–substratum attachment. These suggest that fibronectin and vinculin usually have a complementary distribution. This is in contrast to other workers who suggest that fibronectin and vinculin often lie directly across the membrane from one another. Many factors affect the interpretation of data on both sides of this issue (see text for discussion). (From Chen and Singer, 1982, with permission.)

and a subset of ECM contacts, while fibronectin is found at close contacts and the subset of ECM contacts lacking vinculin (Fig. 7).

The conflicting results on this point may result in part because of an assumption that focal contacts observed by IRM are always equivalent to adhesion plaques observed by electron microscopy. Furthermore, electron microscopy has the potential to resolve variability within focal or close contact areas that would be considered homogeneous by either IRM or immunofluorescence. In addition, discrepancies may arise from differences in the serum content of growth medium, the growth state of cells, length of time in culture, and/or the composition and pH of fixatives, all of which have been reported to affect the morphology of adhesion sites (Singer, 1982; Avnur and Geiger, 1981; Chen and Singer, 1982). Whether or not it is actually present in focal contacts, a recent study suggests that cellular fibronectin may be required for the final coordinated development of focal contacts and well-organized MFB. Couchman *et al.* (1982) showed that over a 48-hr period, cardiac fibroblasts, migrating out of fresh cardiac tissue explants, converted from a highly motile nonreplicative population with close contacts but no focal contacts or MFB (stage I), to a less motile replicative population with prominent focal contacts and MFB (stage II). Cellular fibronectin as well as other cellular glycoproteins were required for this conversion process. The presence of anti-cellular fibronectin prevented the development of the focal-contact MFB arrays (transition to stage II), whereas addition of exogenous cellular fibronectin in late stage I accelerated the transition to stage II. In apparent contrast to these results, Couchman *et al.* (1983) report that human fibroblasts can attach and spread on laminin-coated substrates and can form MFB and focal contacts when plated on laminin in fibronectin-depleted serum even if they have been pretreated with cycloheximide and are plated in the presence of cycloheximide and antifibronectin. Thus, under these conditions, neither plasma nor cellular fibronectin seems to be required for the elaboration of the focal-contact/MFB arrays, and although cellular fibronectin may in some way promote their formation, it is not unique in its ability to do so. This suggests that some intermediate cellular events that can be triggered by more than one mechanism or matrix ligand are ultimately responsible for the organization of MFB/focal-contact arrays. It is important to keep in mind that these studies have been conducted on cells growing *in vitro* and that close contact and focal contact are *in vitro* terms. Now that specific antibodies to matrix and cytoskeletal components have been obtained, investigation of molecular relationships *in vivo* can be pursued more vigorously. Whatever the nature of the relationship of fibronectin to adhesion plaques and subsurface MFBs, it is clear that the organization of ECM molecules and cytoskeletal components at sites of cell–substratum adhesion is controlled in a coordinate fashion. Presumably one or more transmembrane glycoproteins are involved. Efforts to identify these molecules are described in the next sections.

3.2. Cell Surface Receptors for Matrix Molecules

Several basic approaches have been used over the past few years to identify integral membrane glycoprotein receptors for matrix molecules relevant to cell–substratum adhesion resulting in a rather long list of candidate membrane molecules at M_r from 47 to 265K (Table 2). A few years ago, there would not have been much to say on the subject. Now the waters are muddy indeed. Thus, in one sense, rapid progress is being made. However, it will be clear from the subsequent sections that in no case has there been a definitive identification of an integral membrane molecule as a receptor for a particular substratum-associated ligand. No receptor has been purified to homogeneity and shown to directly affect

Table 2. Integral Membrane Glycoproteins Thought to Be Involved in Cell–Substratum Adhesion

Surface membrane glycoprotein	Source	Speculated role in adhesion	How identified	Associated antibody (see text)	Additonal characteristics of antigens	References
140K-region glycoproteins	Hamster fibroblasts and adherent melanoma	Unknown	Broad-spectrum, adhesion-disrupting antisera, followed by biochemical isolation	Anti-GP	Neither species- nor cell type-specific. Present on all mammalian adherent cells tested except macrophages. Reduced or undetectable in nonadherent melanoma, thymocytes.	Wylie *et al.* (1979), Damsky *et al.* (1979), Knudsen *et al.* (1981, 1982a)
	MMTE	Unknown	As above	Anti-SFM I	As above	Damsky *et al.* (1981, 1982)
140K-region glycoproteins	Chick myoblasts	Unknown	Adhesion-disrupting monoclonal antibody	CSAT		Neff *et al.* (1982)
	Chick fibroblasts		Immunoaffinity chromatography using immobilized monoclonal antibody CSAT		SDS–PAGE pattern: Nonreduced → Reduced 160 → 140 135 → 115 114 → 125	Neff *et al.* 1982, Knudsen *et al.* (1982b)
					Appears to surround focal-contact sites by immunofluorescence	Damsky *et al.* (1983b)
					Also present on other adherent cell types that are not rounded by CSAT but whose morphology is affected by the antibody	Decker and Horwitz (1982)
140K-region glycoproteins	Chick myoblasts and fibroblasts	Unknown	Immunoaffinity chromatography using immobilized monoclonal antibody JG22	JG22	Likely to be same antigen as that recognized by CSAT	Greve and Gottlieb (1982)
					Localized in fibroblasts to close but not focal contacts using immunoelectron microscopy	Chen (1982)

(Continued)

and Singer, 1982; Chen, 1982) took advantage of the much greater resolution provided by frozen thin-section double-immunoelectron microscopy to localize the JG22 antigen on fibroblasts with respect to α-actinin, vinculin, and fibronectin. Those results suggested that the JG22 antigen was located both adjacent to focal contacts, in regions termed close contacts, and with the subset of ECM contacts containing fibronectin and α-actinin, but not vinculin. The similar localization pattern and the fact that JG22 interferes with the ability of purified CSAT antigen to block CSAT monoclonal antibody-induced changes in adhesiveness (Knudsen and Horwitz, personal communication) suggest that these two monoclonal antibodies recognize the same antigen.

Determining why some cell types are rounded and detached by monoclonal antibodies while other cell types are not, even though they clearly express a similar antigen, may prove to be very difficult. CSAT antibody clearly has the greatest effect on cells of skeletal myogenic lineage. It was even able to detach BudR-treated myoblasts which continue to proliferate, do not fuse, and resemble fibroblasts in morphology. The response to antibody by BudR-treated myoblasts, however, could be modulated by calcium levels, time in culture, and the presence of cytochalasin (Decker and Horwitz, 1982). If BudR myoblasts were plated in the presence of low calcium, they were detached by antibody even after several days in culture. If normal levels of calcium were present in the medium, the BudR myoblasts became more and more refractile to the antibody with culture age. If cytochalasin D (CD) was included in the medium along with the antibody, the older BudR-treated myoblasts were more easily rounded and detached by the antibody. Most chick embryo fibroblasts were not detached by the antibody even in the presence of low calcium, CD, and antibody (Decker and Horwitz, 1982). Thus, different cell types clearly have different strategies for substrate adhesion. The adhesion mechanism for skeletal myogenic cells can be subverted by the binding of one monoclonal antibody. This antibody binds equally well to fibroblasts (Neff *et al.*, 1982), but in most cases does not cause detachment even at 50-fold higher concentrations of monoclonal antibody. Fibroblasts either have more than one distinct adhesion mechanism, or the antigen recognized by the monoclonal antibody is complexed with auxiliary molecules in a manner not found in myoblasts. In either case, a more complex antiserum would seem to be required to detach fibroblasts. The fact that fibroblasts display several morphologically distinguishable types of adhesion sites in culture supports this suggestion, as do the experiments described previously suggesting that fibroblasts are able to use different membrane receptors to attach and spread on different adhesion ligands. Of great interest in this regard is the recent report of Oesch and Birchmeier (1982). They isolated a monoclonal antibody (FC-1) that delayed the adhesion of chick fibroblasts to culture dishes. In double-immunofluorescence studies, FC-1 and antivinculin staining were superimposable, suggesting that the FC-1 antigen resides at focal-contact sites, as opposed to CSAT and JG22 antigens, which reside adjacent to focal-contact sites. Biochemical studies suggest that the FC-1 antigen has an M_r of about 60K (Oesch and Birchmeier, 1982).

The morphology of adhesion specializations of myoblasts has not been well studied. The fact that their adhesion can be manipulated by a single monoclonal antibody suggests that myoblasts may contain only a single type of matrix–membrane complex for substratum adhesion. The fact that the monoclonal antibody binds to other cells as well suggests that this adhesion mechanism represents part of the adhesion strategy found in other cell types. The existence of multiple adhesion strategies for different cells of a given tissue may be of great importance *in vivo*. Such variability would permit a graded response by different cell types within a tissue to environmental stimuli that affect adhesiveness, e.g., during development when selected groups of cells are stimulated to migrate while others nearby remain stationary. Although the mechanisms to account for the phenomena described above are unknown at this time, it is clear that monoclonal antibodies recognizing different domains of various adhesion molecules may be extremely useful in dissecting the process into its component parts.

Studies in several laboratories using adhesion-perturbing broad-spectrum antisera, in combination with either antibody absorption studies using adherent and nonadherent cells, selective proteolysis, or both, have implicated other candidate molecules in the process of cell–substratum adhesion. Hsieh and Sueoka (1980) prepared antiserum against whole rat neuronal cells that inhibited spreading of these cells. Using antibody absorption by different adherent and nonadherent cell lines and SDS–PAGE analysis, these workers showed that adherence correlated with the presence of 120 and 80K surface glycoproteins. Tarone *et al.* (1982), Oppenheimer-Marks and Grinnell (1982), and Oppenheimer-Marks *et al.* (1982) have utilized broad-spectrum antisera in conjunction with selective proteolysis and antibody absorption experiments to identify membrane molecules that affect cell–substratum adhesion on fibronectin-coated dishes. Adhesion and spreading onto fibronectin-coated dishes in the absence of serum were not impaired by treating cells with trypsin in the presence of Ca^{2+}. Because some surface components were degraded under these conditions, correlations could be drawn between resistant polypeptides and the ability of cells to adhere to fibronectin. Treatment of cells with trypsin–Ca^{2+} and subsequent immunoprecipitation of nonionic detergent extracts using an anti-BHK serum that rounds and detaches cells from fibronectin coated surfaces revealed polypeptides of 130, 120, and 80K

(Tarone *et al.*, 1982). If cells were treated with chymotrypsin or trypsin in the presence of Ca^{2+} and then used to absorb cell–substratum detachment activity from an antiserum directed against BHK wheat germ receptors (anti-WG), only cells that retained glycoproteins of 47, 83, and 105K after proteolysis absorbed the adhesion-disrupting activity from the anti-WG receptor serum (Oppenheimer-Marks *et al.*, 1982). Both sets of experiments present a highly restricted number of glycoproteins as candidates for a fibronectin receptor.

3.2.2. Use of Adhesion Variants

Use of adhesion variants has produced some interesting results over the last several years (reviewed by Briles, 1982). The power of this approach should increase greatly with the advent of monospecific antisera to several different adhesion-related molecules. Unfortunately, in many of the variants used to date, the nature of the defect is unknown or the defect affects a biosynthetic pathway such as protein glycosylation and therefore affects many other molecules than just those involved in adhesion.

The AD6 mutant of a mouse fibroblast line adheres poorly to various substrata. These cells are unable to acetylate glucosamine and all of their glycoproteins are poorly glycosylated. Normal glycosylation and normal adhesion interactions can be restored by supplementing the culture medium of AD6 cells with *N*-acetylglucosamine (Pouyssegur and Pastan, 1977).

Hughes and his colleagues (Edwards *et al.*, 1976; Pena *et al.*, 1979) have produced cell lines resistant to the toxic lectin *Ricinus communis*. Some of these variants lack certain enzymes in the glycosylation pathway and therefore do not have exposed β-galactose residues, which are required for binding to ricin. The polypeptides synthesized have altered oligosaccharide chains presumably causing the cells to adhere poorly to fibronectin-coated dishes. However, these cells were able to attach and spread on surfaces coated with anti-BHK serum or with Con A, suggesting that the cytoskeletal machinery required for spreading was normal. As described later, these workers have used such mutants, along with normal BHK cells, in chemical cross-linking studies designed to identify the cell surface fibronectin receptor (Aplin *et al.*, 1981).

Harper and Juliano (1980, 1981a,b) have isolated a variant of CHO cells defective in its adhesion to fibronectin-coated substrates. These cells attached and spread on SAM produced by human diploid fibroblasts, and as such displayed an adhesion mechanism that appeared not to involve fibronectin. The wild-type CHO cells had receptors for both SAM and fibronectin, for they spread on SAM in the presence of antifibronectin, which inhibited completely their attachment and spreading of fibronectin-coated dishes. The two adhesion systems differ in their protease sensitivity, the adhesion to SAM being much more sensitive to trypsinization in the absence of Ca^{2+} than adhesion to fibronectin. In correlating the loss of adhesion to SAM following mild trypsinization with changes in the cell surface glycoprotein pattern as analyzed by SDS–PAGE, the best candidate for the cell surface mediator of adhesion to the ligand in SAM appeared to be a 265K surface glycoprotein. However, a role for a 140K component, which is also trypsin-sensitive, was not ruled out. The matrix-associated ligand component in SAM for this cell surface glycoprotein is not known at this time. These experiments clearly point out a fibronectin-independent adhesion system in CHO cells.

Schubert and La Courbiere (1982) have studied adhesion variants of the rat skeletal muscle cell line L6. Both lines shed complex aggregates of glycoproteins and GAGs termed "adherons." Adherons from the control L6 line will support substratum adhesion in the absence of serum of both L6 cells and the nonadherent variant M3A in a Ca^{2+}-dependent manner. M3A adherons will not support the adhesion of either cell line. The electrophoretic patterns of the adherons from the two sources are very different. The wild-type adherons have significantly more fibronectin and collagenous proteins as well as several other components of unknown identity. The M3A adherons, on the other hand, are enriched in a 50K component. As the gel patterns are very complex, it is not yet possible to point to a particular set of molecules as being responsible for the altered adhesion activity. However, the wild-type adherons clearly contain an isolated supramolecular complex that is capable of promoting adhesion. Careful dissection of this complex and comparison with the adhesion-defective particles should contribute to our understanding of intermolecular associations in adhesion.

3.2.3. Looking at What Cells Leave Behind

3.2.3a. SAM. Adherent cells can be removed from their substrata by mechanical force, by chelating agents, or they can be extracted *in situ* using low concentrations of nonionic or zwitterionic detergents. In all cases, material remains behind that contains both known matrix or pericellular components such as fibronectin, collagens, and proteoglycans, as well as intracellular cytoskeletal proteins. In light of considerable evidence that the distribution and organization of fibronectin and cytoskeletal proteins can be manipulated in a coordinate fashion, it might be expected that the integral membrane molecule(s) mediating these and other transmembrane adhesion interactions would be enriched in SAM. This reasoning has led several investigators to look for integral membrane adhesion-related molecules in SAM. The extensive studies by Culp and colleagues on SAM (reviewed in Culp, 1978; Rollins *et al.*, 1982), Rees and colleagues (Badley *et al.*, 1978, 1980; Bayley and Rees, 1982), and Avnur and Geiger (1981) have led to the now accepted notion that substrate-adhesion sites represent a biochemically as well as structurally differentiated area of the cell surface. The work of Culp and colleagues has concentrated primarily on the role of GAGs and proteoglycans in adhesion. Recent work of Kjellén *et al.* (1981), Laterra *et al.* (1983), and Rapraeger and Bernfield (1982, 1983) has implicated a surface membrane-associated heparan sulfate proteoglycan in substratum adhesion. Rapraeger and Bernfield (1983a,b) have reported that the membrane-bound heparan sulfate proteoglycan can be inserted into lipid vesicles and can associate with cytoskeletal elements *in vitro*. The relationship of the membrane-associated proteoglycan to putative integral membrane glycoprotein receptors for adhesion matrix ligands is not understood but is of considerable interest. Badley *et al.* (1978, 1980) and Bayley and Rees (1982) have focused on cytoskeletal molecules and ricin-binding membrane glycoproteins that remain behind after cells are mechanically removed from the substratum with a vigorous stream of buffer. A class of ricin receptors distinct from fibronectin were found to be enhanced at isolated adhesion sites and at the leading edges of cells. The distribution of these receptors on cells could be distinguished from that of fibronectin by monitoring both FITC–ricin and antifibronectin staining. Areas uniquely stained by the FITC–ricin correlated better than antifibronectin staining with the termini of stress fibers and with focal adhesions visualized by IRM. The behavior of these ricin receptors was also more strongly correlated with the integrity of focal contacts in response to mild trypsinization and cytochalasin B than the behavior of fibronectin. Comparison of electrophoretic profiles from focal adhesions and whole plasma membranes reveals two ricin receptors of cellular origin with M_r of 80 and 160K enriched in focal adhesions. By contrast, Con A receptors were present to a lesser extent in adhesions than in whole membranes. These authors conclude from their combined biochemical and ultrastructural analyses that the 80 and 160K membrane ricin receptors are potential candidates for mediating the transmembrane association of the cytoskeleton at focal adhesions with ligands in the substratum.

3.2.3b. Detergent-Insoluble Matrices. Another approach to looking for surface membrane receptors for matrix molecules is to extract cells *in situ* with low concentrations of nonionic or zwitterionic detergents. When this is done, a detergent-insoluble framework is left behind with the same overall shape as the intact cell. This framework contains the nucleus in a network of fibrous cytoskeletal elements and an extracellular or pericellular matrix that contains GAGs, collagenous and noncollagenous glycoproteins. Carter and Hakomori (1981) have reported the presence of four major glycoproteins in this matrix, gp250, gp220 (fibronectin), gp170, and gp140. Similar studies by Lehto *et al.* (1980, 1981, 1983) and Virtanen *et al.* (1982) have focused on a 140K glycoprotein as the major noncollagenous glycoprotein in the detergent-resistant surface lamina of normal cells. Studies on the 140K glycoprotein by both groups have shown it to be quite trypsin-resistant. Like collagen, gp140 has significant amounts of both hydroxylysine and hydroxyproline (Carter, 1982). However, gp140 is not susceptible to collagenase. The presence of these amino acids may enable this molecule to bind to fibronectin in the matrix via the latter's collagen-binding site (Carter, 1982). The presence of hydroxylysine and hydroxyproline in this 140K glycoprotein distinguishes it from those identified in chick fibroblasts using the CSAT monoclonal antibody by Knudsen *et al.* (1982b). Amino acid analysis has shown that the latter glycoproteins contain no hydroxylated amino acids (Knudsen, personal communication).

The presence of gp140 as a detergent-insoluble molecule was sensitive to both cell density and transformation. Whether gp140 was produced at the same rate in sparse and transformed cells has not been reported. Gp140 has been partially purified under nonreducing conditions from the detergent-insoluble matrix and in that form will support the attachment of cells when immobilized on the substratum (Carter, 1982). However, as gp140 was not the only component present in the partially purified material, some other component may have been responsible for the attachment and spreading activity. Gp140 has been purified more extensively under reducing conditions, though it loses its cell attachment-promoting activity as a result (Carter, 1982). Lehto (1983) has reported an association of gp140 with vimentin-containing intermediate filaments in detergent-resistant cytoskeletal preparations.

3.2.4. Chemical Cross-Linking Agents

Chemical cross-linking agents are potentially very useful for analyzing nearest-neighbor relationships among molecules, and could shed light on the identity of cell surface receptors for adhesion ligands. The most useful cross-linking agents have been hetero-bifunctional molecules that carry a photoactivatable group. With these agents, one reactive group can be preattached to a ligand in the dark and then the ligand–cross-linker complex exposed to the cell surface in the light. A second important feature is that the photoactivatable group should be potentially reactive with a variety of chemical groups, so there is no bias against a nearby molecule that presents a molecular configuration for which the reagent has low affinity. Finally, the cross-linker should be cleavable so that fol-

cells that inhibit their cell–substratum adhesion *in vitro* and their metastatic behavior *in vivo*. These antibodies do not react with normal mouse tissues.

The use of monoclonal antibodies has also enabled us to broaden our concept of the organization of membrane molecules at sites of substratum adhesion. For example, studies with the monoclonal antibody CSAT (Neff *et al.*, 1982; Knudsen *et al.*, 1982b) indicate that the membrane components of close-contact adhesion sites may exist as a complex of three glycoproteins rather than as a single transmembrane molecule. If so, then working models such as those proposed by Chen and Singer (1982) may need to be modified to show such complexes. Not all the molecules in these complexes need span the membrane to link cytoskeletal and extracellular elements. Instead, one may serve to interact with the cytoskeleton, one with the extracellular environment, and one to coordinate the interaction of the other two glycoproteins. Obviously, any modification of this scheme is possible. The important point is that we must begin to consider adhesion models in which more than one adhesion strategy exists and in which more than one membrane molecule can be involved at any one site. It may also be that the composition of the various contact sites does not remain static and is not limited to the family of 140K glycoproteins or the 60K proteins, but may be modified to include other elements depending, as stated earlier, on the substratum with which the cell is interacting or the state of growth or differentiation of the cell. Hence, these must be considered dynamic structures that can be modified in response of the cell to its environment.

An important point to keep in mind is that most of the work described here has been done with cultured cells. Whether or not these same adhesion complexes function *in vivo* has yet to be directly demonstrated. Immunofluorescence and ultrastructural studies show that many of these molecules are expressed by cells *in vivo* at presumptive adhesion sites, but no studies exist demonstrating that masking these molecules *in vivo* with antibodies prevents cellular adhesion or that these antibodies can affect *in vivo* adhesion in a demonstrable manner.

Many adhesion-related membrane glycoproteins and many matrix-associated molecules that are clearly involved in cell–substratum adhesion have been identified. In general, it has not yet been possible to associate a particular membrane glycoprotein with a particular matrix molecule. Two very recent reports, however, have identified possible receptors for two different matrix molecules using immobilized matrix molecules and detergent extracts of cells. Mollenhauer and van der Mark (1983) have isolated a 31K integral membrane glycoprotein from extracts of chondrocyte membranes using a type II collagen–Sepharose affinity chromatography. The 31K molecule can be inserted into liposomes and interacts with the ends of type II collagen molecules *in vitro*. Using a laminin–Sepharose column, Malinoff and Wicha (1983) have identified a protein from extracts of mouse fibrosarcoma cells with a high binding affinity for laminin (2nM) that migrates as a 69K molecule on reduced SDS–polyacrylamide gels. So far, this approach has not been successful in identifying a fibronectin "receptor." While it is clear that the cellular adhesion is promoted by many different molecules (fibronectin, laminin, GAGs, SAM, etc.), precisely how these molecules and membrane components interact with one another is not known.

3.3. Looking to the Future

The identification of adhesion-related integral membrane glycoproteins is, of course, only the first step. How these molecules function and the nature of their interactions with one another and with components in the matrix and the cytoskeleton must then be elucidated. These efforts will take many different directions and will include approaches that use both specific antibodies to membrane glycoproteins and the purified glycoproteins themselves. At all points efforts must be made to relate observations made *in vitro* to *in vivo* adhesion.

Specific antibodies can be used to localize their respective antigens at both the light and electron microscope levels in well-spread cells, in cells during the course of spreading, and in conjunction with matrix- or cytoskeleton-associated antigens. Studying the effects of specific antibodies on living cells in culture and in *in vivo* (particularly in embryonic systems) can help to elucidate the function of the particular antigens. This might be especially interesting in situations where an antibody to one adhesion component binds, but cannot itself round and detach cells. This is the case with the monoclonal antibody that detaches chick myoblasts but not chick embryo fibroblasts (Neff *et al.*, 1982). It is conceivable that such an antibody may induce subtle changes in morphology, motility, or the distribution of adhesion contacts in cells.

One area beginning to receive attention involves dissecting the sequence of molecular events in the adhesion process and determining exactly when a particular molecule is involved. One would predict that if a component were a matrix ligand receptor, it would be involved at a very early time point in adhesion, whereas if it were involved in spreading or locomotion, a later point would be implicated. To distinguish such events, an assay would be needed that is able to dissect the timing of adhesion events precisely, as well as either a specific antibody against a particular component or a mutant lacking this component. The assay developed by McClay *et al.* (1980) is well suited to the task of analyzing early events in adhesion. It is possible by this technique to quantitate the number of cells attached to a particular substratum, to assess the strength of cell attachment, and to measure the degree of cell spreading. This assay also allows one to control the time of contact between cells and substratum and the temperature at which the interaction is carried out.

The isolation and characterization of integral membrane adhesion glycoproteins themselves will enable investigators to study the direct effects of these antigens on adhesion. So far, most studies of these antigens have been indirect, using antibody inhibition rather than antigen promotion or inhibition of adhesion. Because the antigens are hydrophobic and have a membrane as their natural environment, it may be necessary to incorporate an isolated antigen into lipid vesicles to study its effects on adhesion. These kinds of studies have been done with the neural cell–cell adhesion molecule [N-CAM (Rutishauser *et al.*, 1982); see section on cell–cell adhesion].

Isolation of the antigen should also permit the generation of proteolytic fragments representing domains that interact only with the substratum, the membrane, or the cytoskeleton. Studies with fibronectin provide an excellent model for the kinds of experiments possible using particular domains and their specific antibodies. Studying interactions between any cytoplasmic domain of the molecule and the cytoskeleton presents a particular challenge because of the inaccessibility of the components in the intact cell. Inside-out vesicles or resealed cell ghosts have been used to study the interaction of the integral membrane glycoprotein (Band 3) with the cytoskeleton in the erythrocyte (reviewed in Branton *et al.*, 1981). Some features of the erythrocyte model may be transposable to other cells. For instance, Goodman *et al.* (1981), Bennett *et al.* (1982), and Burridge *et al.* (1982) have found spectrinlike molecules in other cell types. Finally, models for studying membrane–cytoskeleton interactions in *Dictyostelium* (Luna *et al.*, 1981) and isolated intestinal microvilli (Coudrier *et al.*, 1981; Matsudaira and Burgess, 1981) are being established. Cell–substratum adhesion might lend itself to these same approaches as surface membrane preparations enriched in focal-contact sites, and also detergent-insoluble matrices of adherent cells are easily obtained.

3.4. Relationships between Cell–Cell and Cell–Substratum Adhesion

As one can see from the data presented in this review, cell–substratum and cell–cell adhesion have been studied using similar approaches. In fact, it is possible to think of the two types of adhesion within the same conceptual framework. Each involves the recognition of molecules on an adhesive surface, whether that surface is a cell or a substratum. The transmission of information in each case results in the reorganization of the cytoskeleton and of surface membrane molecules. As each type of adhesion appears to require some specificity, there must be many molecules unique to each system especially in the recognition and early attachment phases of these processes. This is illustrated in the MMTE system by the fact that disruption of cell–substratum and cell–cell interactions can be inhibited selectively by different sets of glycoproteins (Fig. 2) and in studies on N-CAM where anti-N-CAM will block neurite fasciculation and the interaction of neurites with myoblasts, but has no effect of the interaction of the neurites with the substratum (Rutishauser *et al.*, 1978b, 1983). In addition, the monoclonal antibody CSAT, which disrupts the cell–substratum adhesion of skeletal myoblasts, has no effect on myoblast fusion (Neff *et al.*, 1982). However, cell–cell and cell–substratum adhesion may share some molecules involved in the organization of adhesion sites. Chen and Singer (1982) suggest that there are sites of cell–cell adhesion in fibroblasts that are analogous to certain types of cell–substratum adhesion sites. This is based not only on morphological similarity, but also on the presence or absence of fibronectin and vinculin or α-actinin at these cell–cell interaction sites. In addition, in epithelial cells, vinculin has been localized to the zonula adherens of the junctional complex as well as to sites of cell–substratum adhesion (Geiger *et al.*, 1980). The extent to which overlap exists in the sets of molecules required to carry out cell–cell and cell–substratum adhesion will be learned only after careful localization of the various molecules thought to be involved and a greater understanding of their molecular properties.

The fields of cell–cell and cell–substratum adhesion are currently at an exciting stage. Thanks to the extensive study of tissue culture cells and other *in vitro* adhesion models, antibodies are being developed that are specific for integral membrane glycoproteins involved in adhesion. Together with similar probes specific for cytoskeletal and ECM molecules, these can now be used to dissect the transmembrane molecular complexes involved in adhesion and to understand the control of expression of these molecules *in vivo* in normal, malignant, and embryonic systems.

ACKNOWLEDGMENTS. Work originating from the authors' laboratory was supported by Grants CA-21319, CA-19144, CA-27909, CA-32311, HD-15663, and CA-10815. In preparing this review, we appreciate the helpful discussions and comments of Drs. A. F. Horwitz, S. Roth, and A. Ross.

REFERENCES

Abercrombie, M. J., Heaysman, J. E. M., and Pegrum, S. M., 1971, Locomotion of fibroblasts in culture. IV. Electron microscopy of the leading lamella, *Exp. Cell Res.* **67**:359–367.

Aplin, J. D., and Hughes, R. C., 1982, Complex carbohydrates of the extracellular matrix: Structures, interactions and biological roles, *Biochim. Biophys. Acta* **694**:375–418.

Aplin, J. D., Hughes, R. C., Jaffe, C. L., and Sharon, N., 1981, Reversible crosslinking of cellular components of adherent fibroblasts to fibronectin and lectin coated substrata, *Exp. Cell Res.* **134**:488–494.

Avnur, Z., and Geiger, B., 1981, Substrate-attached membranes of cultured cells: Isolation and characterization of ventral cell membranes and the associated cytoskeleton, *J. Mol. Biol.* **53**:361–379.

Badley, R. A., Lloyd, C. W., Woods, A., Carruthers, L., Allcock, C., and Rees, D. A., 1978, Mechanisms of cellular adhesion. III. Preparation and preliminary characterization of adhesions, *Exp. Cell Res.* **117**:231–244.

Badley, R. A., Woods, A., Smith, C. G., and Rees, D. A., 1980, Actomyosin relationships with surface features in fibroblast adhesion, *Exp. Cell Res.* **126**:263–272.

Bayley, S., and Rees, D., 1982, Analysis of proteins, glycoprotein and glycosaminoglycans of fibroblast adhesions to substratum, *Biochim. Biophys. Acta* **689**:351–362.

Bennett, V., Davis, J., and Fowler, W. E., 1982, Brain spectrin: A membrane-associated protein related in structure and function to erythrocyte spectrin, *Nature (London)* **299**:126–128.

Bertolotti, R., Rutishauser, U., and Edelman, G. M., 1980, A cell surface molecule involved in aggregation of embryonic liver cells, *Proc. Natl. Acad. Sci. USA* **77**:4831–4835.

Birchmeier, C., Kreis, T. E., Eppenberger, H. M., Winterhalter, K. H., and Birchmeier, W., 1980, Corrugated attachment membrane on WI-38 fibroblasts: Alternating fibronectin fibers and actin-containing focal contacts, *Proc. Natl. Acad. Sci. USA* **77**:4108–4112.

Birchmeier, W., Liebermann, T. A., Imhof, B. A., and Kreis, T. E., 1981, Intracellular and extracellular components involved in the formation of ventral surfaces of fibroblasts, *Cold Spring Harbor Symp. Quant. Biol.* **46**:755–767.

Birdwell, C. R., Gospodarowicz, D., and Nicolson, G. L., 1978, Identification, localization and role of fibronectin in cultured bovine endothelial cells, *Proc. Natl. Acad. Sci. USA* **75**:3273–3277.

Brackenbury, R., Thiery, J.-P., Rutishauser, U., and Edelman, G. M., 1977, Adhesion among neural cells of the chick embryo. I. An immunological assay for molecules involved in cell–cell binding, *J. Biol. Chem.* **252**:6835–6840.

Brackenbury, R., Rutishauser, U., and Edelman, G., 1981, Distinct calcium-independent and calcium-dependent adhesion systems of chicken embryo cells, *Proc. Natl. Acad. Sci. USA* **78**:387–391.

Branton, D., Cohen, C. M., and Tyler, J., 1981, Interaction of cytoskeletal proteins on the human erythrocyte membrane, *Cell* **24**:24–33.

Briles, E. B., 1982, Lectin-resistant cell surface variants of eukaryotic cells, *Int. Rev. Cytol.* **75**:101–166.

Burridge, K., and Feramisco, J., 1980, Microinjection and localization of a 130K protein in living fibroblasts: A relationship to actin and fibronectin, *Cell* **19**:587–595.

Burridge, K., Kelly, T., and Mangeat, P., 1982, Nonerythrocyte spectrins: Actin–membrane attachment problems occurring in many cell types, *J. Cell Biol.* **95**:478–487.

Buskirk, D. R., Thiery, J.-P., Rutishauser, U., and Edelman, G. M., 1980, Antibodies to a neural cell adhesion molecule disrupt histogenesis in cultured chick retina, *Nature (London)* **285**:488–489.

Carter, W. G., 1982, The cooperative role of the transformation-sensitive glycoproteins gp140 and fibronectin in cell attachment and spreading, *J. Biol. Chem.* **257**:3249–3257.

Carter, W. G., and Hakomori, S.-I., 1981, A new cell surface detergent-insoluble glycoprotein matrix of human and hamster fibroblasts, *J. Biol. Chem.* **256**:6953–6960.

Chen, W.-T., 1982, Development of the attachment sites between the cell surface and the extracellular matrix in cultured fibroblasts, *J. Cell Biol.* **95:**100a.

Chen, W.-T., and Singer, J., 1980, Fibronectin is not present in the focal adhesions formed between normal cultured fibroblasts and their substrate, *Proc. Natl. Acad. Sci. USA* **77:**7318–7322.

Chen, W.-T., and Singer, J., 1982, Immunoelectron microscopic studies of the sites of cell–substratum and cell–cell contact in cultured fibroblasts, *J. Cell Biol.* **95:**205–223.

Cohen, S. M., Gorbsky, G., and Steinberg, M., 1983, Immunochemical characterization of related families of glycoproteins in desmosomes, *J. Biol. Chem.* **258:**2621–2627.

Cook, J. H., and Lilien, J., 1982, The accessibility of certain proteins on embryonic chick neural retina cells in iodination and tryptic removal is altered by calcium, *J. Cell Sci.* **55:**85–103.

Couchman, J. R., Rees, D. A., Green, M. R., and Smith, C. G., 1982, Fibronectin has a dual role in locomotion and anchorage of primary chick fibroblasts, *J. Cell Biol.* **93:**402–410.

Couchman, J. R., Höök, M., Rees, D. A., and Timpl, R., 1983, Adhesion, growth and matrix production of fibroblasts on laminin substrates, *J. Cell Biol.* **96:**177–183.

Coudrier, E., Reggio, H., and Louvard, D., 1981, The cytoskeleton of intestinal microvilli contains two polypeptides immunologically related to proteins of skeletal muscle, *Cold Spring Harbor Symp. Quant. Biol.* **46:**881–892.

Courtoy, P. J., Kanwar, Y. S., Hynes, R. O., and Farquhar, M. G., 1980, Fibronectin localization in the rat glomerulus, *J. Cell Biol.* **87:**691–698.

Culp, L., 1978, Biochemical determination of cell adhesion, *Curr. Top. Membr. Transp.* **11:**327–395.

Curtis, A. S. G., 1964, The mechanism of adhesion of cells to glass, *J. Cell Biol.* **20:**199–215.

Damsky, C. H., Wylie, D., and Buck, C. A., 1979, Studies on the function of cell surface glycoproteins. II. Possible role of surface glycoproteins in the control of cytoskeletal organization and surface morphology, *J. Cell Biol.* **80:**403–415.

Damsky, C. H., Knudsen, K. A., Dorio, R. J., and Buck, C. A., 1981, Manipulation of cell–cell and cell–substratum interactions in mouse mammary tumor epithelial cells using broad spectrum antisera, *J. Cell Biol.* **89:**173–184.

Damsky, C. H., Knudsen, K. A., and Buck, C. A., 1982, Integral membrane glycoproteins related to cell–substratum adhesion in mammalian cells, *J. Cell. Biochem.* **18:**1–13.

Damsky, C. H., Richa, J., Solter, D., Knudsen, K. A., and Buck, C. A., 1983a, Identification and purification of a cell-surface glycoprotein-mediated intercellular adhesion in embryonic and adult tissue, *Cell* **34:**455–466.

Damsky, C. H., Buck, C. A., and Horwitz, A. F., 1983b, Localization of a surface membrane substratum adhesion glycoprotein using a monoclonal antibody, *J. Cell Biol.* **97:**99a.

Damsky, C. H., Richa, J., Solter, D., and Buck, C. A., 1983c, Cell-CAM 120/80; a cell–cell adhesion molecule present in the early embryo and adult epithelia, *J. Cell Biol.* **97:**251a.

Decker, C., and Horwitz, A. F., 1982, Adhesive and morphologic hierarchies determined using a monoclonal antibody, *J. Cell Biol.* **95:**111a.

Diaz, L. A., and Marcello, C. L., 1978, Pemphigoid and pemphigus antigens in cultured epidermal cells, *Br. J. Dermatol.* **98:**631–637.

Ducibella, T., 1980, Divalent antibodies to mouse embryonal carcinoma cells inhibit compaction in the mouse embryo, *Dev. Biol.* **79:**356–366.

Ducibella, T., and Anderson, E., 1975, Cell shape and membrane changes in the eight-cell mouse embryo: Prerequisites for morphogenesis of the blastocyst, *Dev. Biol.* **47:**45–58.

Ducibella, T., Albertini, D. F., Anderson, E., and Biggers, J. D., 1975, The preimplantation mammalian embryos: Characterization of intercellular junctions and their appearance during development, *Dev. Biol.* **45**:231–250.

Dunia, I., Nicholas, J. F., Jakob, H., Benedetti, E. L., and Jacob, F., 1979, Junctional modulation in mouse embryonal carcinoma cell serum, *Proc. Natl. Acad. Sci. USA* **76**:3387–3391.

Edelman, G., 1983, Cell adhesion molecules, *Science* **219**:450–457.

Edelman, G., and Chuong, C.-M., 1982, Embryonic to adult conversion of neural cell adhesion molecules in normal and staggerer mice, *Proc. Natl. Acad. Sci. USA* **79**:7036–7040.

Edelman, G., Gallin, W. J., Delouvee, A., Cunningham, B. A., and Thiery, J.-P., 1983, Early epochal maps of two different cell adhesion molecules, *Proc. Natl. Acad. Sci. USA* **80**:4384–4388.

Edwards, J. G., Dysart, J., and Hughes, R. C., 1976, Cellular adhesiveness reduced in ricin-resistant hamster fibroblasts, *Nature (London)* **264**:66–68.

Foidart, J. M., Bere, G. W., Yaer, M., Rennard, S. I., Gullino, M., Martin, G. R., and Katz, S. I., 1980, Distribution and immunoelectron microscope localization of laminin, a non-collagenous basement membrane glycoprotein, *Lab. Invest.* **42**:336–342.

Fox, C. H., Cottler-Fox, M. H., and Yamada, K. M., 1981, Distribution of fibronectin in attachment sites of chick fibroblasts, *Exp. Cell Res.* **130**:477–481.

Frazier, W., and Glaser, L., 1979, Surface components and cell recognition, *Annu. Rev. Biochem.* **48**:491–523.

Gallin, W. J., Edelman, G., and Cunningham, B., 1983, Characterization of L-CAM, a major cell adhesion molecule from embryonic liver cells, *Proc. Natl. Acad. Sci. USA* **80**:1038–1042.

Geiger, B., 1979, A 130K protein from gizzard: Its localization at the termini of microfilament bundles in cultured chicken cells, *Cell* **18**:193–205.

Geiger, B., 1981, Involvement of vinculin in contact-induced cytoskeletal interactions, *Cold Spring Harbor Symp. Quant. Biol.* **46**:671–682.

Geiger, B., Tokuyasu, K. T., Dutton, A. H., and Singer, S. J., 1980, Vinculin, an intracellular protein localized at specialized sites where microfilament bundles terminate at cell membranes, *Proc. Natl. Acad. Sci. USA* **77**:4127–4131.

Gerisch, G., 1977, Univalent antibody fragments as tools for the analysis of cell interactions in *Dictyostelium*, *Curr. Top. Dev. Biol.* **14**:243–270.

Goodman, S. R., Zagon, I. S., and Kulikowski, R., 1981, Identification of a spectrin-like protein in non-erythroid cells, *Proc. Natl. Acad. Sci. USA* **78**:7570–7574.

Gorbsky, G., and Steinberg, M. S., 1981, Isolation of the intercellular glycoproteins of desmosomes, *J. Cell Biol.* **90**:243–248.

Greve, J. M., and Gottlieb, D. I., 1982, Monoclonal antibodies which alter the morphology of cultured chick myogenic cells, *J. Cell. Biochem.* **18**:221–230.

Grinnell, F., 1978, Cellular adhesiveness and extracellular substrate, *Int. Rev. Cytol.* **53**:65–143.

Grinnell, F., 1980, Fibroblast receptor for cell substratum adhesion: Studies on the interaction of baby hamster kidney cells with latex beads coated by cold insoluble globulin (plasma fibronectin), *J. Cell Biol.* **86**:104–112.

Grinnell, F., 1982, Cell attachment and spreading factors, in: *Growth and Maturation Factors* (G. Guroff, ed.), pp. 267–292, Wiley, New York.

Grinnell, F., and Minter, D., 1978, Attachment and spreading of baby hamster kidney cells to collagen substrata: Effect of cold insoluble globulin, *Proc. Natl. Acad. Sci. USA* **75**:4408–4412.

Grumet, M., Rutishauser, U., and Edelman, G. M., 1982, Neural cell adhesion molecule is on embryonic muscle cells and mediates adhesion to nerve cells *in vitro*, *Nature (London)* **295**:693–695.

Grunwald, G. B., Geller, R. L., and Lilien, J., 1980, Enzymatic dissection of embryonic cell adhesive mechanisms, *J. Cell Biol.* **85**:766–776.

Grunwald, G. B., Pratt, R. S., and Lilien, J., 1982, Enzymatic dissection of embryonic cell adhesive mechanisms. III. Immunological identification of a component of the calcium-dependent adhesive system of embryonic chick neural retina cells, *J. Cell Sci.* **55**:69–87.

Harper, P. A., and Juliano, R. L., 1980, Isolation and characterization of Chinese hamster ovary cell variants defective in adhesion to fibronectin coated collagen, *J. Cell Biol.* **87**:755–763.

Harper, P. A., and Juliano, R. L., 1981a, Fibronectin-independent adhesion of fibroblasts to the extracellular matrix: Mediation by a high molecular weight membrane glycoprotein, *J. Cell Biol.* **91**:647–653.

Harper, P. A., and Juliano, R. L., 1981b, Two distinct mechanisms of fibroblast adhesion, *Nature (London)* **290**:136–138.

Hascall, V. C., and Hascall, G. K., 1981, Proteoglycans, in: *Cell Biology of Extracellular Matrix* (E. D. Hay, ed.), pp. 39–64, Plenum Press, New York.

Hausman, R. E., and Moscona, A. A., 1976, Isolation of retina-specific cell-aggregating factor from membranes of embryonic neural retina tissue, *Proc. Natl. Acad. Sci. USA* **73**:3594–3598.

Hay, E., 1981, Extracellular matrix, *J. Cell Biol.* **91**:205S–223S.

Hayman, E., Engvall, E., A'Hearn, E., Barnes, D., Perschbacher, M., and Ruoslahti, E., 1982, Cell attachment on replicas of SDS-polyacrylamide gel reveals two adhesive plasma proteins, *J. Cell Biol.* **95**:20–24.

Heath, J. P., 1982, Adhesions to substratum and locomotory behavior of fibroblastic and epithelial cells in culture, in: *Cell Behavior* (R. Bellairs, A. Curtis, and G. Dunn, eds.), pp. 77–103, Cambridge University Press, London.

Heath, J. P., and Dunn, G. A., 1978, Cell to substratum contacts of chick fibroblasts and their relation to the microfilament system: A correlated interference–reflection and high-voltage electron microscope study, *J. Cell Sci.* **29**:197–212.

Hedman, K., Kurkinen, M., Alitalo, K., Vaheri, A., Johansson, S., and Höök, M., 1979, Isolation of the pericellular matrix of human fibroblast cultures, *J. Cell Biol.* **81**:83–91.

Hedman, K., Johansson, S., Vartio, T., Kjellén, L., Vaheri, A., and Höök, M., 1982, Structure of the pericellular matrix: Association of heparin and chondroitin sulfates with fibronectin-procollagen fibers, *Cell* **28**:663–671.

Hoffman, S., Sorkin, B. C., White, P. C., Brackenbury, R., Mailhammer, R., Rutishauser, U., Cunningham, R. A., and Edelman, G. M., 1982, Chemical characterization of a neural cell adhesion molecule purified from embryonic brain membranes, *J. Biol. Chem.* **257**:7720–7729.

Höök, M., Rubin, K., Oldberg, A., Öbrink, B., and Vaheri, A., 1977, Cold insoluble globulin mediates the adhesion of rat liver cells to plastic petri dishes, *Biochem. Biophys. Res. Commun.* **79**:726–733.

Hsieh, P., and Sueoka, N., 1980, Antisera inhibiting mammalian spreading and possible cell-surface antigens involved, *J. Cell Biol.* **86**:866–873.

Hughes, R. C., Pena, S. D., Clark, J., and Dourmashkin, R. R., 1979, Molecular requirements for the adhesion and spreading of hamster fibroblasts, *Exp. Cell Res.* **121**:307–314.

Hull, B., and Staehelin, L. A., 1979, The terminal web, a re-evaluation of its structure and function, *J. Cell Biol.* **81**:67–82.

Hyafil, F., Morello, D., Babinet, C., and Jacob, F., 1980, A cell-surface glycoprotein involved in the compaction of embryonal carcinoma cells and cleavage stage embryos, *Cell* **21**:927–934.

Hyafil, F., Babinet, C., and Jacob, F., 1981, Cell–cell interaction in early embryogenesis: A molecular approach to the role of calcium, *Cell* **26**:447–454.

Hynes, R. O., 1981a, Fibronectin and its relation to cellular structure and behavior, in: *Cell Biology of Extracellular Matrix* (E. D. Hay, ed.), pp. 295–334, Plenum Press, New York.

Hynes, R. O., 1981b, Relationships between fibronectin and the cytoskeleton, in: *Cytoskeletal Elements and Plasma Membrane Organization* (G. Poste and G. L. Nicolson, eds.), pp. 100–137, Elsevier/North-Holland, Amsterdam.

Hynes, R. O., and Destree, A., 1978, Relationships between fibronectin (LETS protein) and actin, *Cell* **15**:875–886.

Hynes, R. O., and Yamada, K. M., 1982, Fibronectins: Multifunctional modular glycoproteins, *J. Cell Biol.* **95**:369–378.

Hynes, R. O., Destree, A. T., and Wagner, D., 1981, Relationships between microfilaments, cell–substratum adhesion and fibronectin, *Cold Spring Harbor Symp. Quant. Biol.* **46**:659–669.

Izzard, C., and Lochner, L. R., 1976, Cell to substrate contacts in lung fibroblasts: An interference reflection study with an evaluation of the technique, *J. Cell Sci.* **21**:129–159.

Jakoi, E. R., and Marchase, R., 1979, Ligatin from embryonic chick neural retina, *J. Cell Biol.* **80**:624–633.

Jöhansson, E., Kjellén, L., Höök, M., and Timpl, R., 1981, Substrate adhesion of rat hepatocytes: A comparison of laminin and fibronectin as attachment proteins, *J. Cell Biol.* **90**:260–264.

Jørgensen, O., Steen, O., Delouvie, A., Thiery, J.-P., and Edelman, G. M., 1980, The nervous specific protein D2 is involved in adhesion assay neuritis from cultured rat ganglia, *FEBS Lett.* **111**:39–42.

Kemler, R., Babinet, C., Eisen, H., and Jacob, F., 1977, Surface antigens in early differentiation, *Proc. Natl. Acad. Sci. USA* **74**:4449–4452.

Kjellén, L., Pettersson, I., and Höök, M., 1981, Cell-surface proteoglycan: An intercalated membrane proteoglycan, *Proc. Natl. Acad. Sci. USA* **78**:5371–5375.

Kleinman, H., Klebe, R. J., and Martin, G. R., 1981, Role of collagenous matrices in the adhesion and growth of cells, *J. Cell Biol.* **88**:473–487.

Knox, P., and Griffiths, S., 1979, A cell-spreading factor in human serum that is not cold insoluble globulin, *Exp. Cell Res.* **123**:421–424.

Knudsen, K. A., Rao, P. E., Damsky, C. H., and Buck, C. A., 1981, Membrane glycoproteins involved in cell–substratum adhesion, *Proc. Natl. Acad. Sci. USA* **78**:6071–6075.

Knudsen, K. A., Damsky, C. H., and Buck, C. A., 1982a, Expression of adhesion-related membrane components in adherent versus non-adherent hamster melanoma cells, *J. Cell. Biochem.* **18**:157–167.

Knudsen, K. A., Buck, C. A., Damsky, C. H., and Horwitz, A. F., 1982b, Adhesion-related glycoproteins isolated using a monoclonal antibody, *J. Cell Biol.* **95**:111a.

Knudsen, K. A., Buck, C. A., and Horwitz, A. F., 1983, Characterization of cell-matrix adhesion-related integral membrane glycoproteins, *J. Cell Biol.* **97**:99a.

Lahav, J., Schwartz, M. A., and Hynes, R. O., 1982, Analysis of platelet adhesion with a radioactive chemical crosslinking reagent: Interaction of thrombospondin with fibronectin and collagen, *Cell* **31**:253–262.

Laterra, J., Silbert, J. E., and Culp, L. A., 1983, Cell-surface heparan sulfate mediates some adhesive responses to glycosaminoglycan binding matrices including fibronectin, *J. Cell Biol.* **96:**112–123.

Lehto, V.-P., 1983, 140,000 dalton surface glycoprotein: A plasma membrane component of the detergent resistant cytoskeletal preparations of cultured human fibroblasts, *Exp. Cell Res.* **143:**272–286.

Lehto, V.-P., Vartio, T., and Virtanen, I., 1980, Enrichment of a 140-kd surface glycoprotein in adherent, detergent-resistant cytoskeletons of cultured human fibroblasts, *Biochem. Biophys. Res. Commun.* **95:**909–916.

Lehto, V.-P., Vartio, T., and Virtanen, I., 1981, Persistence of a 140,000 M_r surface glycoprotein in cell-free matrices of cultured human fibroblasts, *FEBS Lett.* **124:**289–292.

Lehto, V.-P., Vartio, T., Badley, R. A., and Virtanen, I., 1983, Characterization of a detergent-resistant surface lamina in cultured human fibroblasts, *Exp. Cell Res.* **143:**287–295.

Lilien, J., Balsamo, J., McDonough, J., Hermolin, J., Cook, J., and Rutz, R., 1979, Adhesive specificity among embryonic cells, in: *Surfaces of Normal and Malignant Cells* (R. O. Hynes, ed.), pp. 389–419, Wiley, New York.

Luna, E. J., Fowler, V., Swanson, J., Branton, D., and Taylor, D. L., 1981, Membrane cytoskeleton from *Dictyostelium discoideum*. I. Identification and partial characterization of an actin binding activity, *J. Cell Biol.* **88:**396–409.

McClain, D., and Edelman, G., 1982, A neural cell adhesion molecule from human brain, *Proc. Natl. Acad. Sci. USA* **79:**6380–6384.

McClay, D. R., Wessel, G. M., and Marchase, R. B., 1980, Intercellular recognition: Quantitation of initial binding events, *Proc. Natl. Acad. Sci. USA* **78:**4975–4979.

McPherson, J., Sage, H., and Bornstein, P., 1981, Isolation and characterization of a glycoprotein secreted by aortic endothelial cells in culture: Apparent identity with platelet thrombospondin, *J. Biol. Chem.* **256:**11330–11336.

Madri, J. A., Roll, F. J., Furthmayr, H., and Foidart, J.-M., 1980, Ultrastructural localization of fibronectin and laminin in the basement membrane of the murine kidney, *J. Cell Biol.* **86:**682–687.

Magnani, J. L., Thomas, W. A., and Steinberg, M. S., 1981, Two distinct adhesion mechanisms in embryonic neural retina cells, *Dev. Biol.* **81:**96–105.

Malinoff, H. L., and Wicha, M. S., 1983, Isolation of a cell surface receptor for a laminin from murine fibrosarcoma cells, *J. Cell Biol.* **96:**1475–1479.

Marchase, R. B., Koro, L. A., Kelly, C. M., and McClay, D. R., 1982, Retinal ligatin recognizes glycoproteins bearing oligosaccharides terminating in phosphodiester-linked glucose, *Cell* **28:**813–820.

Matsudaira, P. T., and Burgess, D. R., 1981, Structure and function of brush border cytoskeleton, *Cold Spring Harbor Symp. Quant. Biol.* **46:**845–854.

Mayer, B. W., Hay, E., and Hynes, R. O., 1981, Immunocytochemical localization of fibronectin in embryonic chick trunk and area vasculosa, *Dev. Biol.* **82:**267–286.

Mollenhauer, J., and van der Mark, K., 1983, Isolation and characterization of a collagen-binding glycoprotein from chondrocyte membrane, *EMBO J.* **2:**45–50.

Moscona, A. A. (ed.), 1974, in: *Cell Surface in Development*; Surface specification of embryonic cells: lectin receptors, cell recognition and specific cell ligands. pp. 67–99, Wiley, New York.

Neff, N. T., Lowrey, C., Decker, C., Tovar, A., Damsky, C., Buck, C., and Horwitz, A. F., 1982, A monoclonal antibody detaches embryonic skeletal muscle from extracellular matrices, *J. Cell Biol.* **95:**654–666.

Newgreen, D., and Thiery, J.-P., 1980, Fibronectin in early avian embryos: Synthesis and distribution along the migration pathways of neural crest cells, *Cell Tissue Res.* **211:**269–291.

Nicolas, J.-F., Kemler, R., and Jacob, F., 1981, Effects of anti-embryonal carcinoma serum on aggregation and metabolic cooperation between teratocarcinoma cells, *Dev. Biol.* **81:**127–132.

Nielsen, L. D., Pitts, M., Grady, S. R., and McGuire, E. J., 1981, Cell–cell adhesion in the embryonic chick: Partial purification of liver adhesion molecules from liver membranes, *J. Cell Biol.* **86:**315–326.

Ocklind, C., and Öbrink, B., 1982, Intercellular adhesion of rat hepatocytes: Identification of a cell-surface glycoprotein involved in the initial adhesion process, *J. Biol. Chem.* **257:**6788–6795.

Ocklind, C., Forsum, U., and Öbrink, B., 1983, Cell-surface localization and tissue distribution of a hepatocyte cell–cell adhesion glycoprotein (cell-CAM 105), *J. Cell Biol.* **96:**1168–1171.

Ogou, S.-I., Yashida-Noro, C., and Takechi, M., 1983, Calcium dependent cell–cell adhesion molecules common to hepatocytes and teratocarcinoma stem cells, *J. Cell Biol.* **97:**944–948.

Oesch, B., and Birchmeier, W., 1982, New surface component of fibroblasts focal contacts identified by a monoclonal antibody, *Cell* **31:**671–679.

Oppenheimer-Marks, N., and Grinnell, F., 1982, Inhibition of fibronectin receptor function by antibodies against BHK wheat germ agglutinin receptors, *J. Cell Biol.* **95:**876–884.

Oppenheimer-Marks, N., Marshall, J., and Grinnell, F., 1982, Cell surface glycoproteins in fibronectin receptor function, *J. Cell Biol.* **95:**246a.

Pena, S. D. J., Mills, G., and Hughes, R. C., 1979, Two-dimensional electrophoresis of surface glycoproteins of normal BHK cells and ricin-resistant mutants, *Biochim. Biophys. Acta* **550:**100–109.

Pouyssegur, J., and Pastan, I., 1977, Mutants of mouse fibroblasts altered in the synthesis of cell-surface glycoproteins, *J. Biol. Chem.* **252:**1639–1646.

Rapraeger A. C., and Bernfield, M., 1983a, Heparan sulfate proteoglycan from mouse mammary epithelial cells: A putative membrane proteoglycan associates quantitatively with lipid vesicles, *J. Biol. Chem.* **258:**3632–3636.

Rapraeger, A. C., and Bernfield, M., 1982, An integral membrane proteoglycan is capable of binding components of the cytoskeleton and the extracellular matrix, *14th Michigan Molecular Institute Symposium: Extracellular Matrix* (S. Hawkes and J. L. Wang, eds.), p. 265–270, Academic Press, New York.

Rees, D. A., Badley, R. A., Bayley, S. A., Couchman, J. R., Smith, C. G., and Woods, A., 1981, Surface components in fibroblast adhesion and movement, in: *Cellular Interactions* (J. T. Dingle and J. L. Gordon, eds.), pp. 67–80, Elsevier/North-Holland, Amsterdam.

Revel, J., and Wolken, K., 1973, Electron microscope investigation of the underside of cells in culture, *Exp. Cell Res.* **78:**1–14.

Rollins, B. S., Cathcart, M. K., and Culp, L. A., 1982, Fibronectin–proteoglycan binding as the molecular basis for fibroblast adhesion to extracellular matrices, in: *Glycoconjugates*, Vol. III (M. Horowitz, ed.), pp. 289–329, Academic Press, New York.

Rorschneider, L., Rosok, M., and Shriver, K., 1981, Mechanism of transformation by Rous sarcoma virus: Events within adhesion plaques, *Cold Spring Harbor Symp. Quant. Biol.* **46:**953–967.

Roth, S., 1983, Biochemistry of cell adhesion in vertebrates, in: *Cell Interactions and Development: Molecular Mechanisms* (K. Yamada, ed.), pp. 77–98, John Wiley & Sons Inc., New York, New York.

Rothbard, J. B., Brackenbury, R., Cunningham, B., and Edelman, G. M., 1982, Differences in the carbohydrate structure of neural cell adhesion molecule from adult and embryonic chicken brain, *J. Biol. Chem.* **257:**11064–11069.

Rubin, K., Höök, M., Öbrink, B., and Timpl, R., 1981, Substrate adhesion of rat hepatocytes: Mechanism of attachment to collagen substrates, *Cell* **24:**463–470.

Ruoslahti, E., Engvall, E., and Hayman, E. G., 1981, Fibronectin: Current concepts of structure and function, *Cell Res.* **1:**1–36.

Rutishauser, U., Thiery, J.-P., Brackenbury, R., Sela, B.-A., and Edelman, G., 1976, Mechanisms of adhesion among cells from the neural tissues of the chick embyro, *Proc. Natl. Acad. Sci. USA* **73:**577–581.

Rutishauser, U., Thiery, J.-P., Brackenbury, R., and Edelman, G. M., 1978a, Adhesive assay neural cells of the chick embyro. III. Relationship of the surface molecule CAM to cell adhesion and the development of histotypic patterns, *J. Cell Biol.* **79:**371–381.

Rutishauser, U., Gall, W. E., and Edelman, G. M., 1978b, Adhesion among neural cells of the chick embryo. IV. Role of cell surface molecule CAM in the formation of neurite bundles in culture of spinal ganglia, *J. Cell Biol.* **79:**382–393.

Rutishauser, U., Hoffman, S., and Edelman, G. M., 1982, Binding properties of a cell-adhesive molecule from neural tissue, *Proc. Natl. Acad. Sci. USA* **79:**685–689.

Rutishauser, U., Grumer, M., and Edelman, G., 1983, Neural cell adhesion molecule mediates initial interactions between spinal cord neurons and muscle cells in culture, *J. Cell Biol.* **97:**145–152.

Schor, S., and Court, J., 1979, Different mechanisms in the attachment of cells to native and denatured collagen, *J. Cell Sci.* **38:**267–281.

Schubert, D., and La Courbiere, M., 1982, Properties of extracellular adhesion-mediating particles in a myoblast clone and its adhesion-deficient variant, *J. Cell Biol.* **94:**108–114.

Schwartz, M. A., Das, O. P., and Hynes, R. O., 1982, A new radioactive crosslinking reagent for studying the interactions of proteins, *J. Biol. Chem.* **257:**2343–2349.

Singer, I., 1979, The fibronexus: A transmembrane association of fibronectin-containing fibers and bundles of 5-μm microfilaments in hamster and human fibroblasts, *Cell* **16:**675–685.

Singer, I. I., 1982, Association of fibronectin and vinculin with focal contacts and stress fibers in stationary hamster fibroblasts, *J. Cell Biol.* **92:**398–409.

Solter, D., and Knowles, B. B., 1975, Immunosurgery of mouse blastocysts, *Proc. Natl. Acad. Sci. USA* **72:**5099–5102.

Staehelin, L. A., 1974, Structure and function of intercellular junctions, *Int. Rev. Cytol.* **39:**191–283.

Stanley, J. R., Alvarez, O. M., Bere, E. W., Eaglestein, W. H., and Katz, S. I., 1981a, Detection of basement membrane zone antigens during epidermal wound healing in pigs, *J. Invest. Dermatol.* **77:**240–243.

Stanley, J. R., Hawley-Nelson, P., Yuspa, S. H., Shevach, E. M., and Katz, S. I., 1981b, Characterization of bullous pemphigoid antigen: A unique basement membrane protein of stratified squamous epithelia, *Cell* **24:**897–903.

Steinberg, M., 1978, in: Cell–cell recognition in multicellular assembly: levels of assembly, *Soc. Exp. Biol. Symp.* **XXXII:**25–29.

Sugrue, S. P., and Hay, E., 1982, Interaction of embryonic corneal epithelium with exogenous collagen, laminin and fibronectin: role of endogenous protein synthesis, *Dev. Biol.* **92:**97–106.

Takeichi, M., 1977, Functional correlation between cell adhesive properties and some cell-surface proteins, *J. Cell Biol.* **75:**464–474.

Takeichi, M., Ozaki, H. S., Takunaga, K., and Okada, T. S., 1979, Experimental manipulation of the cell surface to affect cellular recognition mechanisms, *Dev. Biol.* **70:**195–205.

Takeichi, M., Atsumi, J., Yoshida, K. U., and Okada, T. S., 1981, Selective adhesion of embryonal carcinoma cells and differentiated cells by Ca^{2+}-dependent sites, *Dev. Biol.* **87**:340–350.

Tarone, G., Galletto, G., Prat, M., and Comoglio, P., 1982, Cell-surface molecules and fibronectin mediated cell adhesion: Effect of proteolytic digestion of membrane proteins, *J. Cell Biol.* **94**:179–186.

Thiery, J.-P., Brackenbury, R., Rutishauser, U., and Edelman, G. M., 1977, Adhesion among neural cells of the chick embryo. II. Purification and characterization of a cell adhesion molecule from neural retina, *J. Biol. Chem.* **252**:6841–6845.

Thiery, J.-P., Duband, J.-L., Rutishauser, U., and Edelman, G. M., 1982, Cell adhesion molecules in early chicken embryogenesis, *Proc. Natl. Acad. Sci. USA* **79**:6737–6741.

Thomas, W. A., and Steinberg, M. S., 1981, Two distinct adhesion mechanisms in embryonic neural retina cells. II. An immunological analysis, *Dev. Biol.* **81**:106–114.

Tokuyasu, K. T., and Singer, S. J., 1976, Improved procedures for immunoferritin labeling of ultra-thin frozen sections, *J. Cell Biol.* **71**:894–906.

Urushihara, H., and Takeichi, M., 1980, Cell–cell adhesion molecule: Identification of a glycoprotein relevant to the Ca^{2+}-independent aggregation of Chinese hamster fibroblasts, *Cell* **20**:363–371.

Urushihara, H., Ueda, M. J., Ockeda, T. S., and Takeichi, M., 1977, Calcium-dependent and independent adhesion of normal and transformed BHK cells, *Cell Struct. Funct.* **2**:289–296.

Urushihara, H., Ozaki, H. S., and Takeichi, M., 1979, Immunological detection of cell-surface components related with aggregations of Chinese hamster and chick embryonic cells, *Dev. Biol.* **70**:206–216.

Virtanen, I., Vartio, T., Badley, R. A., and Lehto, V.-P., 1982, Fibronectin in adhesion, spreading and cytoskeletal organization of cultured fibroblasts, *Nature (London)* **298**:660–663.

Vollmers, H. P., and Birchmeier, W., 1983, Monoclonal antibodies inhibit the adhesion of mouse B16 melanoma cells in vitro and block lung metastasis in vivo, *Proc. Natl. Acad. Sci. USA* **80**:3729–3733.

Wehland, J., Osborn, M., and Weber, K., 1979, Cell to substratum contacts in living cells: A direct correlation between interference reflection and indirect immunofluorescence microscopy using antibodies against α-actinin, *J. Cell Sci.* **37**:257–273.

Wylie, D. E., Damsky, C. H., and Buck, C. A., 1979, Studies on the function of cell-surface glycoproteins. I. Use of antisera to surface membranes in the identification of membrane components relevant to cell–substrate adhesion, *J. Cell Biol.* **80**:385–402.

Yamada, K. M., 1981, Fibronectin and other structural proteins, in: *Extracellular Matrix* (E. D. Hay, ed.), pp. 95–114, Plenum Press, New York.

Yamada, K. M., 1983, Cell-surface interactions with extracellular materials, *Annu. Rev. Biochem.* **52**:761–800.

Yamada, K. M., and Olden, K., 1978, Fibronectin: Adhesive glycoproteins of the cell surface and blood, *Nature (London)* **275**:179–184.

Yoshida, C., and Takeichi, M., 1982, Teratocarcinoma cell adhesion: Identification of a cell surface protein involved in calcium-dependent cell aggregation, *Cell* **28**:217–224.

Attachment Proteins and Their Role in Extracellular Matrices

A. Tyl Hewitt and George R. Martin

1. COMPOSITION AND STRUCTURE OF EXTRACELLULAR MATRICES

It has long been appreciated that extracellular matrices provide tissues with their strength, stability, and structure (Gross, 1974; Hay, 1981a,b). However, the extracellular matrix itself was considered to be rather inert and to show little specificity. More recently, it has been established that different tissues contain unique matrices generated by the resident cells, such as those associated with fibrous tissues, cartilage, and basement membranes. The components of these different matrices, which include collagens, proteoglycans, and glycoproteins, are different in each tissue both in terms of type and amount (Bornstein and Sage, 1980; Burgeson *et al.*, 1976; Chung and Miller, 1974; Levitt and Dorfman, 1974; Levitt *et al.*, 1975; Goetinck *et al.*, 1974; Royal and Goetinck, 1977; Kefalides *et al.*, 1979; Hay, 1981a,b; Kleinman *et al.*, 1981, 1982a; Miller *et al.*, 1971; Miller and Matukas, 1969, 1974). Further, it appears likely that the proteins produced for the matrix of a given tissue show specific interactions that generate a supramolecular complex of defined stoichiometry.

A. Tyl Hewitt • Wynn Center for the Study of Retinal Degenerations, The Wilmer Ophthalmological Institute, The Johns Hopkins Hospital, Baltimore, Maryland 21205.   *George R. Martin* • Laboratory of Developmental Biology and Anomalies, National Institute of Dental Research, National Institutes of Health, Bethesda, Maryland 20205.

This supramolecular complex not only determines the physical properties of the tissue but also defines the phenotype of the cells in contact with it.

Although the components of the extracellular matrix interact to form a complex, each material carries out specific functions. Collagens provide the major structural element (Bornstein and Sage, 1980), whereas glycoproteins, such as fibronectin, laminin, and chondronectin, bind cells to the matrix (Kleinman *et al.*, 1981). Proteoglycans bind to the glycoproteins forming the ternary element in the complex, and determine the spacing between collagen fibers and the permeability of the matrix to macromolecules (Hascall and Hascall, 1981). This chapter will describe the components of the matrices, particularly the tissue-specific cell attachment glycoproteins, and relate these components to the structures they form and the influences they exert on the growth and differentiation of tissue.

1.1. Collagen

Collagens have several characteristics that distinguish them from other proteins, but within the genetically distinct collagens there are many common chemical factors. Each collagen contains glycine as every third residue as well as hydroxyproline and hydroxylysine, which are not present in most proteins. Collagens exhibit a distinctive triple-helical structure that produces characteristic X-ray diffraction patterns. The native proteins resist degradation by most proteases but are degraded by very specific enzymes, the collagenases. The collagenases produced by certain bacteria are used *in vitro* to assay for collagen, and other proteins are not cleaved by them (Peterkofsky and Diegelmann, 1971; Bornstein and Sage, 1980; Kühn, 1982).

In recent years, it has become clear that there are many different collagens (Table 1). The major collagen of skin, tendon, and bone, type I collagen, was the first to be well characterized. As other collagens were discovered, they were assigned an identifying number based on the order in which they were discovered. Type II collagen is from cartilage (Miller and Matukas, 1969), type III from fetal tissues, parenchyma of internal organs, and blood vessels (Chung and Miller, 1974; Epstein, 1974; Trelstad, 1974; McLees *et al.*, 1977), type IV from basement membranes (Timpl *et al.*, 1978; Kefalides *et al.*, 1979; Kresina and Miller, 1979; Tryggvason *et al.*, 1980; Gehron Robey and Martin, 1981), and type V from smooth muscle, placenta, blood vessels, and other sites (Burgeson *et al.*, 1976; Chung *et al.*, 1976; Mayne *et al.*, 1978; Rhodes and Miller, 1978; Bentz *et al.*, 1978; Sage and Bornstein, 1979). Quite a number of other collagens have been described that are likely to be distinct proteins

Table 1. Collagens

Designation	Tissue distribution	Associated cell type	Associated attachment protein	Enzyme susceptibility
Type I	Skin, bone, tendon, cornea, ligament, widespread in interstitial connective tissue	Fibroblasts, osteoblasts, stromacytes	Fibronectin	Collagenase I (interstitial)
Type II	Cartilage, vitreous, chick cornea, notochord	Chondrocytes, embryonic neuroepithelium, notochord cells, vitreocytes	Chondronectin	Collagenase I (interstitial)
Type III	Sclera, fetal skin, blood vessels, parenchymal organs	Fibroblasts, muscle cells	Fibronectin	Collagenase I (interstitial), trypsin
Type IV	Basement membranes	Epithelial and endothelial cells	Laminin	Collagenase IV
Type V	Placental membranes, blood vessels, smooth muscle	Smooth muscle cells, chondrocytes under certain conditions	fibronectin; integral membrane protein of smooth muscle cells	Collagenase V, α-thrombin, plasmin

and the total number of collagens in vertebrates could exceed 20. Some 50 distinct collagen genes have been estimated to occur in the nematode *Caenorhabditis elegans* (Kramer *et al.*, 1982). Interestingly, certain nonstructural proteins, including Clq of the complement cascade (Reid and Porter, 1976) and acetylcholinesterase (Rosenberry and Richardson, 1977), which is present in neuromuscular junction, contain collagenous segments. The function of these regions is not known but could relate to the binding of the molecule to matrix or to cellular receptors.

While the fibers they form in the tissue differ in appearance, collagen types I, II, III, and V have similar dimensions (300 × 1.5 nm) and molecular weights. These collagen molecules also have similar physical properties and are largely insoluble under physiological conditions (Bornstein and Sage, 1980; Kühn, 1982). They arise from soluble precursor forms, the procollagens, which are enzymatically processed to give rise to the collagen molecules (Lapière *et al.*, 1971; Prockop *et al.*, 1979). Type IV collagen is longer and more flexible than the other collagens and it is soluble under physiological conditions. It does not undergo major enzymatic shortening but is incorporated as secreted into basement membranes (Orkin *et al.*, 1977; Timpl *et al.*, 1978, 1981; Tryggvason *et al.*, 1980; Gehron Robey and Martin, 1981). This matrix is a thin continuous sheet composed of a nonfibrous network of type IV collagen molecules.

The type of collagen in a tissue and its conformation determine the structure and the functional characteristics of the tissue. The fibers formed by type I collagen are large, highly ordered, and extensively crosslinked to ensure maximal tensile strength (Gross, 1974; Gay and Miller, 1978). The type I collagen fibers are found arrayed in large parallel bundles. Type II (cartilage) collagen occurs in thin, short fibers that are randomly arrayed and widely separated in the matrix (Hay *et al.*, 1978; Kühn and van der Mark, 1978). Presumably, proteoglycan and fluid occupy the space between the fibers thereby generating a matrix with less tensile strength but more compressibility (Hascall and Hascall, 1981). Type III collagen forms thin fibers that are usually associated with cells. They are thought to be the reticular fibers defined by classical histological stains (Nowack *et al.*, 1976; Gay and Miller, 1978). Type IV collagen occurs as a continuous open network lacking fibrillar structures (Timpl *et al.*, 1981; Kühn, 1982). The network structure creates a matrix with some tensile strength, high elasticity, and an open structure that allows ready passage of fluid. The appearance of type V collagen in tissues is uncertain. It is often found pericellularly (Gay *et al.*, 1981) but in amounts too small for defined structures to be visible in the microscope. As noted in Table 1 and Section 2, these collagens contain specific binding sites for other matrix constituents and possibly receptors on the cell surface.

Table 2. Examples of Proteoglycans

Tissue/type	Core protein	Glycosaminoglycans
Cartilage		
Chondroitin sulfate proteoglycan, $M_r \sim 2 \times 10^6$	$M_r \sim 350,000$	Predominantly chondroitin sulfate: chain size, $M_r \sim 20,000$ (keratan sulfate also present on same core protein), ~ 100 chains/molecule
Cornea		
Keratan sulfate proteoglycan, $M_r \sim 75,000$	$M_r \sim 40,000$	Keratan sulfate: $M_r \sim 10,000$, 2–3 chains/molecule
Chondroitin sulfate proteoglycan $M_r \sim 150,000$	$M_r \sim 95,000$	Chondroitin sulfate: $M_r \sim 55,000$, 1–2 chains/molecule
Basement membrane		
(From EHS-sarcoma) $M_r \sim 0.5–1.0 \times 10^6$	$M_r \sim 350,000$	Heparan sulfate: $M_r \sim 70,000$, 6–12 chains/molecule
Liver		
(From hepatocyte plasma membrane) $M_r \sim 80,000$	$M_r \sim 24,000$	Heparan sulfate: $M_r \sim 14,000$, 4 chains/molecule

The turnover of collagen is thought to require an initial attack by very specific enzymes (Table 1). Collagen types I–III are attacked by interstitial (type I) collagenase, which cleaves peptide bonds at a single site on these molecules (Gross and Nagai, 1965; Gross *et al.*, 1974; Gross, 1981; Miller *et al.*, 1976; Horwitz *et al.*, 1977; Woolley *et al.*, 1978). Types IV and V collagen are not attacked by this enzyme but appear to be attacked by other collagenases that are in turn specific for them (Liotta *et al.*, 1979, 1981; Mainardi *et al.*, 1980). This specificity may allow certain collagen matrices to be degraded while others remain intact and maintain tissue form even during remodeling and repair.

1.2. Proteoglycans

Proteoglycans are unusual macromolecules consisting in large part of sulfated glycosaminoglycans. The proteoglycans arise biosynthetically from specific proteins that serve as the core to which the carbohydrate residues are subsequently added and sulfated to form the intact proteoglycan (Hascall and Hascall, 1981). The common glycosaminoglycans, including heparan sulfate, chondroitin sulfate and keratan sulfate, are widely distributed among different tissues and cells, whereas the protein core to which they are attached shows cell- and tissue-specific differences (Table 2). Thus, both liver cells and basement membranes contain heparan sulfate proteoglycans but these are chemically and genetically distinct molecules (Hassell *et al.*, 1980a; Kjellén *et al.*, 1980, 1981).

Although ubiquitous, their functions are not well defined. More information is available about the large chondroitin sulfate proteoglycan of cartilage than other forms. This proteoglycan binds at specific intervals along strands of hyaluronic acid creating a very large complex or aggregate (Hascall, 1977; Tengblad, 1981). The presence of sulfate on the glycosaminoglycan side chains gives them a high negative charge and causes them to be maximally extended due to repulsive forces (Hascall and Sajdera, 1970; Pasternack *et al.*, 1974; Hascall, 1977). This large complex creates and maintains spaces between the collagen fibers in the matrix. The inability of certain mouse mutants to synthesize cartilage proteoglycan results in small cartilages with tightly packed cells and collagen fibers (Kimata *et al.*, 1981). Similar effects can be produced in developing animals by xylosides, which interfere with the biosynthesis of the proteoglycan by competitively inhibiting the addition of glycosaminoglycan side chains to the core protein.

There appear to be several types of heparan sulfate proteoglycan of widely different size. Some heparan sulfate proteoglycans appear to be integral components of cell membranes (Kjellén *et al.*, 1980) (Table 2). These may be involved in cell–cell interactions through cell surface receptors for heparan sulfate or in cell–matrix interactions with the various attachment proteins. The heparan sulfate proteoglycan of basement membranes is found in the matrix and is much larger than the other heparan sulfate proteoglycans (Hassell *et al.*, 1980a). Histological stains show that sulfated chains of the proteoglycan are arranged along the surface of the basement membrane in a regular array (Trelstad *et al.*, 1974; Kanwar and Farquhar, 1979a). The proteoglycan is bound to the basement membrane proper creating a negatively charged shield that repels negatively charged macromolecules. It is well known that basement membranes restrict the passage of negatively charged proteins greater than 4.0 nm in diameter. The enzymatic removal of the heparan sulfate proteoglycan from glomerular basement membrane makes this matrix much more permeable to macromolecules (Kanwar and Farquhar, 1979b; Kanwar *et al.*, 1980). Further, glomerular basement membranes from animals made nephrotic by treatment with the aminonucleoside of puromycin pass protein freely and lack the heparan sulfate proteoglycan but retain laminin and type IV collagen (Mynderse *et al.*, 1983). These studies suggest that the heparan sulfate proteoglycan is responsible for this filtration function in basement membranes. In other sites, heparan sulfate proteoglycans may have other roles possibly acting through cell surface receptors.

Both chondroitin sulfate and keratan sulfate are found in the cornea. However, unlike the glycosaminoglycans on the cartilage proteoglycan, these glycosaminoglycans are attached to different core proteins specific for the cornea (Table 2) and both are smaller than and distinct from the cartilage proteoglycan (Hassell *et al.*, 1979). Due to their small size and the reduced number of glycosaminoglycan chains, these molecules fit within the highly ordered array of collagen in the cornea (Trelstad and Coulombre, 1971; Hassell *et al.*, 1979; Hascall and Hascall, 1981) and allow for the tissue to be hydrated. Errors in the synthesis of these proteoglycans alter this matrix and can result in opacities, which impair vision. For example, in corneal macular dystrophy, where the cornea becomes opaque, the keratan sulfate proteoglycan is absent, whereas the synthesis of the chondroitin sulfate proteoglycan is normal. However, a glycoprotein that is recognized by antibodies to the keratan sulfate proteoglycan core protein is present in extracellular and intracellular deposits (Hassell *et al.*, 1980b, 1982). This indicates that the keratan sulfate proteoglycan core protein is not processed properly so that the glycosaminoglycan chains are not added. These studies also suggest that, in normal vision, the properties, arrangements, and interactions of the components of this unique extracellular matrix are required for the transparency of the cornea.

1.3. Glycoproteins

1.3.1. Fibronectin: The Fibroblast Attachment Protein

In recent years, it has become clear that certain glycoproteins are significant components of these matrices and have a major role in the interaction of cells with extracellular matrices (Kleinman *et al.*, 1981). Fibronectin was the first of these glycoproteins to be identified. Fibronectin is produced by fibroblasts, endothelial cells, hepatocytes, and other cells and is found in many tissues (Linder *et al.*, 1975; Vaheri *et al.*, 1976; Birdwell *et al.*, 1978; Chen *et al.*, 1978; Jaffe and Mosher, 1978; Stenman and Vaheri, 1978; Vaheri *et al.*, 1978; Voss *et al.*, 1979; Foidart *et al.*, 1980a; Villiger *et al.*, 1981). It is a prominent component of the proteins associated with the cell surface (Hynes, 1973; Ruoslahti *et al.*, 1973; Yamada and Weston, 1974), and is also found along collagen fibers (Furcht *et al.*, 1980), in basement membranes (Stenman and Vaheri, 1978; Wartiovaara *et al.*, 1979; Madri *et al.*, 1980), and in fibers of its own (Chen *et al.*, 1978; Vuento *et al.*, 1980). It is also present in serum at 0.3–1 mg/ml (Yamada and Olden, 1978; Hynes and Yamada, 1982). Fibronectin is a large protein (M_r 440,000) (Fig. 1) containing two chains joined by disulfide bonds at the carboxy terminus (Mosesson *et al.*, 1975; Hynes and Yamada, 1982). In rotary shadowing using electron microscopy, it is a V-shaped protein when extended (Engel *et al.*, 1981) (Table 3).

Table 4. Localization and Binding Sites of Attachment Proteins

Attachment protein	Histological localization	Binding sites		
		Collagen type	Glycosaminoglycans	Cell type
Fibronectin	Pericellular and matrix	I–V	Heparan sulfate; heparin	Fibroblasts
Chondronectin	Pericellular	II	Chondroitin sulfate; heparin	Chondrocytes
Laminin	Basal surface of cell; basement membrane	IV	Heparan sulfate; heparin	Epithelial cells; endothelial cells

Like fibronectin, laminin also binds to various components. For example, laminin has been found to bind to cell surfaces (Terranova *et al.*, 1980, 1983a), type IV collagen (Terranova *et al.*, 1980, 1983a), and heparan sulfate (Sakashita *et al.*, 1980; Del Rosso *et al.*, 1981; Woodley *et al.*, 1983) (Table 4). Laminin is active as an attachment protein for epithelial and endothelial cells and stimulates their attachment to type IV collagen (Terranova *et al.*, 1980). Laminin does not bind to other collagens (Terranova *et al.*, 1980; Kleinman *et al.*, 1982b) and prefers the native over the denatured protein. Preliminary studies suggest that the binding activities reside in separate protease-resistant domains analogous to those in fibronectin (Rao *et al.*, 1982; Terranova *et al.*, 1983a).

Immunolocalization studies show laminin to be adjacent to the basal surface of epithelial cells and in the basement membrane (Table 4). In many basement membranes, laminin is localized adjacent to the cells in the lamina lucida, although in others it has a more even distribution (Rohde *et al.*, 1979; Timpl *et al.*, 1978, 1979; Foidart *et al.*, 1980b; Madri *et al.*, 1980; Laurie *et al.*, 1982). Interestingly, laminin appears in the developing embryo at the morula stage prior to the appearance of other known components of extracellular matrices (Leivo *et al.*, 1980) and may be associated with the subsequent appearance of distinct structures.

1.3.3. Chondronectin: The Chondrocyte Attachment Protein

Serum was found to contain a protein distinct from fibronectin that mediated the attachment of chondrocytes to type II collagen substrates (Hewitt *et al.*, 1980). This protein (M_r 175,000) (Fig. 1) was isolated from serum and found to be distinct from fibronectin, laminin, or other known factors (Hewitt *et al.*, 1982a). It was named *chondronectin* (Table 3). Unlike fibronectin and laminin, chondronectin is a compact molecule containing two or possibly three disulfide-linked chains (Hewitt *et al.*, 1982a; Varner *et al.*, submitted). Chondronectin contains distinct binding domains for type II collagen, for chondrocyte cell surfaces, and for cartilage proteoglycan (Table 4). The cell binding domain resists tryptic digestion (Varner *et al.*, submitted). It is produced by chondrocytes and is found in cartilage, serum, and the vitreous. Recent studies show that the chondronectin-mediated binding of chondrocytes to type II collagen requires cartilage proteoglycan (Table 5), which is usually produced endogenously by the cells (Hewitt *et al.*, 1982b). Immunohistological studies suggest that chondronectin is not uniformly distributed in cartilage but is restricted to the regions around the cells at the interface between the chondrocyte membrane and the matrix (Hewitt *et al.*, 1982a,b) (Table 4).

Table 5. Proteoglycan Requirement for Attachment Protein Activity

Attachment protein	Proteoglycan requirement
Fibronectin	Not required; but collagen–fibronectin interaction stabilized by heparin
Laminin	Not required; but laminin does bind heparan sulfate and heparin
Chondronectin	Required (cartilage chondroitin sulfate proteoglycan)

1.4. Specificity and Structure of Extracellular Matrices

Each matrix discussed here contains different collagens, glycoprotein attachment factors, and proteoglycans. The attachment proteins play a central role in the matrix, for they bind to collagen and to proteoglycan and generate a unique macromolecular complex that in turn binds to the surface of the resident cells. Various studies are consistent with specific and limited numbers of interactions of the components, which result in matrices of defined stoichiometry and composition (Kleinman *et al.*, 1982b; Woodley *et al.*, 1983). Further, these matrices interact in a precisely organized fashion with cell types specific for the particular matrix. Presumably, this interaction depends on the presence of specific receptors on the cell surface that bind to the attachment protein and to the other components.

Recent studies would suggest that the attachment protein plus the other matrix proteins have a collegial influence on the cell. In culture, the presence of an attachment factor specific to the cell type enhances the growth and survival of the cell (Baron-Van Evercooren *et al.*, 1982; McGarvey *et al.*, submitted; Terranova *et al.*, submitted) and stabilizes its phenotype (Grotendorst *et al.*, 1982). In addition, it has been observed that the matrix from a given tissue has the strongest positive effects in culture as a matrix for the cells taken from that tissue. For example, hepatocytes, which usually survive only 2 or 3 weeks in culture, have been maintained *in vitro* for several months on a "biomatrix" obtained from liver (Rojkind *et al.*, 1980). Thus, the environment of the cell influences its growth, survival, and phenotype.

2. CELL INTERACTIONS WITH COLLAGEN

The interaction of cells with an extracellular matrix is essential for the normal growth and function of cells in tissue. An interaction with collagen is a key factor in the maintenance of growth potential and phe-

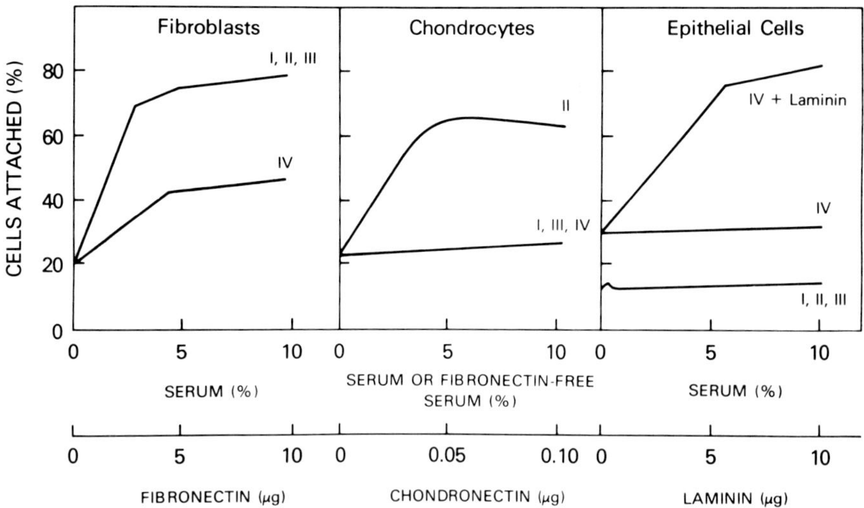

Figure 2. Comparison of attachment properties of fibroblasts, chondrocytes, and epithelial cells to collagen types I, II, III, and IV. Cells (10^5) were added to 35-mm collagen-coated bacteriologic plastic petri dishes in the presence of serum or the appropriate attachment factor. After 90 min, the percentages of attached cells were determined.

notypic expression. It has now become apparent that most cells do not interact directly with collagen but rather that a glycoprotein is required. In studies using fibroblasts and CHO cells, it was demonstrated that attachment to a collagen-coated substrate would not take place in the absence of serum (Klebe, 1974). Fibronectin was subsequently determined to be the serum factor responsible for this activity (Pearlstein, 1976) (Fig. 2). While these studies were performed *in vitro* using an exogenous source of fibronectin, these findings have validity *in vivo* because biosynthetic and immunfluorescence studies have confirmed that fibronectin is found in many tissues particularly those containing fibroblasts (e.g., Linder *et al.*, 1973; Chen *et al.*, 1978; Stenman and Vaheri, 1978; Vaheri *et al.*, 1978; Foidart *et al.*, 1980a; Weiss and Reddi, 1980). These studies led to the concept that the interaction of cells with collagen is mediated by other matrix components.

Fibronectin cannot be the sole attachment protein as it is not present in all tissues as, for example, cartilage (Dessau *et al.*, 1978) and basement membrane at certain stages and in some locations (Stenman and Vaheri, 1978; Wartiovaara *et al.*, 1979; Ekblom, 1981). Further, studies with epithelial and endothelial cells demonstrated that, in contrast to fibroblasts, these cells attached preferentially to type IV (basement membrane) collagen and that this interaction was not stimulated by serum or fibronectin

(Murray *et al.*, 1979; Wicha *et al.*, 1979) (Fig. 2). Because much time was required for attachment, these studies suggested that epidermal cells produced their own attachment factor. Laminin was subsequently demonstrated to be responsible for the attachment of epidermal cells as well as of other epithelial and endothelial cells. Furthermore, antibodies directed against laminin inhibit the attachment of these cells to type IV collagen (Terranova *et al.*, 1980). The mechanism for attachment is similar to that of fibronectin in that both interact with collagen prior to cell attachment. However, the laminin interaction is specific for type IV collagen (Terranova *et al.*, 1980). There is no proteoglycan requirement for laminin-mediated attachment (Table 5), although binding sites for heparin and heparan sulfate are present on laminin (Sakashita *et al.*, 1980; Del Rosso *et al.*, 1981; Woodley *et al.*, 1983) (Table 4).

Some cells can use more than one attachment factor. For example, it has been demonstrated that hepatocytes (Berman *et al.*, 1980; Johansson *et al.*, 1981) as well as some epithelial and endothelial cells (Gold and Pearlstein, 1980; Burrill *et al.*, 1981) can use both laminin and fibronectin for attachment under the appropriate conditions. This versatility may be important in wound healing where the cells interact with fibronectin and interstitial collagens in order to effect repair.

Cartilage also lacks fibronectin (Vaheri *et al.*, 1976; Dessau *et al.*, 1978; Lewis *et al.*, 1978; Stenman and Vaheri, 1978; Kimata *et al.*, 1982). It has been demonstrated that serum and fibronectin-free serum stimulate chondrocyte attachment (Fig. 2) whereas fibronectin does not (Hewitt *et al.*, 1980). Chondronectin was subsequently isolated from serum and shown to be specific for the attachment of chondrocytes, but not of fibroblasts or epithelial cells. Antibodies directed against chondronectin block chondrocyte attachment mediated by exogenously added chondronectin and inhibit the spontaneous attachment of chondrocytes that occurs over extended periods of time without added chondronectin (Hewitt *et al.*, 1982a). This attachment in the absence of added chondronectin indicates that chondrocytes produce a molecule immunologically and functionally similar to the serum-derived protein.

Unlike fibronectin and laminin, chondronectin does not bind directly to collagen, but also requires the chondroitin sulfate proteoglycan. Chondrocyte attachment is inhibited by xylosides, which interfere with proteoglycan formation. This inhibition of attachment can be overcome by the addition of intact cartilage proteoglycan monomer (Hewitt *et al.*, 1982b). These findings show that chondronectin binds to glycosaminoglycans and that intact cartilage proteoglycan monomer is required for attachment (Table 5). A schematic representation of chondrocyte attachment is shown in Fig. 3.

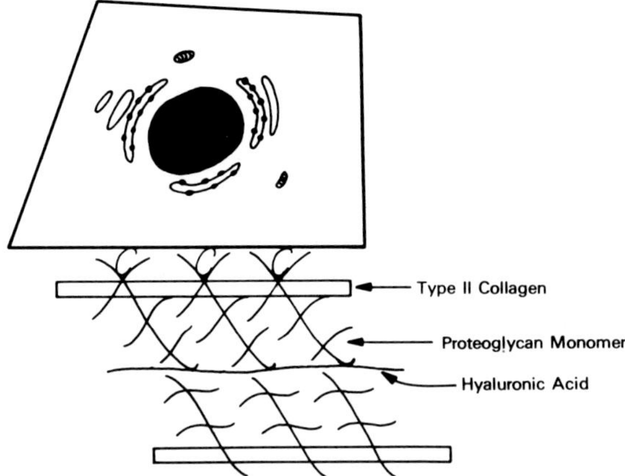

Figure 3. Schematic representation of chondrocyte attachment. Unlike fibronectin- or laminin-mediated attachment, chondronectin requires an interaction with proteoglycan in order to mediate chondrocyte attachment to collagen.

Considering the complexity of extracellular matrices, there may be more than one mechanism by which cells interact with components of the matrix. Smooth muscle cells, for example, exhibit multiple methods of interaction. Fibronectin mediates their attachment to all the interstitial collagens (Gold and Pearlstein, 1980; Grotendorst et al., 1981) and laminin supports attachment to type IV collagen (Grotendorst et al., 1982). However, these cells attach maximally to type V collagen in the absence of serum or any other soluble factor probably through an integral cell surface glycoconjugate that interacts directly with type V collagen (Grotendorst et al., 1981). Hepatocytes can also interact directly with collagen (Seglen and Fosså, 1978; Rubin et al., 1978; 1979). Their direct interaction with collagen is not specific for any one collagen type or for any one region within the collagen molecule (Rubin et al., 1981) suggesting that the repeating amino acid sequence common in collagen may be recognized by a cell surface receptor.

3. ROLE OF EXTRACELLULAR MATRIX AND ATTACHMENT PROTEINS IN DEVELOPMENT

The roles of extracellular matrices in morphogenesis, differentiation, matrix production, and growth have been studied in numerous systems and have been the subject of several recent reviews (Hay, 1981a,b; Kleinman et al., 1981, 1982a). Some of these studies have suggested that collagen by itself can direct differentiation. For example, collagenous substrates will act as a promoter and stabilizer of epithelial stroma production during corneal differentiation (Dodson and Hay, 1974; Meier and Hay, 1974). Cartilage formation by embryonic chick somites is also stimulated by procollagen and collagen, with type II collagen being the most effective (Kosher and Church; 1975; Lash and Vasan, 1978). The branched morphology of salivary gland and lung rudiments *in vitro* is disrupted by collagenase suggesting a role for collagen in these morphogenetic processes (Grobstein and Cohen, 1965; Spooner and Faubion, 1980). Collagen alone, however, is not always able to elicit differentiation. The formation of cartilage by cranial neural crest cells is not stimulated by collagenous substrates alone but is elicited by the extracellular matrix produced by retinal pigmented epithelial cells (Newsome, 1976). Demineralized bone powder, which is predominantly type I collagen, will induce endochondral ossification when implanted subcutaneously. As type I collagen by itself will not (Reddi and Anderson, 1976), it is apparent that other factors in the bone powder are responsible for recruiting cells of mesenchymal origin to the site of the implant and triggering their differentiation.

The role of attachment proteins in tissue morphogenesis is becoming more apparent. Interactions of attachment proteins with collagen, other matrix components, and cells of the same and adjacent tissues may be in part responsible for stimulation of growth and inductive processes commonly associated with tissue interactions and differentiation. Laminin, for example, has been shown to stimulate the growth of cells that use it as an attachment protein, such as epithelial cells, while having an inhibitory effect on the growth of fibroblasts (Terranova *et al.*, submitted). Schwann cells, which also use laminin for attachment, exhibit a dose-dependent stimulation of growth in the presence of laminin and also have a more stellate appearance than cells grown in the absence of laminin (McGarvey *et al.*, submitted). Although this suggests a correlation between cell shape and proliferation, it is possible that laminin exerts its effect indirectly by stimulating the synthesis of a growth promoter. Interestingly, laminin will also promote neurite outgrowth in cultures of fetal sensory ganglia (Baron-Van Evercooren *et al.*, 1982). Neurons themselves do not produce laminin. However, the production of basement membrane components by Schwann cells may provide a permissive environment that can influence the development of neurons.

The same attachment proteins are not always present throughout the entire developmental history of a particular tissue. This is evident in such processes as limb development (Dessau *et al.*, 1978, 1980; Silver *et al.*, 1981; Tomasek *et al.*, 1982) when muscle and cartilage develop from apparently similar mesenchymal cells. The appearance or disappearance of attachment proteins appears to follow a distinct developmental timetable and it is possible that the modulation of one attachment protein acts as a cue for the next stage of the differentiation process. Myogenesis provides an example of such a programmed switch. When cultured *in vitro*, myoblasts require a collagenous substrate in order to attach and differentiate (Hauschka and Konigsberg, 1966; Hauschka and White, 1972). The attachment process requires a serum protein (Hauschka and White, 1972), which has been shown to be fibronectin (Chicquet *et al.*, 1979). *In vivo*, at early stages of limb development, fibronectin has been demonstrated throughout the mesenchymal extracellular matrix by immunofluorescence (Silver *et al.*, 1981; Tomasek *et al.*, 1982). When fusion of myoblasts into myotubes commences, however, there is a decrease of fibronectin both *in vivo* (Tomesek *et al.*, 1982) and *in vitro* (Chen, 1977; Furcht *et al.*, 1978). The fusion of myoblasts can be delayed *in vitro* by exogenous fibronectin, whereas precocious fusion results from the addition of antibodies against fibronectin to the cultures (Podleski *et al.*, 1979). These findings indicate that the removal of fibronectin from the surfaces of myoblasts is essential for the fusion process and, presumably, for the appearance of laminin (Foidart *et al.*, 1980b; Grotendorst *et al.*, 1982). It is possible that the initial presence of fibronectin may prevent myoblast fusion until these cells have reached the appropriate developmental stage.

During chondrogenesis, there are also changes in the extracellular attachment glycoproteins. While there is a marked reduction of staining for fibronectin in regions undergoing myogenesis, immunofluorescence studies indicate that adjacent areas undergoing prechondrogenic condensation exhibit an increase in staining intensity (Silver *et al.*, 1981; Tomasek *et al.*, 1982). However, as differentiation progresses and large amounts of cartilagenous matrix are elaborated, staining for fibronectin is lost (Dessau *et al.*, 1978, 1980; Tomasek *et al.*, 1982). Immunofluorescence staining of cartilagenous limb rudiments also demonstrates the pericellular appearance of chondronectin (Hewitt *et al.*, 1982b). Attachment proteins, when added exogenously to cultured mesenchyme, influence chondrogenic expression. When fibronectin is added to micromass cultures of chick limb mesenchyme under serum-free conditions, chondrogenesis is inhibited. On the other hand, addition of chondronectin to the cultures will stimulate the accumulation of proteoglycans in the cartilage matrix (Hassell and Horigan, unpublished observations; Grotendorst *et al.*, 1982). These data confirm that chondronectin promotes the chondrocyte phenotype and suggest that chondrogenesis may be regulated by receptors for chondronectin.

Thus, association with the "proper" attachment protein may be important for the maintenance of normal phenotypic expression (Fig. 4). Disruption of these interactions due to disease or injury may expose the cells to a "foreign" attachment protein. This could result in an abnormal

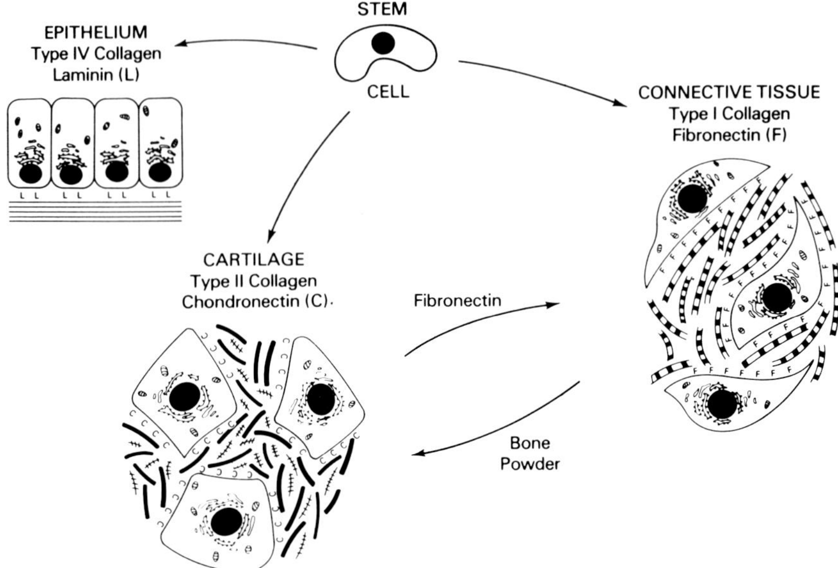

Figure 4. Model for the influence of attachment proteins on the phenotypic expression of cells. Cells in a tissue are associated with a specific attachment protein and other matrix components (collagen and proteoglycans) that are characteristic of the tissue. Other attachment proteins can influence the phenotypic expression of the cells as is seen by the effect of fibronectin on the chondrocytes. (From Kleinman *et al.*, 1980, with permission.)

condition. For example, chondrocytes, when liberated from their matrix, synthesize fibronectin until they are again surrounded by a matrix (Dessau *et al.*, 1978). Continued exposure to increased levels of fibronectin, however, will eventually result in the loss of the chondrocyte phenotype (Pennypacker *et al.*, 1979; West *et al.*, 1979). Chondrocytes cultured in fibronectin-free serum will not undergo such a transition.

4. ATTACHMENT PROTEINS IN PATHOLOGICAL CONDITIONS

The importance of attachment proteins and their interaction with other matrix components is suggested by changes in the attachment properties of cells in various pathological conditions. While fibroblasts require collagen and fibronectin for attachment, transformed fibroblasts have decreased levels of fibronectin on their surfaces and synthesize reduced levels of collagen (Hynes, 1973; Vaheri *et al.*, 1976, 1978; Kamine and Rubin, 1977; Adams *et al.*, 1977) (Table 6). Transformed fibroblasts lose their requirement for a collagenous substrate for attachment and are able

Table 6. Changes in Attachment Protein Requirements

Cell	Normal	Transformed	Metastatic
Fibroblasts	Fibronectin	Fibronectin decreased	Laminin (to type IV)
Chondrocytes	Chondronectin	Fibronectin	?
Epithelial cells	Laminin	Laminin decreased	Laminin

to attach to plastic substrates and to proliferate in the presence of *cis*-hydroxyproline (Liotta *et al.*, 1978; Vembu *et al.*, 1979; Kleinman *et al.*, 1981), a proline analog that prevents collagen secretion when incorporated into collagen (Uitto and Prockop, 1974). Normal fibroblasts do not attach or proliferate under these conditions (Liotta *et al.*, 1978). Thus, there appears to be a correlation between tumorigenicity, increased resistance to *cis*-hydroxyproline, and decreased reliance on collagen and fibronectin for attachment. Transformed epithelial cells similarly show reduced levels of laminin compared to normal epithelial cells (Hayman *et al.*, 1981, 1982). Chondrocytes from a nonmetastatic rat chondrosarcoma were found not to use chondronectin to bind to collagen but to use fibronectin (Kimata *et al.*, 1982). However, the cells still produce the normal cartilage macromolecules including type II collagen and the cartilage type of proteoglycan.

In the process of metastasis, tumor cells must cross organ boundaries and capillary walls in order to initiate new colonies at sites some distance from the primary tumor. It has been demonstrated that highly metastatic tumor cells bind to subendothelial basement membranes more avidly than cells with less metastatic potential (Poste and Fidler, 1980). Laminin selectively promotes the attachment of metastatic, tumorigenic cells to type IV collagen regardless of the origin of the cells (Terranova *et al.*, 1982) (Table 6). Highly metastatic cells also have specific, high-affinity receptors for laminin on their surfaces whereas nontumorigenic cells and fibroblasts do not (Terranova *et al.*, 1983a). Once attached to subendothelial basement membranes, metastasis is completed by degradation of basement membrane collagen by type IV collagenase produced by the cells (Liotta *et al.*, 1979, 1980). This suggests that metastasis is correlated both with an affinity for laminin and the ability to degrade basement membrane components.

Neutrophils, like metastatic cells, are highly motile cells that must also traverse the vascular endothelium to reach sites of infection. Recent studies have also demonstrated that neutrophils use laminin to bind to type IV collagen. Neutrophils may produce laminin as a cell surface component in response to chemotactic signals released during the initial stages of an infection (Terranova *et al.*, 1983b).

Attachment proteins may also play a role in certain parasitic conditions. For example, the chronic stages of Chagas' disease are characterized by circulating antibodies against laminin (Szarfman *et al.*, 1982). The cause for the production of these antibodies is uncertain. Interestingly, affinity-purified antilaminin antibodies from the sera of patients with Chagas' disease interact with the cell surface of the invasive form of *Trypanosoma cruzi* (Szarfman *et al.*, 1982). This implies that the parasite produces a molecule that is immunologically similar to mammalian laminin. It is not known if this cross-reactive molecule also has a function similar to laminin during the invasive stages of the disease. How the antibodies relate to the degenerative and fibrotic changes associated with Chagas' disease is not clear. It is possible that the antilaminin antibodies block the inhibitory activity of laminin on fibroblast growth and thus allow fibroblasts to proliferate at a faster rate.

5. SUMMARY

The studies described in this chapter have pointed out the importance of cellular interactions with their extracellular matrices. Of particular significance is the role played by attachment proteins. These glycoproteins are involved in anchoring cells to the matrix and are able to influence morphology, growth, and phenotypic expression. Chondronectin, laminin, and fibronectin are probably just a few examples of a family of proteins that are involved in mediating the association of specific cells with their respective matrices. Such interactions could in turn explain the localization and morphogenesis of cells in tissues.

ACKNOWLEDGMENTS. The authors thank Dr. Hynda Kleinman for her helpful suggestions and Ms. Sandra Ward and Ms. Ginny Floyd for the preparation of the manuscript.

REFERENCES

Adams, S. E., Sobel, M. E., Howard, B. H., Olden, K., Yamada, K. M., DeCrombrugghe, B., and Pastan, I., 1977, Levels of translatable mRNAs for cell surface protein, collagen precursors, and two membrane proteins are altered in Rous sarcoma virus-transformed chick embryo fibroblasts, *Proc. Natl. Acad. Sci. USA* **74**:3399–3403.

Ali, I. U., Mautner, V. M., Lanza, R. P., and Hynes, R. O., 1977, Restoration of normal morphology, adhesion and cytoskeleton in transformed cells by addition of a transformation-sensitive surface protein, *Cell* **11**:115–126.

Baron-Van Evercooren, A., Kleinman, H. K., Ohno, S., Marangos, P., Schwartz, J. P., and Dubois-Dalcq, M., 1982, Nerve growth factor, laminin, and fibronectin promote neurite growth in human fetal sensory ganglia cultures, *J. Neurosci. Res.* **8:**179–194.

Bentz, H., Bächinger, H. P., Glanville, R., and Kühn, K., 1978, Physical evidence for the assembly of A and B chains of human placental collagen in a single triple helix, *Eur. J. Biochem.* **92:**563–567.

Berman, M. D., Waggoner, J. G., Foidart, J.-M., and Kleinman, H. K., 1980, Attachment to collagen by isolated hepatocytes from rats with induced hepatic fibrosis, *J. Lab. Clin. Med.* **95:**660–671.

Birdwell, C. R., Gospodarowicz, D., and Nicolson, G. L., 1978, Identification, localization and role of fibronectin in cultured bovine endothelial cells, *Proc. Natl. Acad. Sci. USA* **75:**3273–3277.

Blumenstock, F. A., Saba, P., Weber, P., and Cho, E., 1976, Purification and biochemical characterization of a macrophage-stimulating alpha-2-globulin opsonic protein, *J. Reticuloendothel. Soc.* **19:**157–172.

Bornstein, P., and Sage, H., 1980, Structurally distinct collagen types, *Annu. Rev. Biochem.* **49:**957–1003.

Burgeson, R. E., El Adli, F. A., Kaitela, I. I., and Hollister, D. W., 1976, Fetal membrane collagens: Identification of two new collagen alpha chains, *Proc. Natl. Acad. Sci. USA* **73:**2579–2583.

Burrill, P. H., Bernardini, I., Kleinman, H. K., and Kretchmer, N., 1981, Effects of serum, fibronectin, and laminin on adhesion of rabbit intestinal epithelial cells in culture, *J. Supramol. Struct. Cell. Biochem.* **16:**385–392.

Chen, L. B., 1977, Alteration in cell surface LETS protein during myogenesis, *Cell* **10:**393–400.

Chen, L. B., Murray, A., Figal, R. A., Bushnell, A., and Walsh, M. L., 1978, Studies on intercellular LETS glycoprotein matrices, *Cell* **14:**377–391.

Chicquet, M., Puri, E. C., and Turner, D. C., 1979, Fibronectin mediates attachment of chicken myoblasts to a gelatin-coated substratum, *J. Biol. Chem.* **254:**5475–5482.

Chung, E., and Miller, E. J., 1974, Collagen polymorphism: Characterization of molecules with the chain composition $[\alpha 1(III)]_3$ in human tissues, *Science* **183:**1200–1203.

Chung, E., Rhodes, R. K., and Miller, E. J., 1976, Isolation of three collagenous components of probable basement membrane origin from several tissues, *Biochem. Biophys. Res. Commun.* **71:**1167–1174.

Del Rosso, M., Cappelletti, R., Viti, M., Vannucchi, S., and Chiarugi, V., 1981, Binding of the basement membrane glycoprotein, laminin, to glycosaminoglycans, *Biochem. J.* **199:**699–704.

Dessau, W., Sasse, J., Timpl, R., Jilek, F., and von der Mark, K., 1978, Synthesis and extracellular deposition of fibronectin in chondrocyte cultures, *J. Cell Biol.* **79:**342–355.

Dessau, W., von der Mark, H., von der Mark, K., and Fischer, S., 1980, Changes in patterns of collagens and fibronectin during limb bud chondrogenesis, *J. Embryol. Exp. Morphol.* **57:**51–60.

Dodson, J. W., and Hay, E. D., 1974, Secretion of collagen by corneal epithelium: Effect of the underlying substratum on secretion and polymerization of epithelial products, *J. Exp. Zool.* **189:**51–72.

Ekblom, P., 1981, Formation of basement membranes in the embryonic kidney: An immunohistological study, *J. Cell Biol.* **91:**1–10.

Engel, J., Odermatt, E., Engel, A., Madri, J. A., Furthmayr, H., Rohde, H., and Timpl, R., 1981, Shapes, domain organization and flexibility of laminin and fibronectin: Two multifunctional extracellular matrix proteins, *J. Mol. Biol.* **150:**97–120.

Engvall, E., and Ruoslahti, E., 1977, Binding of soluble form of fibroblasts surface protein, fibronectin, to collagen, *Int. J. Cancer* **20**:1–5.

Engvall, E., Ruoslahti, E., and Miller, E. J., 1978, Affinity of fibronectin to collagens of different genetic types and to fibrinogen, *J. Exp. Med.* **147**:1584–1595.

Epstein, E. H., Jr., 1974, [$\alpha 1(III)$]$_3$ human skin collagen: Release by pepsin digestion and preponderance in fetal life, *J. Biol. Chem.* **249**:3225–3231.

Foidart, J.-M., Berman, J. J., Paglia, L., Rennard, S. I., Abe, S., Perantoni, A., and Martin, G. R., 1980a, Synthesis of fibronectin, laminin, and several collagens by a liver-derived epithelial cell line, *Lab. Invest.* **42**:525–532.

Foidart, J.-M., Bere, E. N., Yaar, M., Rennard, S. I., Gullino, M., Martin, G. R., and Katz, S. I., 1980b, Distribution and immunoelectron microscopic localization of laminin, a noncollagenous basement membrane glycoprotein, *Lab. Invest.* **42**:336–342.

Furcht, L. T., Mosher, D. F., and Wendelschafer-Crabb, G., 1978, Immunocytochemical localization of fibronectin (LETS protein) on the surface of L_6 myoblasts: Light and electron-microscopic studies, *Cell* **13**:263–271.

Furcht, L. T., Smith, D., Wendelschafer-Crabb, G., Mosher, D. F., and Foidart, J.-M., 1980, Fibronectin presence in native collagen fibrils of human fibroblasts: Immunoperoxidase and immunoferritin localization, *J. Histochem. Cytochem.* **28**:1319–1333.

Gauss-Müller, V., Kleinman, H. K., Martin, G. R., and Schiffmann, E., 1980, Role of attachment and attractants in fibroblast chemotaxis, *J. Lab. Clin. Med.* **96**:1071–1080.

Gay, S., and Miller, E. J., 1978, *Collagen in the Physiology and Pathology of Connective Tissue*, Fisher Verlag, Stuttgart.

Gay, S., Rhodes, R. K., Gay, R. E., and Miller, E. J., 1981, Collagen molecules comprised of $\alpha_1(V)$-chains (B-chains): An apparent localization in the exoskeleton, *Coll. Res.* **1**:53–58.

Gehron Robey, P., and Martin, G. R., 1981, Type IV collagen contains two distinct chains in separate molecules, *Coll. Res.* **1**:27–38.

Goetinck, P. F., Pennypacker, J. P., and Royal, P. D., 1974, Proteochondroitin sulfate synthesis and chondrogenic expression, *Exp. Cell Res.* **87**:241–248.

Gold, L. I., and Pearlstein, E., 1980, Fibronectin–collagen binding and requirement during cellular adhesion, *Biochem. J.* **186**:551–554.

Greenberg, J. H., Seppä, S., Seppä, H., and Hewitt, A. T., 1981, Role of collagen and fibronectin in neural crest cell adhesion and migration, *Dev. Biol.* **87**:259–266.

Grinnell, F., Harp, D. G., and Minter, D., 1977, Cell adhesion and spreading factor, *Exp. Cell Res.* **110**:175–190.

Grobstein, C., and Cohen, J., 1965, Collagenase: Effect on the morphogenesis of embryonic salivary epithelium *in vitro, Science* **150**:626–628.

Gross, J., 1974, Collagen biology: Structure, degradation, and disease, *Harvey Lect.* **68**:351–432.

Gross, J., 1981, An essay on biological degradation of collagen, in: *Cell Biology of Extracellular Matrix* (E.D. Hay, ed.), pp. 217–258, Plenum Press, New York.

Gross, J., and Nagai, Y., 1965, Specific degradation of the collagen molecule by tadpole collagenolytic enzymes, *Proc. Natl. Acad. Sci. USA* **54**:1197–1201.

Gross, J., Harper, E., Harris, E. D., Jr., McCroskery, P., Highberger, J. H., Corbett, C., and Kang, A. H., 1974, Animal collagenases, specificity of action and structure of the substrate cleavage site, *Biochem. Biophys. Res. Commun.* **61**:605–612.

Grotendorst, G. R., Seppä, H. E. J., Kleinman, H. K., and Martin, G. R., 1981, Attachment of smooth muscle cells to collagen and their migration toward platelet-derived growth factors, *Proc. Natl. Acad. Sci. USA* **78**:3669–3672.

Grotendorst, G. R., Kleinman, H. K., Rohrbach, D. H., Hewitt, A. T., Varner, H. H., Horigan, E. A., Hassell, J. R., Terranova, V. P., and Martin, G. R., 1982, Role of attachment factors in mediating the attachment, distribution, and differentiation of cells, in: *Growth of Cells in Hormonally Defined Media* (G. H. Sato, A. B. Pardee, and D. A. Sirbasku, eds.), pp. 403–413, Cold Spring Harbor Laboratory, New York.

Hascall, V. C., 1977, Interaction of cartilage proteoglycans with hyaluronic acid, *J. Supramol. Struct.* **7:**101–120.

Hascall, V. C., and Hascall, G. K., 1981, Proteoglycans, in: *Cell Biology of Extracellular Matrix* (E. D. Hay, ed.), pp. 39–63, Plenum Press, New York.

Hascall, V. C., and Sajdera, S. W., 1970, Physical properties and polydispersity of proteoglycan from bovine nasal cartilage, *J. Biol. Chem.* **245:**4920–4930.

Hassell, J. R., Newsome, D. A., and Hascall, V. C., 1979, Characterization and biosynthesis of proteoglycans of corneal stroma from rhesus monkey, *J. Biol. Chem.* **254:**12346–12354.

Hassell, J. R., Gehron Robey, P., Barrach, H.-J., Wikzek, J., Rennard, S. I., and Martin, G. R., 1980a, Isolation of a heparan sulfate-containing proteoglycan from basement membrane, *Proc. Natl. Acad. Sci. USA* **77:**4494–4498.

Hassell, J. R., Newsome, D. A., Krachmer, J. H., and Rodriques, M. M., 1980b, Macular corneal dystrophy: Failure to synthesize a mature keratan sulfate proteoglycan, *Proc. Natl. Acad. Sci. USA* **77:**3705–3709.

Hassell, J. R., Newsome, D. A., Nakazawa, K., Rodriques, M., and Krachmer, J., 1982, Defective conversion of a glycoprotein precursor to keratan sulfate proteoglycan in macular corneal dystrophy, in: *Extracellular Matrix* (S. Hawkes and J. L. Wang, eds.), pp. 397–406, Academic Press, New York.

Hauschka, S. D., and Konigsberg, I. R., 1966, The influence of collagen on the development of muscle colonies, *Proc. Natl. Acad. Sci. USA* **55:**119–126.

Hauschka, S. D., and White, N. K., 1972, Studies of myogenesis *in vitro*, in: *Research in Muscle Development and the Muscle Spindle* (B. Q. Banker, R. J. Przybylski, J. P. van der Meulen, and M. Victor, eds.), pp. 53–71, Excerpta Medica, Amsterdam.

Hay, E. D. (ed.), 1981a, *Cell Biology of Extracellular Matrix*, Plenum Press, New York.

Hay, E. D., 1981b, Extracellular matrix, *J. Cell Biol.* **91:**205S–223S.

Hay, E. D., Hasty, D. L., and Kiehnau, K. L., 1978, Fine structure of collagens and their relation to glycosaminoglycans, in: *Collagen–Platelet Interactions* (H. Gastpar, K. Kühn, and R. Marx, eds.), pp. 129–151, Schattauer Verlag, Stuttgart.

Hayman, E. G., Engvall, E., and Ruoslahti, E., 1981, Concomitant loss of cell surface fibronectin and laminin from transformed rat kidney cells, *J. Cell Biol.* **88:**352–357.

Hayman, E. G., Oldberg, Å, Martin, G. R., and Ruoslahti, E., 1982, Codistribution of heparan sulfate proteoglycan, laminin, and fibronectin in the extracellular matrix of normal rat kidney cells and their coordinate absence in transformed cells, *J. Cell Biol.* **94:**28–35.

Hewitt, A. T., Kleinman, H. K., Pennypacker, J. P., and Martin, G. R., 1980, Identification of an adhesion factor for chondrocytes, *Proc. Natl. Acad. Sci. USA* **77:**385–388.

Hewitt, A. T., Varner, H. H., Silver, M. H., Dessau, W., Wilkes, C. M., and Martin, G. R., 1982a, The isolation and partial characterization of chondronectin, an attachment factor for chondrocytes, *J. Biol. Chem.* **257:**2330–2334.

Hewitt, A. T., Varner, H. H., Silver, M. H., and Martin, G. R., 1982b, The role of chondronectin and cartilage proteoglycan in the attachment of chondrocytes to collagen, in: *Limb Development and Regeneration*, Part B (R. O. Kelley, P. F. Goetinck, and J. A. MacCabe, eds.), pp. 25–33, Liss, New York.

Hopper, D. E., Adelmann, B. C., Gentner, G., and Gay, S., 1976, Recognition by guinea pig peritoneal exudate cells of conformationally different states of the collagen molecule, *Immunology* **30:**249–259.

Horwitz, A. L., Hance, A. J., and Crystal, R. G., 1977, Granulocyte collagenase: Selective degradation of type I relative to type III collagen, *Proc. Natl. Acad. Sci. USA* **74**:897–901.

Hynes, R. O., 1973, Alteration of cell-surface proteins by viral transformation and by proteolysis, *Proc. Natl. Acad. Sci. USA* **70**:3170–3174.

Hynes, R. O., and Yamada, K. M., 1982, Fibronectins: Multifunctional modular glycoproteins, *J. Cell Biol.* **95**:369–377.

Jaffe, E. A., and Mosher, D. F., 1978, Synthesis of fibronectin by cultured human endothelial cells, *J. Exp. Med.* **147**:1779–1791.

Jilek, F., and Hörmann, H., 1978, Cold-insoluble globulin (fibronectin). IV. Affinity to soluble collagen of various types, *Hoppe-Seyler's Z. Physiol. Chem.* **359**:247–250.

Johansson, S., Kjellén, L., Höök, M., and Timpl, R., 1981, Substrate adhesion of rat hepatocytes: A comparison of laminin and fibronectin as attachment proteins, *J. Cell Biol.* **90**:260–264.

Kamine, J., and Rubin, H., 1977, Coordinate control of collagen synthesis and cell growth in chick embryo fibroblasts and the effect of viral transformation on collagen synthesis, *J. Cell Physiol.* **92**:1–12.

Kanwar, Y. S., and Farquhar, M. G., 1979a, Anionic sites in the glomerular basement membrane: *In vivo* and *in vitro* localization to the laminae rarae by cationic probes, *J. Cell Biol.* **81**:137–153.

Kanwar, Y. S., and Farquhar, M. G., 1979b, Presence of heparan sulfate in the glomerular basement membrane, *Proc. Natl. Acad. Sci. USA* **76**:1303–1307.

Kanwar, Y. S., Linker, A., and Farquhar, M. G., 1980, Increased permeability of the glomerular basement membrane to ferritin after removal of glycosaminoglycans (heparan sulfate) by enzyme digestion, *J. Cell Biol.* **86**:688–693.

Kefalides, N. A., Alper, R., and Clark, C. C., 1979, Biochemistry and metabolism of basement membranes, *Int. Rev. Cytol.* **61**:167–228.

Kimata, K., Barrach, H. J., Brown, K. S., and Pennypacker, J. P., 1981, Absence of proteoglycan core protein in cartilage from *cmd/cmd* (cartilage matrix deficiency) mouse, *J. Biol. Chem.* **256**:6961–6968.

Kimata, K., Foidart, J.-M., Pennypacker, J. P., Kleinman, H. K., Martin, G. R., and Hewitt, A. T., 1982, Immunofluorescence localization of fibronectin in chondrosarcoma cartilage matrix, *Cancer Res.* **42**:2384–2391.

Kjellén, L., Oldberg, Å, and Höök, M., 1980, Cell-surface heparan sulfate: Mechanisms of proteoglycan–cell association, *J. Biol. Chem.* **255**:10407–10413.

Kjellén, L., Pettersson, I., and Höök, M., 1981, Cell surface heparan sulfate: An intercalated membrane proteoglycan, *Proc. Natl. Acad. Sci. USA* **78**:5371–5375.

Klebe, R. J., 1974, Isolation of a collagen-dependent cell attachment factor, *Nature (London)* **250**:248–251.

Kleinman, H. K., McGoodwin, E. B., Martin, G. R., Klebe, R. J., Fietzek, P. P., and Woolley, D. E., 1978, Localization of the binding site for cell attachment in the α1(I) chain of collagen, *J. Biol. Chem.* **253**:5642–5646.

Kleinman, H. K., Hewitt, A. T., Grotendorst, G. R., Martin, G. R., Murray, J. C., Rohrbach, D. H., Terranova, V. P., Rennard, S. I., Varner, H. H., and Wilkes, C. M., 1980, Role of matrix components in adhesion and growth of cells, in: *Current Research Trends in Prenatal Craniofacial Development* (Pratt and Christiansen, eds.), pp. 277–295, Elsevier North Holland, New York.

Kleinman, H. K., Klebe, R. J., and Martin, G. R., 1981, Role of collagenous matrices in the adhesion and growth of cells, *J. Cell Biol.* **88**:473–485.

Kleinman, H. K., Rohrbach, D. H., Terranova, V. P., Varner, H. H., Hewitt, A. T., Grotendorst, G. R., Wilkes, C. M., Martin, G. R., Seppä, H., and Schiffman, E., 1982a,

Collagenous matrices as determinants of cell function, in: *Immunochemistry of the Extracellular Matrix*, Vol. II (H. Furthmayr, ed.), pp. 151–174, CRC Press, Boca Raton, Fla.

Kleinman, H. K., Woodley, D. T., McGarvey, M. L., Gehron Robey, P., Hassell, J. R., and Martin, G. R., 1982b, Interactions and assembly of basement membrane components, in: *Extracellular Matrix* (S. Hawkes and J. Wang, eds.), pp. 45–52, Academic Press, New York.

Kosher, R. A., and Church, R. L., 1975, Stimulation of *in vitro* chondrogenesis by procollagen and collagen, *Nature (London)* **258:**327–330.

Kramer, J. M., Cox, G. N., and Hirsh, D., 1982, Comparisons of the complete sequences of two collagen genes from *Caenorhabditis elegans*, *Cell* **30:**599–606.

Kresina, T., and Miller, E. J., 1979, Isolation and characterization of basement membrane collagen from human placental tissues—Evidence for the presence of two genetically distinct chains, *Biochemistry* **18:**3089–3097.

Kühn, K., 1982, Chemical properties of collagen, in: *Immunochemistry of the Extracellular Matrix*, Vol. I (H. Furthmayr, ed.), pp. 1–29, CRC Press, Boca Raton, Fla.

Kühn, K., and van der Mark, K., 1978, The influence of proteoglycans on the macromolecular structure of collagen, in: *Collagen–Platelet Interaction* (H. Gastpar, K. Kühn, and R. Marx, eds.), pp. 123–127, Schattauer Verlag, Stuttgart.

Lapière, C. M., Lenaers, A., and Kohn, L. D., 1971, Procollagen peptidase: An enzyme excising the coordination peptides of procollagen, *Proc. Natl. Acad. Sci. USA* **68:**3054–3058.

Lash, J. W., and Vasan, N. S., 1978, Somite chondrogenesis *in vitro*: Stimulation by exogenous extracellular matrix components, *Dev. Biol.* **66:**151–171.

Laurie, G. W., Leblond, C. P., and Martin, G. R., 1982, Localization of type IV collagen, laminin, heparan sulfate proteoglycan and fibronectin to the basal lamina of basement membranes, *J. Cell Biol.* **95:**340–344.

Leivo, I., Vaheri, A., Timpl, R., and Wartiovaara, J., 1980, Appearance and distribution of collagens and laminin in the early mouse embryo, *Dev. Biol.* **76:**100–114.

Levitt, D., and Dorfman, A., 1974, Concepts and mechanisms of cartilage differentiation, *Curr. Top. Dev. Biol.* **8:**103–145.

Levitt, D., Ho, P., and Dorfman, A., 1975, Effects of 5-bromodeoxyuridine on ultrastructure of developing limb bud cells *in vitro*, *Dev. Biol.* **43:**75–90.

Lewis, C. A., Pratt, R. M., Pennypacker, J. P., and Hassell, J. R., 1978, Inhibition of limb chondrogenesis *in vitro* by vitamin A: Alterations in cell surface characteristics, *Dev. Biol.* **64:**31–47.

Linder, E. A., Vaheri, A., Ruoslahti, E., and Wartiovaara, J., 1975, Distribution of fibroblast surface antigen in the developing chick embryo, *J. Exp. Med.* **142:**41–49.

Liotta, L. A., Vembu, D., Kleinman, H. K., Martin, G. R., and Boone, C., 1978, Collagen required for proliferation of cultured connective tissue cells but not their transformed counterparts, *Nature (London)* **272:**622–624.

Liotta, L. A., Abe, S., Gehron Robey, P., and Martin, G. R., 1979, Preferential digestion of basement membrane collagen by an enzyme derived from a metastatic murine tumor, *Proc. Natl. Acad. Sci. USA* **76:**2268–2272.

Liotta, L. A., Tryggvason, K., Garbisa, S., Hart, I., Foltz, C. M., and Shafie, S., 1980, Metastatic potential correlates with enzymatic degradation of basement membrane collagen, *Nature (London)* **284:**67–68.

Liotta, L. A., Lanzer, W. L., and Garbisa, S., 1981, Identification of a type V collagenolytic enzyme, *Biochem. Biophys. Res. Commun.* **98:**184–190.

McDonald, J. A., Kelley, D. G., and Broekelmann, T. J., 1982, Role of fibronectin in collagen deposition: Fab' to the gelatin-binding domain of fibronectin inhibits both fibronectin and collagen organization in fibroblast extracellular matrix, *J. Cell Biol.* **92:**485–492.

McGarvey, M. L., Baron-Van Evercooren, A., Dubois-Dalcq, M., and Kleinman, H. K., 1983, Role of laminin in the adhesion and growth of rat Schwann cells (submitted).

McLees, B. D., Schleiter, G., and Pinnell, S. R., 1977, Isolation of type III collagen from human adult parenchymal lung tissue, *Biochemistry* **16**:185–190.

Madri, J. A., Roll, J., Furthmayr, H., and Foidart, J.-M., 1980, Ultrastructural localization of fibronectin and laminin in the basement membranes of the murine kidney, *J. Cell Biol.* **86**:682–687.

Mainardi, C. L., Seyer, J. N., and Kang, A. H., 1980, Type specific collagenolysis of type V collagen degrading enzyme from macrophages, *Biochem. Biophys. Res. Commun.* **97**:1108–1115.

Marquette, D. J., Molnar, K. M., Yamada, K. M., Schlesinger, D., Darby, S., and Van Alten, P., 1981, Phagocytosis-promoting activity of avian plasma and fibroblastic cell surface fibronectin, *Mol. Cell. Biochem.* **36**:147–155.

Mayne, R., Vail, M. S., and Miller, E. J., 1978, Characterization of collagen chains synthesized by cultured smooth muscle cells derived from rhesus monkey thoracic aorta, *Biochemistry* **17**:446–452.

Meier, S., and Hay, E. D., 1974, Control of corneal differentiation by extracellular materials: Collagen as a promoter and stabilizer of epithelial stroma production, *Dev. Biol.* **38**:249–270.

Miller, E. J., and Matukas, V. J., 1969, Chick cartilage collagen: A new type of α1 chain not present in bone or skin of the species, *Proc. Natl. Acad. Sci. USA* **64**:1264–1268.

Miller, E. J., and Matukas, V. J., 1974, Biosynthesis of collagen: The biochemist's view, *Fed. Proc.* **33**:1197–1201.

Miller, E. J., Epstein, E. H., and Piez, K. A., 1971, Identification of three genetically distinct collagens by cyanogen bromide cleavage of insoluble human skin and cartilage collagen, *Biochem. Biophys. Res. Commun.* **42**:1024–1029.

Miller, E. J., Harris, E. D., Chung, E., Finch, J. E., Jr., McCroskery, P. A., and Butler, W. T., 1976, Cleavage of type II and III collagens with mammalian collagenase: Site of cleavage and primary structure at NH_2 terminal position of the smaller fragments released from both collagens, *Biochemistry* **15**:787–792.

Mosesson, M. V., Chen, A. B., and Huseby, R. M., 1975, The cold-insoluble globulin of human plasma: Studies of its essential structural features, *Biochim. Biophys. Acta* **386**:509–524.

Murray, J. C., Stingl, G., Kleinman, H. K., Martin, G. R., and Katz, S. I., 1979, Epidermal cells adhere preferentially to type IV (basement membrane) collagen, *J. Cell Biol.* **80**:197–202.

Mynderse, L. A., Martinez-Hernandez, A., Hassell, J. R., Kleinman, H. K., and Martin, G. R., 1983, Loss of heparan sulfate proteoglycan from glomerular basement membrane of nephrotic rats, *Lab. Invest.*, **48**:292–302.

Newsome, D. A., 1976, *In vitro* stimulation of cartilage in embryonic chick neural crest cells by products of retinal pigmented epithelium, *Dev. Biol.* **49**:496–507.

Nowack, H., Gay, S., Wick, G., Becker, U., and Timpl, R., 1976, Preparation and use of antibodies specific for type I and type III collagen and procollagen, *J. Immunol. Methods* **12**:117–124.

Orkin, R. W., Gehron, P., McGoodwin, E. B., Martin, G. R., Valentine, T., and Swarm, R., 1977, A murine tumor producing a matrix of basement membrane, *J. Exp. Med.* **145**:204–220.

Pasternack, S. G., Veis, A., and Breen, M., 1974, Solvent-dependent changes in proteoglycan subunit conformation in aqueous guanidine hydrochloride solutions, *J. Biol. Chem.* **249**:2206–2211.

Pearlstein, E., 1976, Plasma membrane glycoprotein which mediates adhesion of fibroblasts to collagen, *Nature (London)* **262**:497–500.

Pennypacker, J. P., Hassell, J. R., Yamada, K. M., and Pratt, R. M., 1979, The influence of an adhesive cell surface protein on chondrogenic expression *in vitro*, *Exp. Cell Res.* **121**:411–415.

Peterkofsky, B., and Diegelmann, R. F., 1971, Use of a mixture of proteinase-free collagenases for the specific assay of radioactive collagen in the presence of other proteins, *Biochemistry* **10**:988–994.

Podleski, T. R., Greenberg, I., Schlesinger, J., and Yamada, K. M., 1979, Fibronectin delays the fusion of L$_6$ myoblasts, *Exp. Cell Res.* **122**:317–326.

Poste, G., and Fidler, I. J., 1980, The pathogenesis of cancer metastasis, *Nature (London)* **283**:139–146.

Prockop, D. J., Kivirikko, K. I., Tuderman, L., and Guzman, N. A., 1979, The biosynthesis of collagen and its disorders, *N. Engl. J. Med.* **301**:13–23, 77–85.

Rao, C. N., Margulies, I. M. K., Tralka, T. S., Terranova, V. P., Madri, J. A., and Liotta, L. A., 1982, Isolation of a subunit of laminin and its role in molecular structure and tumor cell attachment, *J. Biol. Chem.* **257**:9740–9744.

Reddi, A. H., and Anderson, W. A., 1976, Collagenous bone matrix-induced endochondral ossification and hemopoiesis, *J. Cell Biol.* **69**:557–572.

Reid, K. B. M., and Porter, R. R., 1976, Subunit composition and structure of subcomponent Clq of the first component of human complement, *Biochem. J.* **155**:19–23.

Rhodes, R. K., and Miller, E. J., 1978, Physiochemical characterization and molecular organization of the collagen A and B chains, *Biochemistry* **17**:3442–3448.

Rohde, H., Wick, G., and Timpl, R., 1979, Immunochemical characterization of the basement membrane glycoprotein, laminin, *Eur. J. Biochem.* **102**:195–201.

Rojkind, M., Gatmaitan, Z., Mackensen, S., Giambrone, M.-A., Ponce, P., and Reid, L., 1980, Connective tissue biomatrix: Its isolation and utilization for long-term cultures of normal rat hepatocytes, *J. Cell Biol.* **87**:255–263.

Rosenberry, T., and Richardson, J., 1977, Structure of 18S and 14S acetylcholinesterase: Identification of collagen-like subunits that are linked by disulfide bonds to catalytic subunits, *Biochemistry* **16**:3550–3558.

Royal, P. D., and Goetinck, P. F., 1977, *In vitro* chondrogenesis in mouse limb mesenchymal cells: Changes in ultrastructure and proteoglycan synthesis, *J. Embryol. Exp. Morphol.* **39**:79–95.

Rubin, K., Oldberg, A., Höök, M., and Öbrink, B., 1978, Adhesion of rat hepatocytes to collagen, *Exp. Cell Res.* **117**:165–177.

Rubin, K., Johansson, S., Pettersson, I., Ocklind, C., Öbrink, B., and Höök, M., 1979, Attachment of rat hepatocytes to collagen and fibronectin: A study using antibodies directed against cell surface components, *Biochem. Biophys. Res. Commun.* **91**:86–94.

Rubin, K., Höök, M., Öbrink, B., and Timpl, R., 1981, Substrate adhesion of rat hepatocytes: Mechanism of attachment to collagen substrates, *Cell* **24**:463–470.

Ruoslahti, E., Vaheri, A., Kuusela, P., and Linder, E., 1973, Fibroblast surface antigen: A new serum protein, *Biochim. Biophys. Acta* **322**:352–358.

Ruoslahti, E., Engvall, E., and Hayman, E. G., 1981, Fibronectin: Current concepts of its structure and functions, *Coll. Res.* **1**:95–128.

Sage, H., and Bornstein, P., 1979, Characterization of a novel collagen chain in human placenta and its relation to AB collagen, *Biochemistry* **18**:3815–3822.

Sakashita, S., Engvall, E., and Ruoslahti, E., 1980, Basement membrane glycoprotein, laminin, binds to heparin, *FEBS Lett.* **116**:243–246.

Seglen, P. O., and Fosså, J., 1978, Attachment of rat hepatocytes in vitro to substrata of serum protein, collagen, or concanavalin A, *Exp. Cell Res.* **116**:199–206.

Seppä, H. E. J., Yamada, K. M., Seppä, S. T., Silver, M. H., Kleinman, H. K., and Schiffmann, E., 1981, The cell binding fragment of fibronectin is chemotactic for fibroblasts, *Cell Biol. Int. Rep.* **5**:813–820.

Silver, M. H., Foidart, J.-M., and Pratt, R. M., 1981, Distribution of fibronectin and collagen during mouse limb and palate development, *Differentiation* **18:**141–149.

Spooner, B. S., and Faubion, J. M., 1980, Collagen involvement in branching morphogenesis of embryonic lung and salivary gland, *Dev. Biol.* **77:**84–102.

Stenman, S., and Vaheri, A., 1978, Distribution of a major connective tissue protein, fibronectin, in normal human tissues, *J. Exp. Med.* **147:**1054–1064.

Szarfman, A., Terranova, V. P., Rennard, S. I., Foidart, J.-M., DeFatima Lima, M., Scheinman, J. I., and Martin, G. R., 1982, Antibodies to laminin in Chagas' disease, *J. Exp. Med.* **155:**1161–1171.

Tengblad, A., 1981, A comparative study of the binding of cartilage link protein and hyaluronate binding region of the cartilage proteoglycan to hyaluronate-substituted Sepharose gel, *Biochem. J.* **199:**297–305.

Terranova, V. P., Rohrbach, D. H., and Martin, G. R., 1980, Role of laminin in the attachment of PAM 212 (epithelial) cells to basement membrane collagen, *Cell* **22:**719–726.

Terranova, V. P., Liotta, L. A., Russo, R. G., and Martin, G. R., 1982, Role of laminin in the attachment and metastasis of murine tumor cells, *Cancer Res.* **42:**2265–2269.

Terranova, V. P., Rao, C. N., Kalebic, T., Margulies, I. M., and Liotta, L. A., 1983a, Laminin receptor on human breast carcinoma cells, *Proc. Natl. Acad. Sci. USA* **80:**444–448.

Terranova, V. P., Liotta, L. A., Vasanthakumar, G., Thorgeirsson, U., Siegal, G. P., and Schiffmann, E., 1983b, The role of laminin in the adherence and chemotaxis of neutrophils, *Fed. Proc. FASEB* **42:**2851–2852.

Terranova, V. P., Kleinman, H. K., Sultan, L. H., and Martin, G. R., Regulation of cell attachment and cell number by fibronectin and laminin (submitted).

Timpl, R., Martin, G. R., Bruckner, P., Wick, G., and Wiedemann, H., 1978, Nature of the collagenous protein in a tumor basement membrane, *Eur. J. Biochem.* **84:**43–52.

Timpl, R., Rohde, H., Gehron Robey, P., Rennard, S. I., Foidart, J.-M., and Martin, G. R., 1979, Laminin—A glycoprotein from basement membranes, *J. Biol. Chem.* **254:**9933–9937.

Timpl, R., Wiedermann, H., Van Delden, V., Furthmayr, H., and Kühn, K., 1981, A network model for the organization of type IV collagen molecules in basement membranes, *Eur. J. Biochem.* **120:**203–211.

Tomasek, J. J., Mazurkiewicz, J. E., and Newman, S. A., 1982, Nonuniform distribution of fibronectin during avian limb development, *Dev. Biol.* **90:**118–126.

Trelstad, R. L., 1974, Human aorta collagens: Evidence for three distinct species, *Biochem. Biophys. Res. Commun.* **57:**717–725.

Trelstad, R. L., and Coulombre, A. J., 1971, Morphogenesis of the collagenous stroma in the chick cornea, *J. Cell Biol.* **50:**840–858.

Trelstad, R. L., Hayashi, K., and Toole, B. P., 1974, Epithelial collagens and glycosaminoglycans in the embryonic cornea: Macromolecular order and morphogenesis in the basement membrane, *J. Cell Biol.* **62:**815–830.

Tryggvason, K., Gehron Robey, P., and Martin, G. R., 1980, Biosynthesis of type IV procollagens, *Biochemistry* **19:**1284–1289.

Uitto, J., and Prockop, D. J., 1974, Incorporation of proline analogues into collagen polypeptides: Effects on the production of extracellular procollagen and on the stability of the triple-helical structure of the molecule, *Biochim. Biophys. Acta* **336:**234–251.

Vaheri, A., Ruoslahti, B., Westermark, B., and Pontén, J., 1976, A common cell-type specific surface antigen in cultured human glial cells and fibroblasts: Loss in malignant cells, *J. Exp. Med.* **143:**64–72.

Vaheri, A., Kurkinen, M., Lehto, V.-P., Linder, E., and Timpl, R., 1978, Codistribution of pericellular matrix proteins in cultured fibroblasts and loss with transformation: Fibronectin and procollagen, *Proc. Natl. Acad. Sci. USA* **75**:4944–4948.

Varner, H. H., Hewitt, A. T., Furthmayr, H., Fietzek, P. P., Nilsson, B. B., Osborne, J. C., Jr., DeLuca, S., and Martin, G. R., Further characterization of chondronectin, the chondrocyte attachment factor, (submitted).

Vembu, D., Liotta, L. A., Paranjpe, N., and Boone, C. W., 1979, Correlation of tumorigenicity with resistance to growth inhibition by *cis*-hydroxyproline, *Exp. Cell Res.* **124**:247–252.

Villiger, B., Kelley, D. G., Engleman, W., Kuhn, C., and McDonald, J. A., 1981, Human alveolar macrophage fibronectin: Synthesis, secretion, and ultrastructural localization during gelatin-coated latex particle binding, *J. Cell Biol.* **90**:711–720.

Voss, B., Allah, S., Rauterberg, J., Ullrich, K., Gieselmann, V., and Von Figura, K., 1979, Primary cultures of rat hepatocytes synthesize fibronectin, *Biochem. Biophys. Res. Commun.* **90**:1348–1354.

Vuento, M., Vartio, T., Saraste, M., von Bonsdorff, C. H., and Vaheri, A., 1980, Spontaneous and polyamine-induced formation of filamentous polymers from soluble fibronectin, *Eur. J. Biochem.* **105**:33–42.

Wartiovaara, J., Leivo, I., and Vaheri, A., 1979, Expression of the cell surface-associated glycoprotein, fibronectin, in the early mouse embryo, *Dev. Biol.* **69**:247–257.

Weiss, R. E., and Reddi, A. H., 1980, Synthesis and localization of fibronectin during collagenous matrix–mesenchymal cell interaction and differentiation of cartilage and bone in vivo, *Proc. Natl. Acad. Sci. USA* **77**:2074–2078.

West, C. M., Lanza, R., Rosenbloom, J., Lowe, M., Holtzer, H., and Avdalovic, N., 1979, Fibronectin alters the phenotypic properties of cultured chick embryo chondroblasts, *Cell* **17**:491–501.

Wicha, M., Liotta, L. A., Garbisa, S., and Kidwell, W. R., 1979, Basement membrane collagen requirements for attachment and growth of mammary epithelium, *Exp. Cell Res.* **124**:181–190.

Woodley, D. T., Rao, C. N., Hassell, J. R., Liotta, L. A., Martin, G. R., and Kleinman, H. K., 1983, Interactions of basement membrane components, *Biochem. Biophys. Acta.* **761**:278–283.

Woolley, D. E., Glanville, R. W., Roberts, D. R., and Evanson, J. M., 1978, Purification, characterization and inhibition of human skin collagenase, *Biochem. J.* **169**:265–276.

Yamada, K. M., and Olden, K., 1978, Fibronectins—Adhesive glycoproteins of cell surface and blood, *Nature (London)* **275**:179–184.

Yamada, K. M., and Weston, J. A., 1974, Isolation of a major cell surface glycoprotein from fibroblasts, *Proc. Natl. Acad. Sci. USA* **71**:3492–3496.

Yamada, K. M., Yamada, S. S., and Pastan, I., 1976, Cell surface protein partially restores morphology, adhesiveness, and contact inhibition of movement to transformed fibroblasts, *Proc. Natl. Acad. Sci. USA* **73**:1217–1221.

3

Role of Glycoproteins during Early Mammalian Embryogenesis

Raymond J. Ivatt

1. INTRODUCTION

The focus of this chapter is to identify the cellular interactions that occur during early mammalian embryogenesis and to summarize the current understanding of the molecular events at the cell surface that are involved in regulating these interactions. This description is restricted to short-range cellular interactions, for example, between neighboring cells, and between cells and their local environment, such as the early basement membranes, and does not address longer-range interactions such as the role of growth factors during early development.

During early mammalian embryogenesis, particular cell surface carbohydrates have been implicated in cellular recognition. I have explored the nature of these components and how their expression is regulated, paying close attention to the roles of cell surface determinants in regulating recognition during fertilization, during compaction of the embryo and during its subsequent reorganization. There has been considerable progress in identifying specific glycoproteins that may be crucial recognition determinants in these situations. Here, monoclonal antibody technology and classical biochemistry have provided a powerful combination of approaches.

Raymond J. Ivatt • Department of Tumor Biology, University of Texas, M.D. Anderson Hospital and Tumor Institute at Houston, Houston, Texas 77030.

The cellular interactions that regulate the manner in which the early basement membranes are laid down and the cellular consequences of these extracellular materials are explored. Throughout this chapter there has been an attempt to correlate the biochemical descriptions obtained with the embryonal carcinoma system with the phenomenology observed with the normal embryo. Sometimes this correlation has been striking, as in the case of the glycoprotein(s) putatively involved in compaction and in embryonal carcinoma aggregation. The use of tissue culture model systems, e.g., the yolk sac carcinoma, has provided information that could only have come from model systems, either because of the logistics, or because of the ability to manipulate the tissue culture system, and has provided insights that are now being explored in the normal embryo.

2. ROLE OF GLYCOPROTEINS DURING FERTILIZATION

The initial event in mammalian fertilization is the interaction of sperm with the egg's zona pellucida, a glycoprotein envelope surrounding the egg's plasma membrane (Hartmann *et al.*, 1972; Gwatkin, 1976, 1977). Subsequently, the plasma membrane and outer acrosomal membrane of the sperm head fuse to expose acrosin, a protease associated with the inner acrosomal membrane (McRorie and Williams, 1974; Brown *et al.*, 1975; Hartree, 1977; Brown and Hartree, 1978). This is known as the acrosomal reaction, and the release of acrosin enables sperm to penetrate the zona pellucida (McRorie and Williams, 1974). After penetrating the zona pellucida, fusion of a sperm with the egg's plasma membrane triggers a series of events that prevent polyspermy. Sperm–egg fusion probably results in depolarization of the egg's plasma membrane and is associated with the fast block to polyspermy, which occurs within a minute (Sato, 1979). Depolarization is correlated with a release of calcium from the egg's cortex (Vacquier, 1975; Steinhardt *et al.*, 1977) followed by fusion of cortical granules with the plasma membrane and discharge of the cortical granules into the perivitelline space (Szollosi, 1976; Gwatkin *et al.*, 1973; Gulyas, 1980). These contents have been implicated in the slow block to polyspermy that occurs in mammals, namely the zona reaction (Braden *et al.*, 1954; Gwatkin, 1976, 1977; Scheul, 1978; Gulyas, 1980). This reaction occurs in the mouse about 10 min after the fusion of the sperm and egg (Sato, 1979), and results in the alteration of the zona so that additional sperm may attach loosely but do not bind (Hartmann *et al.*, 1972), and sperm already penetrating the zona are prevented from further penetration (Sato, 1979). This section will review the changes in cervical mucin that occur during estrus and affect sperm accessibility for fertilization, the properties of the sperm cell surface, and the roles of the zona pellucida as the sperm receptor and in the slow block to polyspermy.

2.1. Cervical Mucin Changes during Estrus

Cervical mucus, a secretory product of the epithelial lining of the endocervical canal, is a complex fluid with distinctive rheological properties. This fluid performs a critical role in regulating access of spermatozoa to the upper reproductive tract where fertilization occurs. Abnormalities in this secretion can be a significant factor in infertility in humans (Blasco, 1977), and alterations in this secretion mediate the action of certain contraceptive agents (Moghissi and Marks, 1971). The rheological properties of this fluid, and thereby sperm receptivity, vary during the menstrual cycle and are controlled by hormones. For example, while estrogens induce secretion of copious amounts of a clear acellular watery mucus of decreased viscoelasticity (Wolf *et al.*, 1978) through which sperm readily migrate, progesterones induce secretion of a scanty opaque viscous mucus that acts as an effective barrier to sperm migration through the cervix. Hormonal control therefore limits sperm access to the uterus to a period just prior to ovulation.

The mucus can be partitioned into an aqueous phase and a gel phase by centrifugation. The principal component of the gel phase is a glycoprotein termed "cervical mucin." This protein has physicochemical properties similar to typical epithelial mucin (Spiro, 1970).

Mucins exist as extended, tangled (Lee *et al.*, 1977), or cross-linked (Rose *et al.*, 1979) macromolecules that form networks. They are most probably the major determinants responsible for the rheological properties of the mucus (Wolf *et al.*, 1977). For example, the mucin secreted by the ovine submaxillary gland associates and has an apparent M_r greater than 10^6 in its native form (Hill *et al.*, 1977a). This mucin contains a large number of disaccharide chains, structure NeuNAcα2-6GalNAcα1-Ser/Thr, and is 63% carbohydrate by weight (Hill *et al.*, 1977b). Removal of just the sialic acid from this protein results in substantial dissociation of the mucin chains, while removal of both the sialic acid and the GalNAc residues results in complete dissociation to the monomer unit. The unglycosylated protein has a random coil structure rather than the extended structure of the glycosylated molecule (Hill *et al.*, 1977a). This transition sets a clear precedent for the carbohydrate moiety determining the physical and therefore the physiological properties of the mucin.

Detailed structural data are becoming available for bovine (Bhushana Rao and Masson, 1977; Gelman and Vered, 1976) and monkey (Hatcher and Jeanloz, 1974; Hatcher *et al.*, 1977) cervical mucins. We have very little structural or chemical data for rodent mucins at present. The data for human mucin are limited to amino acid and carbohydrate compositions (Iacobelli *et al.*, 1977; Wolf *et al.*, 1980; Van Kooij *et al.*, 1980), and a recent structural characterization of the glycan components of human midcycle cervical mucin (Yorewicz and Moghissi, 1981).

2.2. Sperm Cell Surface Determinants

Spermatozoa progressively gain the ability to fertilize ova during their passage through the epididymis (review: Bedford and Cooper, 1978). This phenomenon is generally referred to as "epididymal maturation." The epididymis is divided anatomically into three regions—*caput* (head), *corpus* (body), and *cauda* (tail)—in order of sperm passage. Spermatozoa acquire the ability to fertilize as they descend the middle corpus epididymis (Bedford, 1966; Orgebin-Christ, 1969). This functional maturation is accompanied by a variety of morphological and biochemical modifications of the spermatozoa, including the size, shape, and internal structure of the acrosome (Bedford, 1963, 1965; Fawcett and Phillips, 1969), and the cohesiveness between the outer acrosomal membrane and the overlying sperm plasma membrane (Bedford, 1965; Fawcett and Phillips, 1969). These changes are accompanied by biochemical changes in the sperm plasma membrane during epididymal maturation, involving changes in surface antigens, lipids, intramembrane particles, fluidity, and ion permeability (Yanagimachi, 1981; Shapiro and Eddy, 1980). There is a change in the electrophoretic mobility of the sperm (Bedford, 1963), which is correlated with changes in the distribution of certain anionic surface charges, as detected by the binding of positively charged colloidal iron hydroxide particles (Bedford *et al.*, 1972; Yanagimachi *et al.*, 1972; Flechon, 1973).

During their passage through the epididymis, sperm shed some glycoconjugates known as "coating" or "decapacitation" factors (Oliphant and Brachet, 1973; Gwatkin and Anderson, 1973), and adsorb other materials that are secreted by the vas deferens and by the epididymis (Edwards *et al.*, 1964; Weil, 1965). Some of these adsorbed components are tightly bound and not readily released by saline washes. Recently, sperm have been shown to become positive for an antigen designated "stage-specific embryonic antigen 1" (SSEA-1) (Fox *et al.*, 1982). This antigen was defined by monoclonal antibodies raised against teratocarcinoma stem cells (Solter and Knowles, 1978, 1979) and is discussed in detail in subsequent sections. It is not yet clear whether this antigen, which is carbohydrate in nature, is adsorbed, or if passage through the epididymis results in modification or exposure or preexisting antigen. Interestingly, a carbohydrate that may have similar properties to the carrier glycan for the SSEA-1 antigen has been identified in *caudal* epididymal fluid (Shur and Hall, 1982a). This glycan can be galactosylated by a sperm cell surface galactosyltransferase activity. During capacitation, it has been proposed that these (epididymal fluid) glycans, which bind the galactosyltransferase and thereby act as competitive inhibitors for the enzyme, are shed by sperm with a resultant activation of the enzyme (Shur and Hall, 1982a). Such glycoconjugates have the properties of the decapacitation factors described earlier (Oliphant and Brachet, 1973; Gwatkin and Anderson, 1973). Receptors on the surface of sperm for wheat germ agglutinin (WGA) are extensively modified after contact of the sperm with seminal plasma (Nicolson and Yanagimachi, 1972). Receptors for other lectins such as Con A and *Ricinus communis* agglutinin (RCA) change during epididymal maturation and upon ejaculation. A comparison (Nicolson *et al.*, 1977) of *caput* and *caudal* epididymal sperm has shown that the distributions of these lectin receptors change independently. For example, with ferritin–WGA, the acrosomal region of *caput* epididymal sperm is densely labeled, while *caudal* epididymal sperm are very sparsely labeled except for the apical acrosomal region (ejaculated sperm are not labeled); with ferritin–RCA, binding to the head is reduced during epididymal maturation, while binding to the tail is not affected; and with ferritin–Con A, the receptor distribution is not substantially altered during epididymal maturation.

The major components of the cell surface of rabbit sperm have been identified by cell surface labelling (Nicolson *et al.*, 1979). There appear to be eight major bands (regions) when the labeled components are examined by polyacrylamide gel electrophoresis in SDS. A band with an apparent M_r of 86K is very prominent on *caudal* epididymal sperm and decreases during epididymal maturation, while the exposure of four bands with M_r of 35, 39, 50, and 78K markedly increases. The functional consequences of these cell surface changes are being investigated.

2.3. The Zona Pellucida

The zona pellucida is a glycoprotein envelope surrounding the egg's plasma membrane. The zona reaction, which occurs shortly after penetration of the first sperm, results in alteration of the zona pellucida such that additional sperm cannot bind (Hartmann *et al.*, 1972), and sperm that have already partially penetrated are prevented from penetrating further (Sato, 1979). There is evidence that the zona has roles as sperm receptor and in blocking polyspermy.

There are three major glycoprotein species in the zona pellucida of mouse oocytes and eggs, which have been designed ZP1, ZP2, and ZP3. These glycoproteins have apparent M_r of 200, 120, and 83K, respectively

(Bleil and Wassarman, 1980a,b). All three of these components are synthesized and secreted by mouse oocytes during their growth phase (Bleil and Wassarman, 1980b). The best evidence that the zona pellucida contains a sperm receptor has been presented by Bleil and Wassarman (1980c). These authors purified the three glycoprotein components of the zona pellucida and used a competition assay to evaluate their effects on the binding of sperm to eggs. Only ZP3, the smallest component, was found to interfere with the binding of sperm to eggs *in vitro*, and only if the ZP3 was prepared from unfertilized eggs. If the ZP3 was prepared from two-cell embryos, then it had no effect on binding of sperm to eggs *in vitro*. These results implicate ZP3 as a receptor for sperm and suggest that during the acrosome reaction it is modified so that it no longer acts as a sperm receptor.

The zona pellucida receptor on the sperm has not been identified. However, a very interesting series of observations suggests that a galactosyltransferase on the surface of sperm is responsible for binding to the zona pellucida (Shur and Hall, 1982b). Inhibitors of the galactosyltransferase inhibit sperm–zona pellucida interaction. Zona pellucida-derived glycopeptides inhibit sperm–zona pellucida interaction and can be galactosylated in the presence of *exogenous* UDP-Gal. Pretreatment of eggs with an enzyme that removes GlcNAc, the acceptor for the galactosyltransferase, reduces sperm–egg binding by 86%, while pretreatment of the eggs with an enzyme that removes Gal (thereby generating an increased number of potential acceptors for the galactosyltransferase) increases sperm–egg binding by 55%. Whether or not ZP3 is the component that acts as receptor for the sperm galactosyltransferase has not been established. It would be interesting to test this directly and to see if the difference between the ZP3's isolated from unfertilized and fertilized eggs is in their glycosylation, either gain of galactose, or loss of GlcNAc (the galactose acceptor). The precedent for sperm binding to glycans with terminal GlcNAc was established by Gwatkin and Anderson (1973) who demonstrated the release of epididymal coating glycoconjugates after treatment with β-*N*-acetylglucosaminidase. These various studies may be describing the two sides of a complementary interaction.

Following fertilization, ZP2 also undergoes modification (Bleil *et al.*, 1981). This modification has been suggested to be responsible for the inability of a sperm that has already partially penetrated the zona pellucida prior to fertilization from penetrating further. The modification of ZP2 appears to be a proteolytic cleavage as the protein retains its apparent molecular weight under nonreducing conditions but migrates on SDS–polyacrylamide gels under reducing conditions with an apparent M_r of 90K. This modification is observed in embryos but not in ZP2 obtained from unfertilized eggs or oocytes. It can be caused *in vitro* by the addition of the calcium ionophore A23187, which artificially activates the eggs and is therefore not necessarily caused by a sperm component. This is important as ZP2 is found to be much less susceptible to proteolytic digestion following fertilization (Gwatkin, 1976, 1977). A reduced susceptibility to proteases such as acrosin would thereby reduce the ability of sperm to penetrate. Alternatively, there may be a major structural reorganization of the zona pellucida and this may block further penetration. Alterations in the ultrastructure of the zona pellucida have been noted by some (Jacowski and Dumont, 1979; Baranska *et al.*, 1970) but not by other investigators (Meyerhofer *et al.*, 1977). It is an aesthetically pleasing mechanism for a proteolytic enzyme of the egg to cut ZP2 and in the resultant conformational change make it less susceptible to further cleavage by proteolytic enzymes of the sperm.

3. ROLE OF GLYCOPROTEINS DURING PREIMPLANTATION DEVELOPMENT

The preimplantation development of the mouse has been described in a number of excellent reviews (Rugh, 1968; Herbert and Graham, 1974; Garnder and Papaioannou, 1975; Sherman and Wudl, 1976; Graham, 1977; Rossant and Papaioannou, 1977; Sherman, 1979). The purpose of this section is to identify the cellular interactions implicated in directing early differentiation. There are a number of cell surface components that have been implicated as playing major roles in intercellular interactions in the early embryo and this brief cellular description should provide a basis for examining their functions.

After fertilization, the mouse egg undergoes a number of cleavages, the first taking 24 hr and subsequent divisions taking 12 hr. Divisions do not occur synchronously and therefore mouse embryos with any integral number of blastomeres occur. The events that occur during preimplantation development are illustrated schematically in Fig. 1. At the 8- to 16-cell stage, the cluster of cells undergo an event known as compaction in which the cells become very closely associated and lose their distinct cell boundaries. The appearance of the embryo before and after compaction is shown in the scanning electron micrographs of Fig. 2. The cellular interactions occurring during this event are critical and will be discussed in detail, for in the rearrangement that occurs subsequent to compaction two distinct cell types with fundamentally different fates segregate. These cell types form the trophectoderm and the inner cell mass. The trophectoderm is an outer shell of cells that resembles an organized epithelium. The cells within this epithelium are extraembryonic in their fate, initially giving rise to the trophoblast, an invasive organ responsible for the im-

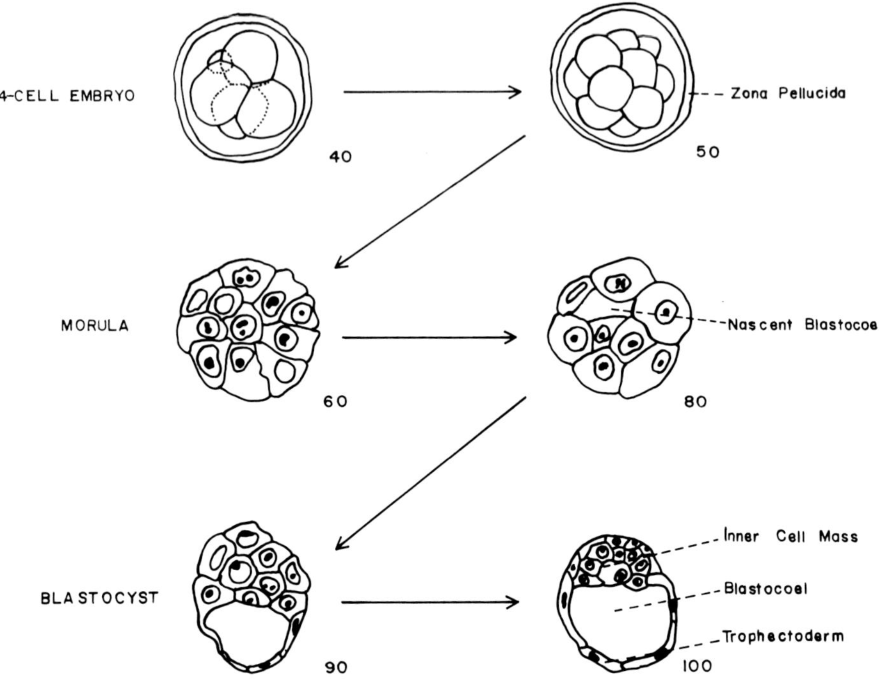

Figure 1. Stages during the preimplantation development of the mouse embryo. The time in hours is presented at the bottom right-hand corner of each schematic. After fertilization, the first cell division takes 24 hr and subsequent cell divisions occur every 12 hr. At the 8- to 16-cell stage, compaction occurs and the resultant cell aggregate is known as the morula. The blastocoel cavity begins to form and two different cell types can be distinguished: the outer epithelial cells of the trophectoderm, and the inner pluripotent cells of the inner cell mass. Implantation occurs during the fifth day. Prior to implantation, the blastocyst hatches from a glycoprotein-rich shell called the zona pellucida. (Taken from Rugh, 1968.)

plantation of the early embryo in the uterine wall, and subsequently contributing to nutritive and protective structures. In contrast, the inner cell mass, which is a cluster of cells within the trophectodermal cavity, will give rise to the embryo proper. The pluripotent stem cells from an unusual tumor, the teratocarcinoma, have many morphological (Pierce *et al.*, 1964; Bernstine *et al.*, 1973), biochemical (Muramatsu *et al.*, 1978), serological (Artzt *et al.*, 1973; Jacob, 1977), and biological (Kahan and Ephrussi, 1970; Rosenthal *et al.*, 1970) properties in common with these inner cell mass cells. The biological properties of these cells within the environment of the inner cell mass have been explored and reveal an ability to contribute to normal fetal tissue derived from all three of the primary germ layers (Kleinsmith and Pierce, 1964; Brinster, 1974; Mintz

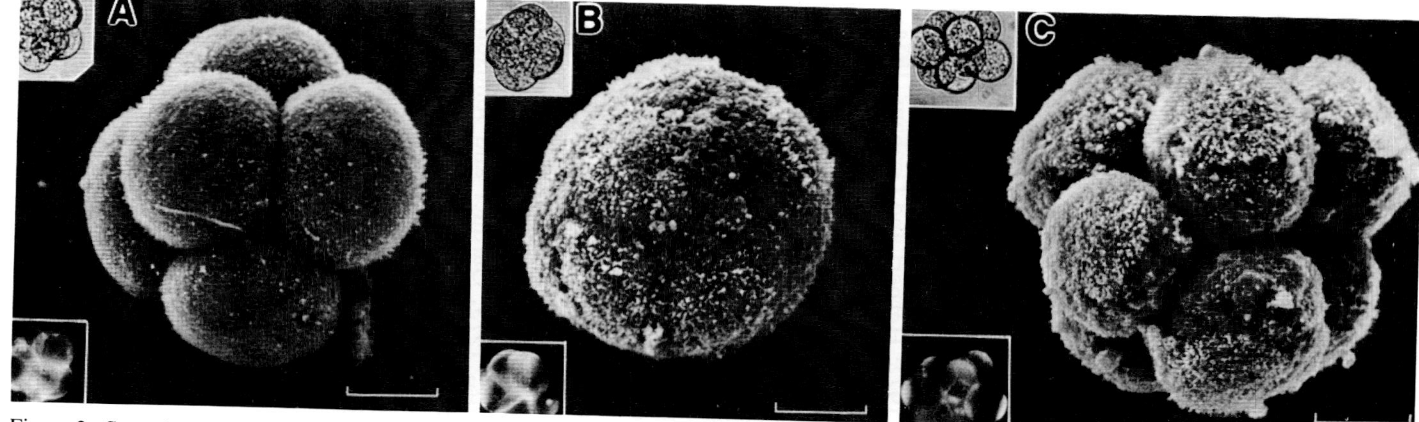

Figure 2. Scanning electron micrographs of the preimplantation embryo. (A, B) The 8- to 16-cell embryo before and after compaction. (C) A morula that has been decompacted by removal of calcium. Note the difference in appearance of the aggregates in (A) and (C)—the homogeneous distribution of microvilli in (A), in contrast to the very localized distribution of microvilli in (C). This localization of ultrastructural detail correlates with the localization of lectin receptors. (Taken from Reeve and Ziomek, 1981.)

cision suggests that maintenance of these different fates requires sustained cell–cell contacts.

3.2. Teratocarcinomas as a Model System for Early Embryogenesis

Teratomas are unusual tumors in that they can contain a wide variety of cell types, usually chaotically arranged but sometimes highly organized. They were first observed in mice by Stevens and Little (1954) and have been studied in detail by Stevens (1967a) and Pierce (1967). Testicular teratomas arise spontaneously in strain 129 mice and appear to be derived from primitive germ cells (Stevens, 1967b), while spontaneous parthenogenesis appears to cause the appearance of ovarian teratomas in strain LT mice (Stevens and Varnum, 1974). Teratomas can be induced experimentally by grafting either entire early embryos or the genital ridge into the testes of syngeneic mice (Stevens, 1970). When transplantable, these tumors are called "teratocarcinomas" and contain stem cells with remarkable properties. These stem cells are called "embryonal carcinoma cells." They can be pluripotent, with single cells injected into syngeneic mice giving rise to tumors containing differentiated derivatives of all three of the primary germ layers (Illmensee and Mintz, 1976). This extensive differentiative capacity has been demonstrated for single cells in culture (Nicolas et al., 1975; Martin, 1975, 1978, 1980). The pluripotency of the teratocarcinoma stem cells suggests an analogy to very early embryonic cells and currently these stem cells are thought to be analogs of the cells of the inner cell mass. This resemblance is emphasized by the following remarkable observation. When embryonal carcinomas are placed in the blastocyst cavity of an early embryo, instead of forming a tumor these cells will usually contribute to the formation of an otherwise normal embryo and even to normal but chimeric mice. Similarly, when early embryos are transplanted to extrauterine sites such as the kidney capsule, they give rise to teratocarcinomas. The availability of a cell line in tissue culture that mimics the early events of embryogenesis is, of course, of great interest to developmental biologists. The strong influence of the cellular environment upon the malignancy of these cells is of great interest to tumor biologists.

The developmental potential of teratocarcinoma stem cells and their embryonic equivalence to the cells of the inner cell mass have been reviewed (Martin, 1975 1980; Jacob, 1975; Graham, 1977; Hogan, 1977). For the purposes of this discussion, they will be briefly summarized. Embryonal carcinomas will differentiate readily in culture, and indeed frequent subculturing on a fibroblast feeder layer is required to maintain an undifferentiated population (Martin, 1975). Some pluripotent embryonal carcinomas readily form aggregates, referred to as "embryoid bodies'' upon which develops a surface layer of endoderm. This differentiation recapitulates the first differentiation to occur on the inner cell mass. Subsequently, a cavity will form and be lined by a layer of columnar epithelial cells. After this stage, the organization becomes progressively more chaotic. However, there is a striking similarity to normal development in these early events (Rossant and Papaioannou, 1977). This developmental pattern, coupled with the observation that trophectodermal derivatives are never obtained, identifies the (murine) embryonal carcinoma cells as later than the morula stage and probably the inner cell mass stage. There is some debate as to whether they represent the initial cells that form the inner cell mass or neurectoderm. This is because the inner cell mass is capable of *regenerating* endoderm. Therefore, the cells that have given rise to differentiated endoderm remain fairly flexible at this early stage and do not lose the potential to differentiate in this direction. The protein synthetic patterns of embryonal carcinomas more closely resemble those of the inner cell mass cells from the expanded blastocysts than from day 3 blastocysts (Howe *et al.*, 1980).

Other teratocarcinoma cell lines (e.g., F9) appear to be nullipotent when cultured *in vitro* or *in vivo*, although at a very low frequency they can differentiate into endodermal cells (Sherman and Miller, 1978). However, F9 cells can be induced by retinoic acid to differentiate into endoderm (Strickland and Mahdavi, 1978). This chemical inducibility makes them a convenient model for laboratories that are not so dedicated to the rigorous tissue culture that the pluripotent cells demand. There are various embryonal carcinomas with different development potentials available. In the presence of DMSO, aggregates of P19 embryonal carcinoma cells (McBurney and Rogers, 1982) will form either skeletal or cardiac muscle depending on the culture conditions, but not neuronal or glial cells. However, in the presence of retinoic acid, aggregates of these cells will form neurons and glial cells but not muscle. Thus, drugs can be used to generate two quite different spectra of embryonic tissue types from the same population of embryonal carcinoma cells (McBurney *et al.*, 1982). Other cells will form either fibroblastic cells or neuronal cells depending on the culture conditions. For example, when cultured in the absence of serum, line 1003 embryonal carcinoma cells form neuronal cells, and on the addition of fetal calf serum to the culture medium, the cells form fibroblastic cell types (Darmon and Serrero, 1983). These and other examples make embryonal carcinoma differentiation an unusual opportunity for studying such varied aspects as cellular recognition and gene expression. The usefulness of such tissue culture models is clear and they are beginning to yield important results. Some of these are summarized below and demonstrate that studies originally undertaken to examine embryonal carcinoma cellular recognition may have identified components involved in maintaining the integrity of the inner cell mass and the compacted morula.

3.3. Programmed Changes in Glycoprotein Expression during Preimplantation Development

The programmed expression of proteins and glycoproteins has been examined both serologically, and by one- and two-dimensional SDS–polyacrylamide gel electrophoresis, with proteins labeled either metabolically or by surface radioiodination. The general picture to emerge (Epstein and Smith, 1973; Van Blerkom and Manes, 1974; Van Blerkom and Brockway, 1975; Dewey *et al.*, 1978; Levinson *et al.*, 1978; Howe and Solter, 1979; Johnson and Calarco, 1980a) is that there are only a few changes in the relative abundance of polypeptides following fertilization and that between the 2-cell and 8- to 16-cell stages a large change in the protein synthetic patterns occurs. These patterns then remain relatively stable for some time afterwards although there are proteins specific for either the inner cell mass or the trophectoderm.

More detailed examination of the proteins labeled by one-cell embryos during the period immediately following fertilization demonstrated the assembly of a group of proteins whose expression is dependent on fertilization and whose expression is transient (Cullen *et al.*, 1980). In this study, eggs were removed, incubated *in vitro*, and labeled at 2-hr intervals after fertilization. There are a number of changes in the expression of a group of proteins with apparent M_r of 35K. These changes are not observed with unfertilized eggs cultured and labeled over the same period. This group of proteins is not observed in eight-cell embryos. Cascio and Wassarman (1982) have extended these studies by showing that a similar group of proteins, which they designate FP1–FP6 for fertilization proteins, are regulated posttranscriptionally. They found that proteins FP1–FP4 were assembled by mouse oocytes though in very low amounts. In contrast, these proteins (with tubulin, actin, and lactate and dehydrogenase) were found to be the major products when oocyte mRNA was translated in an mRNA-dependent cell-free translation system and that there was a marked stimulation in translation following fertilization.

Levinson *et al.* (1978) in a comprehensive examination of the proteins assembled during preimplantation development resolved 400–600 protein spots by two-dimensional polyacrylamide gel electrophoresis and identified 36 of these spots as being specific for a particular stage. Howe *et al.* (1980) have examined the nuclear and cytoplasmic proteins assembled during preimplantation development and find that both classes of proteins show similar trends in their changes, namely, a major period of change between the two-cell and eight-cell stage and a period of relative stability from the eight-cell to early blastocyst stage. They also observed minor differences between trophectoderm and inner cell mass protein patterns. The patterns of protein expression of the cells of the inner cell mass and the cells presumed to be their progenitors, the inside cells of the compacted morula, have been found to be very similar (Handyside and Johnson, 1978). The protein synthetic patterns of the inner cell mass cells and embryonal carcinomas have been compared and the embryonal carcinoma cell pattern was found to be more advanced than the inner cell mass cells at 4 days (Howe *et al.*, 1980). Dewey *et al.* (1973, 1977) compared the patterns of embryonal carcinomas and embryonic cells. They found in addition that the environment of the embryonal carcinoma affected its pattern of protein expression. Specifically, they compared embryoid body (aggregated stem cells) that had either a surface layer of endoderm or no endoderm, and found the pattern of expression of the embryonal carcinomas in the cores to be different. This was in spite of the finding that the cells from these cores formed very similar tumors.

Various antisera have identified different patterns of expression. For example, the *H-2* locus appears not to be expressed on unfertilized eggs nor on the preimplantation embryo. The inner cell mass has been suggested to assemble H-2-reactive material as determined by immunoprecipitation of metabolically labeled material (Webb *et al.*, 1977). Other minor histocompatibility antigens such as H3, H6, and HY are present at all stages of preimplantation development and on all cell types (Heyner *et al.*, 1969; Palm *et al.*, 1971; Searle *et al.*, 1974; Muggleton-Harris and Johnson, 1976; Krco and Goldberg 1976). Two monoclonal antibody preparations that recognize defined carbohydrate determinants react in different patterns. Willison and Stern (1978) prepared monoclonal antibodies against the Forssman antigen, a lipid-linked carbohydrate determinant. These antibodies do not react with the preimplantation embryo until the late morula stage when they react weakly with the outer cells. Later they react with the inner cell mass ($3\frac{1}{2}$ days) and with endoderm ($4\frac{1}{2}$ days), but curiously, with reference to their reaction with the outer cells of the morula, not with trophectoderm. The other antibodies were part of a series of monoclonal antibodies prepared by Solter and Knowles (1978, 1979). The antigens recognized by several of these antibodies showed temporally regulated expression and were designated "stage-specific embryonic antigens." The properties of one of these antibodies, anti-SSEA-1, have been studied extensively, and a very effective competing hapten is the trisaccharide Galβ1-4-(Fucα1-3)-GlcNAcβ (Gooi *et al.*, 1981). The endogenous antigen has not been characterized although glycoconjugates immunoprecipitated by anti-SSEA-1 have been isolated (Andrews *et al.*, 1982). This antigen is not expressed by the unfertilized egg and is not expressed after fertilization until the eight-cell stage. It is expressed by embryos beyond this stage, although the trophectoderm is variably weak and then negative (Solter and Knowles, 1978). Polyclonal antisera raised against totipotent teratocarcinoma cells from the OTT6050 tumor do not react with the unfertilized egg but do react from the two-cell stage onwards (Dewey *et al.*, 1977). Antiserum raised against a nullipotent cell line, F9, behaves similarly, reacting maximally at the eight-cell stage and then more weakly. The inner cell mass and trophectoderm are both positive (Artzt, 1973; Artzt *et al.*, 1973; Artzt and Bennett, 1975; Jacob, 1977). There has been considerable study of the antigen(s) recognized by these antisera (reviewed by Gachelin *et al.*, 1982). One surprising finding is that these antisera recognize determinants common to the sperm of several mammalian species tested (Buc-Caron *et al.*, 1974; Jacob, 1977) while these antisera did not detect determinants on the sperm from several nonmammalian species. The antigen may also be on the same large glycoconjugate that can be immunoprecipitated by anti-SSEA-1 (Andrews *et al.*, 1982). These glycoconjugates are large and highly branched and therefore potentially carry many determinants. It is not yet established whether or not F9 and SSEA-1 antigens overlap in their structural characteristics. There is, however, no report of a cell surface that is SSEA-1 positive and F9 negative or vice versa. In contrast to the above antisera raised against the above teratocarcinoma cell lines, an antiserum raised against the teratocarcinoma cell line PCC4 does not react with the preimplantation embryo until day $4\frac{1}{2}$ (Gachelin *et al.*, 1977).

Various rat anti-mouse antisera react with distinct and different patterns. For example, antisera against the blastocyst do not react with unfertilized eggs, but show a weak reaction after fertilization. This reactivity increases reaching a maximum at the 8- to 12-cell stage then falling off so that the blastocyst is curiously unreactive (Wiley and Calarco, 1975). Antisera raised against the ectoplacental cone do not react with the unfertilized egg, and show a patchy reaction at the 8-cell stage. However, they react strongly at the 16- to 32-cell stage, and then lose their reactivity so that the blastocyst is weakly positive or negative (Searle and Jenkinson, 1978). Although antisera raised against the placenta do not react with the unfertilized egg, they react at all stages of preimplantation development including the inner cell mass (Wiley and Calarco, 1975). Curiously, antisera against T/t determinants, a locus defined by a series of developmental mutations (reviewed, Bennett, 1975, 1978) appear to react at the 8-cell stage regardless of the time of their lethality (Kemler *et al.*, 1976).

The biochemical characterization of the cell surface of embryonic cells is just beginning. Usually it involves taking an antiserum that shows stage specificity in its binding and immunoprecipitating cellular material that has been labeled either metabolically or by surface radioiodination. For example, antiserum against a pluripotent mouse teratocarcinoma cell line (Webb, 1980) immunoprecipitates five major bands with M_r of 150, 115, 82, 48, and 12K. The relative abundance of these proteins was found to vary among different cell lines examined. Rabbit anti-mouse blastocyst antiserum brings down two proteins with M_r of 65 and 70K (Johnson and Calarco, 1980b). These proteins are reported to occur only during the two-cell to morula stage. These proteins are metabolically labeled by radioactive methionine and by glucosamine and are presumed to be glycoproteins. In support of this presumption is the finding that the antiserum reacts with only a single protein species when immunoprecipitated from cells cultured in the presence of the glycosylation inhibitor tunicamycin (Johnson and Calarco, 1980c). The M_r of this new species is 60K. Another interesting finding in this study was the observation that binding of the antiserum to the cells is much reduced by treatment of the cell surface with a glycohydrolase that removes terminal *N*-acetylglucosamine residues. As the antiserum can bind to the lower-molecular-weight component assembled in the presence of tunicamycin—the unglycosylated component—it may be that glycohydrolase digestion causes the component to be removed from the cell surface rather than affecting binding. As will be discussed in detail below, Shur (1982a) has identified a cell surface component that is retained on the surface apparently by its interaction with a cell surface galactosyltransferase. Removal of terminal *N*-acetylglucosamine from this compound released it from the galactosyltransferase and caused it to be shed into the medium.

Rowinski *et al.* (1976) studied the agglutinability of embryos during preimplantation development and found the zygote to eight-cell stage readily agglutinated at 10 μg/ml Con A, the morula to be less readily agglutinated requiring 100 μg/ml, while the blastocyst was not agglutinable even at 5000 μg/ml. Interestingly, the inner cell mass prepared by immunosurgery was readily agglutinated at 10 μg/ml. Changes in the agglutinability of the oocyte have also been observed during meiosis (De Felici and Siracusa, 1980) and after fertilization (Magnuson and Stackpole, 1978). These studies with lectin agglutinability have recently been extended by Magnuson and Epstein (1981) who have examined the expression of Con A receptors during preimplantation development. They used Con A binding, followed by immunoprecipitation with anti-Con A antibodies to isolate endogenously labeled proteins. Upon examination by two-dimensional polyacrylamide gel electrophoresis, they observed distinct, stage-specific patterns of glycoprotein expression.

Immunosurgery of the blastocyst has been used to provide a very convenient method for preparing inner cell masses (Solter and Knowles, 1975). In principle, zona-free embryos are treated with rabbit anti-mouse antisera, washed, and then treated with guinea pig complement. The outer trophectodermal cells are quantitatively lysed and the inner cell mass cells which were not exposed to the original rabbit antisera are spared. After pipetting, the inner cell masses are freed of the trophectodermal debris.

One specific glycoprotein that has been much studied, initially because of its decreased expression on transformation and subsequently because of the very important roles it plays in cellular adhesion, is fibronectin. The biological properties and biochemical characteristics of this protein have been reviewed, and are discussed in detail in a later section. Zetter and Martin (1978) have examined the expression of fibronectin during preimplantation development and by several established cell lines that resemble early embryonic cells. Their conclusions are as follows. Fibronectin is made by embryonal carcinomas and by preimplantation embryos. It was not found on cleavage-stage or compacted embryos or on the trophectoderm of either early or late blastocysts. However, it was found on the inner cell mass prepared by immunosurgery and the level of this expression was much higher on expanded blastocysts at day 4 than at day 3.

3.4. Role of Glycoproteins during Early Embryonic Development

The clear demonstration that cell surface components change during a period of preimplantation development where there are dramatic changes in intercellular contacts now leads to studies that address the functional consequences of these changes. An important step in this process has been the demonstration that an inhibitor of protein glycosylation inhibits the process of compaction and therefore blastocyst formation (Surani, 1979; Azin and Surani, 1979; Atienza-Samols *et al.*, 1980). Cleavage and aggregation of embryos are not affected. Two components, one less than M_r 68K and one greater than 165K, were not expressed in the presence of tunicamycin as detected after radioiodination of the cell surface (Surani *et al.*, 1981).

This section will examine the reaggregation of teratocarcinoma cells and the cell surface components that have been identified in intercellular recognition between these cells and will then explore the possible role that these components may have in the process of compaction and in the segregation of the inner cell mass.

There are several components that have been identified in the intercellular recognition of embryonal carcinomas. They may each represent distinct systems, some of them may be discovered to be overlapping, or indeed it may turn out that they each represent a different aspect of the same system. Five different approaches will be described that identify roles in intercellular recognition for the following: a cell surface component of 140K and a proteolytic fragment of 85K (antibodies against either will prevent embryonal carcinoma cells from reaggregating and cause the decompaction of the normal early embryo); a cell surface carbohydrate-binding protein that recognizes fucose and mannose; an adhesion factor that requires terminal galactose residues for its function; and a cell surface glycosyltransferase that will add galactose to a class of glycans that are very prominent on the embryonal carcinoma cell surface.

3.4.1. A Cell Surface Glycoprotein of 140K

Takeichi *et al.* (1982) have identified a cell surface component of 140K essentially by comparing the protein profiles of cells trypsinized either in the presence or absence of calcium. The rationale for this is as follows. F9 teratocarcinoma cells trypsinized in the presence of 0.01% trypsin and 1 mM EGTA are unable to aggregate until certain cell surface components are restored. In contrast, when F9 cells are trypsinized in the presence of 1 mM calcium, they reaggregate in the presence of calcium without a lag period. Antibodies have been raised against F9 cells trypsinized in the presence of calcium. Fab fragments prepared from these antibodies inhibit the aggregation of cells trypsinized in the presence of calcium. Consistent with this finding is the observation that cells trypsinized in the presence of calcium are able to absorb out this reaggregation-inhibiting effect, whereas cells trypsinized in the absence of calcium are not. Subsequent experiments have shown that the antiserum prepared against cells trypsinized in the presence of calcium, when absorbed against cells trypsinized in the absence of calcium, immunoprecipitates relatively few components compared to the unabsorbed antiserum. In fact, there is a single obvious band with an apparent M_r of 140K. Monoclonal antibodies against this component prevent reaggregation in suspension, and when added to monolayer cultures cause the cells to round up and separate from each other. The antibody is not cytotoxic and does not cause the cells to detach from the substratum.

3.4.2. A Cell Surface-Derived Tryptic Glycopeptide of 84K

A cell surface-derived component that may or may not be related to the 140K protein described above is a tryptic fragment released from the surface of embryonal carcinoma cells. This fragment, with M_r of 84K, was identified as follows. Kemler *et al.* (1977) prepared antisera against F9 embryonal carcinoma cells and demonstrated that Fab fragments prepared from the anti-F9 serum could cause decompaction of the early (8- to 16-cell stage) embryo. These fragments also were found to prevent cell–cell interactions between PCC4 Aza R1 embryonal carcinoma cells. Specifically, the ability of these cells to cooperate metabolically in culture was measured by observing the rescue of HRPT-negative embryonal carcinoma cells by HRPT-positive cells (Nicolas *et al.*, 1981). This cooperation was abolished in the presence of anti-F9 Fab fragments. These Fab fragments are not normally cytotoxic, for early embryos cultured in their presence will continue to divide, and the effects of the Fab fragments are reversibly lost on their removal. The ability of this antiserum to affect compaction was lost after absorption by embryonal carcinoma cells but not by a variety of other cells, including fibroblastic cells, lymphocytes, and splenocytes. Antisera prepared against a number of mouse cell types, including liver, brain, and lymphocytes, were demonstrated to bind to the early embryo but had no effect on compaction. A component with M_r of 84K, prepared from the cell surface of embryonal carcinoma cells by trypsin digestion, would block the anticompaction effect of the Fab fragments (Hyafil *et al.*, 1980). Trypsinates prepared from cells that do not effect the immunoprecipitation of the 84K component also do not inhibit the effect of the anti-F9 Fab fragments on compaction. Monoclonal antibodies against the 84K component were unable to disrupt compaction. However, the binding of these antibody molecules was able to prevent anti-F9-induced decompaction (Hyafil *et al.*, 1981). This requires the monoclonal antibodies to obscure the binding of antibodies in the anti-F9 antisera without obscuring the site on the 84K component putatively involved in intercellular adhesion. Other antisera have been shown to disrupt compaction (Ducibella, 1980) but the components recognized by the antisera have not yet been identified.

The evidence for a role for the 84K component in the compaction process is convincing. Strangely, work in a very different system has identified a polypeptide of this size as a functional component in cellular interactions. This work is summarized in detail in Chapter 1, and began as an examination of the components released into (serum-free) medium by human epithelial mammary tumor cells (Damsky *et al.*, 1981; Knudsen *et al.*, 1981). Antiserum against these components was designated anti-SFM II, and, in contrast to a similar antiserum prepared against mouse mammary tumor cells (designated anti-SFM I), interferred with cell–cell interactions without blocking cell–substratum interaction. Material from the 80–85K region of SDS–PAGE gels blocked the effect of anti-SFM II, and antiserum prepared against material in this region of SDS–PAGE gels disrupts cell–cell interactions. Like the anti-F9 antisera described above, anti-SFM II also prevents compaction of the normal embryo (discussed in Chapter 1) and trypsinates from F9 cells prevent this inhibition. The cross-reactivity of the 84 and 80–85K components suggests that an adhesive mechanism similar to the one that maintains the cohesion of the morula may be expressed by mature tissue. An examination of the distribution of the component detected by anti-SFM II reveals that it is expressed on epithelioid cells from a variety of normal tissues (Damsky *et al.*, 1981). It is not a component of junctional complexes as its expression both precedes junction formation in the mouse embryo and is observed on teratocarcinoma cells that do not have junctions, although in both of these cellular situations the cells are forming very close cell–cell contacts. This suggests the exciting possibility that this component is part of a recognition system involved in the establishment of junctional complexes. Understanding the function of this component is very likely to make a substantial contribution to our understanding of how cellular interactions are regulated. The relationship of this proteolytic fragment of 84K to the 140K component, identified by Takeichi *et al.* above, is unfortunately unclear and awaits the exchange of the respective serological reagents.

3.4.3. A Cell Surface Carbohydrate-Binding Protein (Endogenous Lectin)

An entirely different approach (reviewed by Martin *et al.*, 1982) has exploited the rich spectrum of carbohydrates expressed on the surface of red blood cells from different species. These experiments use the ability of either intact cells to form rosettes with red blood cells or cell extracts to cause red blood cells to agglutinate. The specificity of this aggregation is then tested by the ability of defined cell fractions or, for example, complex carbohydrates of known composition, to inhibit rosette formation. Using this approach, Grabel *et al.* (1979) have identified a cell surface carbohydrate-binding protein on embryonal carcinoma cells. Teratocarcinoma stem cells were found to form rosettes with glutaraldehyde-fixed trypsinized red blood cells. These rosettes were found to be inhibited by mannose-rich polymers, suggesting that a carbohydrate-dependent recognition was being observed (Grabel *et al.*, 1979). More recently, fucose-rich polymers have been demonstrated to inhibit rosette formation and to be much more effective in inhibiting the aggregation of stem cells (Grabel *et al.*, 1981). The success of this approach is demonstrated by the observation that carbohydrates that will inhibit rosette formation will also inhibit the aggregation of embryonal carcinoma cells, and will cause the cells in performed aggregates to form much looser interactions. The component responsible is expressed on embryonal carcinoma cells prepared by disaggregation in calcium-free medium and is lost on trypsinization.

3.4.4. A Galactose-Dependent Adhesion Factor

Another approach has demonstrated the presence of a teratoma cell adhesion factor in the ascites fluid of mice carrying teratoma cells (Oppenheimer, 1975). This factor enhances the aggregation of teratoma cells when tested in *in vitro* reaggregation assays. When this factor is treated with β-galactosidase, its activity is substantially reduced. Treatment of the cells prior to addition of the factor does not affect the ability of the factor to enhance aggregation, suggesting that the adhesion factor itself contains the galactose group. Galactose monosaccharide is effective in inhibiting aggregation mediated by this factor.

3.4.5. A Cell Surface Galactosyltransferase

Cell surface glycosyltransferases have been perceived in very different ways over the last decade since Roseman's (1970) suggestion that they act as receptors for glycoconjugates during cellular recognition. This history is similar to the Russian proverb where a woman complains that her neighbor cracked a bowl lent to her; at first the neighbor denies having borrowed the bowl, but during the ensuing argument she concedes that she borrowed the bowl but adds that it was not cracked when she returned it; however, as the lender is insistent, she finally remembers that the bowl was cracked when she borrowed it. So with cell surface glycosyltransferases the argument has shifted from "Are they really on the outer surface of intact, healthy cells," to "Are they really active when on the outside of cells?," and now finally to "O.K. So they are outside and potentially active but is there any functional importance to them?"

Briefly, the control experiments that have been painstakingly performed and have caused this general turnaround in attitude have demonstrated that glycosyl acceptors that are not internalized can be glycosylated by intact cells, that incorporation of radiolabel is not because the sugar nucleotide is internalized or incorporated by broken cells (a large fraction of broken cells depress incorporation due to competition from unlabeled nucleotide sugars), and have controlled for incorporation of radiolabeled sugar released by hydrolysis of nucleotide sugar and internalized (reviewed by Shur, 1982b). My own reservations have decreased as the mechanisms of membrane formation and turnover have been elucidated. In the light of our current knowledge regarding the extensive turnover of Golgi, vesicular, and plasma membranes, it would be very surprising if glycosyltransferases were not found, for at least part of their lifetime, exposed on the cell surface.

Shur (1982a) has demonstrated that teratocarcinoma stem cells have a high level of a glycosyltransferase on their cell surfaces. Appropriate controls have determined that internal galactosyltransferases contribute less than 5% of the observed activity. The galactosyltransferase when presented with *exogenous* UDP-Gal incorporates the radiolabeled galactose into *endogenous* acceptors. These acceptor glycans have been identified as polyactosamine in nature. When galactosylated in the presence of the exogenous sugar donor, some of the endogenous acceptor glycans are released. The author suggested that they were retained by the cell surface because of their association with the galactosyltransferase. This aspect has been explored by demonstrating that the teratocarcinoma cells will adhere to immobilized polylactosamine and that this adhesion is reversed by the addition of *exogenous* UDP-Gal (Shur, 1983). A natural extension of this possible role in adhesion has been the demonstration that *exogenous* UDP-Gal will reverse performed cellular adhesions between embryonal carcinoma cells.

3.4.6. A Consensus?

There are therefore several lines of experimentation, which identify protein-linked complex carbohydrates at the cell surface as recognition molecules in intercellular adhesion. However, it is hard to completely rationalize these different approaches. For example, the experiments that demonstrate the importance of having terminal galactose residues for a functional adhesive component are directly opposed to the experiments that demonstrate that addition of galactose to cell surface components causes disaggregation. As the reaggregation assays are performed under different conditions, it may well be that different systems are being observed and that these various systems operate in parallel. The fucose-dependent and GlcNAc-dependent systems were observed with cells disaggregated by calcium-free medium and EDTA, respectively, while the galactose-dependent adhesion was observed with trypsinized cells. Nonetheless, it is surprising that the galactose-dependent system is not operational in untrypsinized cells. As mentioned above, a precursor–product relationship between the 140K component and the 84K proteolytic fragment has not been established. A glib consensus of these results would be that embryonal carcinoma cells have on their cell surface a glycoprotein of 140K, that this component has a protease-sensitive site that when cleaved releases a major polypeptide of 84K, and that these glycoproteins carry glycans having the terminal fucose, galactose, and GlcNAc residues required by the various carbohydrate-dependent recognition systems described above. As we shall see below, there *is* an unusual class of protein-linked glycan associated with embryonal carcinomas and embryonic cells, and that these very large, heavily branched glycans are rich in terminal fucose, galactose, and GlcNAc. Unfortunately, whether or not the 140 or 84K components bear any relationship to each other, and indeed if either of them carries this particular class of glycan, has, of course, not yet been established.

4. REGULATION OF GLYCOPROTEIN BIOSYNTHESIS IN EARLY EMBRYONIC CELLS

The preceding section documents the evidence for important roles for cell surface glycoproteins in early embryonic intercellular recognition. In several of these cases, the particular carbohydrate carried by the glycoprotein appears to be implicated in intercellular recognition. This section will describe what is know about the carbohydrates on the surface of early embryonic cells, how they change during early embryogenesis, and will explore the mechanisms that may be involved in regulating their programmed expression.

4.1. Carbohydrates Characteristic of Early Embryonic Cells

Our knowledge of the carbohydrates expressed on the surface of early embryonic cells is extremely limited. There have been almost no direct biochemical investigations of early embryonic cells and only a few direct studies with lectins and with antibodies. The current description of early embryonic cell surfaces is derived by analogy from the scant information we have regarding the surfaces of embryonal carcinoma cells. This section therefore will describe the biochemical studies that have been performed to characterize the cell surface carbohydrates of embryonal carcinomas and then to examine the extent of this analogy with the cell surface carbohydrates of early embryonic cells.

The first indication that embryonal carcinoma cell surfaces had unusual carbohydrate determinants came from a study by Muramatsu *et al.* (1978). In this study, embryonal carcinoma cells and early embryos were labeled with fucose and the resultant glycoproteins were exhaustively digested with proteases. When sized by gel-permeation chromatography, it was observed that a prominent fraction of the fucose-labeled glycopeptides were excluded from Sephadex G-50. This result suggested an unusually large size. For comparison, the bulk of the glycopeptides from embryonic fibroblastic cells are well included into this sizing matrix. The glycopeptides prepared from early embryos (cultured from the two-cell stage for 48 hr in the presence of radiolabeled fucose) showed a similar large size. It should again be mentioned that the following description of structural determinants is for the fucose-labeled glycopeptides from embryonal carcinomas and not for the (similarly sized) glycopeptides from normal embryos.

The large protein-linked glycans prepared from either mouse or human embryonal carcinomas have the following structural properties, which are summarized in Fig. 4. They are very rich in galactose and GlcNAc. They are sensitive to the enzyme endogalactosidase, which cuts

Figure 4. Schematic illustration of the carbohydrates present on mammalian cells. The large (upper) scheme illustrates some of the structural combinations possible with embryonic carbohydrates. These include: long linear repeats of galactose and GlcNAc and branch points where galactose is doubly substituted with GlcNAc. This provides a large number of nonreducing terminals that can be decorated with fucose and sialic acid residues or they can terminate in either galactose or GlcNAc. There is thus tremendous heterogeneity both in the skeleton and in the exterior decoration. The lower scheme illustrates the more common glycan found on differentiated cells.

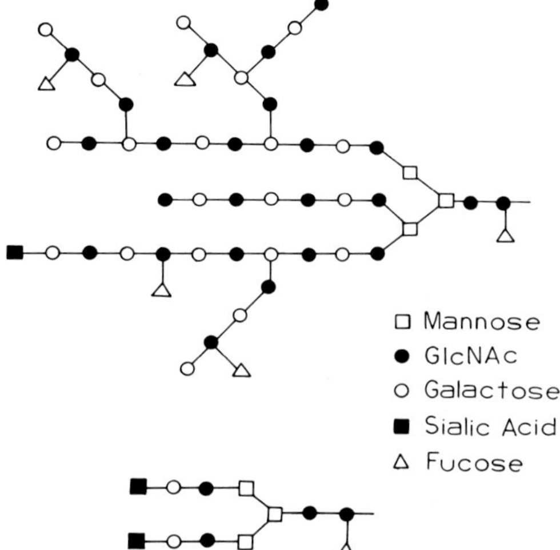

the sugar sequence GlcNAcβ1-3Galβ1-4GlcNAcβ1- (Fukuda and Matsumura, 1976). This enzyme is very appropriately discussed in Chapter 4. This repeating structure is the carbohydrate skeleton of keratan sulfate, the large glycan from erythrocytes termed "erythroglycan" (Finne *et al.*, 1978; Järnefelt *et al.*, 1978; Fukuda *et al.*, 1979), and a very large lipid-linked carbohydrate from ovarian cyst fluid termed the "megalosaccharide" (Lloyd and Kabat, 1968). The poor sensitivity of the glycans from embryonal carcinomas to this enzyme suggests a high degree of GlcNAcβ1-3(GlcNAcβ1-6)Galβ1-4 branching. This is supported by the methylation analysis, which shows a high degree of doubly substituted galactose. These high-molecular-weight glycans have been shown to be on the cell surface by the following approaches: using tryptic digestion of intact cells to release the glycopeptides (Muramatsu *et al.*, 1979); using galactose oxidase and sodium borotritide to label glycans on intact cells that have terminal galactose residues (Prujansky-Jacobovitz *et al.*, 1979); and by demonstrating that a cell surface galactosyltransferase on embryonal carcinoma cells would add Gal to polylactosamine glycans in the presence of exogenous UDP-Gal (Shur, 1982a). Similar glycans have also been identified on the surface of human teratocarcinoma cells (Rasilo, 1980; Rasilo *et al.*, 1980; Muramatsu *et al.*, 1982).

The size distribution of protein-linked glycans assembled by embryonal carcinoma cells is shown in Fig. 5. The glycans from fibroblastic cells are shown for comparison. The high-molecular-weight material excluded from this column has been identified by other structural investigations to have polylactosamine character. The excluded material is

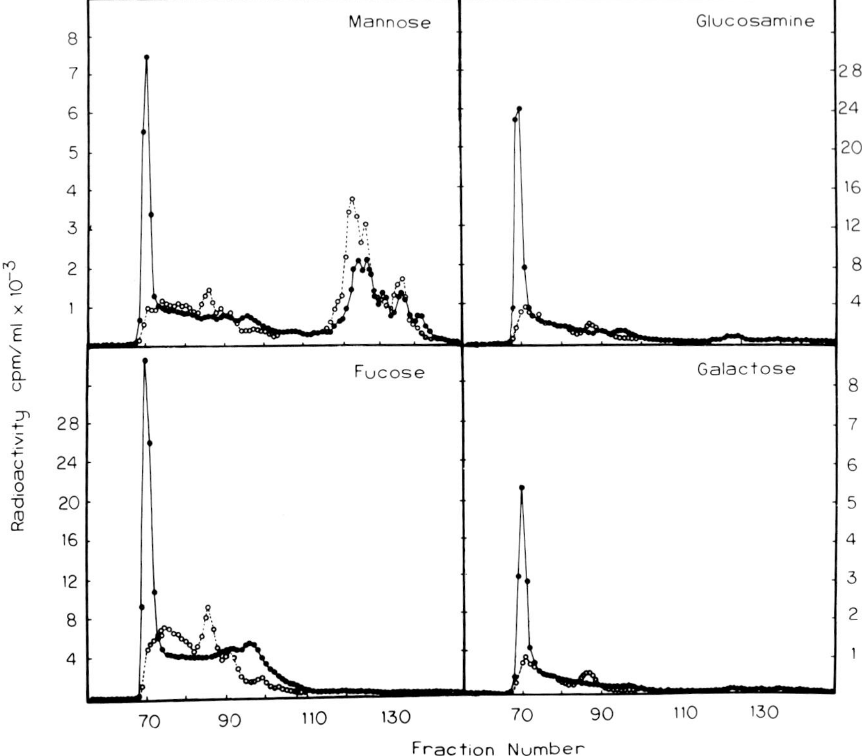

Figure 5. Gel filtration profiles contrasting the large glycopeptides found on embryonal carcinoma cells with the similar glycopeptides found on embryonic fibroblasts. Both cell types were metabolically labeled with mannose, fucose, galactose, and glucosamine and the radiolabeled glycoproteins were exhaustively digested with Pronase. The resultant glycopeptides were separated by size on Bio-Gel P-10 columns. ●, Embryonic carbohydrates; ○, fibroblastic glycans.

clearly a major component of the embryonal carcinoma glycans. In addition to the high-molecular-weight glycans, embryonal carcinomas also express a variety of glycans that coelute with the fibroblastic glycans. Almost no work has been done on these smaller glycans, probably because they are similar in size to glycans that have been identified so frequently previously and were therefore expected to be similar in structure. This expectation has been demonstrated to be unwarranted.

The spectrum of glycans assembled by a cell reflects many factors. One factor is the glycosyltransferase activities available to the cell. Another is the spectrum of presumptive glycoproteins available for glycosylation. As each glycoprotein carries a characteristic spectrum of carbohydrates, the total array of glycans assembled by a cell is a product of these factors and others. One approach to examine the biosynthetic potential of different cells has been to compare the glycans assembled in a single polypeptide chain common to the different cells. Because of the inherent problems in identifying common glycoproteins, an alternative approach has been used. This has been to infect different cells with the same virus and examine the glycosylation of its glycoproteins. In order to compare the biosynthetic potentials of embryonal carcinomas and cells derived from them, Etchison *et al.* (1981) infected the various cell lines with vesicular stomatitis virus (VSV). When the glycopeptides prepared from the viruses grown in different cells were compared by gel filtration, they were found to have similar size distributions. However, when the glycopeptides were digested with a mixture of exoglycohydrolases (neuroaminidase, β-galactosidase, and β-hexosaminidase) and an endoglycohydrolase [endo D, which cleaves the GlcNAc dimer in Manα1-3(Manα1-6)Manβ1-4GlcNAcβ1-4GlcNAcβ1-], they showed substantial differences in their susceptibilities to digestion. The basis for these different susceptibilities has not been identified but it is clear that closer examination of the small glycans from embryonal carcinomas that coelute with usual complex-type glycans is necessary. They may have structural features in common with the cores of the large glycans. As the core regions of the large glycans are a small proportion of the molecules, they are harder to study. Careful examination of the core regions of the small glycans should reveal interesting details which should be tested for in the large glycans. The observation that the G-protein of VSV did not carry polylactosamine glycans is interesting in its own right and also shows the limitation of the viral approach.

The glycans now regarded as characteristic for embryonal carcinomas were originally identified as fucose-containing components and there are two different fucosyl linkages that have subsequently been identified. The most recent is a Fucα1-3Galβ1-4GlcNAcβ sequence (Miyauchi *et al.*, 1982), which was identified in a very curious way. The trisaccharide sequence was chemically synthesized and monoclonal antibodies were prepared against it. The natural distribution of this sequence was then sought and found to be restricted to colon adenocarcinomas and teratocarcinomas.

The other fucosyl decoration is Fucα1-3(Galβ1-4)GlcNAcβ1-3. As described earlier, this determinant was identified by a monoclonal antibody prepared against F9 embryonal carcinomas and followed up because its expression changed during early development (Solter and Knowles, 1978). Quite naturally, this antigen was called "stage-specific embryonic antigen 1" and abbreviated SSEA-1. It has also been identified on maturing sperm (Fox *et al.*, 1982). The binding determinants recognized by this antibody have been characterized by competition with known glycans (Nudelman *et al.*, 1980; Gooi *et al.*, 1981) and by examining the major carbohydrate determinants prepared from embryonal carcinoma cells, which are immunoprecipitated by this antibody (Andrews *et al.*, 1982). These glycans are in the high-molecular-weight class described above. These glycans potentially carry a large number of determinants and they can be immunoprecipitated with peanut agglutinin and anti-peanut agglutinin, fucose-binding protein and anti-fucose-binding protein, and F9 antisera (Muramatsu *et al.*, 1979). Whether or not the same glycan chains carry determinants for more than one lectin or antiserum has not been demonstrated but these glycans are multiply branched and could very easily carry different determinants on different branches.

The relative sizes of these large glycans and the smaller complex-type glycans are also illustrated in Fig. 5. Also emphasized in this figure is the complexity of the nonreducing terminals, with terminals expressing galactose, GlcNAc, sialic acid, and the fucose-containing trisaccharide identified by anti-SSEA-1. These carbohydrate determinants would potentially fulfill the requirements of the three carbohydrate-dependent adhesion systems described above. In fact, Shur (1982a) has identified polylactosamine chains ending in GlcNAc residues as a major acceptor for the galactosyltransferase in the presence of *exogenous* UDP-Gal. The endogenous receptor for the fucose-binding protein has not been identified, nor has the carbohydrate nature of the galactose-containing adhesion factor.

While there is an overall similarity between the large glycan identified on the surface of embryonal carcinomas and the similar glycan on the surface of red blood cells, it is important to identify the following differences.

1. The component on the red blood cell is relatively homogeneous in size, being 8–10K in molecular weight, and it is predominantly carried by a single protein species, Band 3. The component on the embryonal carcinoma cell surface is very heterogeneous in size running, from 3K up to 30K in molecular weight, and may be carried by a number of different protein species.
2. The determinant Fucα1-3(Galβ1-4)GlcNAcβ1-3 (SSEA-1) is found on embryonal carcinoma cells but not on red blood cells, whereas the determinant Fucα1-2Galβ1-4GlcNAcβ1-3 (O blood group) is found on red blood cells but not on embryonal carcinoma cells. Both of these fucose-containing determinants are found on repeating Galβ1-4GlcNAcβ1-3 structures.
3. An interesting, developmentally regulated alteration in the structure of the erythrocyte carbohydrate, namely, the transition from

an unbranched repeating glycan found on fetal erythrocytes to the highly branched glycan found on adult erythrocytes, is not paralleled in the early embryo. Specifically, the glycan is found to be highly branched at the earliest time it is identified during normal embryogenesis and therefore this developmentally regulated expression of the branching enzyme is a feature of the development of erythroid cells and not a general feature of embryonic development.

4.2. Programmed Changes in Carbohydrate Expression during Early Embryogenesis

The above description, which summarizes our current understanding of the carbohydrates expressed on the surfaces of early embryonic cells, emphasizes how much we have yet to learn about them. While we are a long way from having a good understanding of the carbohydrates carried by embryonic cells, we are even further from having a good understanding of how they change during development. However, there are a number of well-characterized carbohydrate-binding proteins, either antibodies or lectins, that bind at particular stages during embryogenesis. The scope of this discussion will therefore be limited to a description of the developmentally regulated changes in the binding of these proteins rather than a direct description of the changes that occur.

The original description of embryoglycan was as a large developmentally regulated fucosylated glycopeptide (Muramatsu *et al.*, 1978), and therefore this molecule will be our starting point. Glycans in this class have been shown to be receptors for peanut agglutinin, the fucose-binding protein from *Lotus tetragonolobus,* and anti-F9 antisera (Muramatsu *et al.*, 1979) and also anti-SSEA-1 antibodies (Andrews *et al.*, 1982). Receptors for peanut agglutinin (Reisner *et al.*, 1977), fucose-binding protein (Gachelin *et al.*, 1976), anti-F9 antibodies (Muramatsu and Gachelin, 1979), and anti-SSEA-1 antibodies (Solter *et al.*, 1979; Knowles *et al.*, 1980) are lost by embryonal carcinoma cells when they differentiate. We have found that the polylactosamine skeleton is also lost on differentiation (Ivatt, Martin, and Robbins). Embryoid bodies were formed and allowed to develop in suspension as described previously (Martin and Evans, 1975). The outer shell of endoderm that formed was removed by dispase digestion and the endodermal cells that stay as a sheet were removed from the embryonal carcinoma aggregates by allowing the latter to settle. The separated cell populations were metabolically radiolabeled with fucose and mannose and the glycoproteins were extracted and exhaustively digested with Pronase and separated by gel filtration. The loss of both fucose and mannose from the high-molecular-weight (polylactosamine) region of the gel was observed for the endodermal glycans. At present we do not know if the fucosyltransferase is lost or if the fucose determinant is lost because of loss of acceptor glycan. However, the sorting out of stem and endodermal cells requires the cell surfaces of these two cell types to express different determinants (Steinberg, 1958, 1963, 1970), and we view the loss of this polylactosamine skeleton as a very efficient mechanism for the coordinated loss of a large number of stem cell determinants. The mechanisms by which polylactosamine synthesis may be regulated are discussed below.

Antibodies against the SSEA-1 determinant are found to interact first with the eight-cell embryo (Solter and Knowles, 1978). All cells of the inner cell mass will react with the antibodies. The trophectoderm reacts transiently. This transient reaction is similar to the finding with monoclonals against the Forssman antigen, a glycolipid determinant, where the trophectoderm on the fourth day was positive but both mural and polar cells on the fifth day were negative (Stern *et al.*, 1978; Stern and Willison, 1982).

The high-molecular-weight fucosylated glycans are lost from the normal embryo during the ninth and tenth days (Muramatsu *et al.*, 1980), i.e., during the peak period of histogenesis. This is an oversimplification, in that unlike the differentiation to form endoderm, there is more correctly a loss of fucose from the high-molecular-weight glycans (Ivatt, Martin, and Robbins). We have examined the glycans made during periods of extensive differentiation by embryonal carcinomas allowed to develop as embryoid bodies. In this situation we observe a loss of fucose from the region of gel filtration columns associated with high-molecular-weight material, but we find substantial incorporation of mannose, galactose, and glucosamine into polylactosamine glycans even at stages where extensive differentiation has occurred. Here, loss of fucose is directly associated with decreased activity of the fucosyltransferase and not because of low levels of acceptor glycan. This observation has been repeated with the normal embryo (Ivatt, Martin, and Robbins). Embryos of different developmental stage were excised and labeled with either mannose or fucose and their protein-linked carbohydrates were examined. During the period of histogenesis described above, we found substantial polylactosamine formation based on the mannose incorporation but substantial loss of fucose from the high-molecular-weight glycans.

The finding that the high-molecular-weight glycans persist long into embryogenesis was demonstrated by a study by Heifetz *et al.* (1980) that examined the high-molecular-weight glycans assembled by mouse embryos cultured *in vitro* (Hsu, 1972), and by immunohistochemical studies using antibodies against polylactosamine structures (Kapadia *et al.*, 1981; Feizi *et al.*, 1982). In the former study, embryos were labeled with glycosamine and the high-molecular-weight materials were examined by specific enzymatic or chemical digestions. When normalized to DNA synthesis, the proportion of glucosamine incorporated into a polysaccharide identified as polylactosamine-containing was found to rise to a peak at the neurula stage and then drop. This pattern was also observed with heparan sulfate, but not with hyaluronic acid or chondroitin sulfates, which rose to a peak at the neurula stage and then remained high. The polysaccharide was insensitive to the digestions used to identify the glycosaminoglycans and was partially sensitive to endogalactosidase. The low susceptibility again was presumed to arise from a high degree of branching.

The examination of polylactosamine glycans *in situ* in the pre- (Feizi *et al.*, 1982) and post- (Kapadia *et al.*,1981) implantation embryo has been achieved using naturally occurring monoclonal antibodies. Some of these antibodies arise in an immune hemolytic disorder known as cold agglutinin disease (Dacie, 1962). These antibodies react with the blood group Ii, and recognize either linear saccharides with the repeating structure GlcNAcβ1-3Galβ1-4 (anti-i sera) or similar structures that also have GlcNAcβ1-3(GlcNAcβ1-6)Galβ1-4 branches (anti-I sera) (Feizi, 1981). The various antibodies recognize slightly different regions of the glycan sequence, as discussed in Chapter 4. Anti-I Ma serum, which recognizes GlcNAc1-6 branches, reacts with all stages of preimplantation development, whereas anti-i Den serum, which recognizes linear sequences, does not react with the early embryo. This lack of reactivity of an antibody that recognizes unbranched sections of polylactosamine is in accord with the suggestion that low sensitivity of the glycans to endogalactosidase is the result of extensive branching. Undifferentiated cells were found to be positive for anti-I Ma and negative for anti-i Den. In contrast, the endoderms derived either during the differentiation of embryonal carcinoma cells or during differentiation in the normal embryo are rich in anti-i Den reactivity. At day 6, anti-i Den serum reacts only with the endodermal cells lining the egg cylinder, whereas anti-I Ma and anti-I Step sera react with the trophoblast and the cells lining the (outer) blastocoelic cavity and the (inner) proamniotic cavity.

The conclusions from these observations are that particular carbohydrate determinants are localized within the normal embryo both temporally and topographically. The next section addresses how these determinants are assembled and how this localized expression may be regulated.

4.3. Mechanisms That Regulate the Expression of Glycoprotein Determinants

The purpose of this section is to summarize what we know about the assembly of glycoproteins in early embryonic cells and to discuss the mechanisms that may operate to regulate this assembly. This section is unavoidably speculative because of our present ignorance regarding some fundamental aspects concerning the assembly of glycoproteins. For example, we know that particular glycoproteins have certain carbohydrates characteristically associated with them; however, in no single instance do we know why a particular glycoprotein carries the carbohydrate chain(s) that it does. Consequently, without understanding the mechanisms that ensure the efficient operation of individual glycosylation pathways, we have no basis for describing how competition between different pathways is regulated, and we are even further from understanding how changes in the prominence of particular pathways are achieved during development. Nonetheless, we know enough to eliminate some mechanisms and to design experimental approaches that will more closely define the mechanisms that regulate glycoprotein biosynthesis. This section will briefly summarize the general mechanisms of glycoprotein biosynthesis and identify some critical aspects of the glycosylation process that proposed regulatory mechanisms must be able to explain.

4.3.1. General Outline of the Glycosylation Process

The glycosylation of asparagine-linked carbohydrates begins with the assembly of a large mannose-rich carbohydrate on a lipid carrier and its en bloc transfer to nascent proteins in the rough endoplasmic reticulum and ends with the expression on the cell surface of a wide variety of complex glycans that are assembled on cores derived from this common precursor (reviewed by Waechter and Lennarz, 1978; Elbein, 1979; Parodi and Leloir, 1979; Spiro and Spiro, 1979; Kornfeld and Kornfeld, 1980; Struck and Lennarz, 1980; Hubbard and Ivatt, 1981). The cellular locations of the various reactions are identified in Fig. 6.

The role of the lipid-linked precursor glycan was identified several years ago (Parodi *et al.*, 1972). The structure of the major lipid-linked glycan has been characterized (Li *et al.*, 1978). The glycan consists of 14 sugars attached through a pyrophosphate linkage to a large lipid. The lipid has 18–22 isoprene units and is termed "dolichol." The first two sugar units are *N*-acetylglucosamine. The first sugar is α-linked to phosphate and is subsequently inverted on transfer. The second sugar is β1-4-linked. The next sugar is a β1-4-linked mannose. This sugar is doubly substituted with both α1-3 and α1-6 mannose residues attached to it. The α1-3 mannose has two α1-2 mannose residues attached to it and then three α-glucose residues. The α1-6 mannose also is doubly substituted with α1-3 and α1-6 mannose residues. Both of these residues have an α1-2 mannose residue attached to them. The glucose residues rarely appear in mature glycoproteins. They are the last sugars added during the assembly

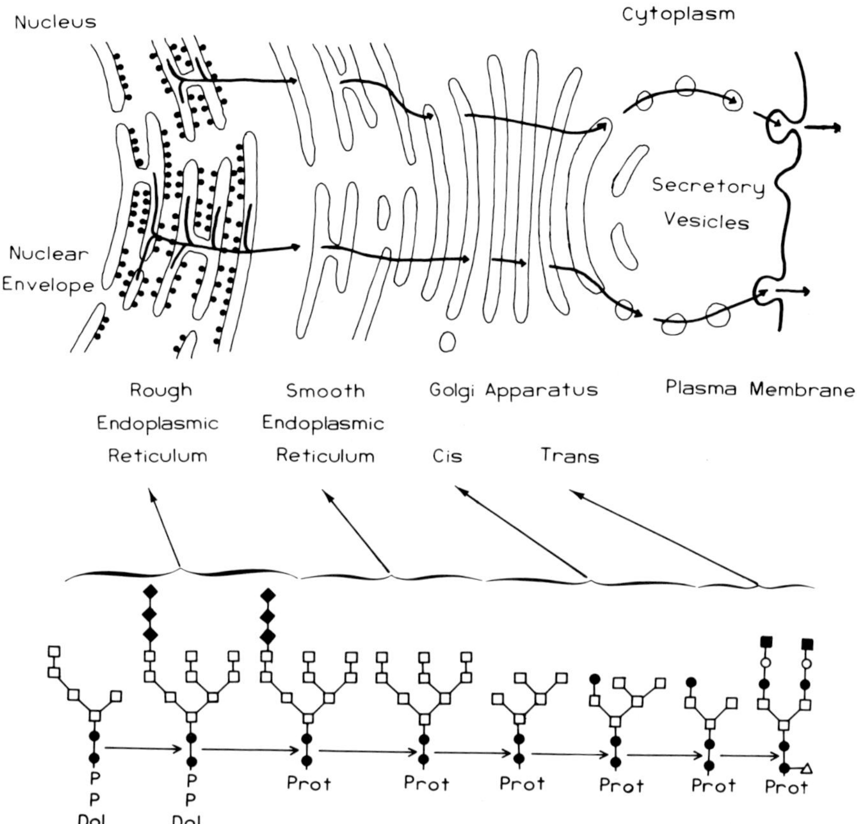

Figure 6. The posttranslation modifications occurring during the maturation of glycoproteins, and their subcellular localizations. The lipid-linked precursor oligosaccharide is assembled in the rough endoplasmic reticulum where it is transferred to nascent proteins. During their passage through the smooth endoplasmic reticulum, glucose residues are removed from the glycoproteins and in the Golgi apparatus mannose residues are removed from them. The extent of removal of these mannose residues dictates whether or not a carbohydrate will retain its high-mannose character or be more extensively processed to form complex-type glycans. The addition of an outer GlcNAc residue acts as a signal for a distinct mannosidase that trims the mannose core from five to three residues. Subsequently, the outer GlcNAc, Gal, and sialic acid residues are added. Apparently, the number of outer branches is determined by the GlcNAc transferases prior to the addition of Gal.

of the lipid-linked precursor glycan (Chapman *et al.*, 1979b) and the first sugars to be removed during the subsequent processing of the protein-linked glycan (Kornfeld and Tabas, 1978; Hubbard and Robbins, 1979). In cell-free systems, these glycose residues have been demonstrated to enhance the rate of glycan transfer from carrier lipid to acceptor protein

(Turco and Robbins, 1977; Spiro *et al.*, 1979). It is very tempting to speculate that their role is to provide a recognition signal for the transferase that transfers the glycan from carrier lipid to acceptor protein so that the precursor glycan is appropriately assembled. Indeed, in mutant cell lines unable to assemble the mannose-9 core [mutants with structures truncated at mannose-7 (Hunt, 1980) and mannose-5 (Chapman *et al.*, 1979a) have been described], glucosylated glycans are a very small fraction of the lipid-linked glycan pool. However, the glucosylated structures that are made are transferred to protein in very obvious preference to the more prevalent unglucosylated glycans (Kornfeld *et al.*, 1979; Hunt, 1980).

The enzymology of the assembly of this precursor glycan is beginning to be described. The mannose residues have been demonstrated to be added in a very strictly controlled sequence (Chapman *et al.*, 1979b) and the sugar donors for each of the mannose residues have been identified (Chapman *et al.*, 1980). The mannose residues of the outer α1-3 and α1-6-linked branch point are derived from lipid-linked mannose (dolichol-phosphate-Man) while those of the inner branch point are derived directly from GDP-Man. These and other data suggest that each mannose residue is added by a different mannosyltransferase. The assembly of this precursor glycan is being studied in cell-free systems and the enzymes are being examined in partially purified form after detergent extraction using defined glycosyl donors and acceptors.

A point that is worth raising here and will recur is that the assembly of this precursor glycan is an efficiently coordinated process. The partially completed (intermediate) structures are present in very small amounts at steady state (Chapman *et al.*, 1979b; Hubbard and Robbins, 1980). Once a glycan chain is initiated, it is very rapidly elongated and completed. There appears to be a slight pause at the mannose-5 stage. This is where the lipid-linked donor takes over from the nucleotide-linked donor; however, this intermediate is still a very minor component relative to the completed structure. I make the suggestion that such efficient assembly is processive and that the coordination of the enzymes is the result of complex formation. Perhaps, after the glycan chain is initiated it is passed to a complex of mannosyltransferases that extend to the mannose-5 stage and it is then transferred to a second complex that completes the structure. This would be compatible with the kinetic discontinuity and with some apparent discrepancies between the various experiments designed to elucidate which side of the rough endoplasmic reticulum the precursor glycan is assembled (Hanover and Lennarz, 1978; Nilsson *et al.*, 1978; Snider *et al.*, 1980). Complex formation in this system is, I reemphasize, complete speculation. The hard experimental facts are that the precursor glycan is assembled efficiently with very low levels of intermediate structures, and, as far as one can tell, with very few mistakes.

After transfer to protein, the glucose residues are removed as the glycoproteins migrate through the endoplasmic reticulum to the Golgi apparatus (Ugalde *et al.*, 1979; Grinna and Robbins, 1979, 1980). They are removed with such different kinetics that the enzyme activities must be at different levels or in different locations. On entering the Golgi apparatus, the outer α1-2 mannose residues are removed (Tulsiani *et al.*, 1977; Opheim and Touster, 1978; Tabas and Kornfeld, 1978). This removal is incomplete at some glycosylation sites, and these sites will, in the mature glycoprotein, carry high-mannose-type glycans. The appearance of such high-mannose glycans is not random and is restricted to particular sites. The removal of mannose residues from sites that carry high-mannose structures on mature glycoproteins occurs with very different kinetics from the removal of mannose residues from sites that carry complex-type glycans in the mature glycoprotein. This difference in kinetics may reflect differences in accessibility of the carbohydrate chain. The glycans at sites that will be processed to complex-type glycans may be readily accessible and rapidly trimmed while the sites that contain high-mannose structures in the mature glycoprotein may be poorly accessible and therefore trimmed very slowly and indeed incompletely. These trimming processes may also occur at different cellular locations. The rapid trimming is thought to occur in the Golgi apparatus while the slow trimming may occur during membrane recycling in either the Golgi apparatus or the lysosomes.

On different glycoproteins, these α1-2 mannose residues are removed in different sequences. On IgM molecules, which contain two high-mannose glycan chains on the mature glycoprotein, the surprising implication from the partially trimmed structures is that at the different glycosylation sites, the mannose residues are removed in strict but distinct orders (Chapman and Kornfeld, 1979a,b). The presumptive α-mannosidases are being characterized.

The various pathways for the removal of the α1-2 mannose residues yield the same mannose-5-containing structure. This glycan contains two mannose residues additional to the mannose-3 core found in complex-type glycans. The removal of these two surplus mannose residues occurs during a very strictly controlled sequence of events, namely, the addition of an outer *N*-acetylglucosamine residue prior to the trimming to the mannose-3 core. This addition is essential for the action of the α-mannosidase, which trims the now-superfluous mannose residues. The mannose-3 core with one outer branch begun is now a suitable acceptor for the *N*-acetylglucosaminyltransferase that will begin the second branch. This curious sequence of events has been established with very careful studies of the enzymology of *N*-acetylglucosaminyl addition (Narasimhan *et al.*, 1977; Tabas and Kornfeld, 1978; Harpaz and Schachter, 1980a,b) and of mutant

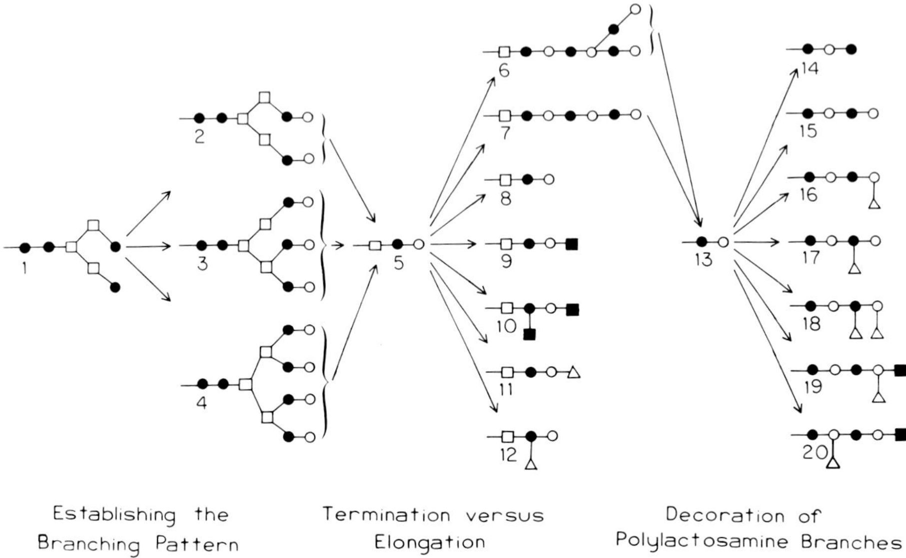

Figure 7. Schematic illustration of some of the possible heterogeneity in embryonic carbohydrate structures. This heterogeneity can, in principle, be considered in three separate phases: the establishment of the branching pattern; the elongation of these branches (to form polylactosamine structures) or their termination with fucose or sialic acid; and the decoration of these polylactosamine branches.

cell lines that were unable to trim beyond the mannose-5 stage (reviewed by Stanley, 1980).

A prevalent group of complex glycans contains additional galactose and sialic acid residues, and not uncommonly the proximal GlcNAc of the core dimer also contains fucose. There are, however, a large number of possible variations in this basic scheme and these are summarized in Fig. 7. The first stage that heterogeneity can occur at is in the number of *N*-acetyllactosamine branches that the core glycan carries (structures 2, 3, and 4). Recent studies have demonstrated that the branching pattern is established very early (Cummings *et al.*, 1982; Gleeson and Schachter, 1983). The enzymes that add the GlcNAcs that will begin the third and probably the fourth branches are inhibited by the presence of galactose. Therefore, the addition of these GlcNAcs must occur before any more external reactions have occurred. The interaction between the galactosyltransferase and the enzymes that add these branching GlcNAcs may play a role in determining the extent of branching.

The next stage during the assembly of complex-type glycans in embryonic cells is whether to terminate the branches with for example sialic acid, or to elongate the branches extensively with repeating units of Gal-GlcNAc (structures 6 and 7). The branch represented in structure 8 can be left undecorated, or be decorated with a variety of substitutions. Structures 9 and 10 represent different sialylations, either terminal or on both GlcNAc and Gal residues. This latter is not common but has been described, for example, for bovine fibronectin (Takasaki *et al.*, 1979). Structures 11 and 12 represent different fucosylations, either α1-2 to Gal or α1-3 to GlcNAc. Usually, glycans that contain several branches will not have both fucose and sialic acid on the same branch and this mutual exclusion appears also to occur between adjacent branches (Beyer *et al.*, 1981). This has been demonstrated by careful examination of the enzymatic properties of the transferases that add sialic acid and fucose to asparagine-linked glycans, and by characterizing the products of these enzymes when they were allowed to compete for the same glycan acceptor. However, these general rules, which were established by careful studies of complex glycans having branches of single lactosamine units, do not seem to apply to glycans that contain branches composed of repeating lactosamine units. For example, the recently reported characterization of human secretory component demonstrated the presence of fucose and sialic acid on sequential lactosamine units of the same branch (Mizoguchi *et al.*, 1982). The variety of decorations found on polylactosamine branches is explored later in this chapter and in Chapter 5.

Structures 6 and 7 represent glycans with either branched or linear chains, respectively. These structures can be repeated many times and the glycans can achieve very high molecular weights. The highly branched structures provide a large number of possible sites for decoration. The possibilities presented in Fig. 7 suggest that some branches may terminate with GlcNAc (structure 14), some with Gal (structure 15), and some with fucose (structure 16). This latter decoration, which converts the Ii blood group to H blood group, is not found on pluripotent stem cells but is found on endodermal cells (Feizi *et al.*, 1982). Some branches may be decorated with a fucose residue on the penultimate GlcNAc residue (structure 17); this forms the antigenic determinant defined by monoclonal antibodies as SSEA-1 (Gooi *et al.*, 1981). Doubly substituted branches yield Lewis-type blood group determinants (structure 18). Recently, both sialylation and fucosylation were demonstrated to occur on either the same *N*-acetyllactosamine unit (structure 19) or on sequential *N*-acetyllactosamine units (structure 20). The fact that these different decorations can occur on different branches of the same large, highly branched glycan means that these glycans carry an enormous amount of potential information for carbohydrate-mediated recognition.

This wide spectrum of products does not arise haphazardly. The observation is that the steady-state levels of intermediates during the formation can be dramatically enhanced by the inclusion of retinoic acid in the culture medium (Strickland and Mahdavi, 1978). When these "nullipotent" cells are cultured at low density on gelatin gels, in the presence of retinoic acid, they form parietal endodermal cells and secrete plasminogen activator and basement membrane components (Strickland and Mahdavi, 1978; Adamson *et al.*, 1979; Jetten *et al.*, 1979). However, when these cells are cultured in the presence of retinoic acid as cell aggregates rather than as sparse monolayers, they secrete large amounts of α-fetoprotein and transferrin and very little of the basement membrane components laminin and type IV collagen (Hogan *et al.*, 1981). There is the possibility that the F9 cells are first induced to differentiate into a multipotent precursor endoderm analogous to the primitive endoderm of the normal embryo and then differentiate into either visceral or parietal endoderm depending on the nature of the intercellular contact signals they receive.

This has been explored by inducing the formation of endoderm by F9 cells cultured as aggregates with retinoic acid in the presence of laminin (Adamson and Grover, 1983). In this situation, the formation of visceral endoderm can be prevented and parietal endoderm forms. Specifically, at low laminin concentrations, visceral endoderm formation is stimulated, possibly as a result of enhanced aggregate formation, while higher concentrations inhibit visceral endoderm formation, possibly by disrupting polarity. This inhibition requires the exogenous laminin to be present during the first 4 days of induction.

The consequences of mixing nullipotent and pluripotent embryonal carcinoma cells and allowing their differentiation in culture have been observed. The initial observation that nullipotent cells restricted the differentiation of pluripotent cells (Rosenstraus and Levine, 1979) is now thought to have been a misinterpretation of the relative rates of division of the two cell types and therefore their relative contributions to the mixed cultures. A recent reexamination (Rosenstraus and Spadoro, 1981) using a genetically marked subclone of F9 cells and modulating the growth rate of the F9 cells yields results suggesting that the pluripotent and nullipotent cells do not affect the differentiation behavior of each other in mixed culture. The possible role of positional effects was not specifically considered in this study. Namely, that if the outer cells on an aggregate of pluripotent cells differentiate into endoderm, surrounding a core of pluripotent cells with a surface layer of nullipotent cells might prevent differentiation. This could occur, either by providing the signals that the *interior* population receive which *prevent* differentiation, or by preventing the pluripotent cells from receiving the signals that the *surface* population receive which *induce* differentiation. This mixed culture would appear to be a useful system to approach the signals that either allow or induce the surface population of cells to differentiate.

It is probable that the retinoic acid-induced differentiation is mediated by a cytoplasmic retinoic acid-binding protein (Sherman *et al.*, 1980). Indeed, mutants that fail to differentiate in response to retinoic acid have no demonstrable cytoplasmic retinoic acid-binding protein and differentiate poorly in tumor form, although they maintain tumorigenic potential (Schindler, 1981). Retinoid acid has been shown to retard tumor growth, decrease mitotic index, decrease extent of necrosis, and increase long-term survival time. Histological and biochemical examination of the tumors suggested induced differentiation had occurred *in vivo* (Strickland and Sawey, 1980; Speers, 1982). The structure–activity relationship of various derivatives of retinoic acid has been examined (Strickland and Sawey, 1980) to identify the key structural features so that more effective agents for inducing differentiation of these tumor cells could be identified. The carboxylic acid function was found to be essential, while the cyclohexenyl ring could be modified. The effectiveness was found to increase with duration of treatment, suggesting the differentiation into endoderm was facilitated at each division.

Nishimune *et al.* (1982) identified reversible and irreversible steps in differentiation of embryonal carcinomas induced by retinoic acid. Ogiso *et al.* (1982) observed disappearance of peanut agglutinin receptors on retinoic acid-induced differentiation of F9 cells. This loss was prevented by culturing the cells with a feeder layer. This loss was reversible if retinoic acid was removed after 2 days, but not if it was present for 4 days, suggesting that there is a reversible and an irreversible phase of this induced differentiation. These findings were also based on secretion of plasminogen activator and expression of the F9 antigen. F9 cells were incubated with retinoic acid for 2 days and the peanut agglutinin (PNA) -positive and -negative cells were separated using coaggregation with rabbit erythrocytes in the presence of PNA. The PNA-negative cells were then examined for various characteristics after culturing either in the continued presence of retinoic acid or in its absence. In the absence of retinoic acid, peanut agglutinin receptors were reexpressed within 5 hr.

5.2. Biochemical and Biological Changes That Accompany Endoderm Formation

There are a number of biochemical and biological changes that accompany endoderm formation in culture. So far, the formation of parietal endoderm has been much more extensively studied than the formation of visceral endoderm. This is because the usual induction conditions with retinoic acid favor the formation of parietal endoderm. The biochemical and biological changes that accompany endoderm formation induced by retinoic acid, and the differences that have been found between undifferentiated embryonal carcinomas and their differentiated endodermal derivatives are discussed in this section. There are a large number of differences that have been cataloged.

The undifferentiated embryonal carcinoma cells are relatively small cells that grow as colonies with indistinct cell boundaries. They have large nuclei and relatively few cytoplasmic organelles. They have gap junctions and modified adherens junctions and are in metabolic cooperation. In contrast, the endodermal cells derived from them are large, have an extensive endoplasmic reticulum, and form epithelial-like junctional complexes with extensive zonula occludens junctions (Lo and Gilula, 1980a). There are also changes in their cell cycle characteristics. The undifferentiated cells have a generation time of 12 hr (10 hr S phase, 2 hr G_2 + M, with little or no G_1). The endodermal cells formed after retinoic acid-induced differentiation have a generation time of 16.8 hr (12.5 hr S phase, 2 hr G_2 + M, and 2.3 hr G_1). The conclusion from these studies was that both G_1 and S lengthen on differentiation (Rosenstraus *et al.*, 1982). The nucleosome repeat distances for embryonal carcinomas, primary endoderm, and long-term parietal endoderm are 196, 205, and 185, respectively (Oshima, 1981), suggesting no obvious systematic relationship between state of differentiation and repeat distance. The ratio of the histones H1a to H1b also changes (Ajiro *et al.* 1980; Oshima, 1981). Howe *et al.* (1980) examined the cytoplasmic and nuclear proteins of embryonal carcinoma cells and inner cell mass cells during their differentiation *in vitro*. Their conclusions from these comparisons were that embryonal carcinoma cells represent a more advanced state than the inner cell mass cells of the 4-day embryo. Specific changes in the cytoplasmic proteins include changes in the isozyme composition of lactate dehydrogenase (Lo and Gilula, 1980b).

Specific cytoskeletal proteins have been identified for the extraembryonic endoderm by Oshima (1981). Two relatively abundant proteins with M_r of 55 and 50K expressed by PYS-2 cells were gel-purified, by SDS–PAGE, and antisera were prepared against them. The larger protein designated endo A was found to be immunologically similar to vimentin and is immunoprecipitated by antivimentin antisera. The smaller protein designated endo B was found to be relatively specific for endodermal cells. Pluripotent (PSA-1) and nullipotent (F9) embryonal carcinoma cells were also negative for these two proteins, if the low level (0.1–1.0%) of endodermal contamination, as assessed by laminin production, is accounted for. The level of these proteins rises dramatically when the F9 cells are induced to differentiate with retinoic acid (Oshima, 1982). The cells showed a strong fibrillar (cytoskeletal) immunofluorescence staining pattern. Endo A but not endo B was found to adsorb the activity of the monoclonal antibody TROMA-1, which also shows endodermal specificity (Brulet *et al.*, 1980).

Howe and Solter (1981) have examined the cell surface proteins of embryonal carcinoma cells and parietal endodermal cells, and have identified a number of developmentally regulated cell surface components. The proteins were radioiodinated using the lastoperoxidase method and analyzed using two-dimensional gel electrophoresis. Not surprisingly, a number of differences were observed between the two cell types. However, several of the spots that were unique to the PYS-2 cells appeared during the retinoic acid-induced differentiation of F9 cells. At present, few of the spots can be identified by more than their molecular weight and isoelectric point, the two subunits of laminin being exceptions. The present focus of this and other studies (Linder *et al.*, 1981) examining the changes in protein expression is to correlate these developmentally regulated cell surface proteins with known antigens and with cell surface-dependent biological phenomena.

Changes in the microviscosity of the plasma membrane (Jetten *et al.*, 1982), in the expression of the large fucosyl glycopeptide described in detail above (Muramatsu and Muramatsu, 1982), and in the expression of the SSEA-1 determinant (Lo and Gilula, 1980b; Solter and Knowles, 1978) also occur. The greatly enhanced expression of basement membrane components has been demonstrated for differentiated endodermal cells (Chung *et al.*, 1977a,b; Hogan *et al.*, 1980). This increase occurs coordinately with increased expression of plasminogen activator during retinoic acid-induced differentiation (Oshima and Linney, 1980). Heath *et al.* (1981) observed the appearance of functional insulin receptors on retinoic acid-induced differentiation of PC13 clone 5 cells, and speculated that this may be part of a general change in growth regulatory mechanisms accompanying embryonal carcinoma cell differentiation and loss of malignancy.

The regulation of the phenotype of somatic cell hybrids formed by fusion of teratocarcinoma cells and either normal splenic lymphocytes or thymoma-derived cells has been examined (Gmur *et al.*, 1981). There was no simple pattern to emerge from an examination of the *in vivo* and *in vitro* morphology of the hybrids or their cell surface properties and karyotypes. Interestingly, transition from embryonal carcinoma to differentiated phenotype and vice versa was observed with clonally derived hybrid cells with passage. In these transitions, a coordinate control of the phenotypic markers of the differentiated state was observed.

Howe and Oshima (1982), examining the phenotype of somatic cell hybrids between embryonal carcinoma cells and parietal endodermal cells, never observed the retention of embryonal carcinoma characteristics (tumorigenicity, stage-specific antigenic marker, and high alkaline phosphatase activity) in hybrids. No hybrid clones, of 1358 examined, retained the embryonal carcinoma morphology. The coordinate expression of endodermal markers (a cytoskeletal marker described above, and basement membrane components described below) with the extinction of embryonal carcinoma functions suggested to the authors that the parietal endodermal cells contained diffusible substances that extinguished embryonal carcinoma functions.

cAMP has been implicated as a modulator of cellular differentiation in a variety of cell types (Weiss and Strada, 1973; Friedman, 1976). In cells treated with retinoic acid, the addition of cAMP increases the production of plasminogen activator (Strickland and Mahdavi, 1978), and the appearance of cells with a neuronal phenotype (Kuff and Fewell, 1980)—having elongated processes and elevated acetylcholinesterase activity. It has been suggested that cAMP serves only to promote the differentiation into endoderm (Strickland *et al.*, 1980); however, cAMP by itself has no effect (Strickland and Mahdavi, 1978). Therefore, retinoic acid must induce changes in the response to cAMP. cAMP-dependent kinases, which mediate response to cAMP, exist in two forms designated types I and II. These two forms have different regulatory subunits R_I and R_{II}. The cAMP dependent kinases also exist either as soluble or as membrane-bound components (cAMP-dependent kinases are reviewed in Strickland *et al.*, 1980; Krebs, 1972; Rubin and Rosen, 1975; Nimmo and Cohen, 1977; Bechtel *et al.*, 1977). Plet *et al.* (1982) have demonstrated that one consequence of exposure to retinoic acid is a marked increase in the ratio of type II cAMP-dependent protein kinase over type I on the plasma membrane. The ratio of type I to type II in the cytosol is 2.5 and remains unaffected by retinoic acid treatment, while the level of type I on the membrane increases transiently (24 hr) 1.6-fold and then declines; the level of type II on the membrane goes up 5-fold and remains elevated for at least 5 days. The ratio of type I to type II thus drops from 7.5 to 1.5. This transient alteration may be interesting in the light of the reversible and irreversible phases of retinoic acid induction described above.

5.3. Programmed Assembly of Basement Membrane Components

Serological examination of embryos at different stages has identified the pattern of expression of the different extracellular matrix components. There is still some controversy surrounding the expression of fibronectin and the cells that actually assemble it. Laminin is found very early, at the morula stage, and this precedes by nearly 2 days the first expression of type IV collagen. This is curious as these two components are almost always found to codistribute in basement membranes. The appearance of interstitial collagens does not occur until the appearance of mesoderm at the primitive streak stage. A major deposition of basement membrane material occurs in a region defined as Reichert's membrane, which underlies the yolk sac. A description of the nature of the components that comprise the extracellular matrix and their appearance in the early basement membranes is presented below.

5.3.1. Expression of Collagens

Collagens are a family of molecules that have common "collagenous" structural features such as a triple-helical nature. However, they also have individual structural features that provide a spectrum of molecular properties. This diversity of molecular properties has been exploited in nature. For example, collagens are major components of tendon, which requires tensile strength, and of articular cartilage, which requires resistance to compression. There are two major classes of collagens, interstitial and basement membrane. Types I, II, and III are regarded as interstitial, while types IV and V are basement membrane associated. These distinctions, however, are generalizations. The interstitial collagens are fibrillar in nature, while the basement membrane collagens form networks with a less obviously regular organization.

The parietal yolk sac carcinoma PYS-2 has been used to study the assembly and shape of type IV collagen (Oberbaumer *et al.*, 1982). Both laminin and type IV collagen were found secreted into the medium and were selectively displaced by a salt gradient during affinity chromatography on heparin–agarose. The various aggregated forms of type IV collagen were examined, including dimers through tetramers joined through the 7 S domain. These forms were visualized by rotary shadowing and electron microscopy. These interactions through the 7 S domain are incorporated into the current model of type IV collagen organization (Fig. 10).

The expression of collagen during early mammalian embryogenesis has been examined by a number of groups (Adamson and Ayers, 1979; Sherman *et al.*, 1980) who have used immunofluorescence to examine the expression of collagens by embryos cultured in defined medium. Under these conditions, the collagen expression reflects endogenously synthesized materials. Antibodies to type I and II collagens did not react with the preimplantation embryo, whereas antibodies to type III collagen and procollagen reacted from the four-cell stage onwards and antibodies to type IV collagen reacted from the two-cell embryos onwards. Both layers of the inner cell mass react with type III collagen and procollagen antibodies, and the outer endodermal layer layer reacts very strongly with type IV antibodies. The authors found no obvious relationship with the

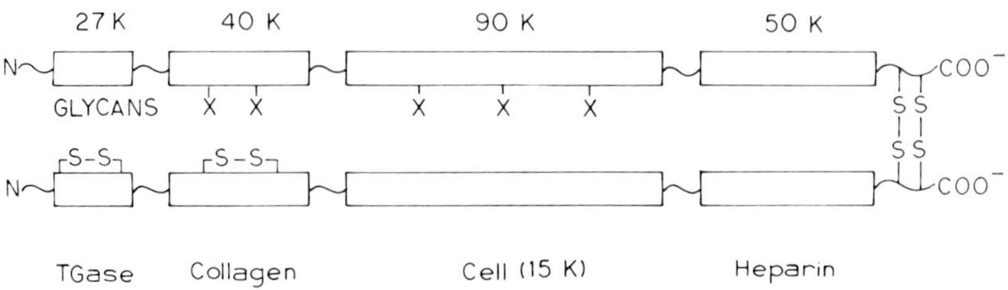

Figure 11. The many binding sites present on fibronectin. Fibronectin is a complex dimeric protein that interacts with itself to form fibrils with proteoglycans such as heparin, with interstitial collagens in the extracellular matrix, and an as yet unidentified receptor on the cell subsurface.

the cell surface and cross-link other fibrous components such as collagen and proteoglycans. These interactions explain why changes in the level of fibronectin have such a dramatic influence on the organization of the extracellular matrix.

During the development of the mouse embryo, fibronectin, as detected by indirect immunofluorescence (Zetter and Martin, 1978), has not been observed until the blastocyst stage. Two-cell, four-cell, eight-cell embryos, and the morula were found to be negative, as was the trophectoderm from either early or late blastocysts. The inner cell mass prepared from early blastocysts by immunosurgery showed a few local patches of fluorescence, while the intensity of the inner cell mass from the late blastocyst was significantly stronger. A similar pattern of expression was also observed by Wartiovaara *et al.* (1978) who were unable to detect fibronectin on the zygote, cleaving or compacted embryo. However, they first detected fibronectin on the implanting embryo coincident with the formation of the primitive endoderm around the inner cell mass. The deposition of fibronectin, which occurs between the primitive endoderm and the ectoderm core, may require the endodermal cells and may not occur on the surface of the pluripotent stem cells.

The synthesis of fibronectin by the pluripotent stem cells of the inner cell mass has not yet been established by metabolic labeling experiments, and so the source of the fibronectin deposited on the inner cell mass is not clear. Embryonal carcinoma cells in culture deposit fibronectin on their surfaces, although in sparse patches; again, the source of this fibronectin has not always been established. Few metabolic labeling studies have demonstrated the synthesis of fibronectin by embryonal carcinoma cells. Visceral endodermal cells derived from embryonal carcinomas have been demonstrated to synthesize and deposit fibronectin (Wolfe *et al.*,

1979). While some parietal endodermal cell lines have been shown to deposit fibronectin (Wartiovaara *et al.*, 1978, 1980; Zetter and Martin, 1978), other parietal endodermal cell lines have been shown not to (Wolfe *et al.*, 1979; Wartiovaara *et al.*, 1980).

At later stages of endoderm formation, fibronectin is readily detectable in the extracellular matrix that forms under both visceral endoderm, which surrounds the egg cylinder, and parietal endoderm, which lines the yolk sac (Wartiovaara *et al.*, 1980). Fibronectin is particularly obvious in the Reichert's membrane that separates the parietal endoderm from the trophoblast. Mesoderm, the third primary germ layer, forms from the primitive streak, and is separated from the ectoderm, from which it forms by a fibronectin-rich layer. In contrast to the ectodermal cells, mesodermal cells stain positively for fibronectin. The mesodermal cells remain separated from both the endoderm and the ectoderm by an extracellular matrix that stains positive for fibronectin (Wartiovaara and Vaheri, 1981). In addition to Reichert's membrane, the allantois, the amnion, and the chorion all stain for fibronectin (Wartiovaara and Vaheri, 1981).

5.3.3. Expression of Laminin

Laminin yields two types of subunit on reduction with apparent M_r of 200–220 and 400–440K (Timpl *et al.*, 1979). Electron microscopic examination of rotary-shadowed specimens of laminin reveal a cruxiform structure with one long branch and three shorter branches (Engel *et al.*, 1981) (Fig. 12). The longer branch is identified as the 440K subunit and the three shorter branches are identified as the 220K subunits. Laminin is more resistant to proteolytic cleavage than fibronectin and this is reflected in its less flexible structure. Seven globular regions have been identified, and recently the biological functions associated with the intact molecule have been assigned to particular regions. This molecule, like fibronectin, has multiple binding sites; these include binding sites for the cell surface, glycosaminoglycans (heparin), and collagen. In the case of laminin, the binding to collagen is specific for type IV (basement membrane specific) collagen (Sakashita and Ruoslahti, 1980; Terranova *et al.*, 1980). A major thrombin-derived fragment of laminin has an M_r of 600K and being composed of three 200K subunits is designated $\alpha 3$ (Rao *et al.*, 1982a,b). This component will bind to cells and will support type IV collagen-dependent attachment (Terranova *et al.*, 1982). Rotary-shadowed specimens of this component under electron microscopy have a "T"-shaped appearance (with three equal arms) with the appropriate globular regions of the intact molecule (Fig. 12). Fragments of laminin prepared either by cathepsin G treatment or by pepsin treatment have M_r of 280 and 350K, respectively, and have lost the ability to bind to type IV col-

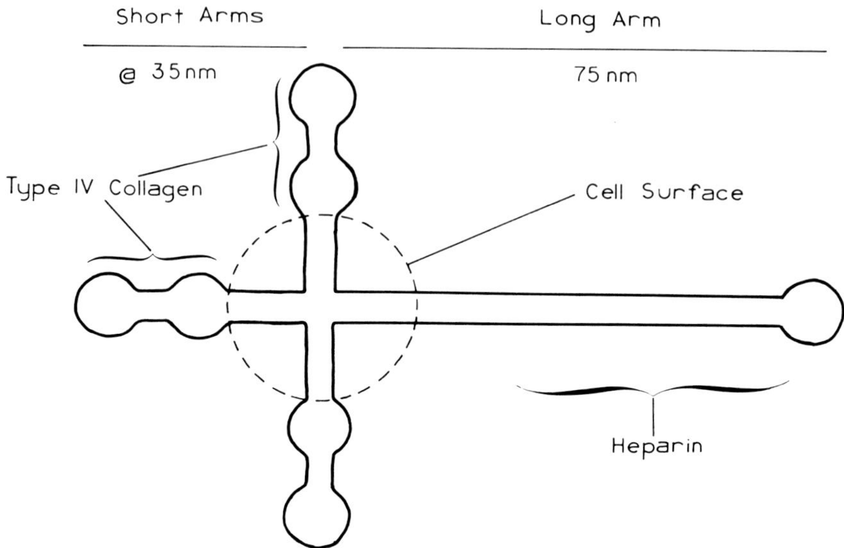

Figure 12. Schematic illustration of the binding domains of laminin. The long (75 nm) arm can be dissociated from the intact laminin molecule, either by mild reducing conditions or by mild proteolysis. Loss of this arm is correlated with loss of heparin binding. The intact laminin molecule contains three shorter arms. Each of these arms is apparently composed of the 220K subunit. This structure is referred to as $\alpha 3$ and will replace intact laminin molecules in laminin-dependent cell attachment to type IV collagen. Digestion of $\alpha 3$ with pepsin removes the outer globular regions and is correlated with loss of type IV collagen binding. However, the abbreviated molecule can inhibit laminin-dependent attachment and apparently retains the cell-binding domain (Terranova *et al.*, 1982).

lagen and have also lost the globular regions so obvious on the 200K subunits. These fragments retain the ability to bind to the cell surface, and competitively inhibit type IV-dependent attachment mediated by either intact laminin or $\alpha 3$. These experiments strongly suggest that the globular regions are the collagen type IV-binding domains and the disulfide-linked trimer region is the cell-binding domain. Recently, a 67K cell surface component was isolated from BL6 murine melanoma cells and identified as the cell receptor to laminin (Rao *et al.*, 1983). As the 400K subunit is lost on thrombin digestion with the ability to bind heparin, the heparin-binding domain is suggested to reside on this unit. Because of its specific association with basement membranes, laminin is a major attachment factor involved in the adhesion of endothelial/epithelial cells to basement membranes (Terranova *et al.*, 1980), and may play an important role in metastasis (Terranova *et al.*, 1982, 1983; Liotta and Hart, 1982).

   A smaller protein of 180K has been identified (Carlin *et al.*, 1981), termed "entactin." This sulfated protein subsequently was found to be very tightly, but not covalently, associated with laminin (Hogan *et al.*, 1982b). The distribution of this component in a variety of subepithelial and vascular basement membranes has been recorded, and it is found at the cell membrane abutting basement membranes and not in the basement membranes proper (Bender *et al.*, 1981). This latter observation suggests a direct role for entactin in attachment.

Laminin is first detectable intracellularly at the 16-cell compacted morula and appears at intercellular contours (Leivo *et al.*, 1980). It subsequently is laid down extensively in Reichert's membrane as described in detail below.

5.3.4. Reichert's Membrane

Reichert's membrane is a basement membrane that forms between the parietal endoderm and the trophectoderm and separates the yolk sac from the trophoblast and the maternal blood supply. The yolk sac is the outermost of the three extraembryonic membranes surrounding the mouse embryo and for a time is the only continuous barrier between the maternal blood sinuses and the embryonic fluid of the yolk sac. The formation of this basement membrane has been examined in yolk sac carcinomas (Pierce, 1965, 1966; Pierce *et al.*, 1963, 1964; Chung *et al.*, 1979; Jetten *et al.*, 1982; Hogan *et al.*, 1980; Leivo *et al.*, 1982), and this basement membrane has been studied as a source of noncollagenous basement membrane glycoproteins. While the functions of this basement membrane are not entirely clear, it probably has an important influence on maternal–fetal interactions. It is begun on the sixth day and increases rapidly in thickness and surface area before being broken down toward the end of gestation (Clark *et al.*, 1975a; Jensh *et al.*, 1977; Dickson, 1979).

The major components of Reichert's membrane are the basement membrane components type IV procollagen and laminin (Timpl *et al.*, 1979; Tryggvason *et al.*, 1980; Hogan *et al.*, 1980; Smith and Strickland, 1981), a heparan sulfate proteoglycan (Hogan *et al.*, 1982a). and two sulfated glycoproteins of 180 and 150K (Hogan *et al.*, 1982a). The latter component is identified as entactin (Carlin *et al.*, 1981), and as described above is probably a noncovalent subunit of laminin, for it is strongly associated with laminin even at high salt (Hogan *et al.*, 1982b).

Reichert's membrane from $13\frac{1}{2}$-day embryos labels uniformly for laminin-entactin (Semoff *et al.*, 1982), in contrast to the nonuniform distribution seen in other basement membranes (Madri *et al.*, 1980; Foidart *et al.*, 1980; Bender *et al.*, 1981). Fibronectin is asymmetrically distributed with its highest concentrations in the outermost layers adjacent to the trophoblast and negligible amounts being present under the parietal endodermal cells (Semoff *et al.*, 1982). Metabolic studies have shown that the components of Reichert's membrane are largely, if not exclusively, synthesized by the parietal endodermal cells (Clark *et al.*, 1975b; Minor *et al.*, 1976a,b). This is consistent with the apparent inability of parietal endodermal cells to synthesize fibronectin (Wolfe *et al.*, 1979; Wartiovaara *et al.*, 1980). The low level of fibronectin on the surface of the trophoblast layer may be the result of a transient expression of fibronectin as the endodermal cells migrate out over the trophectoderm (Enders *et al.*, 1978). While fibronectin is clearly not a major structural component of Reichert's membrane, the more fibrillar morphology of the outer trophoblast layer could be the result of this localized expression of fibronectin. Hogan (in Semoff *et al.*, 1982) reports that trophoblast cells from the $6\frac{1}{2}$-day embryo synthesize fibronectin.

5.4. Assembly of Basement Membrane Components by Endodermal Cultures

The importance of basement membranes to normal functions of kidney, vascular system, eye, gut, and other organs, coupled with the alterations in basement membrane morphology that are associated with various pathological conditions (Kefalides, 1975), has emphasized the need to elucidate their structure, biosynthesis, and molecular basis of action. In spite of the difficulties inherent in isolating pure preparations of basement membrane components, substantial progress has been made in characterizing the collagens found in basement membranes from a variety of sources (Kefalides, 1973). Progress with the noncollagenous proteins has been much slower and has awaited the establishment of simplified systems. The first of these was a mouse parietal yolk sac carcinoma system developed by Pierce *et al.* (1962). This system synthesized large quantities of the extracellular membrane, Reichert's membrane. The immunological and biochemical relatedness of this membrane to normal epithelial basement membrane was carefully established (Midgley and Pierce, 1963; Pierce *et al.*, 1964). Subsequently, a soluble protein obtained from these cultures was shown to be immunologically related to components in the kidney glomerular basement membrane (Johnson and Starcher, 1972). More recently, noncollagenous extracellular matrix components were obtained from a mouse embryonal carcinoma-derived cell line (Chung *et al.*, 1977b). These glycoprotein components were designated GP1 and GP2. The characterization of these components was begun and the smaller component was demonstrated not to be fibronectin (Chung *et al.*, 1979). GP1 and GP2 are now referred to as the large and small subunits of laminin. The embryonal carcinoma-derived endodermal cell line, with a cell line derived from the EHS basement membrane tumor, are the major sources of laminin for physical, biochemical, and biological studies, and the major systems for studying the biosynthesis of laminin.

The ability of embryonal carcinoma cells to synthesize collagen has been examined by Adamson *et al.* (1979). This study used metabolic labeling with leucine and proline followed by either immunoprecipitation or specific fragmentation (collagenase, pepsin, cyanogen bromide degradation) and analysis by SDS–PAGE. The products have been compared to authentic standards and to the components assembled by normal parietal and visceral endoderms. The conclusions that this careful study came to are surprising, namely, that a variety of embryonal carcinoma cells assemble type IV collagen. On differentiation, different embryonal carcinomas yielded different patterns of collagen synthesis. For example, the multipotent line OC 15 synthesized type IV collagen while its differentiated endodermal products were unable to; instead, they synthesized type I collagen. The line PC 13 is unable to differentiate *in vitro*; however, in the presence of retinoic acid, it forms endodermal derivatives that continue to express type IV collagen while expressing type I collagen. The nullipotent line F9 also synthesized type IV collagen, but, in addition, synthesized a large amount of a type I-like collagen. There are no simple conclusions, except that prior to examination it is not clear what collagen types a particular cell line will be expressing.

The expression of high-molecular-weight extracellular proteins by endodermal cells has been examined by Hogan (1980). Using an antiserum raised against teratocarcinoma embryoid bodies, Hogan identified two high-molecular-weight components, 350 and 220K. These proteins designated PYS A and PYS B are now referred to as the large and small subunits of laminin. PYS A and PYS B were found to be expressed by normal parietal endodermal cells from the $10\frac{1}{2}$-day embryo and not by normal visceral endodermal cells from the embryo at the same stage. The work from this laboratory is exceptional in that it compares results obtained with model systems in culture with their counterparts in the normal embryo. This is not a criticism of the model systems. Clearly they are excellent starting points and have provided information that would have come with great difficulty if at all from normal embryo, and they show greater and greater promise as different groups apply them to different problems. However, the object of this research is to understand the normal embryo, and the necessary step that corroborates the *in vitro* and *in vivo* circumstances is not often accomplished.

The assembly and disposition of the basement membrane components laminin and type IV collagen have been examined in a PYS carcinoma cell line (Leivo *et al.*, 1982), using metabolic labeling and immunofluorescence. The materials were assembled into a loose network of fibers with attached dense grains. The cells did not synthesize detectable amounts of fibronectin; however, fibronectin could be adsorbed by the matrix from the culture medium. The structure of the matrix was found to be sensitive to bacterial collagenase. The collagen was deposited without apparent proteolytic cleavage and therefore like the deposition in other systems [lens capsule (Heathcote *et al.*, 1978), parietal yolk sac (Clark *et al.*, 1975a,b; Tryggvason *et al.*, 1980), EHS sarcoma (Tryggvason *et al.*, 1980)] appears to be in the procollagen form. Both pro-α1(IV) and pro-α2(IV) were identified.

The incorporation of laminin into Reichert's membrane in cultures of parietal endoderm still attached to Reichert's membrane has been examined (Hogan *et al.*, 1980). The incorporation of newly assembled laminin into Reichert's membrane was found to be independent of *de novo* type IV procollagen synthesis using an inhibitor (α,α-bipyridyl) of proline and lysine hydroxylation.

The synthesis and secretion of GP2, the small subunit of laminin, and entactin by endodermal derivatives of embryonal carcinoma cells have been examined (Bender *et al.*, 1982). The study using immunological identification and light and electron microscopy described the distribution and fate of these molecules. The proteins were identified in membrane-bound vesicles in the cisternae of the rough endoplasmic reticulum. These vesicles appeared to fuse with the plasma membrane and to transfer their contents to discrete regions of the cell surface. These regions coincided with regions of cell contact, and the authors very reasonably speculate that the cell–cell contacts, in fact, stimulate this localized release. A very unusual finding was the absence of Golgi staining for these proteins. GP2 is known to contain fucose and sialic acid (Chung *et al.*, 1979), and therefore would be expected to be exposed to the glycosyltransferases that reside in the Golgi. This absence of Golgi staining has been observed elsewhere with fibronectin (Hedman *et al.*, 1980), and with the basement membrane produced by parietal yolk sac cells (Martinez-Hernandez *et al.*, 1974). This latter aspect has been reexamined very carefully by Pierce *et al.* (1982a). Using pulse-chase autoradiography, under conditions that clearly demonstrate transport of secretory components through the Golgi in pancreatic cells, it was demonstrated that basement membrane components did not transit through the Golgi in parietal yolk sac cells. We (Ivatt and Oshima) have examined the synthesis and expression of laminin in PFHR-9 endodermal cells. Using metabolic labeling pulse-chase experiments and immunological techniques, we have followed the maturation of the carbohydrate portion of laminin. Our conclusions are that, unlike the bulk of secreted glycoproteins, laminin remains unprocessed for long periods of time and is retained in dense storage vesicles. The absence of Golgi staining observed above could reflect the fact that very little of the laminin is in the Golgi at any one point in time and that the bulk is either in storage vesicles, with immature carbohydrate chains, or in secretory vesicles, with mature carbohydrate chains. Alternatively, it is formally possible that Golgi-like compartments fuse with the storage vesicles prior to secretion and that processing occurs for laminin without transport through the Golgi. The state of maturation of the carbohydrate on laminin in these vesicles is not known, nor is their composition in terms of the enzymes responsible for the maturation of carbohydrates. These are not esoteric points as we have found the intracellular laminin to carry unprocessed carbohydrate while all the glycans carried by extracellular laminin have been extensively processed. Therefore, the processing of the carbohydrate on this glycoprotein is closely correlated temporally with secretion.

The assembly of sulfated glycoproteins by established parietal endodermal cells (PYS and PFHR-9) in culture has been specifically examined by Hogan *et al.* (1982a,b). They described the assembly of two sulfated glycoproteins with M_r of 180 and 150K (entactin), and showed that the large and small (440 and 220K) subunits of laminin are also sulfated, as is fibronectin. These authors suggest that sulfation may be a rather general phenomenon associated with the noncollagenous extracellular matrix proteins.

Heparan sulfate synthesis has been observed in parietal endodermal cells (Hogan *et al.*, 1982a,b). This component has an $M_r > 500K$. Because of the likely importance of the heparan sulfate to the mechanical properties of basement membranes (Kanwar and Farquhar, 1979; Hassel *et al.*, 1979), the availability of a culture system to examine the mechanism and regulation of heparan sulfate synthesis will prove of great value. Leivo *et al.* (1982) have observed the codeposition of a basement membrane heparan sulfate molecule BM-1 (Hassel *et al.*, 1979) with laminin and collagen.

Bonaldo *et al.* (1982) have demonstrated the expression of type IV collagen and heparan sulfate by parietal yolk sac carcinoma cells in culture as a solid tumor, and as an ascites tumor. The unexpected finding from these studies was that if the PYSC was cultured as a solid tumor subcutaneously, the collagen type that was expressed was type I and the glycosaminoglycan that predominated was chondroitin sulfate B. Chondroitin sulfate synthesis is negligible in the ascites form and represents 70% of the glycosaminoglycan in the solid form. In contrast, Oldberg *et al.* (1981) demonstrated that rat yolk sac tumor in ascites culture synthesized significant amounts of a chondroitin sulfate proteoglycan. The glycosaminoglycan chains were completely digested by chondroitinase ABC and substantially (90%) digested by chondroitinase AC.

6. CONCLUSIONS

This chapter has focused on the cell interactions that occur during early mammalian embryogenesis and has examined the current understanding of the molecular events at the cell surface that are involved in regulating these interactions. For obvious reasons, this research has been restricted to rodent embryos with the mouse being, by far, the most important species.

The utility of the embryonal carcinoma system for examining such molecular details has been emphasized. Changes in cell surface glycoproteins have been identified during sperm maturation and during early embryonic development. In some cases it has been possible to associate a functional consequence with these changes. For example, during sperm maturation there is a shedding of complex carbohydrates known as coating or decapacitation factors that may mimic the sperm receptors on the zona pellucida. Particular attention has been paid to the large complex glycans that are lost in a programmed way during postimplantation development and that may turn out to carry the informational determinants involved in cellular recognition. The molecular events involved in the assembly of these complex carbohydrates have been examined and mechanisms that regulate the expression of these components and that operate either at the molecular level or at the cellular level have been discussed.

The restriction of these large carbohydrates to the early embryo and to the reticuloendothelial system in the adult suggests a role for them in cellular interactions that are transient in nature. In the reticuloendothelial system, cells make frequent, readily reversible interactions with other cells and the embryo is characterized by continuous change and reorganization. The most striking correlation is the programmed loss of these carbohydrates during histodifferentiation. This loss correlates with the formation of fetal tissue occurring at different times in different tissues. For example, when liver and thyroid form distinguishable tissues (at 16 and 12 weeks, respectively), the loss of carbohydrates closely correlates with these temporally well-separated events. These large carbohydrates are often highly branched and therefore have a very high information content. They are very hydrophilic and may provide a cell surface-associated mucin that prevents "overtight" interactions between adjacent cells. Tissue formation may be the loss of the cell surface-associated mucin with its potential for readily reversible interactions and its replacement by the interstitial components that characterize the more permanent interactions found within tissues. These interstitial components include the collagens, fibronectins, laminins, and proteoglycans, which appear to have a relatively broad distribution. The specificity with these molecules is achieved by the timing and location of their deposition rather than to intrinsic specificities.

The expression of these components has been correlated with cellular events during early mouse development. However, we have not yet achieved an understanding of the precise roles of these molecules. There are some intriguing possibilities suggested by the transient expression of fibronectin during the initial migration of endodermal cells that will eventually form their parietal yolk sac. The availability of established tissue culture cells that retain the biochemical and biological characteristics of these various embryonic cells provides a unique opportunity to examine and test the roles of extracellular components. It also provides an opportunity to examine the biological significance of these components and to address the mechanisms that regulate their expressions. This is a very exciting phase as it marks the transition from molecular description to functional assignment.

REFERENCES

Adamson, E. D., and Ayers, S. E., 1979, The localization and synthesis of some collagen types in developing mouse embryos, *Cell* **16:**953–965.

Adamson, E. D., and Grover, A., 1983, The production and maintenance of a functioning epithelial layer from embryonal carcinoma cells, in: *Teratocarcinoma Stem Cells* (L. Silver, G. Martin, and S. Strickland, eds.), pp. 69–82, Cold Spring Harbor Laboratory, New York.

Adamson, E. D., Evans, M. J., and Magrane, G. G., 1977, Biochemical markers of the progress of differentiation in cloned teratocarcinoma cell lines, *Eur. J. Biochem.* **79:**607–615.

Adamson, E. D., Gaunt, S. J., and Graham, C. F., 1979, The differentiation of teratocarcinoma stem cells is marked by the types of collagen which are synthesized, *Cell* **17:**469–476.

Ajiro, K., Borun, T. W., and Solter, D., 1980, Quantitative changes in the expression of histone H1 and H2B subtypes and their relationship to the differentiation of mouse embryonal carcinoma cells, *Dev. Biol.* **86:**206–211.

Alexander, S. S., Colonna, G., and Edelhoch, H., 1979, The structure and stability of human cold insoluble globulin, *J. Biol. Chem.* **254:**1501–1505.

Ali, I. V., Mautner, V. M., Lanza, R. P., and Hynes, R. O., 1977, Restoration of normal morphology, adhesion and cytoskeleton in transformed cells by addition of a transformation-sensitive surface protein, *Cell* **11:**115–26.

Alitalo, K., Vaheri, A., Krieg, T., and Timpl, R., 1980, Biosynthesis of two subunits of type IV procollagen and of other basement-membrane proteins by a human tumor line, Eur. J. Biochem. **109:**247–258.

Andrews, P. W., Knowles, B. B., Cossu, G., and Solter, D., 1982, Teratocarcinoma and mouse embryo cell surface antigens: Characterization of the molecule(s) carrying the SSEA-1 antigenic determinant, *Teratocarcinoma and Embryonic Cell Interactions* (T. Muramatsu, G. Gachelin, A. A., Moscoma, and Y. Ikawa, eds.), pp. 103–119, Japan Scientific Societies Press, Tokyo.

Artzt, K., 1973, Surface antigens common to mouse cleavage embryos and primitive teratocarcinoma cells in culture, *Proc. Natl. Acad. Sci. USA* **70**:2988–2992.

Artzt, K., and Bennett, D., 1975, Analogies between embryonic (T/t) antigens and adult major histocompatibility (H-2) antigens, *Nature (London)* **226**:545–547.

Artzt, K., Dubois, P., Bennett, D., Condamine, H., Babinet, C., and Jacob, F., 1973, Surface antigens common to mouse cleavage embryos and primitive teratocarcinoma cells in culture, *Proc. Natl. Acad. Sci. USA* **70**:2988–2992.

Atienza-Samols, S. B., Pine, P. R., and Sherman, M. I., 1980, Effects of tunicamycin upon glycoprotein biosynthesis and development of early mouse embryo, *Dev. Biol.* **79**:19–32.

Austin, C. R., and Bishop, M. W. H., 1959, Differential fluorescence in living rat eggs treated with acridine orange, *Exp. Cell Res.* **17**:35–43.

Azin, M., and Surani, H., 1979, Glycoprotein synthesis and inhibition of glycosylation by tunicamycin in preimplantation mouse embryos: Compaction and trophoblast adhesion, *Cell* **18**:217–227.

Balakier, H., and Pedersen, R. A., 1982, Allocation of cells to inner cell mass and trophectoderm lineages in preimplantation mouse embryos, *Dev. Biol.* **90**:352–362.

Baranska, W., Koldovsky, P., and Koprowski, H., 1970, Antigenic study of unfertilized mouse eggs-cross reactivity with SV-40 induced antigens, *Proc. Natl. Acad. Sci. USA* **67**:193–199.

Bechtel, P. J., Beavo, J. A., and Krebs, E. G., 1977, Purification and characterization of catalytic subunit of skeletal muscle adenosine-3'-5'-monophosphate dependent protein kinase, *J. Biol. Chem.* **252**:2691–2697.

Bedford, J. M., 1963, Changes in the electrophoretic properties of rabbit spermatozoa during passage through the epididymis, *Nature (London)* **200**:1178–1180.

Bedford, J. M., 1965, Changes in the fine structure of the rabbit sperm head during passage through the epididymis, *J. Anat.* **99**:891–906.

Bedford, J. M., 1966, Development of the fertilizing ability of spermatozoa in the epididymis of the rabbit, *J. Exp. Zool.* **163**:319–329.

Bedford, J. M., and Cooper, G. W., 1978, Membrane fusion events in the fertilization of vertebrate eggs, in: *Membrane Fusion* (G. Poste and G. L. Nicolson, eds.), pp. 65–125, Elsevier/North-Holland, Amsterdam.

Bedford, J. M., Cooper, G. W., and Calvin, H. I., 1972, Post-meiotic changes in the nucleus and membranes of mammalian spermatozoa, in: *The Genetics of the Spermatozoa* (R. A. Beatty and S. Gluecksohn-Waelsch, eds.), pp. 69–89, Bogtrykereit Forum, Copenhagen.

Bender, B. L., Jaffe, R., Carlin, B., and Chung, A. E., 1981, Immunolocalization of entactin, a sulfated basement membrane component, in rodent tissues, and comparison with GP-2 (laminin), *Am. J. Pathol.* **103**:419–426.

Bender, B. L., Carlin, B., Jaffe, R., Temple, T., and Chung, A. E., 1982, Production and distribution of entactin and GP-2 in M1526-B3 cells, *Exp. Cell Res.* **137**:415–425.

Bennett, D., 1975, The T-locus in the mouse, *Cell* **6**:441–454.

Bennett, D., 1978, Genetically programmed abnormalities of cell interactions, *Birth Defects Orig. Artic. Ser.* **14**:285–303.

Bernstine, E. G., Hooper, M. L., Grandchamp, S., and Ephrussi, B., 1973, Alkaline phosphatase activity in mouse teratoma, *Proc. Natl. Acad. Sci. USA* **70**:3899–3903.

Beyer, T. A., Rearick, J. I., Paulson, J. C., Prieels, J. P., Sadler, J. E., and Hill, R. L., 1979, Biosynthesis of mammalian glycoproteins and glycosylation pathways in the synthesis of the nonreducing terminal sequences, *J. Biol. Chem.* **254**:12531–12541.

Beyer, T. A., Sadler, J. E., Rearick, J. I., Paulson, J. C., and Hill, R. L., 1981, Glycosyltransferases and their use in assessing oligosaccharide structure and structure–function relationships, *Adv. Enzymol.* **52**:23–176.

Bhushana Rao, K. S. P., and Masson, P. L., 1977, Study of the primary structures of the peptide core of bovine estrus cervical mucin, *J. Biol. Chem.* **252:**7788–7795.

Birdwell, C. R., Gospodarowicz, D., and Nicolson, G. L., 1978, Identification, localization and role of fibronectin in cultured bovine endothelial cells, *Proc. Natl. Acad. Sci. USA* **75:**3273–3277.

Blasco, L., 1977, Clinical approach to evaluation of sperm cervical mucus interactions, *Fertil. Steril.* **28:**1133–1145.

Bleil, J. D., and Wassarman, P. M., 1980a, Mammalian sperm–egg interaction: Identification of a glycoprotein in mouse egg zonae pellucidae possessing receptor activity for sperm, *Cell* **20:**873–882.

Bleil, J. D., and Wassarman, P. M., 1980b, Structure and function of the zona pellucida: Identification and characterization of the proteins of the mouse oocyte's zona pellucida, *Dev. Biol.* **76:**185–202.

Bleil, J. D., and Wassarman, P. M., 1980c, Synthesis of zona pellucida proteins by denuded and follicle enclosed mouse oocytes during culture *in vitro*, *Proc. Natl. Acad. Sci. USA* **77:**1029–1033.

Bleil, J. D., Beall, C. F., and Wassarman, P. M., 1981, Mammalian sperm–egg interaction: Fertilization of mouse eggs triggers modification of the zona pellucida glycoprotein, ZP2, *Dev. Biol.* **86:**189–197.

Bonaldo, M. D. F., Cassaro-Strunz, C. M., and Machado Santelli, G., 1982, Sulfated glycosaminoglycans and collagen patterns in parietal yolk sac carcinoma (PYSC), *Cell Differ.* **11:**99–106.

Braden, A. H. W., Austin, C. R., and David, H. A., 1954, The reaction of the zona pellucida to sperm penetration, *Aust. J. Biol. Sci.* **7:**391–409.

Brinster, R. L., 1974, The effects of cells transferred into the mouse blastocyst on subsequent development, *J. Exp. Med.* **140:**1049–1056.

Brown, C. R., and Hartree, E. F., 1978, Studies on ram acrosin: Activation or proacrosin accompanying the isolation of acrosin from spermatozoa and purification of the enzyme by affinity chromatography, *Biochem. J.* **175:**227–238.

Brown, C. R., Andani, Z., and Hartree, E. F., 1975, Studies on ram acrosin: Isolation from spermatozoa, activation by cations and organic solvents, and influence of cations on its reaction with inhibitors, *Biochem. J.* **149:**133–146.

Brulet, P., Babinet, C., Kemler, R., and Jacob, F., 1980, Monoclonal antibodies against trophectoderm-specific markers during mouse blastocyst formation, *Proc. Natl. Acad. Sci. USA* **77:**413–4117.

Buc-Caron, M. H., Gachelin, G., Hofnung, M., and Jacob, F., 1974, Presence of a mouse embryonic antigen on human spermatozoa, *Proc. Natl. Acad. Sci. USA* **71:**1730–1734.

Burke, J. M., Balian, G., Ross, R., and Bornstein, P., 1978, Synthesis of type I and III procollagen and collagen by monkey aortic smooth muscle cells *in vitro*, *Biochemistry* **16:**3243–3249.

Cascio, S. M., and Wassarman, P. M., 1982, Program of early development in the mammal: Post-transcriptional control of a class of proteins synthesized by mouse oocytes and early embryos, *Dev. Biol.* **89:**397–408.

Carlin, B., Jaffe, R., Bender, B., and Chung, A. E., 1981, Entactin, a novel basal lamina-associated sulfated glycoprotein, *J. Biol. Chem.* **256:**5209–5214.

Carlson, D. M., 1968, Structure and immunochemical properties of oligosaccharides isolated from pig submaxillary mucins, *J. Biol. Chem.* **243:**616–626.

Chapman, A., and Kornfeld, R., 1979a, Structure of the high mannose oligosaccharides of a human IgM myeloma protein, *J. Biol. Chem.* **254:**816–823.

Chapman, A., and Kornfeld, R., 1979b, Structure of the high mannose oligosaccharides of a human IgM myeloma protein, *J. Biol. Chem.* **254:**824–828.

Chapman, A., Trowbridge, I. S., Hyman, R., and Kornfeld, S., 1979a, Structure of the lipid-linked oligosaccharides in class E Thy-1 mutant lymphomas, *Cell* **17**:509–515.

Chapman, A., Li, E., and Kornfeld, S., 1979b, The biosynthesis of the major lipid-linked oligosaccharide of Chinese hamster ovary cells occurs by the ordered addition of mannose residues, *J. Biol. Chem.* **254**:10243–10249.

Chapman, A., Fujimoto, K., and Kornfeld, S., 1980, The primary glycosylation defect in class E Thy-1-negative mutant mouse lymphoma cells is an inability to synthesize dolichol-phosphate-mannose, *J. Biol. Chem.* **255**:4441–4446.

Chen, L. B., Gallimore, P. H., and McDougall, J. K., 1976, Correlation between tumor induction and the large external transformation-sensitive protein on the cell surface, *Proc. Natl. Acad. Sci. USA* **73**:3570–3574.

Chung, A. E., Estes, L. E., Shinozuka, H., Braginski, J., Lorz, C., and Chung, C. A., 1977a, Morphological and biochemical observations on cells derived from the *in vitro* differentiation of the embryonal carcinoma cell line PCC4-F, *Cancer Res.* **37**:2072–2081.

Chung, A. E., Freeman, I. L., and Braginski, J. E., 1977b, A novel extracellular membrane elaborated by a mouse embryonal carcinoma-derived cell line, *Biochem. Biophys. Res. Commun.* **79**:859–868.

Chung, A. E., Jaffe, R., Freeman, I. L., Vergnes, J. P., Braginski, J. E., and Carlin, B., 1979, Properties of a basement membrane-related glycoprotein synthesized in culture by a mouse embryonal carcinoma-derived cell line, *Cell* **16**:277–287.

Clark, C. C., Tomichek, E. A., Koszalka, T. R., Minor, R. R., and Kefalides, N. A., 1975a, The embryonic rat parietal yolk sac: The role of the parietal endoderm in the biosynthesis of basement membrane collagen and glycoprotein *in vitro*, *J. Biol. Chem.* **250**:5259–5267.

Clark, C. C., Minor, R., Koszalka, T. R., Brent, R. L., and Kefalides, N. A., 1975b, The embryonic rat parietal yolk sac: Changes in the morphology and composition of its basement membrane during development, *Dev. Biol.* **46**:243–261.

Clement, A. C., 1962, Development of ilyanassa following removal of D macromere at successive cleavage stages, *J. Exp. Zool.* **149**:193–215.

Copp, A. J., 1978, Interaction between the inner cell mass and trophectoderm of the mouse blastocyst. 1. Study of proliferation, *J. Embryol. Exp. Morphol.* **48**:109–125.

Copp, A. J., 1979, Interaction between the inner cell mass and trophectoderm of the mouse blastocyst. 2. Fate of the polar trophectoderm, *J. Embryol. Exp. Morphol.* **51**:109–120.

Cullen, B., Emigholz, K., and Monahan, J., 1980, The transient appearance of specific proteins in one-cell mouse embryos, *Dev. Biol.* **76**:215–221.

Cummings, R. D., Trowbridge, I. S., and Kornfeld, S., 1982, A mouse lymphoma cell line resistant to the leukoagglutinating lectin from *Phaseolus vulgaris* is deficient in UDP-GlcNAc:α-D-mannoside β1,6N-acetylglucosaminyl transferase, *J. Biol. Chem.* **257**:13421–13427.

Curtis, A. S. G., 1960, Cortical grafting in *Xenopus laevis*, *J. Embryol. Exp. Morphol.* **8**:167–173.

Dacie, J. V., 1962, *The Haemolytic Anemias, Congenital and Acquired, Part II. The Autoimmune Haemolytic Anemias*, pp. 366–377, Churchill, London.

Damsky, C. H., Knudsen, K. A., Dorio, R. J., and Buck, C. A., 1981, Manipulation of cell–cell and cell–substratum interactions in mouse mammary tumor epithelial cells using broad spectrum antisera, *J. Cell Biol.* **80**:173–184.

Darmon, M., and Serrero, G., 1983, Isolation of two different fibroblastic cell types from the embryonal carcinoma cell line 1003, in: *Teratocarcinoma Stem Cells* (L. Silver, G. Martin, and S. Strickland, eds.), pp. 109–120, Cold Spring Harbor Laboratory, New York.

Davidson, E. H., 1976, *Gene Activity in Early Development*, Academic Press, New York.

De Felici, M., and Siracusa, G., 1980, Changes in concanavalin A-mediated agglutinability of mouse oocytes during meiosis, *Dev. Biol.* **76:**428–434.

Dewey, M. J., Gearhart, J. D., and Mintz, B., 1977, Cell surface antigens of totipotent mouse teratocarcinoma cells grown *in vivo*—Their relation to embryo, adult and tumor antigens, *Dev. Biol.* **55:**359–374.

Dewey, M. J., Filler, R., and Mintz, B., 1978, Protein pattern of developmentally totipotent mouse teratocarcinoma cells and normal early embryo cells, *Dev. Biol.* **65:**171–182.

Dickson, A. D., 1979, Disappearance of the decidua capsularis and Reicherts membrane in the mouse, *J. Anat.* **129:**571–577.

Ducibella, T., 1977, Surface changes of the developing trophoblast cell, in: *Development in Mammals*, Vol. 1 (M. H. Johnson, ed.), pp. 5–30, North-Holland, Amsterdam.

Ducibella, T., 1980, Divalent antibodies to mouse embryonal carcinoma cells inhibit compaction in the mouse embryo, *Dev. Biol.* **79:**356–366.

Ducibella, T., Albertini, D. F., Anderson, E., and Biggers, J. D., 1975, Preimplantation mammalian embryo: Characterization of intercellular functions and their appearances during development, *Dev. Biol.* **45:**231–250.

Dziadek, M., 1978, Modulation of alphafetoprotein synthesis in the early postimplantation mouse embryo, *J. Embryol. Exp. Morphol.* **43:**135–146.

Dziadek, M., 1979, Cell differentiation in isolated inner cell masses of mouse blastocysts in vitro: Onset of specific gene expression, *J. Embryol. Exp. Morphol.* **43:**367–379.

Dziadek, M., and Adamson, E. D., 1978, Localization and synthesis of alphafetoprotein in postimplantation mouse embryos, *J. Embryol. Exp. Morphol.* **43:**289–313.

Edwards, R. G., Ferguson, L. C., and Combs, R. R. A., 1964, Blood group antigens on human spermatozoa, *J. Reprod. Fertil.* **7:**153–161.

Elbein, A. D., 1979, The role of lipid-linked saccharides in the biosynthesis of complex carbohydrates, *Annu. Rev. Plant Physiol.* **30:**239–272.

Enders, A. C., Given, R. L., and Schlafke, S., 1978, Differentiation and migration of endoderm in the rat and mouse at implantation, *Anat. Rec.* **190:**65–78.

Engel, J., Odermatt, E., Engel, A., Madri, J. A., Furthmayr, H., Rhode, H., and Timpl, R., 1981, Shapes, domain, organizations and flexibility of LM and FN, two multifunctional proteins of the extracellular matrix, *J. Mol. Biol.* **150:**97–120.

Epstein, C. J., and Smith, S. A., 1973, Amino acid uptake and protein synthesis in preimplantation mouse embryos, *Dev. Biol.* **33:**171–185.

Etchison, J. R., Summers, D. F., and Georgopoulos, G., 1981, Variations in the size and structure of radiolabeled glycopeptides from the glycoprotein of vesicular stomatitis virus grown in four mouse teratocarcinoma cell lines, *J. Biol. Chem.* **256:**3366–3369.

Fawcett, D. W., and Phillips, D. H., 1969, Observations on the release of spermatozoa and on changes in the head during passage through the epididymis, *J. Reprod. Fertil.* **6**(Suppl.):405–418.

Feizi, T., 1981, The blood group Ii system: A carbohydrate antigen system defined by naturally monoclonal or oligoclonal autoantibodies of man, *Immunol. Commun.* **10:**127–156.

Feizi, T., Kapadia, A., Gooi, H. C., and Evans, M. J., 1982, Human monoclonal autoantibodies detect changes in expression and polarization of the Ii antigens during cell differentiation in early mouse embryos and teratocarcinomas, in: *Teratocarcinoma and Embryonic Cell Interactions* (T. Muramatsu, G. Gachelin, A. A. Moscona, Y. Ikawa, eds.), pp. 201–215, Japan Scientific Societies Press, Tokyo.

Finne, J., Krusius, T., Rauvala, H., and Myllyla, G., 1978, Alkali-stable blood group A- and B-active poly(glycosyl)peptides from human erythrocyte membrane, *FEBS Lett.* **89:**111–114.

Fiser, P. S., and MacPherson, J. W., 1976, Development of embryonic structures from isolated mouse blastomeres, *Can. J. Anim. Sci.* **56:**33–36.

Flechon, J. E., 1973, Modifications ultrastructurales et cytochimiques des spermatozoides de lapin au cours du transit epididymaire, *INSERM Colloq.* **26:**115–140.

Foidart, J. M., Berman, J. J., Paglia, L., Rennard, S., Abe, S., Perantoni, A., and Martin, G. R., 1980, Synthesis of fibronectin, laminin, and several collagens by a liver derived epithelial cell line, *Lab. Invest.* **42:**525–532.

Fox, N., Damjanov, I., Knowles, B. B., and Solter, D., 1982, Teratocarcinoma antigen is secreted by epididymal cells and couple to maturing sperm, *Exp. Cell Res.* **137:**485–488.

Friedman, D., 1976, Role of cyclic nucleotides in cell growth and differentiation, *Physiol. Rev.* **56:**652–685.

Fukuda, M. N., and Matsumura, G., 1976, Endo-β-galactosidase of *E. freundii*: Purification and endocytic action on keratan sulfates, oligosaccharides, and blood group active glycoprotein, *J. Biol. Chem.* **251:**218–6223.

Fukuda, M., Fukuda, M. N., and Hakomori, S.-I., 1979a, Developmental change and genetic defect in the carbohydrate structure of Band 3 glycoprotein of human erythrocyte membrane, *J. Biol. Chem.* **254:**3700–3703.

Fukuda, M. N., Papermaster, D. S., and Hargrave, P. A., 1979b, Rhodopsin carbohydrate: Structure of small oligosaccharides attached at two sites near the NH_2 terminal, *J. Biol. Chem.* **254:**8201–8207.

Gachelin, G., Buc-Caron, M. H., Lis, H., and Sharon, N., 1976, Saccharides on teratocarcinoma cell plasma membranes: Their investigation with radioactively labelled lectins, *Biochim. Biophys. Acta* **436:**825–832.

Gachelin, G., Kemler, R., Kelley, F., and Jacob, F., 1977, A new cell surface antigen common to multipotential embryonal carcinoma cells, spermatozoa and mouse early embryos, *Dev. Biol.* **57:**199–209.

Gachelin, G., Delarbre, C., Coulon-Morelenc, M. J., Keil-Dlouha, V., and Muramatsu, T., 1982, F9 antigens: A reevaluation, in: *Teratocarcinoma and Embryonic Cell Interactions* (T. Muramatsu, G. Gachelin, A. A. Moscona, and Y. Ikawa, eds.), pp. 121–140, Japan Scientific Societies Press, Tokyo.

Gardner, R. L., 1975, in: *The Developmental Biology of Reproduction* (C. J. Markert, ed.), pp. 207–238, Academic Press, New York.

Gardner, R. L., and Papaioannou, V. E., 1975, Differentiation in the trophectoderm and inner cell mass, in: *The Early Development of Mammals* (M. Balls and A. E. Wild, eds.), pp. 107–132, Cambridge University Press, London.

Gardner, R. L., and Rossant, J., 1976, Determination during embryogenesis, in: *Embryogenesis in Mammals* (K. Elliott and M. O'Connor, eds.), pp. 5–25, Associated Scientific Publishers, Amsterdam.

Gardner, R. L., Papaioannou, V. E., and Barton, S. C., 1973, Origin of the ectoplacental cone and secondary giant cells in mouse blastocysts reconstituted from isolated trophoblast and inner cell mass, *J. Embryol. Exp. Morphol.* **30:**561–572.

Gelman, R. A., and Vered, J., 1976, Cyanogen bromide fragments of bovine cervical mucus glycoprotein, *Biochim. Biophys. Acta* **427:**627–633.

Gleeson, P. A., and Schachter, H., 1983, Control of glycoprotein synthesis, *J. Biol. Chem.* **258:**6162–6173.

Gmur, R., Knowles, B. B., and Solter, D., 1981, Regulation of phenotype in somatic cell lines with normal or tumor-derived mouse cells, *Dev. Biol.* **81:**245–254.

Gooi, H. C., Feizi, T., Kapadia, A., Knowles, B. B., Solter, D., and Evans, M. J., 1981, Stage-specific embryonic antigen involves alpha-It[3] fucosylated type 2 blood group chains, *Nature (London)* **292:**156–158.

Grabel, L. B., Rosen, S. D., and Martin, G. R., 1979, Teratocarcinoma stem cells have a cell surface carbohydrate-binding component implicated in cell–cell adhesion, *Cell* **17:**477–484.

Grabel, L. B., Glabe, C. G., Singer, M. S., Martin, G. R., and Rosen, S. D., 1981, A fucan specific lectin on teratocarcinoma stem cells, *Biochem. Biophys. Res. Commun.* **102:**1165–1171.

Graham, C. F., 1977, Teratocarcinoma cells and normal mouse embryogenesis, in: *Concepts in Mammalian Embryogenesis* (M. I. Sherman, ed.), pp. 315–376, MIT Press, Cambridge, Mass.

Grinna, L. S., and Robbins, P. W., 1979, Glycoprotein biosynthesis: Rat liver microsomal glucosidases which process oligosaccharides, *J. Biol. Chem.* **254:**8814–8818.

Grinna, L. S., and Robbins, P. W., 1980, Substrate specificities of rat liver microsomal glucosidases which process glycoproteins, *J. Biol. Chem.* **255:**2255–2258.

Gulyas, B. J., 1980, Cortical granules of mammalian eggs, *Int. Rev. Cytol.* **63:**357–392.

Gwatkin, R. B. L., 1976, Fertilization, in: *The Cell Surface in Embryogenesis and Development* (G. Poste and G. L. Nicolson, eds.), pp. 1–54, North-Holland, Amsterdam.

Gwatkin, R. B. L., 1977, *Fertilization mechanisms in Man and Mammals,* Plenum Press, New York.

Gwatkin, R. B. L., and Anderson, O. F., 1973, Effect of glycosidase inhibitors on the capacitation of hamster spermatozoa by cumulus cells in vitro, *J. Reprod. Fertil.* **35:**565–567.

Gwatkin, R. B. L., Williams, D. T., Hartman, J. F., and Kniazuk, M., 1973, The zona reaction of hamster and mouse eggs: Production in vitro of a trypsin like protease from cortical granules, *J. Reprod. Fertil.* **32:**259–265.

Handyside, A. H., 1978, Time of commitment of inside cells isolated from preimplantation mouse embryos, *J. Embryol. Exp. Morphol.* **45:**37–53.

Handyside, A. H., and Barton, S. C., 1977, Evaluation of the technique of immunosurgery for the isolation of inner cell masses from mouse blastocyst, *J. Embryol. Exp. Morphol.* **37:**217–226.

Handyside, A. H., and Johnson, M. H., 1978, Temporal and spatial patterns of the synthesis of tissue-specific polypeptides in the preimplantation mouse embryo, *J. Embryol. Exp. Morphol.* **44:**191–199.

Hanover, J. A., and Lennarz, W. J., 1978, The topological orientation of N,N'-diacetylchitobiosylpyrophosphoryldolichol in artificial and natural membranes, *J. Biol. Chem.* **254:**9237–9246.

Harpaz, N., and Schachter, H., 1980a, Control of glycoprotein synthesis, *J. Biol. Chem.* **255:**4885–4893.

Harpaz, N., and Schachter, H., 1980b, Control of glycoprotein synthesis, *J. Biol. Chem.* **255:**4894–4902.

Hartmann, J. F., Gwatkin, R. B. L., and Hutchinson, C. F., 1972, Early contact interactions between mammalian gametes *in vitro*: Evidence that the vitellus influences adherence between sperm and zone pellucida, *Proc. Natl. Acad. Sci. USA* **69:**2767–2769.

Hartree, E. F., 1977, Spermatozoa, eggs and proteinases, *Biochem. Soc. Trans.* **5:**375–394.

Hassel, J. R., Newsome, D. A., and Hascall, V. C., 1979, Characterization and biosynthesis of proteoglycans of corneal stroma from rhesus monkey, *J. Biol. Chem.* **254:**12346–12354.

Hatcher, V. B., Schwarzmann, G. O. H., Jeanloz, R. W., and McArthur, J. W., 1977, Purification, properties, and partial structure elucidation of a high molecular weight glycoprotein from cervical mucus of the bonnet monkey (*Macaca radiata*), *Biochemistry* **16:**1518–1524.

Hayashi, M., Schlesinger, D. H., Kennedy, D. W., and Yamada, K. M., 1980, Isolation and characterization of a heparin-binding domain of cellular fibronectin, *J. Biol. Chem.* **255:**10017–10020.

Heath, J., Bell, S., and Rees, A., 1981, Appearance of functional insulin receptors during the differentiation of embryonal carcinoma cells, *J. Cell Biol.* **91:**293–297.

Heathcote, J. G., Sear, C. H. J., and Grant, M. E., 1978, Studies on the assembly of the lens capsule, biosynthesis and partial characterization of the collagenous components, *Biochem. J.* **176:**283–294.

Hedman, K., Kurkinen, M., Alitalo, K., Vaheri, A., Johansson, S., and Höök, M., 1980, Isolation of the pericellular matrix of human fibroblast cultures, *J. Cell Biol.* **81:**83–91.

Heifetz, A., Lennarz, W. J., Libbus, B., and Hsu, Y. C., 1980, Synthesis of glycoconjugates during the development of mouse embryos *in vitro, Dev. Biol.* **80:**398–408.

Herbert, M. C., and Graham, C. F., 1974, Cell determination and biochemical differentiation of the early mammalian embryo, *Curr. Top. Dev. Biol.* **8:**151–178.

Heyner, S., Brinster, R. L., and Palm, J., 1969, Effect of isoantibody on preimplantation mouse embryos, *Nature (London)* **222:**783–784.

Hill, H. D., Reynolds, J. A., and Hill, R. L., 1977a, Purification, composition, molecular weight and subunit structure of ovine submaxillary mucin, *J. Biol. Chem.* **252:**3791–3798.

Hill, H. D., Schwyzer, M., Steinman, H. M., and Hill, R. L., 1977b, Submaxillary mucin. 1. Structure and peptide substrates of UDP-N-acetylgalactosamine-mucin transferase, *J. Biol. Chem.* **252:**3799–3804.

Hillman, N., Sherman, M. I., and Graham, C., 1972, Effect of spatial arrangement on cell determination during mouse development, *J. Embryol. Exp. Morphol.* **28:**263–278.

Hogan, B. L. M., 1977, Teratocarcinoma cells as a model for mammalian development, in: *Biochemistry of Cell Differentiation,* Vol. 15 (J. Paul, ed.), pp. 333–376, University Park Press, Baltimore.

Hogan, B. L. M., 1980, High molecular weight extracellular proteins synthesized by endoderm cells derived from mouse teratocarcinoma cells and normal extraembryonic membranes, *Dev. Biol.* **76:**275–285.

Hogan, B. L. M., and Tilly, R., 1978a, *In vitro* development of inner cell masses isolated immunosurgically from mouse blastocysts. I. Inner cell masses from 3.5 day p.c. blastocysts incubated for 24 h before immunosurgery, *J. Embryol. Exp. Morphol.* **45:**93–105.

Hogan, B. L. M., and Tilly, R., 1978b, *In vitro* development of inner cell masses isolated immunosurgically from mouse blastocysts. II. Inner cell masses from 3.5 day to 4.0 day p.c. blastocysts, *J. Embryol. Exp. Morphol.* **45:**107–121.

Hogan B., and Tilly, R., 1981, Cell interactions and endoderm differentiation in cultured mouse embryos, *J. Embryol. Exp. Morphol.* **62:**379–394.

Hogan, B. L. M., Cooper, A. R., and Kurkinen, M., 1980, Incorporation into Reichert's membrane of laminin-like extracellular proteins synthesized by parietal endoderm cells of the mouse embryo. *Dev. Biol.* **80:**289–300.

Hogan, B. L. M., Taylor, A., and Adamson, E., 1981, Cell interactions modulate embryonal carcinoma cell differentiation into parietal or visceral endoderm, *Nature (London)* **291:**235–237.

Hogan, B. L. M., Taylor, A., and Cooper, A., 1982a, Murine parietal endoderm cells synthesize heparan sulfate and 170K and 145K sulfated glycoproteins as components of Reichert's membrane, *Dev. Biol.* **90:**210–214.

Hogan, B. L. M., Taylor, A., Kurkinen, M., and Couchman, J. R., 1982b, Synthesis and localization of two sulfated glycoproteins associated with basement membranes and the extracellular matrix, *J. Cell Biol.* **95:**197–204.

Horstadius, S., Josefson, L., and Runnstrom, J., 1967, Morphogenetic agents from unfertilized eggs of sea urchin *Paracentrotus lividus, Dev. Biol.* **16:**189–202.

Howe, C. C., and Solter, D., 1979, Cytoplasmic and nuclear protein synthesis in preimplantation mouse embryos, *J. Embryol. Exp. Morphol.* **52:**209–225.

Howe, C. C., and Solter, D., 1981, Changes in cell surface proteins during differentiation of mouse embryonal carcinoma cells, *Dev. Biol.* **84:**239–243.

Howe, C. C., Gmur, R., and Solter, D., 1980, Cytoplasmic and nuclear protein synthesis during *in vitro* differentiation of murine ICM and embryonal carcinoma cells, *Dev. Biol.* **74:**351–363.

Howe, W. E., and Oshima, R., 1982, Coordinate expression of parietal endodermal functions in hybrids of embryonal carcinoma and endodermal cells, *Mol. Cell. Biol.* **2:**331–337.

Hsu, Y. C., 1972, Differentiation in vitro of mouse embryos beyond the implantation stage, *Nature (London)* **239:**200.

Hubbard, S. C., and Ivatt, R. J., 1981, Synthesis and processing of asparagine-linked oligosaccharides, *Annu. Rev. Biochem.* **50:**555–583.

Hubbard, S. C., and Robbins, P. W., 1979, Synthesis and processing of protein-linked oligosaccharides in vivo, *J. Biol. Chem.* **254:**4568–4576.

Hubbard, S. C., and Robbins, P. W., 1980, Synthesis of the N-linked oligosaccharides of glycoproteins, *J. Biol. Chem.* **255:**11782–11793.

Hunt, L. A., 1980, CHO cells selected for phytohemagglutinin and Con A resistance are deficient in both early and late stages of protein glycosylation, *Cell* **21:**407–415.

Hyafil, F., Morello, D., Babinet, C., and Jacob, F., 1980, A cell surface glycoprotein involved in the compaction of embryonal carcinoma cells and cleavage stage embryos, *Cell* **21:**927–934.

Hyafil, F., Babinet, C., and Jacob, F., 1981, Cell–cell interaction in early embryogenesis: A molecular approach to the role of calcium, *Cell* **26:**447–454.

Hynes, R. O., 1981, Fibronectin and its relation to cell structure and behavior, in: *Cell Biology of the Extracellular Matrix* (E. D. Hays, eds.), Plenum Press, New York.

Hynes, R. O., and Yamada, K. M., 1982, Fibronectins: Multifunctional modular glycoproteins, *J. Cell Biol.* **95:**369–377.

Iacobelli, S., Garcea, N., and Angeloni, C., 1977, Biochemistry of cervical mucus: A comparative analysis of the secretion from preovulatory, postovulatory and pregnancy periods, *Fertil. Steril.* **22:**727–734.

Illmensee, K., and Mahowald, A. P., 1974, Transplantation of posterior polar plasms in *Drosophila*: Induction of germ cells at anterior pole of egg, *Proc. Natl. Acad. Sci. USA* **71:**1016–1020.

Illmensee, K., and Mintz, B., 1976, Totipotent and normal differentiation of single teratocarcinoma cells cloned by injection into blastocysts, *Proc. Natl. Acad. Sci. USA* **73:**549–553.

Izquierdo, L., 1955, *Arch. Biol.* **66:**403–438.

Izquierdo, L., Lopez, T., and Marticorena, P., 1980, Cell membrane regions in preimplantation mouse embryos, *J. Embryol. Exp. Morphol.* **59:**89–102.

Jacob, F., 1977, Mouse teratocarcinoma and embryonic antigens, *Immunol. Rev.* **33:**3–32.

Jacowski, S., and Dumont, J. N., 1979, Surface alterations of the mouse zona pellucida and ovum following *in vivo* fertilization: Correlation with the cell cycle, *Biol. Reprod.* **20:**150–161.

Jaffe, L. F., 1968, Localization in the developing fucus egg and the general role of localizing currents, *Adv. Morphol.* **7:**295–328.

Jaffe, E. A., and Mosher, D. F., 1978, Synthesis of fibronectin by cultured endothelial cells, *J. Exp. Med.* **147:**1779–1791.

Järnefelt, J., Rush, J., Li, Y. T., and Laine, R. A., 1978, Erythroglycan, a high molecular weight glycopeptide with the repeating structure (galactosyl(1-4)2-deoxy-2-acetamidoglycosyl(1-3)) comprising more than one-third of the protein bound carbohydrate of the human erythrocyte stroma, *J. Biol. Chem.* **253:**8006–8009.

Jensh, J. P., Koszalka, T. R., Jensen, M., Biddle, L., and Brent, R. L., 1977, Morphological alterations in the parietal yolk sac of the rat from the 12th to 19th day of gestation, *J. Embryol. Exp. Morphol.* **39:**9–21.

Jetten, A. M., Jetten, M. E. R., and Sherman, M. I., 1979, Analyses of cell surface and secreted proteins of primary cultures of mouse extraembryonic membranes, *Dev. Biol.* **70:**89–104.

Jetten, A. M., Deluca, L. M., and Meeks, R. G., 1982, Enhancement in apparent membrane microviscosity during differentiation of embryonal carcinoma cells induced by retinoids, *Exp. Cell Res.* **138:**494–498.

Johnson, L. D., and Starcher, B. C., 1972, Epithelial basement membrane: The isolation and identification of a soluble component, *Biochim. Biophys. Acta* **290:**158–165.

Johnson, L. V., and Calarco, P. G., 1980a, Electrophoretic analysis of cell surface proteins of preimplantation mouse embryos, *Dev. Biol.* **77:**224–227.

Johnson, L. V., and Calarco, P. G., 1980b, Immunological characterization of embryonic cell surface antigens recognized by antiblastocysts serum, *Dev. Biol.* **79:**208–223.

Johnson, L. V., and Calarco, P., 1980c, Stage-specific embryo antigens detected by an antiserum against mouse blastocysts, *Dev. Biol.* **79:**224–231.

Johnson, M. H., 1982, Membrane events associated with the generation of a blastocyst, *Int. Rev. Cytol. Suppl.* **12:**1–37.

Johnson, M. H., Handyside, A. H., and Braude, P. R., 1977, in: *Development in Mammals* (M. H., Johnson, ed.), Vol. 2, pp. 67–98, North-Holland, Amsterdam.

Kahan, B. W., and Ephrussi, B., 1970, Developmental potentialities of clonal *in vitro* cultures of mouse testicular teratoma, *J. Natl. Cancer Inst.* **44:**1015–1023.

Kanwar, Y. S., and Farquhar, M. G., 1979, Isolation of GAGS (heparan sulfate) from glomerular basement membranes, *Proc. Natl. Acad. Sci. USA* **76:**4493–4497.

Kapadia, A., Feizi, T., and Evans, M. J., 1981, Changes in the expression and polarization of blood group i and I antigens in postimplantation embryos and teratocarcinomas of mouse associated with cell differentiation, *Exp. Cell Res.* **131:**185–195.

Kefalides, N. A., 1973, Structure and biosynthesis of basement membranes, *Int. Rev. Connec. Tissue Res.* **6:**63–104.

Kefalides, N. A., 1975, Basement membranes:Current concepts of structure and synthesis, *Dermatologica* **150:**4–15.

Kelly, J. J., 1977, Studies of the potency of early cleavage blastomeres of the mouse, in: *The Early Development of Mammals* (M. Balls and A. E. Wild, eds.), pp. 97–106, Cambridge University Press, London.

Kemler, R., Babinet, C., Condamine, H., Gachelin, G., Guernet, J. L., and Jacob, F., 1976, Embryonal carcinoma antigens and T/t locus of mouse, *Proc. Natl. Acad. Sci. USA* **73:**4080–4084.

Kemler, R., Babinet, C., Eisen, H., and Jacob, F., 1977, Surface antigen in early differentiation, *Proc. Natl. Acad. Sci. USA* **74:**4449–4452.

Kleinsmith, L. J., and Pierce, G. B., 1964, Multipotency of single embryonal carcinoma cells, *Cancer Res.* **24:**1544–1552.

Knowles, B. B., Pan, S., Solter, D., Linnenbach, A., Croce, C., and Huebner, K., 1980, Expression of H-2, laminin, and SV-40T and TASA on differentiated and transformed murine teratocarcinoma cells, *Nature (London)* **288:**615–618.

Knudsen, K. A., Rao, P. E., Damsky, C. H., and Buck, C. A., 1981, Membrane glycoproteins involved in cell–substratum adhesion, *Proc. Natl. Acad. Sci. USA* **78:**6071–6075.

Kornfeld, R., and Kornfeld, S., 1980, Structure of glycoproteins and their oligosaccharide units, in: *The Biochemistry of Glycoproteins and Proteoglycans* (W. J. Lennarz, ed.), pp. 1–34, Plenum Press, New York.

Kornfeld, S., and Tabas, I., 1978, The synthesis of complex type oligosaccharides, *J. Biol. Chem.* **253:**7771–7778.

Kornfeld, S., Gregory, W., and Chapman, A., 1979, The synthesis of complex type oligosaccharides, *J. Biol. Chem.* **254:**11649–11654.

Krco, C. J., and Goldberg, E. H., 1976, H-Y (male) Ag-detection on 8-cell mouse embryos, *Science* **193**:1134–1135.

Krebs, E. G., 1972, Protein kinases, *Curr. Top. Cell. Regul.* **5**:99–133.

Kuff, E. L., and Fewell, J. W., 1980, Induction of neural-like cells and acetylcholinesterase activity in cultures of F9 teratocarcinoma treated with retinoic acid and dibutyryl cyclic adenosine monophosphate, *Dev. Biol.* **77**:103–115.

Lee, W. I., Verdugo, P., Blandau, R. J., and Gaddum Rosse, P., 1977, Molecular arrangement of cervical mucus—Reevaluation based on laser light scattering spectroscopy, *Gynecol. Invest.* **8**:254–266.

Leivo, I., Vaheri, A., Timpl, R., and Wartiovaara, J., 1980, Appearance and distribution of collagens and laminin in the early mouse embryo, *Dev. Biol.* **76**:100–114.

Leivo, I., Alitalo, K., Ristell, L., Vaheri, A., Timpl, R., and Wartiovaara, J., 1982, Basal lamina glycoproteins laminin and type IV collagen are assembled into a fine-fibered matrix in cultures of a teratocarcinoma-derived endodermal cell line, *Exp. Cell Res.* **137**:15–23.

Levinson, J., Goodfellow, P., Vadeboncoeur, M., and McDevitt, H., 1978, Identification of stage-specific polypeptides synthesized during murine preimplantation development, *Proc. Natl. Acad. Sci. USA* **75**:3332–3336.

Li, E., Tabas, I., and Kornfeld, S., 1978, Structure of the lipid-linked precursor of the complex-type oligosaccharides of the vesicular stomatitis virus G protein, *J. Biol. Chem.* **253**:7762–7770.

Linder, S., Krondahl, U., Sennerstam, R., and Ringertz, N. R., 1981, Retinoic acid induced differentiation of F9 embryonal carcinoma cells, *Exp. Cell Res.* **132**:453–460.

Liotta, L. A., and Hart, I. R., 1982, *Tumor Invasion and Metastasis,* Nijhoff, The Hague.

Lloyd, K. O., and Kabat, E. A., 1968, Immunochemical studies on blood groups. XLI. Proposed structure for the carbohydrate portions of blood groups A, B, H, Lewis[a] and Lewis[b] substances, *Proc. Natl. Acad. Sci. USA* **61**:1470–1477.

Lo, C. W., 1980, Gap junctions in development, in: *Development in Mammals* (M. H. Johnson, ed.), Vol. 4, pp. 39–80, North-Holland, Amsterdam.

Lo, C. W., and Gilula, N. B., 1980a, PCC4azal teratocarcinoma stem cell differentiation in culture, *Dev. Biol.* **75**:112–120.

Lo, C. W., and Gilula, N. B., 1980b, PCC4azal teratocarcinoma stem cell differentiation in culture: Biochemical studies, *Dev. Biol.* **75**:78–92.

Macarack, E. J., Kirby, E., Kirk, T., and Kefalides, N. A., 1978, Synthesis of cold insoluble globulin by cultured calf endothelial cells, *Proc. Natl. Acad. Sci. USA* **75**:2621–2625.

McBurney, M. W., and Rogers, B. J., 1982, Isolation of male embryonal carcinoma cells and their chromosome replication patterns, *Dev. Biol.* **89**:503–508.

McBurney, M. W., Jones Villeneuve, E. M. V., Edwards, M. K. S., and Anderson, P. J., 1982, Control of muscle and neuronal differentiation in a cultured embryonal carcinoma cell line, *Nature (London)* **299**:165–167.

McRorie, R. A., and Williams, W. L., 1974, Biochemistry of mammalian fertilization, *Annu. Rev. Biochem.* **43**:777–803.

Madri, J. A., Roll, F. J., Furthmayr, H., and Foidart, J.-M., 1980, Ultrastructural localization of FN and LM in the basement membranes of the murine kidney, *J. Cell Biol.* **86**:682–687.

Magnuson, T., and Epstein, C. J., 1981, Characterization of concanavalin A precipitated proteins from mouse embryos: A 2-dimensional gel electrophoresis study, *Dev. Biol.* **81**:193–199.

Magnuson, T., and Stackpole, C. W., 1978, Lectin-mediated agglutination of preimplantation mouse embryos, *Exp. Cell Res.* **116**:466–469.

Martin, G. R., 1975, Teratocarcinomas as a model system for the study of embryogenesis and neoplasia, *Cell* **5**:229–243.

Martin, G. R., 1978, Advantages and limitations of teratocarcinoma stem cells as models of development, in: *Development in Mammals* (M. Johnson, ed.), Vol. 3, pp. 225–265, Elsevier/North-Holland, Amsterdam.

Martin, G. R., 1980, Teratocarcinomas and mammalian embryogenesis, *Science* **209:**768–776.

Martin, G. R., and Evans, M. J., 1975, in: *Teratomas and Differentiation* (M. I. Sherman and D. Solter, eds.), pp. 167–187, Academic Press, New York.

Martin, G. R., Wiley, L. M., and Damjanov, I., 1977, Development of cystic embryoid bodies in vitro from ceptic teratocarcinoma stem cells, *Dev. Biol.* **61:**230–244.

Martin, G. R., Rosen, S. D., and Grabel, L. B., 1982, Specificity and possible role in intercellular adhesion of a teratocarcinoma stem cell surface lectin, in: *Teratocarcinoma and Embryonic Cell Interactions* (T. Muramatsu, G. Gachelin, A. A. Moscona, and Y. Ikawa, eds.), pp. 295–309, Japan Scientific Societies Press, Tokyo.

Martinez-Hernandez, A., Nakane, P. K., and Pierce, G. B., 1974, Intracellular localization of basement membrane antigen in parietal yolk sac cells, *Am. J. Pathol.* **76:**549–556.

Mazanec, K., and Dvorak, M., 1963, *Cesk. Morfol.* **11:**103–108.

Meyerhofer, M., Anderson, O. F., Marx, B. S., and Gwatkin, R. B. L., 1977, The zona reaction: An SEM and light microscope study employing hamster gametes, *Scanning Electron Microsc.* **2:**343–347.

Midgley, A. R., and Pierce, G. B., 1963, Immunohistochemical analysis of basement membranes of mouse, *Am. J. Pathol.* **43:**929–943.

Minor, R. R., Hoch, P. S., Koszalka, T. R., Brent, R. L. and Kefalides, N. A., 1976a, Organ cultures of embryonic rat parietal yolk sac. I. Morphological and autoradiographic studies of the deposition of the collagen and noncollagen glycoprotein components of basement membrane, *Dev. Biol.* **48:**344–364.

Minor, R. R., Strause, E. L., Koszalka, T. R., Brent, R. L., and Kefalides, N. A., 1976b, Organ cultures of the embryonic rat parietal yolk sac. II. Synthesis, accumulation, and turnover of collagen and noncollagen basement membrane glycoproteins, *Dev. Biol.* **48:**365–376.

Mintz, B., 1964, Synthetic processes in early development in mammalian egg, *J. Exp. Zool.* **157:**273–292.

Mintz, B., and Illmensee, K., 1975, Normal genetically mosaic mice produced from malignant teratocarcinoma cells, *Proc. Natl. Acad. Sci. USA* **72:**3585–3589.

Miyauchi, T., Yonezawa, S., Takamura, T., Chiba, T., Tejima, S., Ozawa, M., Sato, E., and Muramatsu, T., 1982, A new fucosyl antigen expressed on colon adenocarcinoma and embryonal carcinoma cells, *Nature (London)* **299:**168–170.

Mizoguchi, A., Mizuochi, T., and Kobata, A., 1982, Structures of the carbohydrate moieties of secretory component purified from human milk, *J. Biol. Chem.* **257:**9612–9621.

Moghissi, K. S., and Marks, C., 1971, Effects of microdose norgestrel on endogenous gonadotropic and steroid hormones, cervical mucus properties, vaginal cytology and endometrium, *Fertil. Steril.* **22:**424–434.

Mosesson, M. W., and Umfleet, R. A., 1970, The cold insoluble globulin of human plasma, *J. Biol. Chem.* **245:**5728–5736.

Mosher, D. F., 1980, *Fibronectin, Prog. Hemostasis Thromb.* **5:**111–151.

Muggleton-Harris, A. L., and Johnson, M. H., 1976, Nature and distribution of serologically detectable alloantigens on preimplantation mouse embryo, *J. Embryol. Exp. Morphol.* **35:**59–72.

Mulnard, J., and Huygens, R., 1978, Ultrastructural localization of nonspecific alkaline phosphatase during cleavage and blastocyst formation in mouse, *J. Embryol. Exp. Morphol.* **44:**121–131.

Muramatsu, H., and Muramatsu, T., 1982, Decreased synthesis of large fucosyl glycopeptides during differentiation of embryonal carcinoma cells induced by retinoic acid and dibutyryl cyclic AMP, *Dev. Biol.* **90**:441–444.

Muramatsu, H., Muramatsu, T., and Avner, P., 1982, Biochemical properties of the high-molecular-weight glycopeptides released from the cell surface of human teratocarcinoma cells, *Cancer Res.* **42**:1749–1752.

Muramatsu, T., Gachelin, G., Nicolas, J. F., Condamine, H., Jakob, H., and Jacob, F., 1978, Carbohydrate structure and cell differentiation: Unique properties of fucosylglycopeptides isolated from embryonal carcinoma cells, *Proc. Natl. Acad. Sci. USA* **75**:2315–2319.

Muramatsu, T., Gachelin, G., and Jacob, F., 1979, Characterization of glycopeptides isolated from membranes of F9 embryonal carcinoma cells, *Biochim. Biophys. Acta* **587**:392–406.

Muramatsu, T., Condamine, H., Gachelin, G., and Jacob, F., 1980, Changes in fucosylglycopeptides during early post-implantation embryogenesis in the mouse, *J. Embryol. Exp. Morphol.* **57**:25–36.

Nadijcka, M., and Hillman, N., 1974, Ultrastructural studies of the mouse blastocyst substages, *J. Embryol. Exp. Morphol.* **32**:675–695.

Narasimhan, S., Stanley, P., and Schachter, H., 1977, Control of glycoprotein synthesis: Lectin resistant mutant containing only one of two distinct N-acetylglucosaminyl transferase activities present in wild type Chinese hamster ovary cells, *J. Biol. Chem.* **252**:3926–3933.

Nicolas, J. F., Dubois, P., Jakob, H., Gaillard, J., and Jacob, F., 1975, Teratocarcinome de la souris: Differenciation en culture d'une lignee de cellules primitives a potentialites multiples, *Ann. Microbiol. (Inst. Pasteur)* 126 A, 3-22.

Nicolas, J. F., Kemler, R., and Jacob, F., 1981, Effects of anti-embryonal carcinoma sera on aggregation and metabolic cooperation between teratocarcinoma cells, *Dev. Biol.* **81**:127–132.

Nicolson, G. L., and Yanagimachi, R., 1972, Terminal saccharides on sperm plasma membranes: Identification with specific agglutinins, *Science* **177**:276–279.

Nicolson, G. L., Usuni, N., Yanagimachi, R., Yanagimachi, H., and Smith, J. R., 1977, Lectin binding sites on the plasma membranes of rabbit spermatozoa: Changes in surface receptors during epididymal migration and after ejaculation, *J. Cell Biol.* **74**:950–962.

Nicolson, G. L., Brodginski, A. B., Beattie, G., and Yanagimachi, R., 1979, Cell surface changes in the properties of rabbit spermatozoa during epididymal passage, *Gam. Res.* **2**:153–162.

Nilsson, O. S., De Tomas, M. E., Peterson, E., Bergman, A., Dallner, G., and Hemming, F. W., 1978, Mannosylation of endogenous proteins of rough and smooth endoplasmic reticulum and of Golgi membranes, *Eur. J. Biochem.* **89**:619–628.

Nimmo, H. G., and Cohen, P., 1977, *Adv. Cyclic Nucleotide Res.* **8**:146–266.

Nishimune, Y., Ogiso, Y., Kume, A., Matsushiro, A., and Noguchi, T., 1982, Identification of reversible and irreversible stages during the differentiation of pluripotent teratocarcinoma cell line, in: *Teratocarcinoma and Embryonic Cell Interactions* (T. Muramatsu, G., Gachelin, A. A. Moscona, and Y. Ikawa, eds.), Japan Scientific Societies Press, Tokyo.

Nuccitelli, R., and Jaffe, L. F., 1974, Spontaneous current pulses through developing fucoid eggs, *Proc. Natl. Acad. Sci. USA* **71**:4855–4859.

Nudelman, E., Hakomori, S.-I., Knowles, B. B., Solter, D., Nowinski, R. C., Tam, M. R., and Young, W. W., 1980, Monoclonal antibodies directed to the stage specific embryonic antigen (SSEA-1) react with a branched glycosphingolipid similar in structure to Ii antigen, *Biochem. Biophys. Res. Commun.* **97**:443–451.

Oberbaumer, I., Wiedermann, H., Timpl, R., and Kühn, K., 1982, Shape and assembly of type IV procollagen from cell culture, *EMBO J* **1**:805–810.

O'Brien, D. A., and Bellve, A. R., 1980a, Protein constituents of the mouse spermatozoon, *Dev. Biol.* **75**:405–418.

O'Brien, D. A., and Bellve, A. R., 1980b, Protein constituents of the mouse spermatozoon: An electrophoretic characterization, *Dev. Biol.* **75**:386–404.

Ogiso, Y., Akinori, A., Nishimune, Y., and Matsushiro, A., 1982, Reversible and irreversible stage in the transition of cell surface marker during the differentiation of pluripotent teratocarcinoma cell induced with retinoic acid, *Exp. Cell Res.* **137**:365–372.

Oldberg, A., Hayman, E. G., and Ruoslahti, E., 1981, Isolation of a chondroitin sulfate proteoglycan from a rat yolk sac tumor and immunochemical demonstration of its cell surface localization, *J. Biol. Chem.* **256**:10847–10852.

Oliphant, G., and Brachet, B. G., 1973, Capacitation of mouse spermatozoa in media with elevated ionic strength and reversible decapacitation with epididymal extract, *Fertil. Steril.* **24**:948–955.

Opheim, D. J., and Touster, O., 1978, Lysosomal α-D-mannosidase of rat liver: Purification and comparison with the Golgi and cytosolic α-D-mannosidases, *J. Biol. Chem.* **253**:1017–1023.

Oppenheimer, S. B., 1975, Functional involvement of specific carbohydrate in teratoma cell adhesion factor, *Exp. Cell Res.* **92**:122–126.

Orgebin-Christ, M. C., 1969, Maturation of spermatozoa in the rabbit epididymis: Fertilizating ability and embryonic mortality in does inseminated with epididymal spermatozoa, *Ann. Biol. Anim. Biochim. Biophys.* **7**:373–389.

Oshima, R. G., 1981, Identification and immunoprecipitation of cytoskeletal proteins from murine extra-embryonic endodermal cells, *J. Biol. Chem.* **256**:8124–8133.

Oshima, R. G., 1982, Developmental expression of murine extra-embryonic endodermal cytoskeletal proteins, *J. Biol. Chem.* **257**:3414–3421.

Oshima, R. G., and Linney, E., 1980, Identification of murine extraembryonic endodermal cells by reaction with teratocarcinoma basement membrane antiserum, *Exp. Cell Res.* **126**:485–490.

Palm, J., Heyner, S., and Brinster, R. L., 1971, Differential immunofluorescence of fertilized mouse eggs with H-2 and non H-2 antibody, *J. Exp. Med.* **133**:1282–1293.

Papaioannou, V. E., McBurney, M. W., and Gardner, R. L., 1975, Fate of teratocarcinoma cells injected into early mouse embryos *Nature (London)* **258**:70–75.

Parodi, A. J., and Leloir, L. F., 1979, Protein glycosylation through lipid-linked intermediates, *Biochim. Biophys. Acta* **559**:1–37.

Parodi, A. J., Behrens, N. H., Leloir, L. F., and Carminatti, H., 1972, The role of polyisoprenol-bound saccharides as intermediates in glycoprotein biosynthesis in liver, *Proc. Natl. Acad. Sci. USA* **69**:3268–3272.

Paulson, J. C., Prieels, J. P., Glasgow, L. R., and Hill, R. L., 1978, Sialyl and fucosyltransferases in the biosynthesis of asparagine-linked oligosaccharides in glycoproteins, *J. Biol. Chem.* **253**:5617–5624.

Pederson, R. A., Spindle, A. I., and Wiley, L. M., 1977, Regeneration of endoderm by ectoderm isolated from mouse blastocysts, *Nature (London)* **270**:435–437.

Pierce, G. B., 1965, Basement membranes. VI. Synthesis by epithelial tumors of the mouse, *Cancer Res.* **25**:656–663.

Pierce, G. B., 1966, The development of basement membranes in the mouse embryo, *Dev. Biol.* **13**:231–249.

Pierce, G. B., 1967, Teratocarcinoma: Model for a developmental concept of cancer, *Curr. Top. Dev. Biol.* **2**:223–246.

Pierce, G. B., Midgley, A. R., Feldman, J. D., and Sri Ram, J., 1962, Parietal yolk sac carcinoma: Clue to the histogenesis of Reichert's membrane of the mouse embryo, *Am. J. Pathol.* **41**:549–556.

Pierce, G. B., Midgley, A. R., and Sri Ram, J., 1963, The histogenesis of basement membranes, *J. Exp. Med.* **117**:339–348.

Pierce, G. B., Beals, T. F., Sri Ram, J., and Midgley, A. R., 1964, Basement membranes. IV. Epithelial origin and immunologic cross reactions, *Am. J. Pathol.* **45**:929–962.

Pierce, G. B., Jones, A., Orfanakis, N. G., Nakame, P. K., and Lustig, L., 1982a, Biosynthesis of basement membrane by parietal yolk sac cells, *Differentiation* **23**:60–72.

Pierce, G. B., Pantazis, C. G., Caldwell, J. E., and Wells, R. S., 1982b, Specificity of the control of tumor formation by the blastocyst, *Cancer Res.* **42**:1082–1087.

Plet, A., Evain, D., and Anderson, W. B., 1982, Effect of retinoic acid treatment of F9 embryonal carcinoma cells on the activity and distribution of cyclic AMP-dependent protein kinase, *J. Biol. Chem.* **257**:889–893.

Prujansky-Jacobovitz, A., Gachelin, G., Muramatsu, T., Sharon, N., and Jacob, F., 1979, Surface galactosyl glycopeptides of embryonal carcinoma cells, *Biochem. Biophys. Res. Commun.* **89**:448–455.

Quaroni, A., Isselbacher, K. J., and Ruoslahti, E., 1978, Fibronectin synthesis by epithelial crypt cells of rat small intestine, *Proc. Natl. Acad. Sci. USA* **75**:5548–5552.

Rao, C. N., Marguiles, I. M. K., Tralka, T. S., Terranova, V. P., Madri, J. A., and Liotta, L. A., 1982a, Isolation of a subunit of laminin and its role in molecular structure and tumor cell attachment, *J. Biol. Chem.* **257**:9740–9744.

Rao, C. N., Marguiles, I. M. K., Goldfarb, R. H., Madri, J. A., Woodley, D. T., and Liotta, L. A., 1982b, Differential proteolytic susceptibility of laminin α and β subunits, *Arch. Biochem. Biophys.* **219**:65–70.

Rao, C. N., Barsky, S. H., Terranova, V. P., and Liotta, L. A., 1983, Isolation of a tumor cell laminin receptor, *Biochem. Biphys. Res. Commun.* **111**:804–808.

Rasilo, M. L., 1980, Fractionation of large glycopeptides of human teratocarcinoma-derived cells by concanavalin A–Sepharose chromatography, *Can. J. Biochem.* **58**:281–286.

Rasilo, M. L., Wartiovaara, J., and Renkonen, O., 1980, Mannose containing glycopeptides of cells derived from human teratocarcinoma, *Can. J. Biochem.* **58**:384–393.

Reeve, J. W. R., and Ziomek, C. A., 1981, Distribution of microvilli on dissociated blastomeres from mouse embryos: Evidence for surface polarization at compaction, *J. Embryol. Exp. Morphol.* **62**:339–350.

Reeve, W. J. D., 1981, Cytoplasmic polarity develops at compaction in rat and mouse embryos, *J. Embryol. Exp. Morphol.* **62**:351–367.

Reisner, Y., Gachelin, G., Dubois, P., Nicolas, J. F., Sharon, N., and Jacob, F., 1977, Interaction of peanut agglutinin, a lectin specific for nonreducing terminal D-galactosyl residues with embryonal carcinoma cells, *Dev. Biol.* **61**:20–27.

Rose, M. C., Lynn, W. S., and Kaufman, B., 1970, Resolution of the major components of human lung mucosal gel and their capabilities for reaggregation and gel formation, *Biochemistry* **18**:4030–4037.

Roseman, S., 1970, The synthesis of complex carbohydrates by multiglycosyltransferase systems and their potential function in intercellular adhesion, *Chem. Phys. Lipids* **5**:270–297.

Rosenthal, M. D., Wishnow, R. M., and Sato, G. M., 1970, *In vitro* growth and differentiation of clonal populations of multipotential mouse cells derived from a transplantable testicular teratocarcinoma, *J. Natl. Cancer Inst.* **44**:1001–1014.

Rosenstraus, M. J., and Levine, A. J., 1979, Alterations in the developmental potential of embryonal carcinoma cells in mixed aggregates of pluripotent and nullipotent cells, *Cell* **17**:337–346.

Rosenstraus, M. J., and Spadoro, J. P., 1981, Autonomy of "nullipotent" and pluripotent embryonal carcinoma cells in differentiating aggregates, *Dev. Biol.* **85:**190–198.

Rosenstraus, M. J., Sundell, C. L., and Liskay, R. M., 1982, Cell-cycle characteristics of undifferentiated and differentiating embryonal carcinoma cells, *Dev. Biol.* **89:**516–520.

Rossant, J., and Papaioannou, V.E., 1977, The biology of embryogenesis, in: *Concepts In Mammalian Embryogenesis* (M. I. Sherman, ed.), pp. 1–36, MIT Press, Cambridge, Massachusetts.

Rossant, J., and Lis, W. J., 1979, Potential of isolated mouse inner cell masses to form trophectoderm derivatives *in vivo, Dev. Biol.* **70:**255–261.

Rossant, J., and Ofer, L., 1977, Properties of extraembryonic ectoderm isolated from post-implantation mouse embryos, *J. Embryol. Exp. Morphol.* **39:**183–194.

Rossant, J., and Vijh, K. M., 1980, Ability of outside cells from preimplantation mouse embryos to form inner cell mass derivatives, *Dev. Biol.* **76:**475–482.

Rowinski, J., Solter, D., and Koprowski, H., 1976, Change of concanavalin A induced agglutinability during preimplantation mouse development, *Exp. Cell. Res.* **100:**404–408.

Rubin, C. S., and Rosen, O. M., 1975, Protein phosphorylation, *Annu. Rev. Biochem.* **44:**831–887.

Rugh, R., 1968, *The Mouse: Its Reproduction and Development,* Burgess, Minneapolis, Minn.

Ruoslahti, E., and Vaheri, A., 1974, Novel human serum protein from fibroblast plasma membrane, *Nature (London)* **248:**789–791.

Ruoslahti, E., Engvall, E., and Hayman, E. G., 1981, Fibronectin: Current concepts of its structure and function, *Coll. Res.* **1:**95–128.

Sabatini, D. D., Kreibich, G., Morimoto, T., and Adesnick, M., 1982, Mechanisms for the incorporation of proteins in membranes and organelles, *J. Cell Biol.* **92:**1–22.

Sadler, J. E., Rearick, J. I., and Hill, R. L., 1979, Purification to homogeneity and enzymatic characterization of an α-N-acetylgalactosaminide α,2-6 sialyl transferase from porcine submaxillary glands, *J. Biol. Chem.* **254:**5934–5941.

Sakashita, S., and Rouslahti, E., 1980, Laminin-like glycoproteins in extracellular matrix of endodermal cells, *Arch. Biochem. Biophys.* **205:**283–290.

Sato, K., 1979, Polyspermy preventing mechanisms in mouse eggs fertilized *in vitro, J. Exp. Zool.* **210:**353–360.

Schachter, H., McGuire, E. J., and Roseman, S., 1971, Sialic acids. XIII. A uridine di-phosphate D-galactose:mucin galactosyltransferase from porcine submaxillary gland, *J. Biol. Chem.* **246:**5321–5328.

Scheul, H., 1978, Secretory functions of egg cortical granules in fertilization and development: A "critical" review, *Gam. Res.* **1:**299–381.

Schindler, J., Matthaei, K. I., and Sherman, M. I., 1981, Isolation and characterization of mouse mutant embryonal carcinoma cells which fail to differentiate in response to retinoic acid, *Proc. Natl. Acad. Sci. USA* **78:**1077–1080.

Searle, R. F., and Jenkinson, E. J., 1978, Localization of trophoblast defined surface antigens during early mouse embryogenesis, *J. Embryol. Exp. Morphol.* **43:**147–156.

Searle, R. F., Johnson, M. H., Billington, W. D., Elson, J., and Chutterbuck-Jackson, S., 1974, Investigation of H-2 and non H-2 antigens on mouse blastocyst, *Transplantation* **18:**136–141.

Semoff, S., Hogan, B. L. M., and Hopkins, C. R., 1982, Localization of fibronectin, laminin-entactin, and entactin in Reichert's membrane by immunoelectronmicroscopy, *EMBO J.* **1:**1171–1175.

Shapiro, B. M., and Eddy, E. M., 1980, When sperm meets egg: Biochemical mechanisms of gamete interaction, *Int. Rev. Cytol.* **66:**257–302.

Sherman, M. I., 1975, Differentiation of teratoma cell line PCC4: azal *in vitro*, in: *Teratomas and Differentiation* (M. I. Sherman and D. Solter, eds.), pp. 189–205, Academic Press, New York.

Sherman, M. I., 1979, Developmental biochemistry of preimplantation mammalian embryos, *Annu. Rev. Biochem.* **48**:443–470.

Sherman, M. I., and Miller, R. A., 1978, F9-embryonal carcinoma cells can differentiate into endoderm-like cells, *Dev. Biol.* **63**:27–34.

Sherman, M. I., and Wudl, L. W., 1976, The implanting mouse blastocyst, in: *The Cell Surface in Animal Embryogenesis and Development* (G. Poste and G. L. Nicolson, eds.), pp. 81–125, Elsevier/North-Holland, Amsterdam.

Sherman, M. I., Gay, R., Gay, S., and Miller, E. J., 1980, Association of collagen with preimplantation and peri-implantation mouse embryos, *Dev. Biol.* **74**:470–478.

Shur, B. D., 1982a, Evidence that galactosyltransferase is a surface receptor for poly(N)acetyllactosamine glycoconjugates on embryonal carcinoma cells, *J. Biol. Chem.* **257**:6871–6878.

Shur, B. D., 1982b, Cell surface glycosyl transferase activities during fertilization and early embryogenesis, in: *The Glycoconjugates* (M. Horowitz, ed.), Vol. III, pp. 145–185, Academic Press, New York.

Shur, B. D., 1983, The role of cell surface galactosyltransferase in embryonal carcinoma cell adhesion, in: *Teratocarcinoma Stem Cells* (L. Silver, G. Martin, and S. Strickland, eds.), pp. 185–195, Cold Spring Harbor Laboratory, New York.

Shur, B. D., and Hall, N. G., 1982a, A role for mouse sperm surface galactosyl transferases in sperm binding to the egg zona pellucida, *J. Cell Biol.* **95**:574–579.

Shur, B. D., and Hall, N. G., 1982b, Sperm surface galactosyl transferase activity during *in vitro* capacitation, *J. Cell Biol.* **95**:567–573.

Smith, K. K., and Strickland, S., 1981, Structural components and characteristics of Reichert's membrane, an extra-embryonic basement membrane, *J. Biol. Chem.* **256**:4654–4661.

Smith, L. D., 1966, Role of germinal plasm in formation of primordial germ cells in *Rana pipiens*, *Dev. Biol.* **14**:330–347.

Snider, M. D., Sultzman, L. A., and Robbins, P. W., 1980, Transmembrane location of oligosaccharide–lipid synthesis in microsomal vesicles, *Cell* **21**:385–392.

Snow, M. H. L., 1973, Tetraploid mouse embryos produced by cytochalasin B during cleavage, *Nature (London)* **244**:513–515.

Solter, D., and Knowles, B. B., 1975, Immunosurgery of mouse blastocysts, *Proc. Natl. Acad. Sci. USA* **72**:5099–5102.

Solter, D., and Knowles, B. B., 1978, Monoclonal antibody defining a stage-specific mouse embryonic antigen (SSEA-1), *Proc. Natl. Acad. Sci. USA* **75**:5565–5569.

Solter, D., and Knowles, B. B., 1979, Developmental stage-specific antigens during mouse embryogenesis, *Curr. Top. Dev. Biol.* **13**:139–165.

Solter, D., Shevinsky, L., Knowles, B. B., and Strickland, S., 1979, The induction of antigenic changes in a teratocarcinoma stem cell line (F9) by retinoic acid, *Dev. Biol.* **70**:515–521.

Speers, W. C., 1982, Conversion of malignant murine embryonal carcinomas to benign teratomas by chemical induction of differentiation *in vivo*, *Cancer Res.* **42**:1843–1849.

Spindle, A. I., 1978, Trophoblast regeneration by inner cell masses isolated from cultured mouse embryos, *J. Exp. Zool.* **203**:483–489.

Spiro, M. J., Spiro, R. G., and Bhoyroo, V. D., 1979, Glycosylation of proteins by oligosaccharide lipids, *J. Biol. Chem.* **254**:7668–7674.

Spiro, R. G., 1970, Glycoproteins, *Annu. Rev. Biochem.* **39**:599–638.

Spiro, R. G., and Spiro, M. J., 1979, Role of lipid-saccharide intermediates in glycoprotein biosynthesis, in: *Glycoconjugate Research*, Vol. 2 (J. Gregory and R. Jeanloz eds.), pp. 613–636, Academic Press, New York.

Stanley, P., 1980, Surface carbohydrate alterations of mutant mammalian cells selected for resistance to plant lectins in: *The Biochemistry of Glycoproteins and Proteoglycans* (W. J. Lennarz, ed.), pp. 161–189, Plenum Press, New York.

Steinberg, M. S., 1958, On the chemical bonds between animal cells: A mechanism for type specific association, *Am. Nat.* **92**:65–82.

Steinberg, M. S., 1963, Reconstruction of tissues by dissociated cells, *Science* **141**:401–408.

Steinberg, M. S., 1970, Does differential adhesion govern self-assembly processes in histogenesis? Equilibrium configurations and the emergence of a hierarchy among populations of embryonic cells, *J. Exp. Zool.* **173**:395–434.

Steinhardt, R., Zucker, R., and Schatten, G., 1977, Intracellular calcium release at fertilization in the sea urchin egg, *Dev. Biol.* **58**:185–196.

Stern, P. L., and Willison, K. R., 1982, Cell surface markers of teratocarcinomas and embryos: Antibodies to Forssman antigen, in: *Teratocarcinoma and Embryonic Cell Interaction* (T. Muramatsu, G. Gachelin, A. A. Moscona, and Y. Ikawa, eds.), pp. 87–101, Japan Scientific Societies Press, Tokyo.

Stern, P. L., Willison, K. R., Lennox, E., Galfre, G., Milstein, C., Secher, D., Ziegler, A., and Springer, T., 1978, Monoclonal antibodies as probes for differentiation and tumor associated antigens—Forssman specificity on teratocarcinoma stem cells, *Cell* **14**:775–783.

Stevens, L. C., 1967a, The biology of teratomas, *Adv. Morphog.* **6**:1–31.

Stevens, L. C., 1967b, Origin of testicular teratomas from primordial germ cells in mice *J. Natl. Cancer Inst.* **38**:549–552.

Stevens, L. C., 1970, The development of transplantable teratocarcinomas from intratesticular grafts of pre- and postimplantation mouse embryos, *Dev. Biol.* **21**:364–382.

Stevens, L. C., and Little, C. C., 1954, Spontaneous testicular teratomas in an inbred strain of mice, *Proc. Natl. Acad. Sci. USA* **40**:1080–1085.

Stevens, L. C., and Varnum, D. S., 1974, The development of teratomas from parthenogenetically activated ovarian mouse eggs, *Dev. Biol.* **37**:369–380.

Strickland, S., and Mahdavi, V., 1978, Induction of differentiation in teratocarcinoma stem cells by retinoic acid, *Cell* **15**:393–403.

Strickland, S., and Sawey, M. J., 1980, Studies on the effects of retinoids on the differentiation of teratocarcinoma stem cells *in vitro* and *in vivo, Dev. Biol.* **78**:76–85.

Strickland, S., Reich, E., and Sherman, M. I., 1976, Plasminogen activator in early embryogenesis: Enzyme production by trophoblast and parietal endoderm, *Cell* **9**:231–240.

Strickland, S., Smith, K. K., and Marotti, M., 1980, Hormonal induction of differentiation in teratocarcinoma stem cells: Generation of parietal endoderm by retinoic acid and dibutyryl c-AMP, *Cell* **21**:347–356.

Struck, D. K., and Lennarz, W. J., 1980, The function of oligosaccharide lipids in the synthesis of glycoproteins, in: *The Biochemistry of Glycoproteins and Proteoglycans* (W. J. Lennarz, ed.), pp. 35–83, Plenum Press, New York.

Surani, M. A. H., 1979, Glycoprotein synthesis and inhibition of glycosylation by tunicamycin in preimplantation embryos, *Cell* **18**:217–227.

Surani, M. A. H., Kimber, S. J., and Handyside, A. H., 1981, Synthesis and role of cell surface glycoproteins in preimplantation mouse development, *Exp. Cell Res.* **133**:331–339.

Szollosi, D., 1976, Development of cortical granules and the cortical reaction in rat and hamster eggs, *Anat. Rec.* **159**:439–446.

Szulman, A. E., 1964, The histological distribution of the blood group substances in man as disclosed by immunofluorescence, *J. Exp. Med.* **199**:503–516.

Tabas, I., and Kornfeld, S., 1978, Identification of an α-mannosidase activity involved in a late stage of processing of complex-type oligosaccharides, *J. Biol. Chem.* **253**:7779–7783.

Tabas, I., Schlesinger, S., and Kornfeld, S., 1978, The processing of high mannose oligosaccharides to form complex-type oligosaccharides in the newly synthesised polypeptides of the vesicular stomatitis virus G protein and the IgG heavy chain, *J. Biol. Chem.* **253**:716–725.

Takasaki, S., Yamashita, K., Suzuki, K., Iwanaga, S., and Kobata, A., 1979, The sugar chains of cold-insoluble globulin, *J. Biol. Chem.* **254**:8548–8553.

Takeichi, M., Atsumi, T., Yoshida, C., and Ogou, S. I., 1982, Molecular approaches to cell–cell recognition mechanisms in mammalian embryos, in: *Teratocarcinoma and Embryonic Cell Interactions* (T. Muramatsu, G. Gachelin, A. A. Moscona, and Y. Ikawa, eds.), pp. 283–293, Japan Scientific Societies Press, Tokyo.

Tarkowski, A. K., 1961, Mouse chimeras developed from fused eggs, *Nature (London)* **190**:857–859.

Tarkowski, A. K., 1965, Embryonic and postnatal development of mouse chimeras, in: *Preimplantation Stages of Pregnancy*, pp. 183–193, Churchill, London.

Tarkowski, A. K., and Wroblewska, J., 1967, Development of blastomeres of mouse eggs isolated at 4- and 8-cell stage, *J. Embryol. Exp. Morphol.* **18**:155–180.

Terranova, V. P., Rohrbach, D. M., and Martin, G. R., 1980, Role of laminin in the attachment of PAMZIZ (Epithelial) cells to basement membrane collagen, *Cell* **22**:719–726.

Terranova, V. P., Liotta, L. A., Russo, R. G., and Martin, G. R., 1982, Role of laminin in the attachment and metastasis of murine tumor cells, *Cancer Res.* **42**:2265–2269.

Terranova, V. P., Rao, C. N., Kabelic, T., Marguiles, I. M., and Liotta, L. A., 1983, Laminin receptor on human breast carcinoma cells, *Proc. Natl. Acad. Sci. USA* **80**:444–448.

Timpl, R., Rhode, H., Gehron Robey, P., Rennard, S. I., Foidart, J. M., and Martin, G. R., 1979, Laminin—A glycoprotein from basement membranes, *J. Biol. Chem.* **254**:9933–9937.

Tryggvason, K., Gehron Robey, P., and Martin, G. R., 1980, Biosynthesis of type IV procollagens, *Biochemistry* **19**:1284–1289.

Tulsiani, D. R. P., Opheim, D. J., and Touster, O., 1977, Purification and characterization of α-D-mannosidase from rat liver Golgi membranes, *J. Biol. Chem.* **252**:3227–3233.

Turco, S. J., and Robbins, P. W., 1977, The initial stages of processing of protein bound oligosaccharides *in vitro*, *J. Biol. Chem.* **254**:4560–4567.

Ugalde, R. A., Staneloni, R. J., and Leloir, L. F., 1979, Microsomal glucosidases acting on the saccharide moiety of the glucose containing dolichyl diphosphate oligosaccharide, *Biochem. Biophys. Res. Commun.* **91**:1174–1181.

Vacquier, V. D., 1975, The isolation of intact cortical granules from sea urchin eggs: Calcium ions trigger granule discharge, *Dev. Biol.* **43**:62–73.

Vaheri, A., and Mosher, D. F., 1978, High molecular cell surface associated glycoprotein (fibronectin) lost in malignant transformation, *Biochim. Biophys. Acta* **516**:1–25.

Van Blerkom, J., and Brockway, G. O., 1975, Quantitative patterns of protein synthesis in the preimplantation mouse embryo, *Dev. Biol.* **44**:148–157.

Van Blerkom, J., and Manes, C., 1974, Development of preimplantation rabbit embryos *in vivo* and *in vitro*. II. A comparison of quantitative aspects of protein synthesis, *Dev. Biol.* **35**:262–282.

Van Kooij, R. J., Roelofs, H. J. M., Kathman, G. A. M., and Kramer, M. F., 1980, Cervical mucus and its mucous glycoprotein during the menstrual cycle, *Fertil. Steril.* **34**:226–233.

Voss, B., Allam, S., and Rautenberg, J., 1979, Primary cultures of rat hepatocytes synthesize fibronectin, *Biochem. Biophys. Res. Commun.* **90**:1348–1354.

Waechter, C. J., and Lennarz, W. J., 1978, The role of polyprenol-linked sugars in glycoprotein synthesis, *Annu. Rev. Biochem.* **45**:95–124.

Wagner, D. D., Ivatt, R. J., Destree, A. T., and Hynes, R. O., 1981, Similarities and differences between the fibronectins of normal and transformed fibroblasts, *J. Biol. Chem.* **256**:11708–11715.

Wartiovaara, J., and Vaheri, A.,1980, Fibronectin and early mammalian embryogenesis, in: *Development in Mammals* (M. H., Johnson ed.), vol. 4, pp. 233–266, Elsevier North-Holland, Amsterdam.

Wartiovaara, J., Leivo, I., Virtanen, I., Vaheri, A., and Graham, C. F., 1978, Appearance of fibronectin during differentiation of mouse teratocarcinoma *in vitro, Nature* (*London*) **272**:355–356.

Wartiovaara, J., Leivo, I., and Vaheri, A., 1980, Matrix glycoproteins in early mouse development and in differentiation of teratocarcinoma cells, in: *The Cell Surface: Mediator of Developmental Processes* (S. Subtelny and N. K. Wessels, eds.), pp. 305–328, Academic Press, New York.

Webb, C. G., 1980, Characterization of antisera against mouse teratocarcinoma OTT6050: Molecular species recognized on embryoid bodies, preimplantation embryos, and sperm, *Dev. Biol.* **76**:203–214.

Webb, C. G., Gall, E., and Edelman, G. M., 1977, Synthesis distribution of H-2 antigens in preimplantation mouse embryos, *J. Exp. Med.* **146**:923–932.

Weil, A. J., 1965, The spermatozoa coating antigen (SCA) of the seminal vesicle, *Ann N.Y. Acad. Sci.* **124**:267–269.

Weiss, B., and Strada, S. J., 1973, Adenosine 3′,5′-monophosphate during fetal and postnatal development, in: *Fetal Pharmacology* (I. Boreus, ed.), pp. 205–235, Raven Press, New York.

Wiley, L. D., and Calarco, P. G., 1975, Effects of immune sera and their localization to cell surface during preimplantation development, *Dev. Biol.* **47**:407–418.

Wiley, L. M., Spindle, A. I., and Pedersen, R. A., 1978, Morphology of isolated mouse inner cell masses developing *in vitro, Dev. Biol.* **63**:1–10.

Willison, K. R., and Stern, P. L., 1978, Expression of Forssman antigenic specificity in preimplantation mouse embryos, *Cell* **14**:785–793.

Wilson, I. B., Bolton, E., and Cuttler, R. H., 1972, Preimplantation in mouse egg as revealed by microinjection of vital markers, *J. Embryol. Exp. Morphol.* **27**:467–479.

Wolf, D. P., Sokoloski, J., Khan, M. A., and Litt, M., 1977, Human cervical mucus. 3. Isolation and characterization of rheologically active mucin, *Fertil. Steril.* **28**:53–58.

Wolf, D. P., Blasco, L., Khan, M. A., and Litt, M., 1978, Human cervical mucus. 4. Viscoelasticity and sperm penetrability during ovulatory menstrual cycle, *Fertil. Steril.* **30**:163–169.

Wolf, D. P., Sokoloski, J., and Litt, M., 1980, Composition and function of human cervical mucus, *Biochim. Biophys. Acta* **630**:545–558.

Wolfe, J., Mautner, V., Hogan, B. L. M., and Tilly, R., 1979, Synthesis and retention of fibronectin (LETS protein) by mouse teratocarcinoma cells, *Exp. Cell Res.* **118**:63–71.

Yamada, K. M., 1982, Biochemistry of fibronectin in: *The Glycoconjugates* (M. I. Horowitz, ed.), Vol. III, Academic Press, New York.

Yamada, K. M., and Olden, K., 1978, Fibronectins: Adhesive glycoproteins of cell surface and blood, *Nature* (*London*) **275**:179–184.

Yamada, K. M., Kennedy, D. W., Kimata, K., and Pratt, R. M., 1980, Characterization of fibronectin interactions with glycosaminoglycans and identification of active proteolytic fragments, *J. Biol. Chem.* **255**:6055–6063.

Yanagimachi, R., 1981, Mechanisms of fertilization in mammals, in: *Fertilization and Embryonic Development in Vitro* (L. Mastrionni and J. D. Biggers, eds.), pp. 81–182, Plenum Press, New York.

Yanagimachi, R., Noda, Y. D., Fujimoto, M., and Nicolson, G. L., 1972, The distribution of negative surface charges on mammalian spermatozoa, *Am. J. Anat.* **135**:497–520.

Yorewicz, E. C., and Moghissi, K. S., 1981, Purification of human midcycle cervical mucin and characterization of its oligosaccharides with respect to size, composition and microheterogeneity, *J. Biol. Chem.* **256**:11895–11904.

Yoshida, C., and Takeichi, M., 1982, Teratocarcinoma cell adhesion: Identification of a cell-surface protein involved in calcium-dependent cell aggregation, *Cell* **28**:217–224.

Zetter, B. R., and Martin, G. R., 1978, Expression of a high molecular weight cell surface glycoprotein (LETS protein) by preimplantation mouse embryos and teratocarcinoma stem cells, *Proc. Natl. Acad. Sci. USA* **75**:2324–2328.

Ziomek, C. A., and Johnson, M. H., 1980, Cell surface interaction induces polarization of mouse 8-cell blastomeres at compaction, *Cell* **21**:935–942.

4

Cell Surface Glycoproteins and Carbohydrate Antigens in Development and Differentiation of Human Erythroid Cells

Minoru Fukuda and Michiko N. Fukuda

1. INTRODUCTION

In the past two decades, our knowledge of inheritance in biological systems has rapidly advanced. We now know that the information for the development of a mature organism is carried by the chemical structure of deoxyribonucleic acid, which, together with associated proteins, is present in nuclei. The next question is how the expression of this information is regulated in order to produce the appropriate highly organized structure. In this context, we have to face the fact that the processes of development and differentiation are highly complex yet miraculously ordered sequences of reactions. It is almost certain that the regulation of these events is the result not only of the expression of particular proteins but also of the interaction between these molecules in the same cell or in different cells. Thus, it is of critical importance to understand cell–cell

Minoru Fukuda and Michiko N. Fukuda • Cancer Research Center, La Jolla Cancer Research Foundation, La Jolla, California 92037. This article is dedicated to the late Professor Egami, who guided us with spirit and encouragement.

interactions in order to understand the development and differentiation of a living organism. Our major assumption is that these cell–cell interactions are mostly governed by cell surface specificities and our major concern is to describe the cell surface specificities in molecular terms.

In order to understand this complex phenomenon, it is important to choose a good model system. Hematopoietic tissue, particularly the erythroid system, is probably one of the best to work with. It is one of the few tissues that continues to renew itself in adults. This fact enables us to follow differentiation and maturation in rather isolated systems, in contrast to the studies on the differentiation and maturation of most of other tissues, which have to deal with embryogenesis in which the location of each embryonic tissue changes and becomes more complex as development proceeds. In addition, the progression of cells from the stem cell stage through discrete developmental stages has been clearly demonstrated in differentiation of the erythroid lineage (Clarkson *et al.*, 1978). In fact, *in vitro* cell culture systems, in which progenitor cells proliferate and mature to give rise to end cells, have been developed, e.g., BFU-E, CFU-E, and CFU-GM (Metcalf, 1977; see Fig. 1). In addition to these advantages, the erythroid cell lineage is particularly interesting for the following reasons. First, the membrane components of mature erythrocytes are comparatively well characterized (Steck, 1974; Marchesi *et al.*, 1976). Second, cell surface antigens of human erythrocytes have been extensively investigated for practical reasons, such as blood transfusion (Race and Sanger, 1975). They have been shown to possess a wide variety of cell surface specificities, which are encoded by cell surface carbohydrates or proteins. Third, the expression of the erythroid specific molecule, hemoglobin, has been extensively studied during development and differentiation (Clarkson *et al.*, 1978). It is also possible to study cell–cell interactions in this system. Interactions between immature hematopoietic cells and the stroma in bone marrow are evidently important in hematopoiesis, and some experimental systems have recently been developed that enable study of these interactions (Dexter and Testa, 1980). These advantages make the erythroid system one of the best models for studying cell surface specificities and cell–cell interactions.

In this chapter, we first summarize the present knowledge of the cell surface glycoproteins and carbohydrate structures of mature erythrocytes as revealed by newly developed methods. We then describe the changes in the carbohydrate structure of Band 3 during the development from fetus to adult. Section 4 deals with membrane differentiation during erythropoiesis. Sections 5 and 6 deal with surface markers in *in vitro* differentiation of leukemic cells and in hematological disorders including leukemia. Finally, we summarize the significance of studying cell surface markers and discuss future directions. Throughout this chapter, we focus

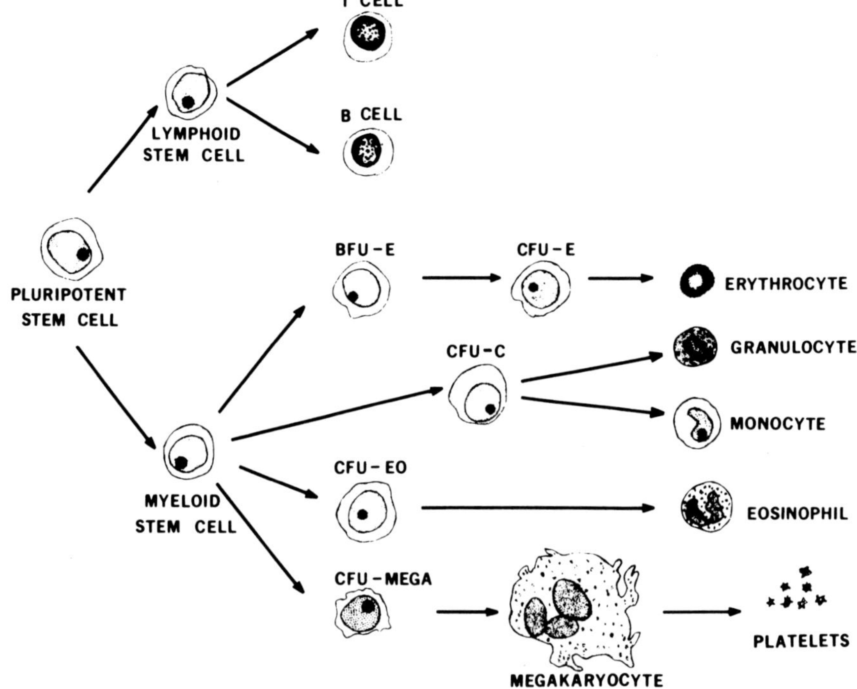

Figure 1. Model of hematopoiesis (from Koeffler and Golde, 1981). CFU-C are also called CFU-GM or GM-CFC. A pluripotent cell capable of self-renewing can be detected in the CFU-Mix and the CFU-S assay.

on human cells to emphasize the importance of understanding the role of cell surface components in cellular interactions of erythropoiesis in humans.

2. CELL SURFACE GLYCOPROTEINS AND CARBOHYDRATE STRUCTURE OF MATURE ERYTHROCYTES

2.1. Two Types of Membrane Proteins (Extrinsic and Intrinsic Proteins)

Mammalian cell membranes are basically composed of protein, lipid, and carbohydrate. In hemoglobin-free red cell membranes, approximately 52% of the membrane mass is protein, 40% is lipid, and 8% is carbohydrate. Carbohydrate is, however, present as conjugates either with protein (glycoproteins) or with lipids (glycosphingolipids) (Steck, 1974). One anal-

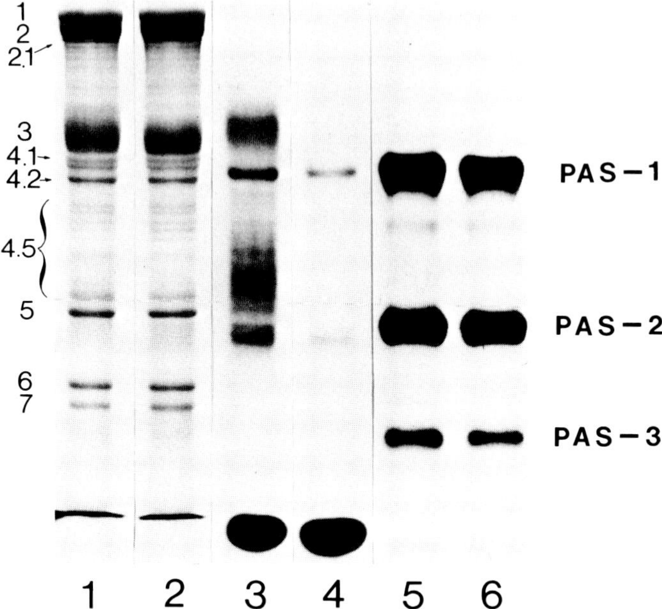

Figure 2. SDS–polyacrylamide gels of adult erythrocyte membrane before and after endo-β-galactosidase digestion. Cell surfaces were labeled by galactose oxidase/NaB³H₄ (gels 3, 4) and by periodate/NaB[³H]₄ (gels 5, 6). 1, 3, 5 are control cells and 2, 4, 6 are cells incubated with endo-β-galactosidase. Gels 3–6 are fluorogram and gels 1, 2 are protein stains of gels 3, 4 before they were treated for fluorography. Fetal erythrocytes showed a similar pattern except Band 3 was sharpened after the enzyme treatment. Electrophoresis was done according to Laemmli (1970). Cells were not pretreated with sialidase.

ysis suggests that only about 7% of this carbohydrate is carried by glycosphingolipids (Sweely and Dawson, 1969); therefore, a predominant portion of carbohydrate is attached to membrane proteins. Analysis of membrane proteins became possible after introduction of SDS–polyacrylamide gel electrophoresis (Weber and Osborn, 1969). This analytical tool revealed several distinct proteins of human erythrocyte membrane as detected by Coomassie blue (see Fig. 2). It has been possible to identify isolated polypeptides and follow one particular protein by SDS–gel electrophoresis. As analytical tools were gradually developed, three most intriguing questions concerning membrane organization were raised. (1) Where are these polypeptides located with respect to the hydrophobic core of the lipid bilayer? (2) To which side of the cell surface (external or cytoplasmic) is the protein attached? (3) Which proteins are glycosylated and where are the carbohydrates located?

In order to answer the first question, isolated membranes are subjected to various solvents in attempts to selectively extract one or the other components. By choosing a particular solvent, it was shown that the proteins of human erythrocytes can be operationally classified into two groups (Singer and Nicolson, 1972). One set of proteins is relatively loosely bound to the membranes and can be removed by varying the pH or ionic strength or by using chelating agents. This group of proteins, therefore, can be extracted witout destroying the lipid bilayer and have been named *extrinsic* proteins as they are external to the lipid bilayer. In contrast, the other set of proteins can be released from membranes with solvents, particularly detergents, that break down the lipid bilayer. The detergents not only disrupt the lipid bilayer but also yield detergent–protein or detergent–lipid–protein complexes that maintain the complexes water-soluble. This class of proteins are called *intrinsic* proteins. They are mostly transmembrane proteins, which penetrate the lipid bilayer (also see below).

With regard to the second question, intact cells are exposed to probes that do not penetrate the bilayer because of their size or chemical characteristics. The membrane constituents that react are judged to have at least part of the molecule residing on the external cell surface. The reactivity toward a probe at the outer surface is compared with that of isolated membranes that are permeable to a given reagent or with that of impermeable inside-out vesicles (Steck and Kant, 1974). If a constituent is reactive only on one surface of the membrane, it indicates the exclusive localization of that constituent on one of the two sides of the membrane. If a membrane component is accessible from both sides, it is probably a transmembrane polypeptide. In the initial stages of studies, protease and sialidase were used as probes to detect molecules. The results showed that acetylcholine esterase could be inactivated or removed by protease and that most of the sialic acid could be removed by sialidase, indicating localization of these components at the outer surface of the cell (Martin, 1970; Eylar *et al.*, 1962). The second, and more sensitive, approach is to introduce radioactive groups into various molecules. The membrane is impermeable to radioactive groups such as formylmethionyl sulfomethylphosphate (Bretscher, 1971), trinitrobentase sulfonate (Arrotti and Garvin, 1972), and stilbenone derivatives (Cabanthik and Rothstein, 1972), that bind covalently to reactive sites on proteins. Another approach is to introduce radioactive groups into reactive sites of protein by enzymes to which the cell membrane is impermeable. This is the case in lactoperoxidase-catalyzed iodination and this technique has frequently been used (Phillips and Morrison, 1970). The technique was originally devised by Phillips and Morrison and improved by adding the glucose oxidase–glucose system to generate peroxide (Hubbard and Cohn, 1972), as enzymatic formation of peroxide is milder. In order to answer the third question, specific labeling of cell surface carbohydrates has been used. A few methods have been developed for labeling cell surface carbohydrates, and will be discussed in the next section. In general, these enzyme-catalyzed methods of cell surface labeling are thought to be more specific and milder compared to the introduction of chemical reagents such as trinitrobenzenesulfonate. The enzyme-catalyzed approach also has the advantage that radioactive reagents are more available, as this type reaction requires only limited radioactive compounds such as ^{125}I, ^{131}I, or $NaB[^3H]_4$. From these experimental approaches, the following conclusions can be obtained. (1) Anisotropy of the erythrocyte membrane clearly exists; none of the proteins are randomly represented at both membrane surfaces and the proteins penetrating the lipid bilayer have asymmetric distribution about their locus. (2) There is absolute asymmetry in the distribution of carbohydrates. All the carbohydrates attached to proteins or lipids are present at the outer surface of erythrocytes. In addition, all integral proteins, which are expressed on the outer surface and associated with the lipid bilayer, are glycoproteins, as far as known.

The properties of human erythrocyte membranes are likely to be applicable to any other cell membrane, and so far no exception has been found to this rule.

2.2. Labeling of Cell Surface Carbohydrates

Two basic methods for specific labeling of cell surface carbohydrates have been developed so far. The first is to label sialic acid by mild periodate oxidation followed by reduction with $NaB[^3H]_4$ (Blumenfeld *et al.*, 1972). It was found that the lipid bilayer is not permeable to periodate at low temperature (0–4°C) (Gahmberg and Andersson, 1977). As sialic acid has adjacent glycols in the linear structure of C-7, C-8, C-9, these glycols can be selectively oxidized by a low concentration of periodate, and aldehydes are formed. The aldehyde groups are then reduced with radioactive $NaB[^3H]_4$ (Fig. 3A). The second approach is to label the nonreducing terminal galactose or *N*-acetylgalactosamine by galactose oxidase followed by $NaB[^3H]_4$ reduction (Gahmberg and Hakomori, 1973). Galactose oxidase oxidizes C-6 of galactose or *N*-acetylgalactosamine and the resultant C-6 aldehyde is reduced with radioactive $NaB[^3H]_4$ (see Fig. 3B). Galactose oxidase does not permeate the cell membrane because of its size and therefore only oxidizes substrate on the external cell membrane (Steck and Dawson, 1974). Most of the galactose residues are substituted with sialic acid, and are shielded from the action of galactose oxidase. Treatment with neuraminidase prior to galactose oxidase has been used to remove the terminal sialic acid and achieve maximum labeling of the

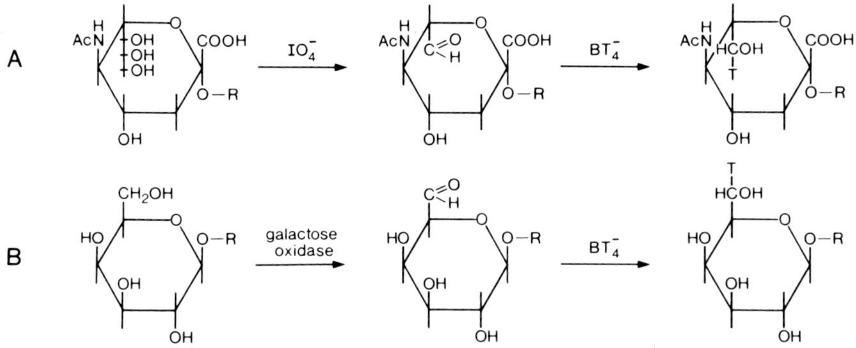

Figure 3. Scheme for labeling sialic acid and galactose. (A) Labeling of sialic acid residues by mild periodate oxidation followed by reduction with NaBT$_4$. (B) Labeling of galactose residues by galactose oxidase treatment followed by reduction with NaBT$_4$. T indicates tritium.

galactose/*N*-acetylgalactosamine residues. It should be noted that specific labeling of isolated glycoproteins and glycolipids had been described before developments of cell surface labeling (Morell *et al.*, 1966; Van Lenten and Ashwell, 1971). The radioactivity incorporated into carbohydrate chains by the methods above is ^{3}H; therefore, the development of fluorography of SDS gels has greatly facilitated the use of these techniques (Bonner and Laskey, 1974), particularly after Gahmberg showed the presence of a number of glycoproteins in human erythrocytes (Gahmberg, 1976).

When human erythrocytes were labeled by periodate oxidation or by the galactose oxidase procedure, the result shown in Fig. 2 was obtained. It was striking that periodate oxidation and galactose oxidase procedures revealed two different sets of cell surface glycoproteins in human erythrocytes. Sialoglycoproteins were specifically revealed by the periodate oxidation procedure, as expected. On the other hand, the galactose oxidase procedure significantly labels Band 3 and Band 4.5 glycoproteins but weakly labels PAS-1 and PAS-2, which are sialoglycoproteins. The selective labeling by using two procedures can only be obtained if the cells are labeled with galactose oxidase in the absence of pretreatment with sialidase. As described later, the differential labeling is actually due to the difference of carbohydrate structure and the two techniques differentiate between two different sets of carbohydrate chains in human erythrocytes.

2.3. Endo-β-galactosidase

Some interesting biological phenomena, such as the homing of lymphocytes, can be diminished by sialidase treatment of lymphocytes (Gesner and Ginsburg, 1964). Although enzymatic modification of the cell surface is an attractive approach to analyze the function of cell surface components, the carbohydrate-degrading enzymes (exo- and endoglycosidase), which can be applied to intact cell surfaces, have been limited. So far, bacterial sialidases, coffee bean α-galactosidase (Harpaz *et al.*, 1977), and *Escherichia freundii* endo-β-galactosidase have been demonstrated to hydrolyze cell surface carbohydrates, without altering the peptide moiety. Recent progress in characterizing the substrate of the endo-β-galactosidase (*E. freundii*) and the use of the enzyme to hydrolyze cell surface glycoconjugates revealed that this enzyme is useful both for the structural analysis of isolated glycoproteins, glycolipids, and oligosaccharides and for the modification of the intact cell surface glycoconjugates (see also Section 3.2).

Endo-β-galactosidase activity produced by *E. freundii* was first noticed as keratan sulfate-degrading enzyme (Kitamikado and Ueno, 1970). Keratan sulfate-degrading enzymes had been found in *Pseudomonas* (Nakazawa *et al.*, 1975) and *Coccobacillus* (Hirano and Meyer, 1971); however, the most important characteristic of the *E. freundii* enzyme is that it hydrolyzes various glycoproteins, glycosphingolipids, and oligosaccharides that have the common structure R→GlcNAcβ1→3Galβ1→4GlcNAc (or Glc) (Fukuda and Matsumura, 1976). Thus, the *E. freundii* enzyme can be called an endo-β-galactosidase. Since then, several endo-β-galactosidases (which hydrolyze both keratan sulfate and other substrates) with a similar substrate specificity as *E. freundii* endo-β-galactose, have been found in *Flavobacterium* (Kitamikado *et al.*, 1981), and *Bacteriodes fragilis* (Doinel *et al.*, 1980).

Table 1 summarizes the substrate specificity of *E. freundii* endo-β-galactosidase (M. N. Fukuda, 1981). The minimum structural requirement for endo-β-galactose is GlcNAcβ1→3Galβ1→4GlcNAc (or Glc). This enzyme can hydrolyze both type 1 chains (R'→Galβ1→*3*GlcNAcβ1→3Galβ1→4R) and type 2 chains (R'→Galβ1→*4*GlcNAcβ1→3Galβ1→4R), which are widely present in blood group-active glycoconjugates. The enzyme can hydrolyze the β-galactosidic residue which has two substitutions at the C-3 and C-4 positions of the neighboring *N*-acetylglucosamine. Substitution on the C-6 position of *N*-acetylglucosamine does not hinder access of the enzyme. In contrast, the substitution at the C-6 position of the galactose residue by sulfate or sugar residues prevents the enzymatic hydrolysis, and hydrolysis does not occur under the conditions described in Table 1 (Fukuda and Matsumura, 1976).

The aglycon moiety (expressed as R in Table 1) also influences the enzymatic hydrolysis. For example, lacto-*N*-tetraitol (Galβ1→4GlcNAcβ1→3Galβ1→4 sorbitol) cannot be hydrolyzed under the enzyme concentration of 125 mU/ml, whereas lacto-*N*-tetraose (Galβ1→

4GlcNAcβ1→3Galβ1→4Glc) can be hydrolyzed under the same conditions (Fukuda and Matsumura, 1975). Reduction of the reducing terminal residue of the oligosaccharide impairs the rate of hydrolysis and similar results were reported for endo-β-*N*-acetylglucosaminidase D (Tarentino and Maley, 1975). The use of endo-β-galactosidase enables us to detect the lactosaminoglycan chain in glycoproteins as seen in Band 3 (Järnefelt *et al.*, 1978; M. Fukuda *et al.*, 1979), glucose transporter (Gorga *et al.*, 1979), glycoproteins in human myelocytic leukemia cells (Fukuda *et al.*, 1981a), and human and mouse embryonal carcinoma cells (Muramatsu *et al.*, 1979, 1982; Rasilo and Renkonen, 1982).

During these studies, two other enzymes distinct from *E. freundii* endo-β-galactosidase have been reported. One enzyme was the keratanase and sulfatase isolated from soil bacterium, *Pseudomonas* (Nakazawa *et al.*, 1975). This enzyme, however, is not a true endo-β-galactosidase, for the enzyme could act only on keratan sulfate but not on other glycoconjugates containing galactose and *N*-acetylglucosamine. It is, therefore, a misnomer that this enzyme is frequently called an endo-β-galactosidase in the literature. If one uses this *Pseudomonas* keratanase, one has to assume that only keratan sulfate is detected by the enzyme. Takasaki and Kobata (1976) reported another type of endo-β-galactosidase from *D. pneumoniae*, which cleaves A and B blood group haptenic structures as follows:

$$\text{Gal(or GalNAc)}\alpha1{\rightarrow}3\text{Gal}\beta1{\rightarrow}4\text{GlcNAc}\beta1{\rightarrow}R$$
$$\underset{\underset{\text{Fuc}\alpha1}{\uparrow}}{2}$$

$$\Rightarrow \quad \text{Gal(or GalNAc)}\alpha1{\rightarrow}3\text{Gal} \ + \ \text{GlcNAc}\beta1{\rightarrow}R$$
$$\underset{\underset{\text{Fuc}\alpha1}{\uparrow}}{2}$$

The enzyme does not act on H-haptenic structure. Recently, it was found that the enzyme fraction initially reported by Takasaki and Kobata also contains endo-β-galactosidase that has a similar substrate specificity to *E. freundii* endo-β-galactosidase (Fukuda, 1982).

2.4. Lactosaminoglycan

Our knowledge of the carbohydrate structures in glycoproteins, particularly those linked to asparagine, has been greatly enriched during the last 10 years and reviewed on several occasions (Kornfeld and Kornfeld,

Table 1. Specificity of E. freundii Endo-β-galactosidase[a]

Structure[b]	Hydrolyzability[c]	Substrates	Ref.[d]
GlcNAcβ1→3Galβ1$\Downarrow$4R	+	Keratan sulfate	1,2
		Lacto-*N*-triaosylceramide	4
		Band 3	3
GalNAcβ1→3Galβ1$\Downarrow$4R	+	X$_2$-glycolipid	5
Galβ1→3GlcNAcβ1→3Galβ1$\Downarrow$4R	+	Lacto-*N*-tetraose	6
Galβ1→4GlcNAcβ1→3Galβ1$\Downarrow$4R	+	Paragloboside	4
Fucα1→2Galβ1→3GlcNAcβ1→3Galβ1$\Downarrow$4R	+	Lacto-*N*-fucopentaose I	7
R′→Galβ1→4GlcNAcβ1→3Galβ1$\Downarrow$4R	+	H$_1$-glycolipid, sialosylparagloboside, etc.	4
Fucα1→2Galβ1→4GlcNAcβ1→3Galβ1$\Downarrow$4R 3 ↑ GalNAcα1	+	A^a- and A^b-glycolipid	4
Galβ1→3GlcNAcβ1→3Galβ1$\Downarrow$4R 4 ↑ Fucα1	+	Lacto-*N*-fucopentaose II (Le^a)	7
Galβ1→4GlcNAcβ1→3Galβ1$\Downarrow$4R 3 ↑ Fucα1	+	Lacto-*N*-fucopentaose III (Le^x)	7

SO_4
↓
6
GlcNAcβ1→3Galβ1⇓→4R   +   Keratan sulfate   2

SO_4
↓
6
GlcNAcβ1→3Galβ1⇓→4R   −   Keratan sulfate   2
R″→Galβ1→4GlcNAcβ1
↘
6
Galβ1⇓→4R   −   H₃-glycolipid   4
3   Band 3   3
↗
R″→Galβ1→4GlcNAcβ1

---

[a] Modified from Table IV in M. N. Fukuda (1981).

[b] R, glucosyl or *N*-acetylglucosaminyl residue. *R′*, Galβ1→3, Galα1→3, NeuAcα2→3, and Fucα1→2. *R″*, Fucα1→2 and (Galβ1→4GlcNAcβ1→3)$_n$.

[c] Hydrolysis under the enzyme concentration at 125 mU/ml. Hydrolysis position is indicated by the open arrows.

[d] References: 1, Kitamikado and Ueno (1970); 2, Fukada and Matsumura (1976); 3, M. Fukuda *et al.* (1979); 4, M. N. Fukuda *et al.* (1978b); 5, Kannagi *et al.* (1982); 6, Fukuda and Matsumura (1975); 7, M. N. Fukuda (1981).

1976, 1980; Montreuil, 1980). The progress in this area has been made possible by improvement in various analytical methods including: (1) separation of glycopeptides after exhaustive digestion of the glycoprotein by Pronase and passage of the glycopeptides through lectin affinity columns, particularly Con A columns, which appear to separate them according to their antennary structure (Ogata *et al.*, 1975; Finne *et al.*, 1980). (2) Sugar sequences and anomeric configuration of each sugar residue have been determined by isolation of exoglycosidases (e.g., Muramatsu and Egami, 1967; Li, 1967) and endoglycosidases (Muramatsu, 1971; Tarentino and Maley, 1974; Fukuda and Matsumura, 1976). (3) Linkage between sugars and sugar sequences have been determined by effective methylation (Hakomori, 1964; Sandford and Conrad, 1966) followed by gas chromatography–mass spectrometric identification of partially *O*-methylated sugars and amino sugars (Björndal *et al.*, 1970; Stellner *et al.*, 1973) or followed by direct-probe mass spectrometry (Karlsson *et al.*, 1974). (4) Hydrazinolysis has been applied in the chemical cleavage of GlcNAc→Asn linkage (Fukuda *et al.*, 1976).

From the accumulated data of the elucidated structures, a few conclusions can be drawn at this point. (1) With a few exceptions, the core portion of asparagine-linked carbohydrate chains has the same structure of Manα1→6(Manα1→3)Manβ1→4GlcNAcβ1→4GlcNAc→Asn. (2) Asparagine-linked carbohydrate chain can be classified into two groups according to the components of the carbohydrate chains: high-mannose and complex type (Kornfeld and Kornfeld, 1976). In addition to the core structure mentioned above, the high-mannose type of carbohydrate chain has additional mannose residues of Manα1→2 linkage or Manα1→6 (Manα1→3)Man branching structure. In the complex type, on the other hand, side chains composed of sialyl→Gal→GlcNAc are attached to the core portion by a variety of linkages and these carbohydrates contain a variable amount of sialic acid. The number of side chains is also variable. In addition, the innermost *N*-acetylglucosamine residues which are linked to asparagine are frequently substituted with fucose through an α1→6 linkage. For other aspects of asparagine-linked oligosaccharides, the reader is referred to recent reviews (Kornfeld and Kornfeld, 1980; Montreuil, 1980; Hakomori *et al.*, 1983).

During the last few years, a novel type of asparagine-linked carbohydrate chain has gradually been recognized. This type of carbohydrate chain, exemplified in the carbohydrate chain of Band 3, is made up of large side chains and a core portion (Finne *et al.*, 1978; Järnefelt *et al.*, 1978; M. Fukuda *et al.*, 1979). The side chains have a characteristic structure composed of a (Galβ1→4GlcNAcβ1→3)$_n$ repeating unit. This side chain is similar to the carbohydrate chain of the ABO blood group antigens from mucus glycoproteins, and part of the terminal regions have ABO blood group antigenic structure. This type of carbohydrate was initially suggested by Tanner and Boxer (1972). They found that human erythrocyte protein E (equivalent to Band 3) and protein F contained carbohydrate chains of which a large portion was galactose and N-acetylglucosamine. Significantly, they also noticed that these carbohydrates contained a small amount of N-acetylgalactosamine when the blood type was A, but not when the blood type was O, suggesting that these carbohydrate chains might carry the ABO blood group determinants. Adair and Kornfeld (1974) found that glycoproteins bound to ricin–Sepharose columns are enriched in galactose and N-acetylglucosamine. Complementary to these studies, Fukuda and Osawa (1973) observed that glycophorin does not bind well to ricin or Con A, and Findlay (1974) isolated Band 3 by Con A absorption. Thus, it became clear at that point that carbohydrate chains of Band 3 and glycophorin are different. Gahmberg *et al.* (1976) then showed for the first time that Band 3 contains one sugar chain of high molecular weight. Drickamer (1978) confirmed the large size of Band 3 carbohydrate chain and pointed out the uniqueness of the carbohydrate moiety of Band 3. Fukuda *et al.* (1978) established the purification procedure of Band 3 and showed that the carbohydrate composition of Band 3 is rich in galactose and N-acetylglucosamine and uniquely different from other glycoproteins. Finne *et al.* (1978) then reported ABO blood group-active glycopeptides of high molecular weight from erythrocytes. Järnefelt *et al.* (1978) also reported glycopeptides of high molecular weight, termed *erythroglycan*, that can be digested with endo-β-galactosidase. As the carbohydrate analysis showed that these glycopeptides were also rich in galactose and N-acetylglucosamine, they suggested that these glycopeptides were derived from Band 3. M. Fukuda *et al.* (1979) confirmed this idea by characterizing glycopeptides from purified Band 3. In addition, they showed a pronounced change in the side chain structure of these glycopeptides during development from fetus to adult (M. Fukuda *et al.*, 1979). From structural analysis by chemical fragmentation, endo-β-galactosidase digestion, and permethylation studies, the carbohydrate chain of Band 3 has been shown to consist of side chains composed of a $(Gal\beta1\rightarrow4GlcNAc\beta1\rightarrow3)_n$ repeating unit and a $(Man)_3(GlcNAc)_2$ core (Krusius *et al.*, 1978; Järnefelt *et al.*, 1978; M. Fukuda *et al.*, 1979) (see Fig. 4A).

The side chains of Band 3 have Ii (Childs *et al.*, 1978; M. N. Fukuda *et al.*, 1979) and ABO antigenic activities (Finne *et al.*, 1978; Järnefelt *et al.*, 1978; M. Fukuda *et al.*, 1979) (see also Section 3). This class of carbohydrate chains has been found not only in human erythroid cells but also in various other cells such as Chinese hamster ovary cells (Li *et al.*, 1979), Friend erythroleukemic cells (Kaizu *et al.*, 1982), mouse embryonic cells (Muramatsu *et al.*, 1979), and possibly human myeloid leu-

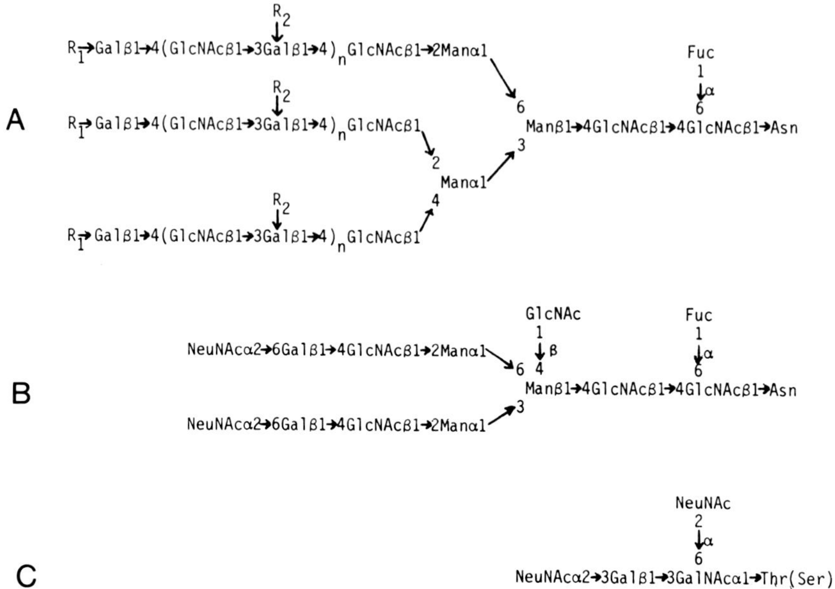

Figure 4. Three types of carbohydrate chains present on mature erythrocytes. (A) Lacto-saminoglycan present in Band 3 and Band 4.5 glycoproteins. R_1, H, Fucα1→2, Neu-NAcα2→3/6; R2, R1→Galβ1→4GlcNAcβ1→6; n = 4–5. (B) Asn-linked oligosaccharides present on glycophorin A. (C) Tetrasaccharide units present in glycophorins. Glycophorin A contains 15 chains of this type.

kemic cells (Fukuda *et al.*, 1981a). For this reason, we named this type of carbohydrate chain *lactosaminoglycan*, as the term *erythroglycan* is only applicable to erythroid cells.

Although the carbohydrate unit of Band 3 is the best characterized oligosaccharide among the lactosaminoglycans, several important aspects of this carbohydrate chain are still unknown. First, it has been shown (Tsuji *et al.*, 1980) that the core structure of Band 3 is essentially the same as other asparagine-linked oligosaccharides, i.e., Manα1→6 (Manα1→3)Manβ1→4GlcNAcβ1→4GlcNAc, but it is not yet known if the β-linked mannose is substituted with additional *N*-acetylglucosamine (bisect N-acetylglucosamine). Second, it is not clear whether the side chains are extended only through α-linked mannose residues as occurs with the other glycoproteins. Third, the number of side chains present in the unit has not been determined, i.e., biantennary or triantennary. Finally, it has not been shown whether any difference exists between gly-cophorin and Band 3 in the core structure of asparagine-linked oligosac-charides (see Fig. 4). Very recently we have determined that the core structure of fetal lactosaminoglycan isolated from cord Band 3 is indistinguishable from that of the complex-type carbohydrate chains (Fukuda *et al.*, 1984).

2.5. Two Types of Glycoproteins with Distinct Functions and Organization

The cell surface of human erythrocytes was labeled by galactose oxidase or by the periodate procedure and followed by endo-β-galactosidase digestion. Membranes were isolated and analyzed by SDS–polyacrylamide gel electrophoresis. As shown in Fig. 2, a significant amount of the glycoproteins labeled by galactose oxidase was digested by the endo-β-galactosidase, whereas periodate-labeled glycoproteins were scarcely affected (M. N. Fukuda *et al.*, 1979). In other words, carbohydrate chains of Band 3 and Band 4.5 and a part of the carbohydrate chains of sialoglycoproteins are susceptible to endo-β-galactosidase. It is interesting that glycoproteins with this type of carbohydrate chain generally have heterogeneity in molecular weight and show a broad band on SDS gel electrophoresis. This is because there is considerable variation of the size and structure of lactosaminoglycans, which contributes to the heterogeneity of molecular size. On the other hand, the majority of oligosaccharides of sialoglycoproteins have a tetrasaccharide structure as shown in Fig. 4C (Thomas and Winzler, 1969). This oligosaccharide cannot be cleaved by endo-β-galactosidase (M. N. Fukuda *et al.*, 1978b). Sialoglycoproteins also have one asparagine-linked oligosaccharide and those structures were elucidated as shown in Fig. 4B (Yoshima *et al.*, 1980; Irimura *et al.*, 1981). Thus, these two types of glycoproteins, Band 3 and Band 4.5 versus sialoglycoproteins, are distinctly different in their oligosaccharide structures.

The major sialoglycoprotein, glycophorin A, has been extensively studied and has been shown to traverse the lipid bilayer once (Marchesi *et al.*, 1976). The function of glycophorin A is not clear, for blood group type En(a−) individuals lack this glycoprotein with no detectable pathological symptoms (Gahmberg *et al.*, 1976). Sialoglycoproteins probably serve as structural elements, whereas Band 3 is the anion transporter for human erythrocytes (Cabanthik and Rothstein, 1972). This protein, therefore, plays an important role in the exchange of carbonate with chloride, in situations when cells are loaded with carbonate. At least part of the Band 4.5 glycoproteins are functional, as the glucose transporter of erythrocytes migrates at the Band 4.5 region on SDS gel electrophoresis. In fact, isolated glucose transporter shows a broad band in SDS gel electrophoresis and is susceptible to endo-β-galactosidase digestion (Gorga *et al.*, 1979). The molecular architecture of Band 3 has also been characterized, and it is assumed that Band 3 traverses the lipid bilayer several

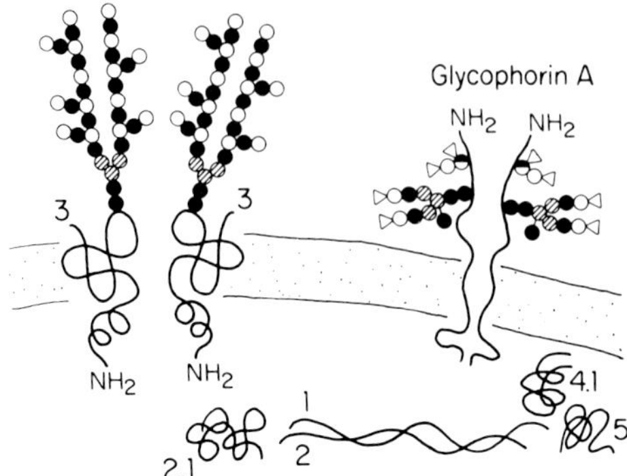

Figure 5. A schematic version of the organization of human erythrocyte membrane proteins. Band 3 traverses the lipid bilayer several times and a long carbohydrate chain is attached to this protein. Glycophorin traverses the lipid bilayer one time, and 15 short oligosaccharides and a relatively short asparagine-linked carbohydrate chain are attached to the protein (only one short oligosaccharide is shown in the figure). Band 2.1 is also called ankyrin, 1, 2 is spectrin, and Band 5 is actin. ●, N-acetylglucosamine; ◉, mannose; ○, galactose; ◐, N-acetylgalactosamine; △, sialic acid. This version is based on the data in Marchesi *et al.* (1976), M. Fukuda *et al.* (1978), Steck *et al.* (1978), Bennet and Stenbuck (1979), Tyler *et al.* (1980), and Morrow *et al.* (1980). It is also suggested that certain populations of Band 3 and glycophorin associate with each other (Nigg *et al.*, 1980).

times, as shown in Fig. 5. Band 4.5 seems to have a similar organization, as it is not very susceptible to the action of protease on the internal or external cell membrane (M. Fukuda and V. T. Marchesi, unpublished data). It is also worth noting that part of the Band 4.5 proteins can be a proteolytic degradation product of Band 3 (Tarone *et al.*, 1979). These two sets of glycoproteins (Band 3, Band 4.5 versus glycophorins) are, therefore, distinctly different in the molecular organization, function, and structure of the attached oligosaccharides (see Fig. 5). Long-chain oligosaccharides, i.e., lactosaminoglycans, may control the polypeptide folding during its biosynthesis and insertion into the lipid bilayer (Fukuda and Fukuda, 1981).

3. DEVELOPMENTAL CHANGES IN CARBOHYDRATE STRUCTURE

3.1. Ii Antigens

Human erythrocytes undergo a pronounced change in hemoglobin content and in cell surface antigens after birth. During the first year of human life, fetal hemoglobin is replaced by adult hemoglobin and fetal antigen (i) is changed into adult (I) antigen. These changes are presumably necessary to adapt the newborn to the external environment (*in utero* versus *outside utero*).

Fetal (i) and adult (I) antigens were initially described as blood group antigens in adults. Wiener *et al.* (1956) reported that a patient with acquired hemolytic anemia produced antibodies (anti-I) that react with the erythrocytes (I antigen) of all normal adults, with the exception of 5 out of 22,000 individuals who exclusively expressed i antigen. Marsh (1961) then found cold agglutinin that reacted with i erythrocytes. Although the majority of adult human erythrocytes belong to the I-positive group, fetal or neonatal erythrocytes react strongly with anti-i antibodies and weakly with anti-I antibodies. Thus, it is assumed that the majority of adults gain I antigen and lose i antigen during development while a very small population of adults lacks such conversion (i-variant adults). After the discovery of anti-I antibodies, it was gradually realized that these antibodies recognize different antigenic determinants and that they might be classified into several groups (Feizi *et al.*, 1971a). A similar classification is also possible for anti-i antibodies. Anti-I or -i antibodies are, therefore, suffixed with the name of the patient who produced the antibodies. A chemical description of these antigens was initiated by Marcus and Kabat (Marcus *et al.*, 1963); they showed that when normal erythrocytes were treated with β-galactosidase and β-N-acetylglucosaminidase, there was a decrease in agglutination of erythrocytes caused by anti-I antibodies, with a concomitant release of galactose and N-acetylglucosamine. Feizi, Kabat, and their colleagues (Feizi *et al.*, 1971b) subsequently found that blood group I antigen is related to blood group ABH by showing that the Smith-degradation products of ABH antigen cross-reacted with anti-I antibodies. They first identified the antigenic determinant for anti-I (Ma) as Galβ1$\rightarrow$4GlcNAcβ1$\rightarrow$6Gal.

Watanabe *et al.* (1975) later observed that a glycolipid fraction with similar TLC behavior as branched glycolipid showed I (Ma) antigenic activity and that a fraction with similar TLC behavior as linear glycolipid showed i-antigenic activity. They isolated two gangliosides (i.e., glycolipids containing sialic acid) and prepared various derivatives of those gangliosides by exoglycosidase digestion. The ganglioside derivatives were tested for i-antigenic and I-antigenic activities by radioimmunoassay. The data clearly indicate that I-antigenic determinants are located in various parts of the branched structures of the glycolipids and that the antibodies can be classified into three groups according to the antigenic determinants that are recognized by the antibodies (Watanabe *et al.*, 1979; Feizi *et al.*, 1979). The antigenic determinants for anti-i antibodies, on the other hand, reside on the linear chain structure of glycolipids (Nieman *et al.*, 1978; Watanabe *et al.*, 1979) (see Fig. 6).

3.2. Developmental Changes in Lactosaminoglycan

In parallel with the progress in studies on i- and I-antigenic determinants, described above, structural distinctions between fetal and adult erythrocytes have been described. This research was prompted by two

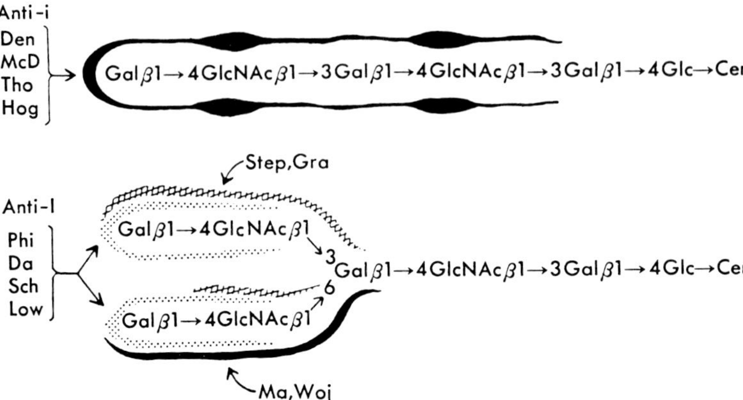

Figure 6. Structures of antigenic determinants for I and i antigens. I- and i-antigenic activities are expressed by branched polylactosamine (I) (Watanabe *et al.*, 1979; Feizi *et al.*, 1979) and linear repeating lactosaminyl structure (i) (Nieman *et al.*, 1978; Watanabe *et al.*, 1979). Endo-β-galactosidase cleaves the linear chain of polylactosamine extensively but poorly hydrolyzes the branched galactose (Fukuda and Matsumura, 1976; M. N. Fukuda *et al.*, 1978b). See also Table 1.

separate lines of evidence that indicated that Band 3 is a major carrier for I and i antigens. First, when intact human erythrocytes were treated with endo-β-galactosidase, there was a considerable decrease in the I-antigenic activities of the cells (M. N. Fukuda *et al.*, 1978a). We have also demonstrated that Band 3 is the major glycoprotein affected by endo-β-galactosidase (Fig. 2). Thus, it could be reasonably assumed that the reduction of Ii-antigenic activity observed following endo-β-galactosidase treatment is the result of hydrolysis of the carbohydrate chains of Band 3 (M. N. Fukuda *et al.*, 1979). The second line of evidence came from affinity adsorption by anti-I antibodies. Band 3 glycoprotein was found to be a major component in erythrocyte glycoproteins that bound to an anti-I antiserum column, when total erythrocyte glycoproteins were applied to the column. Band 3 glycoprotein purified by our method (M. Fukuda *et al.*, 1978) also bound to an anti-I column (Childs *et al.*, 1978). These results clearly indicated that Band 3 glycoprotein is the major carrier for Ii antigens.

We then compared the carbohydrate chains of Band 3 purified from normal adult erythrocytes (OI), an adult variant that fails to express I antigen (Oi) and umbilical cord erythrocytes (Oi) (which are known to have similar cell surface antigens as fetal erythrocytes). It was noticed that glycopeptides of Band 3 from I erythrocytes (I-Band 3) behaved as

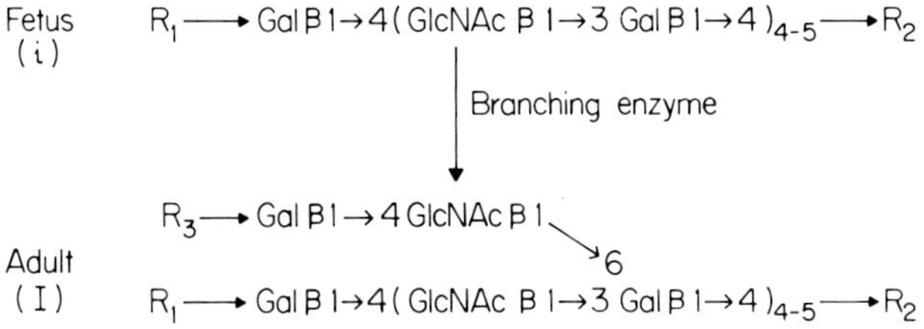

Fetus (i) $R_1 \longrightarrow Gal\beta1\rightarrow4(GlcNAc\ \beta1\rightarrow3\ Gal\beta1\rightarrow4)_{4\text{-}5}\longrightarrow R_2$

Branching enzyme

Adult (I) $R_3 \longrightarrow Gal\ \beta1\rightarrow4\ GlcNAc\ \beta1\searrow_{6}$

$R_1 \longrightarrow Gal\beta1\rightarrow4(GlcNAc\ \beta1\rightarrow3\ Gal\beta1\rightarrow4)_{4\text{-}5}\longrightarrow R_2$

Figure 7. Proposed structural change of Band 3 carbohydrate chain during development from fetus to adult. R_1, H, Fuc$\alpha1\rightarrow2$, NeuNAc$\alpha2\rightarrow3/6$; R_2, (Man)$_3$(GlcNAc) 4–5; R_3, the same as R_1. Two or three chains shown here are linked to the same core portion (R_2).

if the molecular weight were higher than that of Band 3 from i erythrocytes (i-Band 3). Analysis of the chemical compositions indicated that I-Band 3 has more galactose and N-acetylglucosamine than i-Band 3. Permethylation analysis indicated that i-Band 3 has a side chain of linear polylactosaminyl structure, $(Gal\beta1\rightarrow4GlcNAc\beta1\rightarrow3)_n$, whereas I-Band 3 has highly branched polylactosaminyl structure, $R \rightarrow Gal\beta1\rightarrow 4GlcNAc\beta1 \rightarrow 3(Gal\beta1\rightarrow4GlcNAc\beta1 \rightarrow 6)Gal\beta1 \rightarrow 4GlcNAc \rightarrow R'$. Nonetheless, no significant difference between the core structure of the Band 3 carbohydrate from I and i erythrocytes was detected. The results were supported by the characteristics of the oligosaccharides released by endo-β-galactosidase treatment. i-Band 3 produced mostly low-molecular-weight oligosaccharides whereas I-Band 3 produced oligosaccharides of high molecular weight and various sizes. These higher oligosaccharides were found to possess a branched structure. This result was obtained by the fact that endo-β-galactosidase hydrolyzes very little of the branched galactose chain whereas linear-chain polylactosamine can be extensively hydrolyzed, as shown in Section 2.3 (see Table 1). These results clearly indicate that a major event in the change from i to I is the conversion of a linear polylactosamine structure to a branched one as shown in Fig. 7 (M. Fukuda *et al.*, 1979). More recently, Fukuda *et al.* (1984) have shown that fetal lactosaminoglycan contains NeuNAc$\alpha2\rightarrow8$ NeuNAc$\alpha2\rightarrow3$Gal structure whereas such structure is absent in adult lactosaminoglycan. During this analysis, we found that the average number of repeating units in fetal lactosaminoglycan $(Gal\beta1\rightarrow4GlcNAc\beta1\rightarrow3)_n$ is 4 or 5. This calculation is based on the ratio of disaccharides to tri- plus tetrasaccharides released from i-Band 3. The disaccharide $GlcNAc\beta1\rightarrow3Gal$ should be derived from the internal part of the side chain whereas the trisaccharide $Gal\beta1\rightarrow4GlcNAc\beta1\rightarrow3Gal$ and tetrasaccharide Fuc$\alpha1\rightarrow2Gal\beta1\rightarrow4GlcNAc\beta1\rightarrow3Gal$ should be derived from the nonreducing terminus, because very little nonreducing terminal N-acetylglucosamine residue

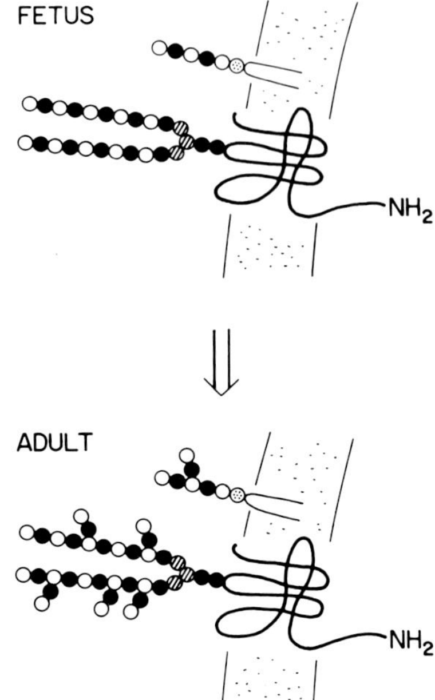

Figure 8. Idealized version of structure change of lactosaminoglycan on glycolipid (paragloboside) and glycoprotein (Band 3). In fetal erythrocytes, a linear unbranched polylactosamine, $(Gal{\rightarrow}GlcNAc)_n$, is linked to $Gal{\rightarrow}Glc{\rightarrow}Cer$ or to $(Man)_3(GlcNAc)_2$ core of Band 3. When fetal erythrocytes develop into adult erythrocytes, the linear chains are converted to those having $Gal{\rightarrow}GlcNAc$ branchings.

is present in intact molecules (M. Fukuda *et al.*, 1979). The molar ratio of disaccharide to tri- plus tetrasaccharide was found to be 3.4.

When oligosaccharides released from surface-labeled cells were fractionated by Sephadex G-50 gel filtration, distinctly different patterns were obtained from I and i erythrocytes: I adult cells released a number of oligosaccharides of various sizes whereas i cells produced oligosaccharides of relatively low molecular weight, similar to the oligosaccharides released from I or i Band 3. The oligosaccharides released by endo-β-galactoside from cell surfaces are derived from lactosaminoglycan of both glycoproteins and glycolipids. The results, therefore, clearly indicate that cell surface glycoconjugates in general change their carbohydrate structure from linear polylactosamine to branched polylactosamine during development from fetus to adult, as depicted in Fig. 8.

The results discussed above demonstrate that the technique of cell surface labeling combined with endo-β-galactosidase digestion can provide the following information. First, it is possible to infer when lactosaminoglycan-type glycoconjugates are carrying a given antigenic determinant, by the reduction of antigenic activity following endo-β-galactosidase digestion, and to identify glycoprotein carrying the polylactosaminyl structure by its susceptibility to cleavage by the enzyme. Furthermore, it is possible to distinguish between adult (I) and fetal (i) antigens by the size distribution of the oligosaccharide structures that are released from the erythrocyte cell surface by endo-β-galactosidase.

These results on Ii antigen structures in Band 3 are consistent with other reports. Ebert *et al.* (1975) found that the I-active glycopeptide fraction prepared by papain digestion was devoid of MN activity, suggesting that it was derived from a glycopeptide other than glycophorin. Watanabe and Hakomori (1976) found that the amount of branched glycolipids is lower in fetal or neonatal erythrocytes compared to that in adult erythrocytes by using techniques of cell surface labeling or agglutination by specific antibodies. Koschielak *et al.* (1979) isolated large glycolipids (''polyglycosyl ceramides'') from both adult and neonatal erythrocytes and found that polyglycosyl ceramides from neonatal erythrocytes contained considerably fewer branching structures than those from adult erythrocytes. Fukuda and Levery (1983) further extended these findings to sialic acid-containing glycolipids (gangliosides) and showed also that the amount of gangliosides with lactosyl structure is much higher in neonatal (cord) cells than in adult cells, although the total amount of glycolipids is indistinguishable between cord and adult cells. These combined results indicate that carbohydrate chains in Band 3 and glycolipids are carriers for Ii antigens and those structures change from linear polylactosamine to branched polylactosamine during development from fetus to adult.

4. CHANGES IN CELL SURFACE GLYCOPROTEINS AND CARBOHYDRATE ANTIGENS DURING ERYTHROPOIESIS

4.1. Erythropoiesis—A Model for Differentiation

Erythropoiesis is the process by which the supply of mature, functional erythrocytes is maintained to the peripheral circulation. In adult life, erythrocytes are ultimately derived from the pluripotent self-renewing hematopoietic stem cell. The stem cells differentiate to produce erythroid-committed progenitor cells that undergo extensive proliferation and mature through a sequence of developmental stages to produce erythrocytes (Fig. 9). The hematopoietic stem cell and the progenitor cell populations cannot be recognized morphologically. However, certain functional assays are available that can be used to detect these cell types. Murine stem cells can be assayed by the CFU-S (colony-forming unit—spleen) technique (Till and McCulloch, 1961). Irradiated mice are injected with a hematopoietic cell population and 8–10 days later macroscopic colonies of developing and mature hematopoietic cells of all myeloid lineages can be seen on the surface of the spleen. It has been shown that

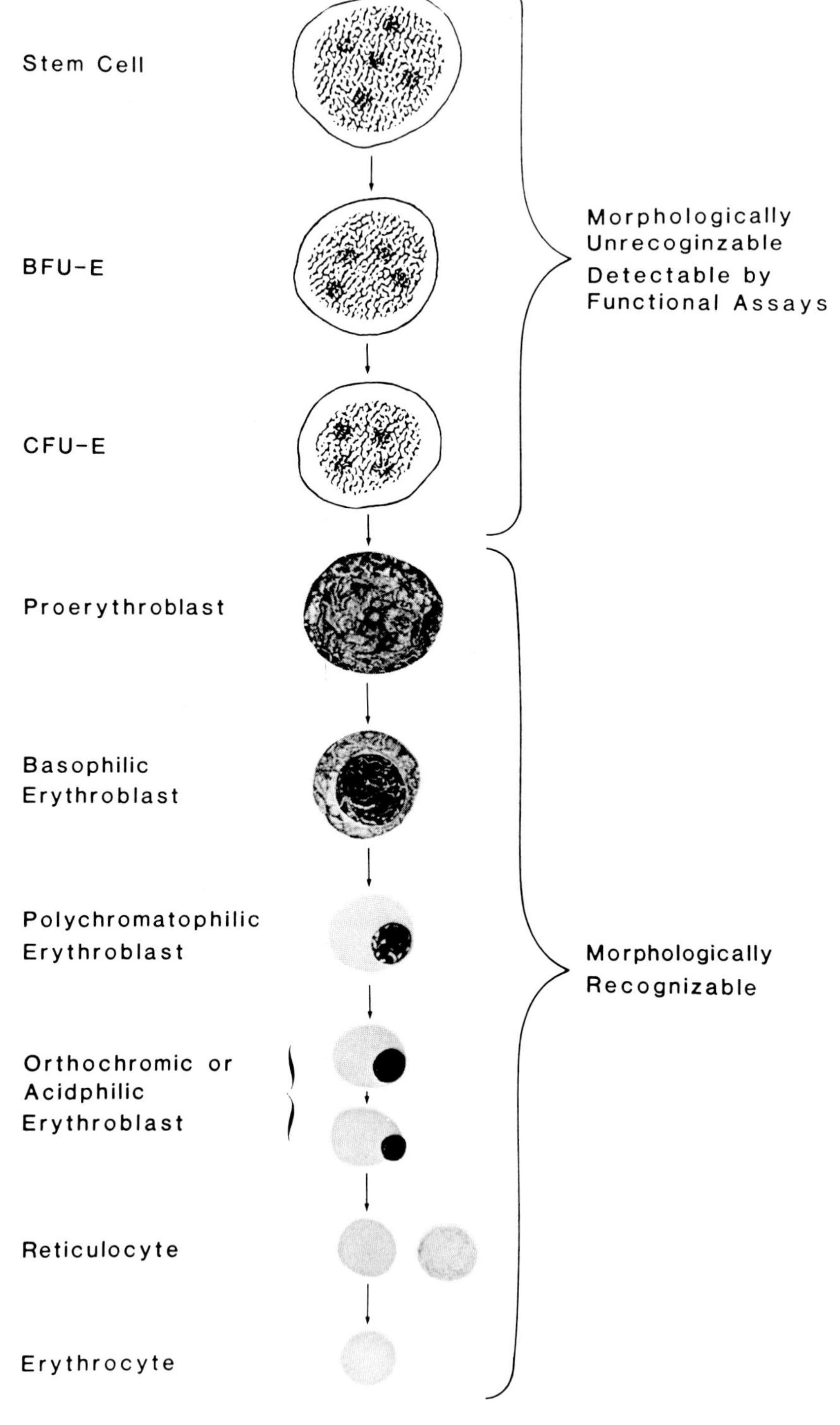

Stem Cell
BFU-E
CFU-E
Proerythroblast
Basophilic
Erythroblast
Polychromatophilic
Erythroblast
Orthochromic or
Acidphilic
Erythroblast
Reticulocyte
Erythrocyte
Morphologically
Unrecoginzable
Detectable by
Functional Assays
Morphologically
Recognizable

these colonies are clones derived from self-renewing pluripotent cells that have the capacity to fully reconstitute hematopoiesis in irradiated mice (i.e., stem cells). Pluripotent and self-renewing cell types of both human and murine origin can also be detected by *in vitro* assays in which the "stem cell" is induced to form clonally derived colonies in semisolid media (Fauser and Messner, 1979; Nakahata and Ogawa, 1982; Metcalf *et al.*, 1979).

In vitro methods have been developed to detect the lineage-restricted progenitor cells that arise by differentiation of the stem cells (Metcalf, 1977). These assays depend on the capacity of progenitor cells to proliferate and form large clones of mature progeny in semisolid media in the presence of specific regulatory molecules or colony-stimulating factors. The progenitor cells of the erythroid lineage that form colonies *in vitro* are classified into two groups. The BFU-E (burst-forming unit—erythroid) is the most primitive erythroid precursor; in fact, the earliest BFU-E are not yet restricted to erythropoiesis. BFU-E produce colonies of hemoglobinized cells in the presence of burst-promoting activity (BPA) and erythropoietin. The CFU-E (colony-forming unit—erythroid) is a more mature erythroid precursor that forms erythrocyte colonies in response to erythropoietin. The progression of BFU-E to CFU-E is independent of erythropoietin. The BFU-E and CFU-E are heterogeneous classes of cells encompassing a range of developmental stages in erythropoiesis, and are subclassified according to how long it takes for a colony of hemoglobinized erythrocytes to develop from the colony-forming cell (Gregory, 1976).

The differentiation and maturation of the other hematopoietic lineages follow a similar course as that described for erythroid cells, i.e., commitment of stem cells into lineage-restricted progenitor cells that, concomitant with extensive proliferation, mature to form nondividing end cells. The availability of *in vitro* methods that functionally distinguish the various stages of erythropoiesis and the well-defined morphological characterization of immediate precursors of erythrocytes make erythropoiesis a valuable system for studying differentiation and development processes (see also Quesenberry and Levitt, 1979).

In the second part of this section, we shall discuss a few points related to erythropoiesis. The first point is the origin of the hematopoietic system. Although we can follow hematopoietic development from the stem cell stage using the experimental systems described above, this approach does

---

Figure 9. Sequences of erythrocyte maturation. Immature erythroid cells at or before the proerythroblast stage are not morphologically recognizable. However, experimental systems are now available to delineate these early stages, i.e., stem cell → BFU-E → CFU-E → proerythroblasts. Stem cells can be assayed by semisolid media colony-forming assays.

not address the question of how hematopoietic stem cells are formed. Although hematopoietic stem cells are pluripotent, they are already committed in that they are restricted to produce only hematopoietic cells. Therefore, it is necessary to consider embryonic developmental processes to understand the formation of hematopoietic stem cells.

The second problem is the role of cell–cell interactions in hematopoiesis. Although it is clear that hematopoiesis will only occur in association with an appropriate stromal environment, it has not yet been conclusively demonstrated that the hematopoietic environment plays a directive role in hematopoiesis. The importance of the microenvironment was initially demonstrated by studies on the embryonic development of hematopoiesis. The studies showed that hematopoiesis may become established only if the stem cells lodged in a cellular matrix suitable for their replication and differentiation (Metcalf and Moore, 1971). The cellular matrix (marrow stroma) is derived from the perichondrial mesenchyme whereas hematopoietic stem cells circulating in the blood come from the yolk sac (Moore and Metcalf, 1970). Morphological examination of bone marrow cells also indicates that there are extensive interactions between hematopoietic cells and the cellular matrix (Lichtman, 1981). The long-term bone marrow culture system developed by Dexter *et al.* (1977) provides an experimental model to study the role of the hematopoietic environment. These mouse bone marrow cultures support the production of hematopoietic stem cells, the whole range of progenitor cells, and certain mature cell types for several months. However, the maintenance of hematopoiesis *in vitro* is absolutely dependent on the prior formation of an adherent layer of marrow-derived cells, which is thought to constitute an *in vitro* counterpart to the hematopoietic environment (Dexter, 1982). The adherent layer is a complex multilayer of cell types including phagocytic mononuclear cells, endothelial-type cells, and giant lipid-laden adipocytes within which extensive interaction between these cell types and hematopoietic cells occur (see Fig. 10). Some success has been achieved in maintaining human long-term bone marrow cultures (Gartner and Kaplan, 1980) although it is likely that the optimal conditions for the growth of human marrow *in vitro* have not yet been determined. Nonetheless, human bone marrow cultures develop an adherent layer of cells that is morphologically similar to that observed in mouse cultures (Toogood *et al.*, 1980). Recently, Keating *et al.* (1982) demonstrated that endothelial cells of the adherent layers of bone marrow cultures derived from marrow transplant recipients were of the marrow transplant *donor* genotype. This observation suggested, for the first time, that the marrow stroma (or certain components thereof) was transplantable. These data considered as a whole indicate the critical role of the marrow stroma in a hematopoiesis. It is important to understand how these cells interact in

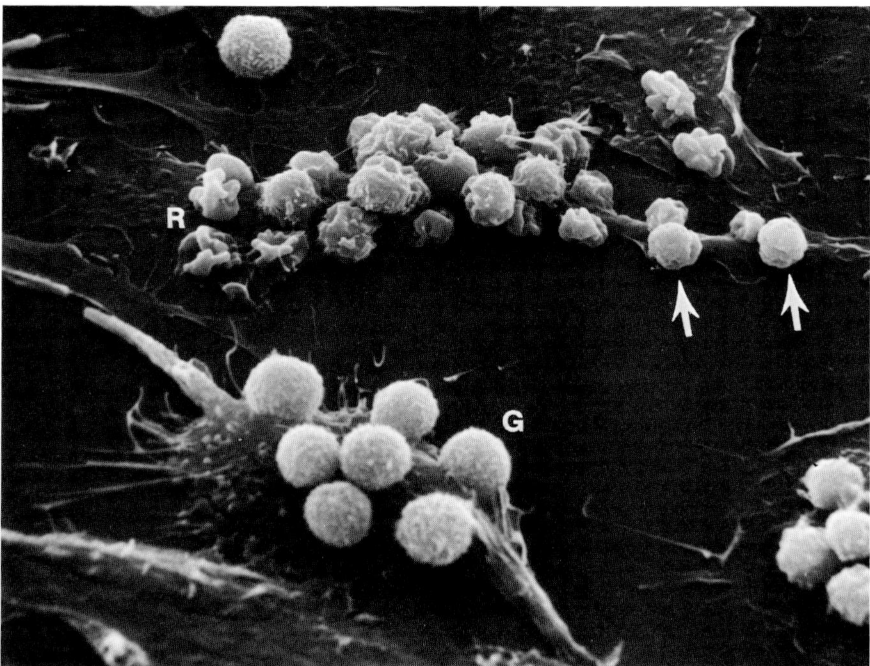

Figure 10. Scanning electron microscopy of cells present in long-term bone marrow culture. Erythroblastic island at the reticulocyte stage (R) with expelled nuclei (arrows) and an adjacent cluster of granulocytes (G) are seen on each adherent layer. (From Allen and Dexter, 1982.)

molecular terms. Another type of cell–cell interaction is illustrated by the role of colony-stimulating factors. It can be reasonably assumed that cells producing various factors such as erythropoietin or colony-stimulating activity play important roles in hematopoiesis. In fact, myeloproliferative viruses induce a large amount of granulocyte–macrophage colony-stimulating activity in fibroblasts and this effect is considered to be a cause of abnormal proliferation of hematopoietic cells (Koury and Pragnell, 1982).

The third question concerns the developmental reason for the various stages of differentiation and maturation. We can reasonably assume that each differentiation and maturation step is necessary to form functionally mature cells, but it is still a very difficult task to assign significance to each stage. This problem might be approached by asking how cells regulate the sequential expression of gene products. If we understand this process, at least to some extent, we may be able to manipulate differentiation and development.

4.2. Cell Surface Glycoproteins and Carbohydrate Structures in Erythroid Precursor Cells (Proerythroblasts and Erythroblasts)

Analysis of cell surface structures of blood precursor cells could be carried out in two ways. One approach is to isolate a pure population of precursor cells and to subject the isolated cells to analysis. This approach is only possible when the required population of cells at a particular stage of erythroid development can be isolated from a heterogeneous population of cells. As stage-specific markers for the majority of differentiation and maturation stages are not yet known and the frequency of progenitor cells is very low, this approach is still only feasible to a limited extent. The second approach is to use antibodies to a given antigen and to identify cells expressing the antigens by morphological criteria. The earliest erythroid precursor cells that are morphologically recognizable are the pro-erythroblasts. This approach is therefore feasible to erythroid cells at and later than the proerythroblast stage. However, by combining serological methods and stem cell culture, it is also possible to investigate the presence or absence of a given antigen on stem cells (which are unrecognizable by morphology). This approach will be described more in the next section.

In initial studies on the biochemical characterization of cell surface structures in erythroid precursor cells, we chose the first approach and studied two types of cells, erythroblasts and K562. Erythroblasts were generated by *in vitro* culture of BFU-E with erythropoietin. The large number of erythroblasts required for this work was kindly supplied by Stamatoyannopoulos and his colleagues. Because the conditions for growing BFU-E colonies also stimulate the development of other types of colonies, e.g., granulocytic, the erythroid colonies were identified microscopically and were isolated from nonerythroid colonies. Erythroid colonies were picked out by a fine pipet and subjected to cell surface labeling. The cell surface glycoprotein pattern of those cells was then compared with that of mature erythroid cells and the results are summarized as follows (Fukuda *et al.*, 1980):

1. Sialoglycoproteins (glycophorin) are present as major components, and can be detected by labeling by the periodate method. The glycoproteins gp105 and gp95, which are not detected in mature erythrocytes, are present as minor components.
2. Band 3 and Band 4.5 glycoproteins, which can be labeled by the galactose oxidase procedure, are present but to a much lesser extent than in mature erythrocytes.
3. The amount of lactosaminoglycan in erythroblasts is also considerably less than that in mature erythroblasts (see also Fig. 11).

These results are not influenced by the ontogenetic origin of the progenitor cells. The extent of branching in the structure of lactosamino-

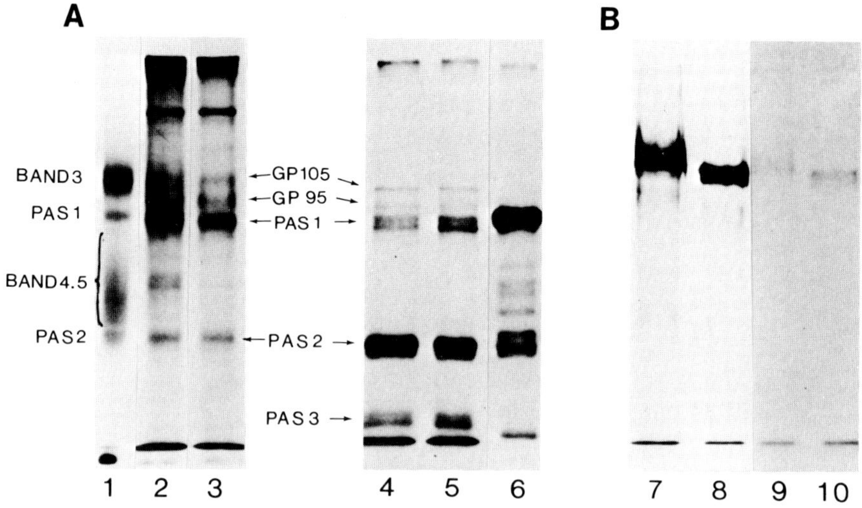

Figure 11. Fluorogram of SDS–polyacrylamide gels of surface-labeled erythroblasts and other cells. 1, Galactose oxidase-labeled or 6, periodate-labeled mature erythrocytes; 2, 3, galactose oxidase-labeled or 4, 5, periodate-labeled erythroblasts; 7, 8, periodate-labeled granulocytes; 9, 10, periodate-labeled monocytes. 1, 2, 4, 6, 7, 9, are control cells and 3, 5, 8, 10 are cells treated with endo-β-galactosidase. Cell numbers used in gels 2, 3 were four times more than those in gels 4, 5 to obtain comparable amount of incorporated radioactivity.

glycan, however, showed dependency on the ontogenetic stage. Erythroblasts derived from adult progenitor cells possess some branching structure in lactosaminoglycan whereas those from newborn or fetal progenitors showed almost undetectable branching structure. Our results also suggest that the branching structure in lactosaminoglycan (I antigen) increases during maturation from erythroblasts to erythrocytes in the adult. Vainchenker *et al.* (1981a) extensively examined this aspect in erythroblasts obtained by *in vitro* culture of BFU-E. They found that the i antigen was preferentially expressed on immature erythroblasts, while the I antigen was expressed at higher levels in mature erythroblasts than in immature ones. As the i and I antigens correspond to linear and branched lactosaminoglycan, respectively, these combined results indicate that the amount of branching in lactosaminoglycan increases during the maturation of adult erythroid cells.

Although proerythroblasts are the earliest precursor erythroid cells that can be recognized by morphology, it is not yet feasible to collect appreciable amounts of proerythroblasts from bone marrow samples. If large numbers of developmentally synchronized BFU-E can be cultured, it may be possible to obtain a relatively pure population of proerythroblasts if cultures are harvested at the appropriate stage. So far, such attempts have not been reported. As the first step to understanding the cell surface structures of proerythroblasts, we analyzed K562 human leukemic cells. K562 cells were isolated by Lozzio and Lozzio (1975) from a patient with chronic myelocytic leukemia in blast crisis. Although the cells were initially regarded as myeloid cells, Gahmberg and Andersson showed the erythroid nature of these cells by demonstrating induced synthesis of hemoglobin (Andersson *et al.*, 1979a) and the presence of glycophorin A (Gahmberg *et al.*, 1979). We have analyzed this cell line by cell surface labeling combined with endo-β-galactosidase digestion and the results can be summarized as follows (Fukuda, 1980):

1. Glycophorin is present as a minor component.
2. Band 3 and Band 4.5 are undetectable.
3. Only very small amounts of lactosaminoglycan (which can be detected by susceptibility to endo-β-galactosidase) are present and its structure is linear (i type).
4. A major glycoprotein is gp105 together with gp95. Both are present as minor components in erythroblasts and almost undetectable in erythrocytes.

Thus, cell surface glycoproteins of K562 cells are found to be significantly different from those of mature erythroid cells. Turco *et al.* (1980) claimed the presence of linear-chain lactosaminoglycan (i type) in gp105 of K562 cells, although we have no evidence that the major oligosaccharide unit of gp105 is lactosaminoglycan. Yoshima *et al.* (1982) confirmed our conclusion by showing that less than 6% of asparagine-linked oligosaccharides are the polylactosaminyl type. As gp105 contains more than 60% of the total carbohydrate chains labeled by [³H]glucosamine, the sugar chains with polyactosamine structure would be minor components of gp105 even if they occur exclusively in the glycoprotein. We have also clearly shown that gp105 is different from Band 3 by immunoprecipitation with specific anti-Band 3 serum. As the antiserum recognizes the amino-terminal cytoplasmic segment of Band 3 (M. Fukuda *et al.*, 1978), the antiserum should also detect Band 3 variants differing in glycosylation, if such variants exist. Our results on glycophorin content appear to be slightly different from those of Gahmberg *et al.* (1979); they reported that K562 cells contain almost the same number of glycophorin as mature erythrocytes. However, the diameter of erythrocytes is 2 to 3 times smaller than that of K562 cells. Thus, the density of glycophorin in K562 cells is 4 to 9 times lower than that of mature erythrocytes. Taken together, these results suggest that K562 cells represent a very immature stage of erythroid cell lineage. Further characteristics of K562 cells will be discussed in Section 5.

The results obtained by cell surface labeling of these cell types are consistent with those obtained by immunological techniques. In particular, two results obtained by Gahmberg *et al.* (1978) and Foxwell and Tanner (1981) are worthy of comment. Gahmberg *et al.* developed a staphylococcal rosette formation technique. This technique involves application of specific antibodies followed by the addition of *Staphylococcus aureus* bacteria. *S. aureus* has protein A, which binds to the Fc portion of a certain class of immunoglobulins. After removal of unbound bacteria, the samples are cytocentrifuged and can be stained by conventional methods. This technique allows us to examine the morphology of the cells that express a given antigen. It also appears that the number of bound bacteria is proportional to the density of antigens present on the cell surface. By using this technique, Gahmberg *et al.* (1978) showed that glycophorin is present only in cells of the erythroid lineage and that it is present in cells beyond the proerythroblast stage. Furthermore, the amount of glycophorin detected progressively increases as the erythroblasts mature to erythrocytes. Foxwell and Tanner (1981) showed that expression of Band 3 is also increased as cells mature beyond the proerythroblast stage. If the number of bound bacteria is normalized according to the cell size, the density of Band 3 is much higher in the later stages of erythropoiesis.

4.3. Cell Surface Glycoproteins and Carbohydrate Antigens in Progenitor Cells

Although antigens on pure populations of mature blood cells can be directly examined, the quantitation of antigen density on stem cells and progenitor cells has not been feasible because these cells cannot yet be obtained as pure populations. The isolation of these cell populations is difficult because they are present in a low frequency in bone marrow and blood and they are not recognizable by morphological criteria. The antigenic structure of hematopoietic progenitors, therefore, must be studied indirectly by combining serologic techniques with functional assays for various classes of stem cells and progenitors (see also review by Fitchen *et al.,* 1981).

The first, and most popular, technique has been antibody- and complement-dependent killing, followed by colony-forming assays. Bone marrow containing stem cells and progenitor cells is incubated with antibody and complement. Surviving stem cells can be measured by long-term culture, the mixed-colony-forming cell assay (Metcalf and Johnson, 1976), or the CFU-S assay and surviving progenitor cells by BFU-E, CFU-E, or CFU-GM colony formation. It is assumed that precursor cells express the antigen if colony formation is decreased by pretreatment with antibodies specific to the antigen and complement. Although this method seems to be sound and simple, it is associated with some drawbacks. Using this technique, it is difficult to distinguish between expression of the antigen by the colony-forming cells and by an accessory cell type whose presence is required for colonies to develop. Another disturbing factor is complement. Although we can use fairly pure antibody preparations (monoclonal antibodies, for example), the source of complement is not pure. The results, therefore, must be carefully examined to detect nonspecific effects.

The second, and more recent, technique is to use fluorescent antibodies and to fractionate the cells by using a fluorescence-activated cell sorter (FACS). Bone marrow cells or target cells are treated with specific antibodies followed by fluorescence-conjugated anti-antibody. Fluorescence-positive and -negative cells are then separated using the FACS, and the antigen-positive and -negative fractions can be cultured to measure stem cells and progenitor cells. This approach may enrich stem cells or progenitor cells. In addition, the effect of "helper cells" may be assessed by reconstruction of two fractions. The latter experiment can be further extended by depleting certain populations of cells by antibody–complement-dependent analysis. The major drawback of this technique is the requirement for a second antibody, for many hematopoietic progenitor cells express Fc receptor. This problem may be overcome to some extent by adding irrelevant antibodies before addition of fluorescence-conjugated antibodies.

The third technique, which is not as common as the others, is to reconstruct hematopoiesis *in vivo* by bone marrow transplantation. Donor bone marrow is fractionated by a FACS or by antibody–complement-dependent depletion. If the fractionated cells are capable of reconstituting hematopoiesis in the hematopoietically ablated recipient, then it may be possible to conclude that the cells removed by the fractionation procedure do not influence hematopoiesis. However, even where hematopoietic cells in the recipient can be shown to be of donor origin, it is possible that cells that "influence" hematopoiesis (e.g., helper cells, stromal cells) have in fact been removed from the graft, but that such cells persist in a functional state in the recipient. Another application of this method may be to exhaustively remove "leukemic stem cells" by selective depletion and to reconstitute leukemia patients with autologous bone marrow transplantation after aggressive cytotoxic therapy. The failure of a marrow transplant can result from various causes and does not necessarily imply a lack of stem cells in the graft. Therefore, this assay only indirectly measures the quantity and quality of stem cells and progenitor cells in a given test population of cells.

Robinson *et al.* (1981) examined glycophorin in BFU-E, and CFU-E by using the FACS. They found that BFU-E or CFU-E were not present in the glycophorin-positive fraction, indicating that glycophorin is present only after the CFU-E stage. Furthermore, almost all the cells of the glycophorin-positive fraction of bone marrow cells can be morphologically identified as erythroid cells at various stages of maturation. Th same authors found that the Ia-like antigen (HLA-DR) is expressed at highest levels by the earliest erythroid precursor (the BFU-E), less in the later precursor (the CFU-E), and lost during further maturation to morphologically recognizable erythroblasts. HLA-ABC antigens are lost more slowly and erythroblasts and reticulocytes may express a low density of these antigens (Brown *et al.*, 1979). Human Ia-like antigens (including HLA-DR) resemble mouse Ia antigens and are believed to have an important role in the generation of the immune response (Klein, 1979). The HLA antigens may also be differentiation antigens which are important in distinguishing self and nonself and in this context play a critical role in cell–cell interactions (see next section also).

In contrast to the clear-cut observations on protein antigens, studies on carbohydrate antigens are still inconclusive. Although the presence of i antigen on erythroid progenitor cells has been reported (O'Hara *et al.*, 1978), this observation has not been confirmed yet. Sieff *et al.* (1982) reported that BFU-E and CFU-E have an almost undetectable amount of i antigen, which was measured by anti-i (Den) antiserum. They also reported the presence of I antigen as detected by anti-I (Step) antiserum on BFU-E and CFU-E. Although anti-I (Step) is a useful reagent for distinguishing fetal and adult erythrocytes, the specificity of the antiserum is not as strict a one as that of anti-I (Ma) antiserum. Anti-I (Ma) antiserum requires a branching structure for antigenic recognition, whereas anti-I (Step) recognizes a linear portion of the I molecule (see Fig. 6). In fact, a unique cell surface structure was detected on erythrocytes of a patient using anti-I (Step) antiserum (Papayannopoulou *et al.*, 1981). As the presence of i antigen has been demonstrated on proerythroblasts and erythroblasts, further studies will be necessary to investigate the presence of Ii antigens on stem cells. Sieff *et al.* (1982) detected a small amount of A antigen on BFU-E and CFU-E by monoclonal antibodies. It is possible that A antigens detected by monoclonal antibodies are "A-like" antigens, containing the GalNAcαl$\rightarrow$3 structure, which could be present on O-glycosidically linked carbohydrate chains. It is not yet known if any carbohydrate antigen specific to stem cells is present.

4.4. A Provisional Model for Membrane Differentiation during Erythropoiesis and Its Implications

Based on the data described in the previous sections, the following points summarize current knowledge of membrane differentiation during erythropoiesis.

1. Glycophorin appears at the proerythroblast stage, significantly increases during further maturation to erythroblasts, and increases slightly with terminal maturation. The amount of glycophorin present at the late erythroblast stage is almost equivalent to that of erythrocytes.

2. Band 3 and Band 4.5 glycoproteins appear slightly later than glycophorin, and basophilic erythroblasts express detectable amounts of these proteins. Their level increases significantly during development from early to late erythroblasts and continues to increase until maturation of the erythrocytes.

3. The amount of lactosaminoglycan increases in parallel to that of Band 3 and Band 4.5. Throughout the period during which there is an increase in the level of lactosaminoglycan, the extent of branching in the structure of lactosaminoglycan is also increased in adult erythroid cells. Thus, immature cells express small amounts of linear lactosaminoglycan, whereas more mature cells express large amounts of branched lactosaminoglycan.

4. Cell surface glycoproteins are highly sialylated in immature cells, and nonsialylated carbohydrate chains, which can be detected by galactose oxidase labeling, are increased with the progress of maturation. The same change was also observed in granulocyte–monocyte lineage (Fukuda *et al.*, 1981a).

5. gp105 and gp95, in contrast, are expressed only by immature cells, although it is not yet clear if early progenitor cells, such as BFU-E or CFU-E, express such glycoproteins.

6. HLA-ABC antigens and HLA-DR antigens are markers for very early stages of erythroid precursor cells. However, stem cells seem to be negative in HLA-DR antigens (Moore *et al.*, 1980); thus, these antigens may be transiently expressed.

7. The appearance and increase of spectrin during erythropoiesis approximately coincides with that of glycophorin (Chang *et al.*, 1976; Eisen *et al.*, 1977; Geidushek and Singer, 1979). Thus, glycophorin and spectrin appear slightly earlier than Band 3 (see Table 2).

It is interesting that glycoproteins specific to erythrocytes such as glycophorin and Band 3 are expressed in the later stages of maturation. Atlhough the function of glycophorin is not yet known, the expression of this glycoprotein appears to be strictly restricted to the erythroid cell lineage (Gahmberg *et al.*, 1978; Fukuda *et al.*, 1980). Band 3, which also seems to be specific to erythroid cells, is the anion transporter for erythrocytes (Cabantchik and Rothstein, 1972); an important function in mature erythrocytes, as it facilitates carbonate–chloride exchange. In the case of Band 3 anion transporter, therefore, it appears that this glycoprotein

Table 2. Membrane Differentiation of Human Erythroid Cells[a]

	Pluripotent stem cell	BFU-E	CFU-E	Proery-throblast	Ery-throblast	Ery-throcyte
HLA-ABC	−	+ +	+	(+)	(+)	(±)
Ia	−	+ +	+	−	−	−
gp105/gp95				+ +	+	−
Glycophorin		−	−	+	+ +	+ + +
Band 3				−	+	+ + +
Spectrin				+	+ +	+ + +
Transferrin receptor	±		+	+ +	+ + +	−
Lactosaminoglycan				+	+ +	+ + +
Blood group						
Ii antigen		(+)?		+ (i)	+ + (Ii)	+ + + (I)
ABH antigen				(+) −	+ +	+ + +
Rh antigen				(+) −	+ +	+ + +
Hemoglobin	−	−	−	−	+ +	+ + +

[a] References are cited in the text. Blanks indicate that studies have not yet been done.

is synthesized in largest amounts in the cell types where its function is required. This notion is supported by the fact that the appearance and increase of Band 3 coincides well with the onset and increase of hemoglobin synthesis. The anion transport properties of K562 cells, which are presumed to be at the proerythroblast stage of development, are distinctly different from those of erythrocytes (Knauf and Law, 1981). This is likely due to the absence of Band 3 in K562 cells. The same cell line, however, seems to have a glucose transporter similar to that in erythrocytes (Dozier *et al.*, 1981). A glucose transport system is a necessary apparatus in many cells. In fact, proteins that are antigenically cross-reactive to the glucose transporter of erythrocytes have been detected in fibroblasts (Salter *et al.*, 1982).

The changes in the expression and structure of lactosaminoglycan that occur during maturation are paralleled with changes in ABO blood group antigens. Karhi *et al.* (1981) reported that blood group A antigen could be detected in basophilic erythroblasts by staphylococcal rosette formation and increases during further maturation. As lactosaminoglycan carries ABO blood group determinants in addition to Ii antigens, it is not surprising that the appearance and increase of these blood group antigens is parallel to that of lactosaminoglycan. The amount of lactosaminoglycan is also in parallel to the expression of Band 3 and Band 4.5. As Band 3 and Band 4.5 are major carriers for lactosaminoglycan in mature erythrocytes, this result implies two possibilities: the lack of lactosaminoglycan in immature cells may be due to a lack of acceptor protein or it may be the result of a lack of glycosyltransferases, which are required for the synthesis of lactosaminoglycan. It will be interesting to test if the glucose transporter in immature cells expresses lactosaminoglycan. Another blood group antigen whose expression follows the same trend as Band 3 and lactosaminoglycan during erythroid development is RhD. Thus, RhD antigens are present in erythroblasts and increase with erythroid maturation (Readern and Masouredis, 1977). The result is consistent with the recent report that Band 3 glycoprotein is a carrier for RhD antigens (Victoria *et al.*, 1981), although conflicting data also exist (Moore *et al.*, 1982).

Structural changes in lactosaminoglycan, which are associated with development and differentiation, are not restricted to human erythroid cells. By analyzing human myeloid cells blocked at various differentiation stages, it was possible to infer that the quantity of lactosaminoglycan in human myeloid cells changes drastically during maturation (Fukuda *et al.*, 1981a). Furthermore, it has been shown in the mouse system that a unique lactosaminoglycan is present in embryonalcarcinoma cells which disappears during differentiation (Muramatsu *et al.*, 1979). In addition, the Galβ1→4 (Fucα1→3)GlcNAcβ1→3Gal structure is present in primitive embryos but is considerably decreased during the differentiation of endodermal cells (Gooi *et al.*, 1981). Thus, it is conceivable that expression of lactosaminoglycan is controlled by differentiation programs in various systems (Hakomori *et al.*, 1982).

More recently, a lot of attention has been focused on another important functional protein, the transferrin receptor. This glycoprotein was initially described by Omary *et al.* (1980) as a proliferation regulator and later identified as the transferrin receptor by several groups (Trowbridge and Omary, 1981; Sutherland *et al.*, 1981). The transferrin receptor has a molecular weight of about 180 kd with a subunit of about 90 kd. It has been shown by Hutchings and Sato (1978) that transferrin is required for the growth of cells in serum-free medium, and the receptor has been identified from placenta (Wada *et al.*, 1979). The amount of transferrin receptor is very high in erythroblasts and reticulocytes but negligible in mature erythrocytes (Hamilton *et al.*, 1979; Sieff *et al.*, 1982). Transferrin is an essential source of iron for hemoglobin synthesis. Expression of the transferrin receptor has been determined by monoclonal antibodies and found to be variable in BFU-E, but increases during maturation through CFU-E to recognizable erythroblasts. This pattern of expression is, therefore, in accordance with the role of the transferrin receptor in both proliferation and hemoglobin synthesis. Although the pattern of expression of the transferrin receptor is quite similar to that of gp105 and gp95, we have immunochemical evidence that they are different molecules. Furthermore, gp105 and gp95 do not have a dimeric molecular weight in the absence of reducing agents (M. Fukuda, unpublished data), as does the transferrin receptor.

We do not yet know the role(s) of the glycoproteins that are specific to the early stages of erythropoiesis such as gp95, gp105, HLA-ABC, and HLA-DR. However, recent studies suggest that these glycoproteins are probably involved in cell–cell interactions at early stages of hematopoiesis. In particular, Torok-Storb *et al.* (1981) found that the formation of colonies from BFU-E (which express Ia antigens) can be influenced by two distinct subpopulations of T cells. One population enhances BFU-E colony formation and is an Ia-negative cell type; the other "limits" the number of BFU-E colonies and is Ia-positive. Furthermore, the presence of Ia-positive monocytes also influences the extent to which BFU-E colonies can be stimulated by Ia-negative T cells (Torok-Storb and Martin, 1982). The modulation of BFU-E colony formation by T-cell subsets and monocytes is to a certain extent restricted by HLA-DR (Ia-like antigen); the maximum extent of BFU-E colony stimulation is only observed when at least one HLA-DR haplotype is identical between the responding BFU-E and the accessory cell population(s). Similarly, gp95 and gp105 may have some role in cell–cell interactions in hematopoiesis. It is possible that these glycoproteins act as receptors for erythroblast-specific agglutinin (Harrison and Chesterton, 1980). One can also speculate that these glycoproteins may be involved in cell proliferation, for human melanoma-associated antigen, p97, has been shown to have transferrin-like molecules and suggested to be a carrier of iron, although it is clearly distinct from the transferrin receptor (Brown *et al.*, 1982).

5. CELL SURFACE MARKERS IN IN VITRO DIFFERENTIATION OF LEUKEMIC CELLS

In mouse systems, the erythroleukemic cell line established by Friend *et al.* (1971) has been a valuable model for the study of the onset of hemoglobin synthesis. Such a cell line for the human system is lacking. Thus far, we have only one human cell line that can be regarded as erythroleukemic. This cell line, K562, was originally established by Lozzio and Lozzio (1975) from the pleural effusion of a patient in the terminal stage of chronic myelogenous leukemia with blast crisis. As this cell line appeared to have almost no markers except some cross-reactivity with granulocytes, the cell line was regarded as myeloblastoid (Lozzio and Lozzio, 1980). Andersson *et al.* (1979b), however, found that surface glycoproteins of K562 cells are different from those of HL-60 promyelocytic cells or undifferentiated myeloblastoid cells. Furthermore, they identified one of the glycoproteins as glycophorin A, which has been shown to be erythroid-specific. In addition, Andersson *et al.* (1979) induced hemoglobin synthesis in K562 cells by treating the cells with butyrate. Rutherford

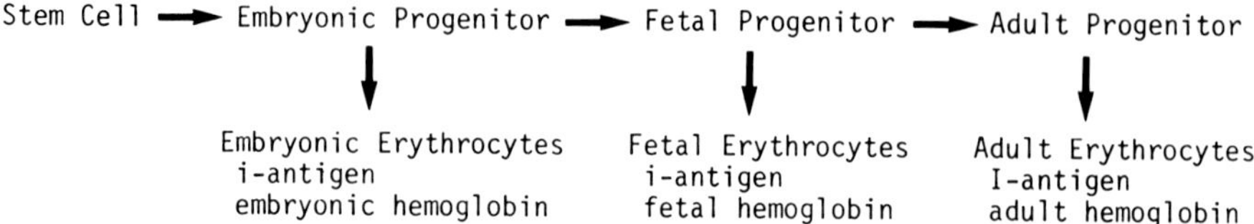

Figure 12. Postulated scheme for ontogenetic dependency in hemoglobin and lactosaminoglycan structure. It is proposed that ontogenetic stage is determined at the level of progenitor cells (Stammatoyannopoulos *et al.*, 1981).

et al. (1979) subsequently demonstrated that hemoglobins synthesized by K562 cells are primarily embryonic hemoglobins. Rutherford *et al.* used hemin (rather than butyrate) as a potent inducer for hemoglobin synthesis in K562 cells and, along with several other groups, could not confirm the observation that sodium butyrate induces hemoglobin synthesis in K562 cells. These studies clearly showed the erythroid nature of K562 cells. However, K562 cell lines may be derived from the very early stages of the erythroid lineage so that some of its characteristics may also resemble those of myeloid cells. We have shown that the cell surface glycoproteins of K562 cells are distinctly different from those of mature erythrocytes and that the major glycoproteins (gp105 and gp95) of K562 cells are similar to those of a myeloblastoid cell line, KGla (Fukuda, 1980; Fukuda *et al.*, 1981a). The same glycoproteins were found as minor components in erythroblasts cultured *in vitro* (Fukuda *et al.*, 1980). Similarly, Marie *et al.* (1981) showed that K562 cells express both a granulocytic marker defined by the My-1 monoclonal antibodies and an erythroid marker, spectrin, in the same cells. Vainchenker *et al.* (1981b) indirectly showed that some K562 cells may be bipotent cells that can undergo differentiation into erythrocytes or megakaryocytes depending on the culture conditions. It is conceivable that cells at such an immature stage still have several characteristics in common with other cell lineages. This notion is supported by the fact that K562 cells can be induced to produce embryonic hemoglobin. Stammatoyannopoulos *et al.* (1981) postulated that the expression of the ontogenetic phenotype is determined before the BFU-E stage, i.e., cells that produce embryonic hemoglobin are derived from BFU-E that are irreversibly programmed for embryonic hemoglobin production and are, by implication, more primitive precursors than those that give rise to adult hemoglobin-producing cells (see Fig. 12). Thus, K562 cells are derived from some point along the developmental pathway of a "primitive" erythroid precursor cell according to this scheme. Friend luekemic cells, the murine erythroleukemic cell line, are probably derived from later stages than K562 cells, as most of the Friend cells produce adult hemoglobin when they are induced.

K562 cells express glycophorin A, blood group N antigen (Gahmberg *et al.*, 1979), and i antigen (Fukuda, 1980; Turco *et al.*, 1980), but Band 3 and ABH antigens are absent (Fukuda, 1980; Horton *et al.*, 1981). When K562 cells are treated with various "inducers," the cells do not generally respond by expressing the cell surface properties of more mature cells. Yurchencho and Furthmayr (1980) reported that inducing agents do not increase either glycophorin or spectrin in K562 cells. Horton *et al.* (1981) obtained similar results by using various monoclonal antibodies and lectins. Thus, it gradually became accepted that cell surface markers and hemoglobin synthesis do not necessarily change in a coordinated fashion. Tonkonow *et al.* (1982) showed that K562 cells induced by hemin increased the expression of globin mRNA by usign a globin cDNA probe but the glycophorin content of the induced cells was not increased. Horton (1982) showed that neither of the red cell enzymes, carbonic anhydrase or pyruvate kinase, was induced with sodium butyrate or hemin, although the conditions caused an increase in hemoglobin synthesis. Similarly, Testa *et al.* (1982) showed that an increase in I antigen was observed in K562 cells treated with sodium butyrate, although hemoglobin was not induced using that agent; conversely, hemin does not increase I-antigen expression but does induce hemoglobin synthesis. Dokhelar *et al.* (1982) reported that butyrate decreases the susceptibility of K562 cells toward natural killer cells, whereas hemin does not influence it. These combined results lead us to the conclusion that "differentiation" of leukemic cells does not necessarily recapitulate the pathway of normal cell maturation in every respect. This characteristic of leukemic cell induction is probably important with respect to the "leukemic nature" of similar cell lines. K562 cells, like Friend cells, also fail to respond to erythropoietin (Guerrasio *et al.*, 1981). It is likely that the abnormal response of these cells to regulatory molecules is an important facet of their leukemic nature.

The analysis of the effect of tumor-promoting phorbol esters on the cell surface glycoproteins of K562 cells is not directly related to this section; however, the results have some relevance to this discussion. The cell surface glycoproteins were changed such that their profile was characteristic of a less mature cell type; these changes included a loss of glycophorin but retention of gp95. Thus, a cell surface marker specific to later stages of maturation was lost, whereas one specific to early stages was maintained. In addition, a new glycoprotein (gp80) appeared after treatment of K562 cells with phorbol esters (M. Fukuda, 1981). These results indicate that the identification of surface markers for cellular differentiation may be of great help in understanding the phenomenon.

6. CELL SURFACE MARKERS IN HEMATOLOGICAL DISORDERS

It has been known for some time that changes in cell surface markers can occur in hematological disorder. Giblett and Crookston (1964) discovered that the number of red cells expressing high levels of i antigen is increased in thalassemia and other hematological disorders. It was subsequently reported that I antigen is decreased and i antigen increased in some cases of leukemia (McGinniss *et al.*, 1964; Jenkins *et al.*, 1965). Hillman and Giblett (1965), then demonstrated that the increased production of hematopoietic cells which can be induced by bleeding was accompanied by an increase in the number of i-expressing red cells. They concluded that stressing bone marrow to produce abnormally high numbers of blood cells was the major cause for the increased expression of i antigen by erythrocytes. We have analyzed the cell surface glycoproteins of erythrocytes from patients with sickle cell anemia. We detected an increased amount of sialylated oligosaccharides in Band 3, which accounts for the increased i antigenicity in this disease (Fukuda *et al.*, 1981b). It has also been reported that the expression of fetal hemoglobin by erythrocytes is increased in hematological disorders (Rochant *et al.*, 1972; Raghavendra *et al.*, 1978). These combined results suggest that the abnormal characteristics of red cells in hematological disorders may be the product of premature terminal differentiation of early erythroid progenitor cells.

However, the abnormal expression of fetal hemoglobin in hematological disorders is not necessarily accompanied by elevation of i-antigen levels on the same cells, and vice versa. Rather, these two markers occur independently of one another (Papayannopoulou *et al.*, 1980). Other markers (such as carbonic anhydrase in red cells) also behave independently from the above two markers in different conditions (Papayannopoulou *et al.*, 1981). Similarly, various erythroid markers in K562 cells are modulated independently as discussed in the previous section Thus, it appears that in certain hematological disorders, uncoordinated and incomplete development programs are expressed in red blood cells. This concept is of critical importance in understanding the character of leukemic cells and tumor cells in general. Although most leukemic cells seem to be derived from a particular stage of differentiation of one cell lineage, it is possible that leukemic cells do not express every characteristic of their normal immature cell counterparts. Further studies are necessary to clarify this point.

In parallel to studies on erythrocyte markers, the surface markers of other hematopoietic cell types have been extensively studied in various disorders, including leukemia (for review see Foon *et al.*, 1982). The cell surface markers specific for T cells are among those that have been studied and well characterized (Reinherz and Schlossman, 1980). Monoclonal antibodies specific to various stages of T-cell maturation were obtained by using thymocytes or T-cell leukemic cells as immunogens, and the cell surface markers that have been identified can be classified into three groups: (1) those present throughout all stages of maturation (Pan T, Leu-1), (2) those specific to early T lymphopoiesis (Thy-1 antigen, transferrin receptor), and (3) those specific to late-stage T lymphopoiesis (Leu 2, Leu 3). The common acute lymphoblastic leukemia antigen (cALLA) was detected in cells of non-T, non-B lymphocytic leukemia. Although this antigen initially seemed to be specific to leukemia, it has now been identified on normal B-cell precursors but not on mature B cells (Janossy *et al.*, 1979). The distribution of this antigen, however, is not restricted to hematopoietic cells. Metzgar *et al.* (1981) found that this antigen, which is defined by J-5 monoclonal antibodies, can be detected in renal tubular and glomerular cells, fetal small intestine epithelial cells, and myoepithelial cells of the adult breast. Furthermore, the molecular weight of the antigen in tissue and leukemic cells was different when measured by SDS gel electrophoresis, probably due to posttranslational modification. These studies provide some clues for the developmental relationship of hematopoietic cells with cells in other tissues and to the functions of the cALLA protein.

7. CONCLUDING REMARKS AND PROSPECTS

The data discussed in the preceding sections demonstrate that distinct changes in the cell surface glycoproteins and carbohydrate antigens accompany the various phases of erythropoiesis—from stem cell to mature erythrocyte. This conclusion has been obtained by a combination of immunological methods and biochemical analysis. Essentially, these changes can be classified in two ways: one is the appearance and disappearance of particular glycoprotein species and the other is changes in the structures of carbohydrate chains attached to proteins. The modifications of these cell surface components are probably operated by two independent mechanisms, which may both modulate the expression of important functional units involved in cell surface specificity and cell–cell interactions.

Among the various changes described, changes of lactosaminoglycan are particularly worth mentioning. First, the carbohydrate chains in Band 3, Band 4.5, and long-chain glycolipids contain a linear unbranched structure in fetal erythrocytes. This structure is replaced by a highly branched structure within several months after birth. Lactosaminoglycan carries blood group ABH determinants in addition to Ii antigens. Thus, the conversion from linear lactosaminoglycan to a branched one determines the general status of blood group determinants. This time of conversion from linear to branched lactosaminoglycan coincides well with the time for switching from fetal to adult hemoglobin. The appearance of presumptive branching enzymes (Fig. 7) is, therefore, a crucial membrane event that may reflect fetal to adult hemoglobin synthesis. Second, the quantity of lactosaminoglycan and the extent of its branching are greatly increased during maturation of adult erythroid cells; immature cells express only a small amount of unbranched lactosaminoglycan, whereas mature erythrocytes express large quantities of branched lactosaminoglycan. This change is also associated with the increase of erythroid-specific glycoproteins such as Band 3, glycophorin, and Band 4.5. These combined results strongly suggest that the expression of lactosaminoglycan plays crucial roles in the processes of development and differentiation, namely in the regulation of expression of cell surface glycoproteins (see also Section 2.5) and possibly in the regulation of gene expression for other erythroid-specific proteins (e.g., hemoglobin). This idea can be supported by the fact that hematological disorders, which are results of abnormal development programs, are associated with the expression of fetal characteristics in adult patients.

It has also been suggested that a deficiency of branched bivalent ABH determinants in the infants may have a protective and compromising effect on ABO-incompatibility pregnancy (Romans *et al.*, 1980). Branched and bivalent determinants could bind to two binding sites of IgG antibody (= "monogamous bivalency" of IgG antibody). The binding of antigen by two combining sites is favored over binding by a single site by a factor of 10^3 or 10^4. As fetal erythrocytes lack such bivalent structures, the fetus has much less reactivity to anti-A or anti-B IgG antibodies. Similarly, it is conceivable that the increase of branched lactosaminoglycan may be associated with the appearance of proteins in stroma cells, which recognize lactosaminoglycan, while unbranched lactosaminoglycan may play some roles in early stages of erythropoiesis. Further studies are necessary to clarify molecular bases underlying these events.

The identification and characterization of cell surface markers can provide information on various aspects of differentiation and oncogenesis. First, the identification of cell surface markers that are specific to discrete stages of differentiation and maturation will allow us to follow the course of these processes in molecular terms. Mechanisms regulating the expression of cell surface markers will lead us to understand how functionally mature cells are formed and what kind of structural elements are necessary for functional maturation. The studies may also enable us to determine how the control of differentiation of tumor cells differs from that of normal cells.

Second, they could be used to test if tumor cells are arrested at, and representative of, a certain stage of maturation along the normal cell development pathway. The availability of monoclonal antibodies specific to cell surface components will permit this hypothesis to be tested. In fact, Reinherz and colleagues have already conducted such studies in T-cell leukemia (Reinherz *et al.*, 1980). The discovery of leukemia cell surface markers may also allow the detection of very minor populations of cells within a normal cell population, which may otherwise be overlooked. For example, B cells characteristic of chronic lymphocytic leukemia were found to be a minor population among normal lymph node cells situated in germinal centers of lymph nodes (Caligaris-Cappio *et al.*, 1982).

Third, these differentiation markers can be directly used for diagnostic and therapeutic purposes. Classification of tumor cells and assessment of prognosis is an immediate application. In addition, these studies may allow us to selectively remove a certain population of cells by using antibodies against such markers. This type of approach has three possible applications. One is to take marrow from the leukemic patient and treat the bone marrow cells with antibodies that are specific to "leukemic" cell surface markers and complement. The treated cells are transplanted back into the patient who, meanwhile, has been treated with intensive cytotoxic therapy in an attempt to delete the leukemic cells remaining in the body. The patient's hematopoietic system should be reconstituted by cells of the nonleukemic autologous marrow graft. Although this approach seems to be very attractive, none of the cell surface markers described so far is specific to leukemic cells and cell surface markers found in leukemic cells have been found in certain populations of immature normal cells. On the other hand, administration of a drug or a toxin bound to antibodies to cell surface markers may be a valuable method to target therapeutic agents (Urdal and Hakomori, 1980; Gilliland *et al.*, 1980). This approach, at least, has more specificity than conventional therapy and enables us to use relatively high levels of drugs, which are, ideally, concentrated around the tumor cells. The third approach is to remove certain functional subpopulations of hematopoietic cells, e.g., those cells that elicit graft-versus-host (GVH) disease in an HLA-mismatched recipient. Indeed, Reinherz *et al.* (1982) recently reported that severe combined immunodeficiency can be treated by grafting the patient with T-lymphocyte-deplated, HLA haplotype-mismatched bone marrow cells. In this case, the T cells responsible for provoking GVH disease were depleted from the graft by treating the marrow cells with T12 monoclonal antibodies and complement. This approach has been experimentally used to prevent GVH disease in mice for some time, but it is only recently that appropriate markers of human T lymphocytes have made it a feasible prospect to use non-HLA-matched donors for human marrow transplants.

In summary, it can be concluded that erythropoiesis provides a valuable model system for studying differentiation and maturation. The progression of cells through the erythroid lineage can be monitored by *in vitro* and *in vivo* functional assays, by morphology, and by the analysis of cell surface glycoproteins. The combination of these methods with biochemical techniques can reveal the existence of cell surface markers that are specific to various stages of differentiation and maturation. Furthermore, the range of assays available for erythroid function will be valuable in determining the role of the cell surface structures in differentiation, e.g., mediating cell–cell interactions and as receptors for various regulatory signals, and hence to understand the abnormalities of differentiation in leukemia. We expect that new information about differentiation and maturation of cells in erythropoiesis will broaden and enrich our knowledge about these fundamental processes, both in relation to normal cell development and leukemogenesis.

Acknowledgments. The authors thank Dr. Elaine Spooncer for useful discussions and pertinent advice on the manuscript. The work done in our laboratories was supported by Grant CA-33000 and in part by Grants CA-33895 and CA-34014 from the National Cancer Institute. M.F. is supported in part by Cancer Center Support Grant CA-30199. The authors thank our colleagues who made our work possible. The authors also thank all those who have provided us with reprints and copies of their unpublished work and Ms. Anna Steve for secretarial assistance.

REFERENCES

Adair, W. L., and Kornfeld, S., 1974, Isolation of the receptors for wheat germ agglutinin and the *Ricinus communis* lectins from human erythrocytes using affinity chromatography, *J. Biol. Chem.* **249**:4696.

Allen, T. D., and Dexter, T. M., 1982, Ultrastructural aspect of erythropoietic differentiation in long-term bone marrow culture, *Differentiation* **21**:86.

Andersson, L. C., Jokinen, M., and Gahmberg, C. G., 1979a, Induction of erythroid differentiation in the human leukemic cell line K562, *Nature (London)* **278**:364.

Andersson, L. C., Nilsson, K., and Gahmberg, C. G., 1979b, K562—A human erythroleukemic cell line, *Int. J. Cancer* **23**:143.

Arrotti, J. J., and Garvin, J. E., 1972, Selective labeling of human erythrocyte components with tritiated trinitrobenzenesulfonic acid and picryl chloride, *Biochem. Biophys. Res. Commun.* **49**:205.

Bennet, V., and Stenbuck, P. J., 1980, Association between ankyrin and the cytoplasmic domain of Band 3 isolated from the human erythrocyte membrane, *J. Biol. Chem.* **255**:6424.

Björndal, H., Hellerqvist, C. G., Lindberg, B., and Svensson, S., 1970, Gas–liquid chromatography and mass spectrometry in methylation analysis of polysaccharides, *Angew. Chem. Int. Ed. Engl.* **9:**610.

Blumenfeld, O. O., Gallop, P. M., and Liao, T. H., 1972, Modification and introduction of a specific radioactive label into the erythrocyte membrane sialoglycoproteins, *Biochem. Biophys. Res. Commun.* **48:**242.

Bonner, W. M., and Laskey, R. A., 1974, A film detection method for tritium-labeled proteins and nucleic acids in polyacrylamide gels, *Eur. J. Biochem.* **46:**83.

Bretscher, M. S., 1971, Human erythrocyte membranes: Specific labeling of surface proteins, *J. Mol. Biol.* **58:**775.

Brown, G., Biberfeld, P., Christensson, B., and Mason, D. Y., 1979, The distribution of HLA on human lymphoid, bone marrow and peripheral blood cells, *Eur. J. Immunol.* **9:**272.

Brown, J. P., Hewick, R. M., Hellstrom, I., Hellstrom, K. E., Doolittle, R. F., and Dreyer, W. J., 1982, Human melanoma-associated antigen p97 is structurally and functionally related to transferrin, *Nature (London)* **296:**171.

Cabantchik, Z. I., and Rothstein, A., 1972, The nature of the membrane sites controlling anion permeability of human red blood cells as determined by studies with disulfonic stilbene derivatives, *J. Membr. Biol.* **15:**207.

Caligaris-Cappio, F., Gobbi, M., Bofill, M., and Janossy, G., 1982, Infrequent normal B lymphocytes express features of B-chronic lymphocytic leukemia, *J. Exp. Med.* **155:**623.

Chang, H., Langer, P. J., and Lodish, H. F., 1976, Asynchronous synthesis of erythrocyte membrane proteins, *Proc. Natl. Acad. Sci. USA* **73:**3206.

Childs, R. A., Feizi, T., Fukuda, M., and Hakomori, S., 1978, Blood group I activity associated with Band 3, the major intrinsic membrane protein of human erythrocytes, *Biochem. J.* **173:**333.

Clarkson, B., Marks, P. A., and Till, J. E. (eds.), 1978, *Differentiation of Normal and Neoplastic Hematopoietic Cells,* Cold Spring Harbor Laboratory, New York.

Dexter, T. M., 1982, Stromal cell associated hemopoiesis, *J. Cell Physiol. Suppl.* **1:**87.

Dexter, T. M., and Testa, N. G., 1980, *In vitro* methods in haemopoiesis and lymphopoiesis, *J. Immunol. Methods* **38:**177.

Dexter, T. M., Allen, T. P., and Lajtha, L. G., 1977, Conditions controlling the proliferation of haemopoietic stem cells *in vitro*, *J. Cell Physiol.* **91:**335.

Doinel, C., Andrew, G., Carton, J. P., Salmon C., and Fukuda, M. N., 1980, TK polyagglutination produced *in vitro* by endo-β-galactosidase, *Vox Sang.* **38:**94.

Dokhelar, M.-C., Testa, U., Vainchenker, W., Finale, Y., Tetaud, C., Salem, P., and Tursz, T., 1982, NK cell sensitivity of the leukemic K562 cells: Effect of sodium butyrate and hemin induction, *J. Immunol.* **128:**211.

Dozier, J. C., Diedrich, D. F., and Turco, S. J., 1981, The hexose transport system in the human K-562 chronic myelogenous leukemia-derived cell, *J. Cell Physiol.* **108:**77.

Drickamer, L. K., 1978, Orientation of the Band 3 polypeptide from human erythrocyte membranes: Identification of NH_2-terminal sequence and site of carbohydrate attachment, *J. Biol. Chem.* **253:**7242.

Ebert, W., Roelcke, D., and Weicker, H., 1975, The I antigen of human red cell membrane, *Eur. J. Biochem.* **53:**505.

Eisen, H., Bach, R., and Embery, R., 1977, Induction of spectrin in erythroleukemic cells transformed by Friend virus, *Proc. Natl. Acad. Sci. USA* **74:**3898.

Eylar, E. H., Madoff, M. A., Brody, U. V., and Oncley, J. L., 1962, The contribution of sialic acid to the surface charge of the erythrocyte, *J. Biol. Chem.* **237:**1992.

Fauser, A. A., and Messner, H. A., 1979, Identification of megakaryocytes, macrophages and eosinophils in colonies of human bone marrow containing neutrophilic granulocytes and erythroblasts, *Blood* **53:**1023.

Feizi, T., Kabat, E. A., Vicari, G., Anderson, B., and Marsh, W. L., 1971a, Immuno-chemical studies in blood groups, XLVII. The I antigen complex precursors in A, B, H, Le^a and Le^b blood group system, *J. Exp. Med.* **133**:39.

Feizi, T., Kabat, E. A., Vicari, G., Anderson, B., and Marsh, W. L., 1971b, Immuno-chemical studies in blood groups. XLIX. The I antigen complex: Specificity differences among anti-I sera revealed by quantitative precipitation studies; partial structure of the I determinant specific for one anti-I serum, *J. Immunol.* **106**:1578.

Feizi, T., Childs, R. A., Watanabe, K., and Hakomori, S., 1979, Three types of blood group I specificity among monoclonal anti-I autoantibodies revealed by analogues of a branched erythrocyte glycolipid, *J. Exp. Med.* **149**:975.

Findlay, J. B. C., 1974, The receptor proteins for concanavalin A and *Lens culinas* phy-tohemagglutinin in the membrane of the human erythrocyte, *J. Biol. Chem.* **249**:4398.

Finne, J., Krusius, T., Rauvala, H., Kekomaki, R., and Myllylä, G., 1978, Alkali-stable blood group A- and B-active poly(glycosyl) peptides from human erythrocyte mem-brane, *FEBS Lett.* **89**:111.

Finne, J., Krusius, T., and Järnefelt, J., 1980, Fractionation of glycopeptides, in: *27th In-ternational Congress of Pure and Applied Chemistry* (A. Varmavouri, ed.), pp. 147–159, Pergamon Press, Elmsford, N.Y.

Fitchen, J. H., Foon, K. A., and Cline, M. J., 1981, The antigenic characteristics of he-matopoietic stem cells, *N. Engl. J. Med.* **305**:17.

Foon, K. A., Schroff, R. W., and Gale, R. P., 1982, Surface markers on leukemia and lymphoma cells: Recent advances, *Blood* **60**:1.

Foxwell, B. M. J., and Tanner, M. J. A., 1981, Synthesis of the erythrocyte anion-transport protein: Immunochemical study of its incorporation into the plasma membrane of eryth-roid cells, *Biochem. J.* **195**:129.

Friend, C., Scher, W., Holland, J. G., and Sato, T., 1971, Hemoblogin synthesis in virus-induced leukemic cells *in vitro*: Stimulation of erythroid differentiation by dimethyl sulfoxide, *Proc. Natl. Acad. Sci. USA* **68**:378.

Fukuda, M., 1980, K562 human leukaemic cells express fetal type (i) antigen on different glycoproteins from circulating erythrocytes, *Nature (London)* **285**:405.

Fukuda, M., 1981, Tumor-promoting phorbol diester-induced specific changes in cell surface glycoprotein profile of K562 human leukemic cells, *Cancer Res.* **41**:4621.

Fukuda, M., and Fukuda, M. N., 1981, Changes in cell surface glycoproteins and carbo-hydrate structures during the development and differentiation of human erythroid cells, *J. Supramol. Struct. Cell. Biochem.* **17**:313.

Fukuda, M., and Osawa, T., 1973, Isolation and characterization of a glycoprotein from human group O erythrocytes, *J. Biol. Chem.* **248**:5100.

Fukuda, M., Kondo, T., and Osawa, T., 1976, Studies on the hydrazinolysis of glycopro-teins, core structures of oligosaccharides obtained from porcine thyroglogulin and pine-apple stem bromelain, *J. Biochem. Tokyo* **80**:1223.

Fukuda, M., Eshdat, Y., Tarone, G., and Marchesi, V. T., 1978, Isolation and character-ization of peptides derived from the cytoplasmic segment of Band 3, the predominant intrinsic membrane protein of the human erythrocyte, *J. Biol. Chem.* **253**:2419.

Fukuda, M., Fukada, M. N., and Hakomori, S., 1979, Developmental change and genetic defect in the carbohydrate structure of Band 3 glycoprotein of human erythrocyte mem-brane, *J. Biol. Chem.* **254**:3700.

Fukuda, M., Fukuda, M. N., Papayannopoulou, T., and Hakomori, S., 1980, Membrane differentiation in human erythroid cells: Unique profiles of cell surface glycoproteins expressed in erythroblasts *in vitro* from three ontogenic stages, *Proc. Natl. Acad. Sci. USA* **77**:3474.

Fukuda, M., Koeffler, H. P., and Minowada, J., 1981a, Membrane differentiation in human myeloid cells: Expression of unique profiles of cell surface glycoproteins in myeloid leukemic cell lines blocked at different stages of differentiation and maturation, *Proc. Natl. Acad. Sci. USA* **78**:6299.

Fukuda, M., Fukuda, M. N., Hakomori, S., and Papayannopoulou, T., 1981b, Anomalous cell surface structure of sickle cell anemia erythrocytes as demonstrated by cell surface labeling and endo-β-galactosidase treatment, *J. Supramol. Struct. Cell. Biochem.* **17**:289.

Fukuda, M., Dell, A., and Fukuda, M. N., 1984, Structure of fetal lactosaminoglycan, the carbohydrate moiety of Band 3 isolated from human umbilical cord erythrocytes, *J. Biol. Chem.*, (in press).

Fukuda, M. N., 1981. Purification and characterization of endo-β-galactosidase from *Escherichia freundii* induced by hog gastric mucin, *J. Biol. Chem.* **256**:3900.

Fukuda, M. N., 1982, Endo-β-galactosidases from *Diplococcus pneumoniae*, *Fed. Proc.* **41**:1160.

Fukuda, M. N., and Levery, S. B., 1983, Glycolipids of fetal, newborn, and adult erythrocytes: glycolipid pattern and structural study of H_3-glycolipid from newborn erythrocytes, *Biochemistry* **22**:5034.

Fukuda, M. N., and Matsumura, G., 1975, Endo-β-galactosidase of *Escherichia freundii*: Hydrolysis of pig colonic mucin and milk oligosaccharides by endoglycosidic action, *Biochem. Biophys. Res. Commun.* **64**:465.

Fukuda, M. N., and Matsumura, G., 1976, Endo-β-galactosidase of *Escherichia freundii*: Purification and endoglycosidic action on keratan sulfates, oligosaccharides, and blood group active glycoprotein, *J. Biol. Chem.* **251**:6218.

Fukuda, M. N., Fukuda, M., Watanabe, K., and Hakomori, S., 1978a, Modification of cell surface antigenicity by endo-β-galactosidase of *E. freundii*, *Fed. Proc.* **37**:1601.

Fukuda, M. N., Watanabe, K., and Hakomori, S., 1978b, Release of oligosaccharides from various glycosphingolipids by endo-β-galactosidase, *J. Biol. Chem.* **253**:6814.

Fukuda, M. N., Fukuda, M., and Hakomori, S., 1979, Cell surface modification by endo-β-galactosidase, change of blood group activities and release of oligosaccharides from glycoproteins and glycosphingolipids of human erythrocytes, *J. Biol. Chem.* **254**:5458.

Gahmberg, C. G., 1976, External labeling of human erythrocyte glycoproteins: Studies with galactose oxidase and fluorography, *J. Biol. Chem.* **251**:510.

Gahmberg, C. G., and Andersson, L. C., 1977, Selective radioactive labeling of cell surface sialoglycoproteins by periodate–tritiated borohydride, *J. Biol. Chem.* **252**:5888.

Gahmberg, C. G., and Hakomori, S., 1973, External labeling of cell surface galactose and galactosamine in glycolipid and glycoprotein of human erythrocytes, *J. Biol. Chem.* **248**:4311.

Gahmberg, C. G., Myllylä, G., Leikola, J., Pirkola, A., and Nordling, S., 1976, Absence of the major sialoglycoprotein in the membrane of human En(a-) erythrocytes and increased glycosylation of Band 3, *J. Biol. Chem.* **251**:6108.

Gahmberg, C. G., Jokinen, M., and Andersson, L. C., 1978, Expression of the major sialoglycoprotein (glycophorin) on erythroid cells in human bone marrow cells, *Blood* **52**:379.

Gahmberg, C. G., Jokinen, M., and Andersson, L. C., 1979, Expression of the major red cell sialoglycoprotein, glycophorin A, in the human leukemia cell line K562, *J. Biol. Chem.* **254**:7442.

Gartner, S., and Kaplan, H. S., 1980, Long-term culture of human bone marrow cells, *Proc. Natl. Acad. Sci. USA* **77**:4756.

Geiduschek, S. B., and Singer, J. J., 1979, Molecular changes in the membranes of mouse erythroid cells accompanying differentiation, *Cell* **16**:149.

Gesner, B. M., and Ginsburg, V., 1964, Effect of glycosidases on the fate of transfused lymphocytes, *Proc. Natl. Acad. Sci. USA* **52**:750.

Giblett, E. R., and Crookston, M. C., 1964, Agglutinability of red cells by anti-i in patients with thalassemia and other haematological disorders, *Nature (London)* **201**:1138.

Gilliland, D. G., Steplewski, Z., Collier, R. J., Mitchell, K. F., Chang, T. H., and Koprowski, H., 1980, Antibody-directed cytotoxic agents: Use of monoclonal antibody to direct the action of toxin A chains to colorectal carcinoma cells, *Proc. Natl. Acad. Sci. USA* **77**:4539.

Gooi, H. C., Feizi, T., Kapadia, A., Knowles, B. B., Solter, D., and Evans, M. J., 1981, Stage-specific embryonic antigen involves $\alpha 1 \rightarrow 3$ fucosylated type 2 blood group chains, *Nature (London)* **292**:156.

Gorga, F. R., Baldwin, S. A., and Lienhard, G. E., 1979, The monosaccharide transporter from human erythrocytes is heterogeneously glycosylated, *Biochem. Biophys. Res. Commun.* **91**:955.

Gregory, C. J., 1976, Erythropoietin sensitivity as a differentiation marker in the hemopoietic system, *J. Cell Biol.* **89**:289.

Guerrasio, A., Vainchenker, W., Breton-Gorius, J., Testa, U., Rosa, R., Thomopoulos, P., Titeux, M., Guichard, J., and Beuzard, Y., 1981, Embryonic and fetal hemoglobin synthesis in K562 cell line, *Blood Cells* **7**:165.

Hakomori, S., 1964, A rapid permethylation of glycolipid and polysaccharide catalyzed by methylsulfinyl carbonium in dimethyl sulfoxide, *J. Biochem. Tokyo* **55**:205.

Hakomori, S., Fukuda, M., and Nudelman, E., 1982, Role of cell surface carbohydrates in differentiation: Behaviour of lactosaminoglycan in glycolipids and glycoproteins, in: *Teratocarcinoma and Embryonic Cell Interactions* (T. Muramatsu, G. Gachelin, A. A., Moscona, and Y. Ikawa, eds.), pp. 179–200, Academic Press, New York.

Hakomori, S., Fukuda, M., Sekiguchi, K., and Carter, W. G., 1984, Chemistry and function of pericellular and intercellular glycoproteins: Fibronectin, laminin and other matrix components, in: *Connective Tissue Biochemistry* (K. Piez and A. H. Reddi, eds.), Elsevier/North-Holland, Amsterdam (in press).

Hamilton, T. A., Wada, H. G., and Sussman, H. H., 1979, Identification of transferrin receptors on the surface of human cultured cells, *Proc. Natl. Acad. Sci. USA* **76**:6406.

Harpaz, N., Flowers, H. M., and Sharon, N., 1977, α-Galactosidase from soybeans destroying blood-group B antigens: Purification by affinity chromatography and properties, *Eur. J. Biochem.* **77**:419.

Harrison, F. L., and Chesterton, C. J., 1980, Erythroid developmental agglutinin is a protein lectin mediating specific cell–cell adhesion between differentiating rabbit erythroblasts, *Nature (London)* **286**:502.

Hillman, R. S., and Giblett, E. R., 1965, Red cell membrane alteration associated with 'marrow stress,' *J. Clin. Invest.* **44**:1730.

Hirano, S., and Meyer, K., 1971, Enzymatic degradation of corneal and cartilaginous keratan sulfates, *Biochem. Biophys. Res. Commun.* **44**:1371.

Horton, M. A., 1982, Analysis of marker enzymes in the K562 erythroleukaemia cell line: No coordinate expression of red cell enzymes in induction of haemoglobin synthesis, *Biomedicine* **36**:213.

Horton, M. A., Cedar, S. H., and Edwards, P. A. W., 1981, Expression of red cell specific determinants during differentiation in the K562 erythroleukemia cell line, *Scand. J. Haematol.* **27**:231.

Hubbard, A. C., and Cohn, Z. A., 1972, The enzymatic iodination of the red cell membrane, *J. Cell. Biol.* **55**:390.

Hutchings, S. E., and Sato, G. H., 1978, Growth and maintenance of HeLa cells in serum-free medium supplemented with hormones, *Proc. Natl. Acad. Sci. USA* **75**:901.

Irimura, T., Tsuji, T., Tagami, S., Yamamoto, K., and Osawa, T., 1981, Structure of a complex-type sugar chain of human glycophorin A, *Biochemistry* **20**:560.

Janossy, G., Bollum, F., Bradstock, K., McMichael, A., Rapson, N., and Greaves, M. F., 1979, Terminal deoxynucleotidyl transferase positive cells in normal human bone marrow have the antigen phenotype of acute lymphoblastic leukemia cells, *J. Immunol.* **123:**1525.

Järnefelt, J., Rush, J., Li, Y.-T., and Laine, R. A., 1978, Erythroglycan, a high molecular weight glycopeptide with the repeating structure [galactosyl(1→4)2-deoxy-2-acetamido-glucosyl(1→3)] comprising more than one-third of the protein bound carbohydrate of human erythrocyte stroma, *J. Biol. Chem.* **253:**8006.

Jenkins, W. J., Marsh, W. L., and Gold, E. R., 1965, Reciprocal relationship of antigens I and i in health and disease, *Nature (London)* **205:**813.

Kaizu, T., Turco, S. J., Rush, J. S., and Laine, R. A., 1982, Synthesis of the branched form of erythroglycan by Friend GM979 erythroleukemic cells, *J. Biol. Chem.* **257:**8272.

Kannagi, R., Fukuda, M. N., and Hakomori, S., 1982, A new glycolipid antigen isolated from human erythrocyte membranes reacting with antibodies directed to globo-N-tetraosyl ceramide (globoside), *J. Biol. Chem.* **257:**4438.

Karhi, K. K., Andersson, L. C., Vuopio, O., and Gahmberg, C. G., 1981, Expression of blood group A antigens in human bone marrow cells, *Blood* **57:**147.

Karlsson, K.-A., Leffler, H., and Samuelson, B. E., 1974, Characterization of the Forssman glycolipid hapten of horse kidney by mass spectrometry, *J. Biol. Chem.* **249:**4819.

Keating, A., Singer, J. W., Killen, P. D., Striker, G. E., Salo, A. C., Sanders, J., Thomas, E. P., Thorning, D., and Fialkow, P. J., 1982, Donor origin of the *in vitro* haematopoietic microenvironment after marrow transplantation in man, *Nature (London)* **298:**280.

Kitamikado, M., and Ueno, R., 1970, Enzymatic degradation of whale cartilage keratosulfate. III. Purification of a bacterial keratosulfate-degrading enzyme, *Bull. Jpn. Soc. Sci. Fish.* **36:**1175.

Kitamikado, M., Ito, M., and Li, Y.-T., 1981, Isolation and characterization of a keratan sulfate-degrading endo-β-galactosidase from *Flavobacterium karatolyticus, J. Biol. Chem.* **256:**3906.

Klein, J., 1979, The major histocompatibility complex of the mouse, *Science* **203:**516.

Knauf, P. A., and Law, F.-Y., 1981, Comparison of anion exchange in K562 erythroleukemic cells and human red blood cells, *J. Supramol. Struct. Cell Biochem.* Supplement 5, p. 124, Alan R. Liss, New York.

Koeffler, H. P., and Golde, D. W., 1981, Chronic myelogenous leukemia—New concepts, *N. Engl. J. Med.* **304:**1201, 1269.

Kornfeld, R., and Kornfeld, S., 1976, Comparative aspects of glycoprotein structure, *Annu. Rev. Biochem.* **45:**217.

Kornfeld, R., and Kornfeld, S., 1980, Structure of glycoproteins and their oligosaccharide units, in: *The Biochemistry of Glycoproteins and Proteoglycans* (W. J. Lennarz, ed.), pp. 1–34, Plenum Press, New York.

Koschielak, J., Zdebska, E., Wilczynska, Z., Miller-Podraza, H., and Dzierzkowa-Borodej, W., 1979, Immunochemistry of Ii-active glycosphingolipids of erythrocytes, *Eur. J. Biochem.* **96:**331.

Koury, J. J., and Pragnell, I. B., 1982, Retroviruses induce granulocyte–macrophage colony stimulating activity in fibroblasts, *Nature (London)* **299:**638.

Krusius, T., Finne, J., and Rauvala, H., 1978, The poly(glycosyl) chains of glycoproteins: Characterization of a novel type of glycoprotein saccharides from human erythrocyte membrane, *Eur. J. Biochem.* **92:**289.

Laemmli, U. K., 1970, Cleavage of structural proteins during assembly of bacteriophage T_4, *Nature (London)* **227:**680.

Li, E., Gibson, R., and Kornfeld, S., 1979, Structure of an unusual complex-type oligosaccharides isolated from Chinese hamster ovary cells, *Arch. Biochem. Biophys.* **199:**393.

Li. Y.-T., 1967, Studies on the glycosidases in jack bean meal. I. Isolation and properties of α-mannosidase, *J. Biol. Chem.* **242**:5474.

Lichtman, M. A., 1981, The ultrastructure of the hemopoietic environment of the marrow: A review, *Exp. Hematol.* **9**:391.

Lozzio, B. B., and Lozzio, C. B., 1980, Properties and usefulness of the original K-562 human myelogenous leukemia cell line, *Leuk. Res.* **3**:363.

Lozzio, C. B., and Lozzio, B. B., 1975, Human chronic myelogenous leukemia cell-line with positive Philadelphia chromosome, *Blood* **45**:321.

McGinniss, M. H., Schmidt, P. J., and Carbone, P. P., 1964, Close association of I blood group and disease, *Nature (London)* **202**:606.

Marchesi, V. T., Furthmayr, H., and Tomita, M., 1976, The red cell membrane, *Annu. Rev. Biochem.* **45**:667.

Marcus, D. M., Kabat, E. A., and Rosenfeld, R. E., 1963, The action of enzymes from *Clostridium teritium* in the I-antigenic determinants of human erythrocytes, *J. Exp. Med.* **118**:175.

Marie, J. P., Izaguirre, C. A., Civin, C. I., Mirro, J., and McCulloch, E. A., 1981, The presence within single K562 cells of erythropoietic and granulopoietic differentiation markers, *Blood* **58**:708.

Marsh, W. L., 1961, Anti-i: Cold antibody defining Ii relationship in human red cells, *Br. J. Haematol.* **7**:200.

Martin, K., 1970, The effect of proteolytic enzymes on acetylocholine-esterase activity, the sodium pump and choline transport in human erythrocytes, *Biochim. Biophys. Acta* **203**:182.

Metcalf, D., 1977, *Hematopoietic Colonies: In Vitro Cloning of Normal and Leukemic Cells,* Springer-Verlag, Berlin.

Metcalf, D., and Moore, M. A. S., 1971, *Hematopoietic Cells,* North-Holland, Amsterdam.

Metcalf, D., and Johnson, C. R., 1978, Mixed hematopoietic colonies *in vitro,* in: *Hematopoietic Cell Differentiation,* ICN-UCLA Symposia on Molecular and Cellular Biology *10,* (D. W. Golde, M. J. Cline, D. Metcalf, C. F. Fox, eds.), pp. 141–151, Academic Press, New York.

Metcalf, D., Johnson, G. R., and Mandel, T. E., 1979, Colony formation in agar by multipotential hemopoietic cells, *J. Cell Physiol.* **98**:401.

Metzgar, R. S., Browitz, M. J., Jones, N. H., and Lowell, B. L., 1981, Distribution of common acute lymphoblastic leukemia antigen in nonhematopoietic tissues, *J. Exp. Med.* **154**:1249.

Montreuil, J., 1980, Primary structure of glycoprotein glycans: Basis for the molecular biology of glycoproteins, *Adv. Carbohydr. Chem. Biochem.* **37**:157.

Moore, M. A. S., and Metcalf, D., 1970, Ontogeny of the hematopoietic system: Yolk sac origin of *in vivo* and *in vitro* colony forming cells in the developing mouse embryo, *Br. J. Haematol.* **18**:279.

Moore, M. A. S., Broxmeyer, H. E., Sheridan, A. P. C., Meyers, P. A., Jacobson, N., and Winchester, R. J., 1980, Continuous human bone marrow cells: Ia antigen characterization of probable pluripotential stem cells, *Blood* **55**:682.

Moore, S., Woodrow, C. H., and McClelland, D. B. L., 1982, Isolation of membrane components associated with human red cell antigens Rh(D), (C), (E) and Fy^a, *Nature (London)* **295**:529.

Morell, A. G., Van Den Hamer, C. J. A., Scheinberg, I. H., and Ashwell, G., 1966, Physical and chemical studies in ceruloplasmin. IV. Preparation of radioactive sialic acid-free ceruloplasmin labeled with tritium in terminal D-galactose residues, *J. Biol. Chem.* **241**:3745.

Morrow, J. S., Speicher, P. W., Knowles, W. J., Hsu, C. J., and Marchesi, V. T., 1980, Identification of functional domains of human erythrocyte spectrin, *Proc. Natl. Acad. Sci. USA* **77:**6592.

Muramatsu, H., Muramatsu, T., and Avner, P., 1982, Biochemical properties of the high-molecular-weight glycopeptides released from the cell surface of human teratocarcinoma cells, *Cancer Res.* **42:**1749.

Muramatsu, T., 1971, Demonstration of an endo-glycosidase acting on a glycoprotein, *J. Biol. Chem.* **246:**5535.

Muramatsu, T., and Egami, F., 1967, α-Mannosidase and β-mannosidase from the liver of *Turbo corrutus:* Purification, properties and application of carbohydrate research, *J. Biochem. Tokyo* **62:**700.

Muramatsu, T., Gachelin, G., Damonneville, M., Delarbre, C., and Jacob, F., 1979, Cell surface carbohydrates of embryonal carcinoma cells: Polysaccharidic side chains of F9 antigens and of receptors to two lectins, FBP and PNA, *Cell* **18:**183.

Nakahata, T., and Ogawa, M., 1982, Identification in culture of a class of hemopoietic colony forming units with extensive capability to self-renew and generate multipotential hemopoietic colonies, *Proc. Natl. Acad. Sci. USA* **79:**3843.

Nakazawa, K., Suzuki, N., and Suzuki, S., 1975, Sequential degradation of keratan sulfate by bacterial enzymes and purification of a sulfatase in the enzymatic system, *J. Biol. Chem.* **250:**905.

Nieman, H., Watanabe, K., Hakomori, S., Childs, R. A., and Feizi, T., 1978, Blood group i and I activities of "lacto-N-norhexaosylceramide" and its analogues: The structural requirement for i-specifities, *Biochem. Biophys. Res. Commun.* **81:**1286.

Nigg, E. A., Bron, C., Girardet, M., and Cherry, R. J., 1980, Band 3–glycophorin A association in erythrocyte membranes demonstrated by combining protein diffusion measurements with antibody-induced cross-linking, *Biochemistry* **19:**1887.

Ogata, S., Muramatsu, T., and Kobata, A., 1975, Fractionation of glycopeptides by affinity column chromatography on concanavalin A-sepharose, *J. Biochem. Tokyo* **78:**687.

O'Hara, C. J., Shumak, K. H., and Price, G. B., 1978, The i antigen on human myeloid progenitors, *Clin. Immunol. Immunopathol.* **10:**420.

Omary, M. B., Trowbridge, G. S., and Minowada, J., 1980, Human cell-surface glycoprotein with unusual properties, *Nature (London)* **286:**888.

Papayannopoulou, T., Chen, P., Maniatis, A., and Stamatoyannopoulos, G., 1980, Simultaneous assessment of i-antigenic expression and fetal hemoglobin in single red cells by immunofluorescence, *Blood* **55:**221.

Papayannopoulou, T., Halfpap, L., Chen, S. H., Fukuda, M., Hoffman, R., Dow, L., and Hill, S., 1981, Fetal red cell markers and their relationships in patients with hematologic malignancies, in: *Hemoglobin in Development and Differentiation* (G. Stamatoyannopoulos and A. W. Nienhuis, eds.), pp. 443–456, Liss, New York.

Phillips, D. R., and Morrison, M., 1970, The arrangement of proteins in the human erythrocyte membrane, *Biochem. Biophys. Res. Commun.* **40:**284.

Quesenberry, P., and Levitt, L., 1979, Hematopoietic stem cells, *N. Engl. J. Med.* **301:**755, 819, 868.

Race, R. R., and Sanger, R., 1975, *Blood Groups in Man*, Blackwell, Oxford.

Raghavendra, R. A. M., Brown, A. K., Rieder, R. F., Clegg, J. G., and Marsh, W. L., 1978, Aplastic anemia with fetal-like erythropoiesis following androgen therapy, *Blood* **51:**711.

Rasilo, M.-L., and Renkonen, O., 1982, Cell-associated glycosaminoglycans of human teratocarcinoma-derived cells of line PA1, *Eur. J. Biochem.* **123:**397.

Rearden, A., and Masouredis, S. P., 1977, Blood group D antigen content of nucleated red cell precursors, *Blood* **50:**981.

Reinherz, E. L., and Schlossman, S. F., 1980, The differentiation and function of human T lymphocytes, *Cell* **19**:821.

Reinherz, E. L., Kung, P. C., Goldstein, G., Levey, R. H., and Schlossman, S. F., 1980, Discrete stages of human intrathymic differentiation: Analysis of normal thymocytes and leukemic lymphoblasts of T-cell lineage, *Proc. Natl. Acad. Sci. USA* **77**:1588.

Reinherz, E. L., Geha, R., Rappeport, J. M., Wilson, M., Penta, A. C., Hussey, R., Fitzgerald, K. A., Daley, J. F., Levine, H., Rosen, F. S., and Schlossman, S. F., 1982, Reconstitution after transplantation with T-lymphocyte-depleted HLA halotype-mismatched bone marrow for severe combined immunodeficiency, *Proc. Natl. Acad. Sci. USA* **79**:6047.

Robinson, J., Sieff, C., Delia, D., Edwards, P. A. W., and Greaves, M., 1981, Expression of cell-surface HLA-DR, HLA-ABC and glycophorin during erythroid differentiation, *Nature (London)* **289**:68.

Rochant, H., Dreyfus, B., Bouguerra, M., and Tonthat, H., 1972, Refractory anemias, preleukemic conditions and fetal erythropoiesis, *Blood* **39**:721.

Romans, D. G., Tilley, C. A., and Dorrington, K. J., 1980, Monogamous bivalency of IgG antibodies. I., Deficiency of branched ABHI-active oligosaccharide chains on red cells of infants causes the weak antiglobulin reactions in hemolytic disease of the newborn due to ABO incompatibility, *J. Immunol.* **124**:2807.

Rutherford, T. R., Clegg, J. B., and Weatherall, D. J., 1979, K562 human leukaemic cells synthesize embryonic haemoglobin in response to haemin, *Nature (London)* **280**:164.

Salter, D. W., Bladwin, S. A., Lienhard, G. E., and Weber, M. J., 1982, Proteins antigenically related to the human erythrocyte glucose transporter in normal and Rous sarcoma virus-transformed chicken embryo fibroblasts, *Proc. Natl. Acad. Sci. USA* **79**:1540.

Sandford, P. A., and Conrad, H. E., 1966, The structure of the *Aerobacter aerogenes* A3(S1) polysaccharide. I. A reexamination using improved procedures for methylation analysis, *Biochemistry* **5**:1508.

Sieff, C., Bicknell, D., Caine, G., Robinson, J., Lam, G., and Greaves, M. F., 1982, Changes in cell surface antigen expression during hemopoietic differentiation, *Blood* **60**:703.

Singer, S. J., and Nicolson, G. L., 1972, The fluid mosaic model of the structure of cell membranes, *Science* **175**:720.

Stammatoyannopoulos, G., Papayannopoulou, T., Brice, M., Kurachi, S., Nakamoto, B., Lim, G., and Farquhar, M., 1981, Cell biology of hemoglobin switching. I. The switch from fetal to adult hemoglobin formation during ontogeny, in: *Hemoglobins in Development and Differentiation* (G. Stamatoyannopoulos and A. W. Nienhuis, eds.), pp. 287–305, Liss, New York.

Steck, T. L., 1974, The organization of proteins in the human red blood cell membranes, *J. Cell Biol.* **62**:1.

Steck, T. L., and Dawson, G., 1974, Topographical distribution of complex carbohydrates in the erythrocyte membrane, *J. Biol. Chem.* **249**:2135.

Steck, T. L., and Kant, J. A., 1974, Preparation of impermeable ghosts and inside-out vesicles from human erythrocyte membranes, *Methods Enzymol.* **31**:172.

Steck, T. L., Koziarz, J. J., Singh, M. K., Reddy, G., and Kohler, H., 1978, Preparation and analysis of seven major, topographically defined fragments of Band 3, the predominant transmembrane polypeptide of human erythrocyte membranes, *Biochemistry* **17**:1216.

Stellner, K., Saito, H., and Hakomori, S., 1973, Determination of aminosugar linkages in glycolipids by methylation: Amino sugar linkage of ceramide pentasaccharide of rabbit erythrocytes and of Forssman antigen, *Arch. Biochem. Biophys.* **155**:464.

Sutherland, R., Delia, D., Schneider, C., Newman, R., Kemshead, J., and Greaves, M., 1981, Ubiquitous, cell surface glycoprotein on tumor cells is proliferation-associated receptor for transferrin, *Proc. Natl. Acad. Sci. USA* **78**:4515.

Sweely, C. C., and Dawson, G., 1969, Lipids of the erythrocyte, in: *Red Cell Membrane Structure and Function* (G. A. Jamieson and T. J. Greenwalt, eds.), p. 172, Lippincott, Philadelphia.

Takasaki, S., and Kobata, A., 1976, Purification and characterization of an endo-β-galactosidase produced by *Diplococcus pneumoniae, J. Biol. Chem.* **251**:3603.

Tanner, M. J. A., and Boxer, D. H., 1972, Separation and some properties of the major proteins of the human erythrocyte membrane, *Biochem. J.* **129**:333.

Tarentino, A. L., and Maley, F., 1974, Purification and properties of an endo-β-N-acetylglucosaminidase from *Streptomyces griseus, J. Biol. Chem.* **249**:811.

Tarentino, A. L., and Maley, F., 1975, A comparison of the substrate specificities of endo-β-N-acetyoglucosaminidases from *Streptomyces griseus* and *Diplococcus pneumoniae, Biochem. Biophys. Res. Commun.* **67**:455.

Tarone, G., Hamasaki, N., Fukuda, M., and Marchesi, V. T., 1979, Proteolytic degradation of human erythrocyte Band 3 by membrane-associated protease activity, *J. Membr. Biol.* **48**:1.

Testa, U., Henri, A., Bettaieb, A., Titeux, M., Vainchenker, W., Tontha, H., Docklear, M. C., and Rochant, H., 1982, Regulation of i- and I-antigen expression in the K562 cell line, *Cancer Res.* **42**:4694.

Thomas, D. B., and Winzler, R. J., 1969, Structural studies on human erythrocyte glycoproteins: Alkali-labile oligosaccharides, *J. Biol. Chem.* **244**:5943.

Till, J. E., and McCulloch, 1961, A direct measurement of the radiation sensitivity of normal mouse bone marrow cells, *Radiat. Res.* **14**:213.

Tonkonow, B. L., Hoffman, R., Burger, D., Elder, J. T., Mazur, E. M., Murnane, M. J., and Benz, E. J., Jr., 1982, Differing responses of globin and glycophorin gene expression to hemin in the human leukemia cell line K562, *Blood* **59**:738.

Toogood, I. R. G., Dexter, T. M., Allen, T. D., Suda, T., and Lajtha, L. G., 1980, The development of a liquid culture system for the growth of human bone marrow, *Leuk. Res.* **4**:449.

Torok-Storb, B., and Martin, P., 1982, Modulation of *in vitro* BFU-E growth by normal Ia-positive T cells is restricted by HLA-DR, *Nature (London)* **298**:473.

Torok-Storb, B., Martin, P., and Hansen, J., 1981, Regulation of *in vitro* erythropoiesis by normal T cells: Evidence for two T-cell subsets with opposing function, *Blood* **58**:171.

Trowbridge, I. S., and Omary, M. B., 1981, Human cell surface glycoprotein related to cell proliferation is the receptor for transferrin, *Proc. Natl. Acad. Sci. USA* **78**:3039.

Tsuji, T., Irimura, T., and Osawa, T., 1980, The carbohydrate moiety of Band-3 glycoprotein of human erythrocyte membranes, *Biochem. J.* **187**:677.

Turco, S. J., Rush, J. S., and Laine, R. A., 1980, Presence of erythroglycan on human K562 chronic myelogeneous leukemia-derived cells, *J. Biol. Chem.* **255**:3266.

Tyler, J. M., Reinhardt, B. N., and Branton, D., 1980, Association of erythrocyte membrane proteins: Binding of purified Bands 2.1 and 4.1 to spectrin, *J. Biol. Chem.* **255**:7034.

Urdal, D. L., and Hakomori, S., 1980, Tumor-associated ganglio-N-triosylceramide, target for antibody-dependent avidin-mediated drug killing of tumor cells, *J. Biol. Chem.* **255**:10509.

Vainchenker, W., Testa, U., Rochant, H., Titeux, M., Henri, A., Bouguet, J., and Breton-Gorius, J., 1981a, Cellular regulation of i and I antigen expressions in human erythroblasts grown *in vitro, Stem Cells* **1**:97.

Vainchenker, W., Testa, U., Guichard, J., Titeux, M., and Breton-Gorius, J., 1981b, Heterogeneity in the cellular commitment of a human leukemic cell line: K562, *Blood Cells* **7**:357.

Van Lenten, L., and Ashwell, G., 1971, Studies in the chemical and enzymatic modification of glycoproteins: A general method for the tritiation of sialic acid containing glycoproteins, *J. Biol. Chem.* **246**:1889.

Victoria, E. J., Mahan, L. C., and Masouredis, S. P., 1981, Anti-rho (D) IgG binds to Band 3 glycoprotein of the human erythrocyte membrane, *Proc. Natl. Acad. Sci. USA* **78**:2898.

Wada, H. G., Hass, P. E., and Sussman, H. H., 1979, Transferrin receptor in human placental brush border membranes: Studies in the binding of transferrin to placental membrane vesicles and the identification of a placental brush border glycoprotein with high affinity for transferrin, *J. Biol. Chem.* **256**:12629.

Watanabe, K., and Hakomori, S., 1976, Status of blood group carbohydrate chains in ontogenesis and in oncogenesis, *J. Exp. Med.* **144**:664.

Watanabe, K., Laine, K. A., and Hakomori, S., 1975, On neutral fucoglycolipids having long, branched carbohydrate chains: H-Active and I-active glycosphingolipids of human erythrocyte membranes, *Biochemistry* **14**:2725.

Watanabe, K., Hakomori, S., Childs, R. A., and Feizi, T., 1979, Characterization of a blood group I-active ganglioside: Structural requirements for I and I specificities, *J. Biol. Chem.* **254**:3221.

Weber, K., and Osborn, M., 1969, The reliability of molecular weight determination by dodecylsulfate-polyacrylamide gel electrophoresis, *J. Biol. Chem.* **244**:4406.

Wiener, A. S., Unger, L. J., Cohen, L., and Feldman, J., 1956, Type-specific cold auto-antibodies as a cause of acquired hemolytic anemia and hemolytic transferrin reactions: Biological test with bovine red cells, *Ann. Intern. Med.* **44**:221.

Yurchencho, P. D., and Furthmayr, H., 1980, Expression of red cell membrane proteins in erythroid precursor cells, *J. Supramol. Struct.* **13**:255.

Yoshima, H., Furthmayr, H., and Kobata, A., 1980, Structures of the asparagine-linked sugar chains of glycophorin, *J. Biol. Chem.* **255**:9713.

Yoshima, H., Shiraishi, N., Matsumoto, A., Maeda, S., Sugiyama, T., and Kobata, A., 1982, The asparagine-linked sugar chains of plasma membrane glycoproteins of K562 human leukaemic cells: A comparative study with human erythrocytes, *J. Biochem. Tokyo* **91**:233.

5

Carbohydrate Structure, Biological Recognition, and Immune Function

Christopher L. Reading

1. INTRODUCTION

Complex carbohydrates are favorable candidates for encoding biological information because of the large number of structures possible in relatively short oligosaccharide sequences. A great variety of complex carbohydrate structures exist in nature, with ample diversity to serve as receptors in recognition phenomena. The reader is referred to reviews for consideration of the carbohydrate structures found in glycolipids (Sweeley *et al.*, 1978), glycoproteins (Walborg, 1978; Sharon and Lis, 1980; Berger *et al.*, 1982) and glycosaminoglycans (Ginsburg and Neufeld, 1969; Heath 1971).

A good deal of evidence indicates that complex carbohydrate structures may play a major role in biological recognition processes (reviews: Sharon and Lis, 1980; Berger *et al.*, 1982). Only in a few cases is there evidence to support this concept based on demonstration of the relationship between structure and function. There are two components to such a recognition system: the carbohydrate-binding protein and the complex carbohydrate receptor. Several soluble and cell surface carbohydrate-binding proteins are known, including antibodies, complement compo-

Christopher L. Reading • The Department of Tumor Biology and the Bone Marrow Transplantation Center, The University of Texas, M. D. Anderson Hospital and Tumor Institute at Houston, Houston, Texas 77030.

nents, hormones, toxins, lectins, enzymes, and carbohydrate-mediated uptake systems. There is also reason to believe that carbohydrate structures on cells change in response to their stage of differentiation and even their position in the cell cycle.

One of the earliest carbohydrate recognition systems was antibodies that recognize blood group structures (Table 1). More recently, evidence has accumulated to implicate recognition of carbohydrate structures in the activation of complement (Table 2). Glycolipids and glycoproteins are capable of serving as receptors for several toxins and hormones. Examples of carbohydrate recognition by soluble mediators are presented in Table 3 (reviews: Critchley and Vicker, 1977; Grollman *et al.*, 1978). Glycolipids and glycoproteins can share terminal sugar sequences (review: Rauvala and Finne, 1979), and both may be important as cellular receptors for the same component.

Lectins are sugar-binding proteins and glycoproteins other than antibodies, which agglutinate cells or precipitate glycoconjugates (Goldstein *et al.*, 1980). The specificities of lectins are usually tested by mono- or oligosaccharide inhibition of agglutination or precipitation reactions, but in some cases, only complex carbohydrates inhibit the reactions. Lectins found in plants (reviews: Sharon and Lis, 1972; Lis and Sharon, 1973, 1977; Goldstein and Hayes, 1978; Brown and Hunt, 1978) are easily purified by using affinity chromatography on carbohydrate gels or immobilized carbohydrates. The plant lectins have been used extensively to characterize cell surface components of mammalian cells (reviews: Etzler, 1974; Nicolson, 1974; Lotan and Nicolson, 1978, 1979; Kornfeld and Kornfeld, 1978). In addition, plant lectins have been used to investigate numerous cell surface phenomena including transmembrane control of cell surface interactions (Ji and Nicolson, 1974), membrane glycoprotein turnover (Karsenti and Avrameas, 1973), lymphocyte mitogenesis (Boldt *et al.*, 1975; Dillner-Centerlind *et al.*, 1980), macrophage phagocytosis (Goldman, 1974), and cell surface glycoprotein changes in experimental metastasis (Reading *et al.*, 1980a,b).

There are few examples of actual recognition roles for plant lectins in nature. Lectins isolated from legumes are believed to be involved in specific cellular interaction between root cells and the symbiotic bacteria of the genus *Rhizobium*, which nodulate the roots and produce the nitrogenase responsible for nitrogen fixation (review: Dazzo, 1980). Thus, soybean agglutinin (SBA), which is inhibitable by *N*-acetylgalactosamine, binds to carbohydrate residues on the bacterium *Rhizobium japonicum*, and the clover lectin trifolium, which is inhibitable by a 2-deoxy-D-glucose, binds carbohydrate residues on the bacterium *R. trifolii*, the clover symbiont.

Table 1. Antibodies to Carbohydrate Determinants

Source	Glycoconjugate recognized	References[a]
Unimmunized serum		
Horse serum	Blood group substances	1
Human serum	Forssman glycolipid	2
Rabbit serum	CDH and CTH	3
Several species' serum	Fetuin glycopeptides	4
Human, guinea pig, and mouse serum	Neuraminidase-treated lymphocytes and lactose	5, 6
Human and mouse serum	T-antigen	7, 8
Multiple sclerosis serum	Glycolipids	9
Lupus erythematosus serum	Glycolipids	10
Serum immunized with:		
Trout immunoglobulin		11
Bacteria, yeast		12–14
Mouse T cells		15
Neisseria meningitidis	Sialic acids	16
E. coli	Sialic acids	17
Group C streptococcus	GalNAc, erythrocytes	18
Rat erythrocytes	Trypsinized bovine RBC, carbohydrate tissue isoantigens	19
Transplanted human organs	Erythrocytes of foreign species	20
Glycolipids		21–25
Proteins conjugated with:		
Diazotized phenyl glycosides		26
Lacto-*N*-difucohexaose I		27
Diazotized *p*-(aminophenyl)ethylamine derivatives of oligosaccharides		28
p-Isothiocyanatophenyl disaccharides		29
Synthetic oligosaccharides		30
Isothiocyanate derivatives of *p*-(aminophenyl)ethylamine oligosaccharides		31
Murine myeloma proteins		
S117	Terminal GlcNAc, highest affinity for GlcNAc-β(1-3)GlcNAc-β(1-6)Gal	32

(Continued)

Table 1. (Continued)

Source	Glycoconjugate recognized	References[a]
MOPC 384	Precipitates LPS; inhibited with β-Me-D-Gal	33
MOPC 406	Precipitates LPS; inhibited with α-D-ManNAc	34
Various	β-D-Gal; highest affinity for galactans	35, 36
Waldenstrom's macroglobulinemia human monoclonal cold agglutinins		
Reactive with:		
Umbilical cord erythrocytes, lacto-N-neotetraose, i blood group		37
SRBC, Pronase-treated human RBC, glycolipids with terminal α- or β-GalNAc		38
Human RBC, G_{M3}, NeuNAc but not NeuNGl		39
I antigen		40
Human lymphoblastoid cell line monoclonal antibodies		
Reactive with:		
Streptococcal group A carbohydrate, GlcNAc		41
Type A erythrocytes, blood A		42
Murine monoclonal antibodies		
CA-1	Broadly tumor-reactive, sialic acid-containing high-M.W. antigen	43
3.1	Melanoma glycolipid	44
O_5	Melanoma glycolipid	45
R_2	Melanoma ganglioside, ganglioside present in bovine eye	45
Antimelanoma	G_{D3}	46
4.2	Melanoma cells, G_{D3}	47
1116NS 19-9	Colorectal carcinoma cells, monosialoganglioside:	48

$$
\begin{array}{c}
\text{G–D–Cer} \\
\diagup \\
\text{F–N} \\
| \\
\text{G} \\
\diagup \\
\text{Sa}
\end{array}
$$

Source	Specificity	References
	high-M.W. mucin	49
1116NS 33a	Colorectal carcinoma cells monosialoganglioside	50
1116NS 10	Lewis B antigen	50
D-B3, C-F8, D-D1, E-E12	Pancreatic ductal carcinoma cells, glycolipids	51
Antineuroblastoma	Carbohydrate portion of glycoprotein	52
SSEA-1	Teratocarcinoma cells, fucosyl antigen	53
SSEA-3	Teratocarcinoma cells	54
Anti-spleen cells	Asialo-G_{M2}	55
Anti-spleen cells	Forssman glycolipid	56
Anti-influenza virus	Forssman glycolipid	57
Anti-tonsil cells	A blood group	58
49H8, 49H24	T antigen, human acute lymphoblastic leukemia cells	59
VeP8, VEP9	3-Fucosyl-N-acetyllactosamine, human promyelomonocytic leukemia cells, granulocytes, monocytes	60
101	Epidermal growth factor receptor (Blood group H type 1)	61

Anti-MHC-Linked Antibodies with Carbohydrate Specificities

Source	Specificity	References
Human anti-HLA	Type XIV pneumococcal polysaccharide, HLA glycopeptides	62
	Deesterified LPS	63
	Glycoproteins and polysaccharides	64
Murine monoclonal	H-2K^k subset, Man, NeuNAc, Gal	65
Allo- and xenogeneic antiserum		66
Ia.1	Man, ManNAc, methyl-α-D-mannopyranoside	
Ia.3	Gal, GalNAc, GalNH$_2$, D-talose, melibiose, raffinose, stachyose, GalUA, phenyl-β-D-galactopyranoside	
Ia.7	L-Fuc, NeuNAc	
Ia.15	GlcNAc, GalNAc	
Monoclonal anti-Ia		67
Ia.2	GlcNH$_2$, glycolipids	
Ia.9	Sialic acid, glycolipids	
Ia.1	Sialic acid, β-Gal, glycolipids	

(*Continued*)

Table 1. (Continued)

Source	Glycoconjugate recognized	References[a]
Murine alloantisera		68
Anti-I-A^k	Neuraminidase-sensitive determinant on I^s cells, D-Gal, glycolipids	
Anti-I-A^s	Neuraminidase-sensitive determinant on I^k cells, D-Gal, glycolipids	
Anti-human Ia-like antibodies	Periodate, neuraminidase- and glycosidase-sensitive antigens, glycolipids	69
Anti-*T*-locus antibodies		70
Anti-T	Sialic acids	
Anti-t^{12}	D-Gal	
Anti-t^{W32}	D-Gal	
Anti-T^0	L-Fuc	
Anti-t^{w18}	L-Fuc	
Anti-t^{w1}	GlcNAc	

[a] (1) Kristiansen (1974); (2) Young *et al.* (1979); (3) Alving and Richards (1977); (4) Sela *et al.* (1975); (5) Johannsen *et al.* (1979); (6) Hughes *et al.* (1973); (7) Springer *et al.* (1982); (8) Kamenov *et al.* (1983); (9) Hirsch and Parks (1976); (10) Hirano *et al.* (1980); (11) Yamaga *et al.* (1978); (12) Itoh and Yamashina (1975); (13) Sarkar and Menge (1977); (14) Trenkner and Sarka (1977); (15) Layton (1980); (16) Winkelhake and Kasper (1972); (17) Kasper *et al.* (1973); (18) Brown and Colling (1982); (19) Horowitz (1978); (20) Kano and Milgrom (1970); (21) Uchida and Nagai (1980); (22) Uemura *et al.* (1980); (23) Naiki *et al.* (1974); (24) Gregson and Hammer (1982); (25) Kundu *et al.* (1980); (26) Bloch *et al.* (1977); (27) Zopf *et al.* (1975); (28) Zopf *et al.* (1978); (29) Reichert and Goldstein (1979); (30) Lemieux *et al.* (1975); (31) Smith and Ginsburg (1980); (32) Vicari *et al.* (1970); (33) Potter (1977); (34) Rovis *et al.* (1972); (35) Glaudemans *et al.* (1978); (36) Jolley *et al.* (1974); (37) Tsai *et al.* (1976); (38) Naiki and Marcus (1977); (39) Tsai *et al.* (1977); (40) Childs *et al.* (1980); (41) Steinitz *et al.* (1979); (42) Koskimies (1980); (43) Ashall *et al.* (1982); (44) Yeh *et al.* (1981); (45) Dippold *et al.* (1980); (46) Pukel *et al.* (1982); (47) Nudelman *et al.* (1982); (48) Koprowski *et al.* (1981); Magnani *et al.* (1981, 1982); (49) T. Klug (personal communication); (50) Z. Steplewski (personal communication); (51) D. Metzgar (personal communication); (52) Momoi *et al.* (1980); (53) Solter and Knowles (1978); Nudelman *et al.* (1980); (54) D. Solter (personal communication); (55) Young *et al.* (1979); (56) Stern *et al.* (1978); (57) Nowinski *et al.* (1980); (58) Voac *et al.* (1980); (59) M. Longnecker, (personal communication); (60) Gooi *et al.* (1983); (61) Fredman *et al.* (1983); (62) Hirata *et al.* (1973); (63) Mittal *et al.* (1973); (64) Sanderson *et al.* (1971); (65) O'Neill and Parish (1981); O'Neill *et al.* (1981); (66) McKenzie *et al.* (1977); (67) Higgins *et al.* (1980a); (68) Parish *et al.* (1981); (69) Sandrin *et al.* (1981a); (70) Cheng and Bennett (1980).

Table 2. Carbohydrate-Binding Complement Components

Reaction	Reactive with	References[a]
Alternate pathway activation	LPS, lipid A	1
	Pneumococcal capsular polysaccharides and cell walls	2
	Sindbis virus	3
Antibody-independent classical pathway activation (via Clq)	LPS, lipid A, Sindbis virus	4
Complement-dependent bactericidal factor	*Salmonella* Ra chemotype, GlcNAc, L-glycero-α-D-mannoheptose	5
C3b covalent complex formation	Cell walls of pneumococci	6
	Plasma membranes, zymosan, glucose polymers, GlcNAc polymers, Sepharose 4B	7
Sialic acid-dependent inhibition of C3b fixation via alternate pathway	*E. coli* K capsular antigen	8
	SRBC	9
	Group B streptococcal Type III capsular polysaccharide	10
	Sindbis virus	11

[a] (1) Gewurz *et al.* (1968); Marcus *et al.* (1971); (2) Winkelstein *et al.* (1976), Winkelstein and Tomasz (1977); (3) Hirsch *et al.* (1980); (4) Loos *et al.* (1974), Morrison and Kline (1977), Cooper and Morrison (1978), Hirsch *et al.* (1980); (5) Ihara *et al.* (1982); (6) Winkelstein *et al.* (1980); (7) Law and Levine (1977); (8) van Dijk *et al.* (1979), Bortolussi *et al.* (1979); (9) Pangburn and Muller-Eberhard (1978), Fearon (1978); (10) Edwards *et al.* (1982); (11) Hirsch *et al.* (1981).

Evidence for an enzymatic role for a plant lectin has been presented for the phytohemagglutinin from mung beans (Hankins and Shannon, 1978). The mung bean lectin was purified and found to possess a strong α-galactosidase activity. These authors provided evidence that the two activities were related to the same protein and suggested that legume lectins may, in general, be plant glycosidases or glycosyltransferases. They suggested that special precautions may be required to preserve enzymatic activity during purification and that binding activity may remain after loss of catalytic activity.

There is considerable evidence that lectins participate in specific cell surface interactions in microorganisms, invertebrates, and vertebrates (reviews: Simpson *et al.*, 1978; Barondes, 1980, 1981). Lectins for which a recognition role has been suggested are presented in Table 4.

Surface glycosyltransferases and glycosidases have also been implicated in biological recognition processes. In 1970, Roseman suggested that cell surface glycosyltransferases could bind to substrate acceptor molecules on adjacent cells and might play a role in intercellular recognition and communication. In the past 14 years, this hypothesis has been explored extensively (reviews: Shur and Roth, 1975; Pierce *et al.*, 1981;

Table 3. Carbohydrate Recognition by Soluble Mediators

Mediator	Inhibited by	References[a]
Cholera toxin	Glycolipids	1
	Glycoproteins	2
E. coli enterotoxin	Glycolipids	3
Tetanus toxin	Glycolipids	4
Botulinum toxin	Glycolipids	5
Plant toxins	Glycolipids	6
	Glycoproteins	7
Serotonin	Glycolipids	8
TSH	Glycolipids	9
	Glycoproteins	10
LH	Glycolipids	11
HCG	Glycolipids	12
Insulin	Glycoproteins	13
Guinea pig macrophage migration inhibition factor (MIF)	Glycoproteins	14
	α-L-Fucose	15
	α-L-Fucose, GalNAc, blood group substances, bovine submaxillary mucin	16
	Fucose-binding lectins, fucose-containing oligosaccharides, α-L-Fuc, L-rhamnose, 6-deoxy-D-glucose	17
	Acidic macrophage fucose-containing glycolipids	18
	Chymotrypsin digestion of macrophages	19
	Trypsin digestion of macrophages	20
Guinea pig and rat MIF	Bovine brain gangliosides	21
Human MIF	α-L-Fucose, L-rhamnose, 6-deoxy-D-glucose	22

Guinea pig macrophage activation factor (MAF)	Bovine brain gangliosides, macrophage glycolipids	23
Mouse MAF	Man, Me-α-D-mannopyranoside, L-Fuc, L-rhamnose, GlcNAc, lectins, digestion of macrophages with α-D-mannosidase, neuraminidase, proteases	24
Guinea pig macrophage migration stimulation factor (MSF)	GalNAc	25
Guinea pig macrophage chemotactic factor, neutrophil chemotactic factor	L-Rhamnose, L-fucose	26
Human leukocyte inhibition factor (LIF)	GlcNAc	27
Suppression of *in vivo* reactions	L-Fucose, L-rhamnose	28
Tuftsin	NeuNAc	29
Mouse lymphotoxin (LT)	Gal, GalNAc, porcine thyroglobulin glycopeptide	30
Guinea pig LT	L cell glycopeptides	31
Human LT	Me-α-D-mannopyranoside, GlcNAc	32
Murine and Human interferon (IF)	Gangliosides G_{M1}, G_{M2}, G_{D1a}, G_{T1}, sialyllactose	33
Soluble immune suppressor factor for T cells (SISS-T)	GlcNAc	34
Soluble immune suppressor factor for B cells (SISS-B)	L-Rhamnose	35
T-cell helper factors for secondary IgG response	GalNAc, Gal, Fuc, ManNAc	36

[a] (1) van Heyningen *et al.* (1971), Holmgren *et al.* (1974), Cuatrecasas (1973a,b), Gill and King (1975), Staerk *et al.* (1974), Basu *et al.* (1976); (2) Morita *et al.* (1980), Strombeck and Harrold (1974), Grollman *et al.* (1978); (3) Pierce (1973); (4) van Heyningen (1974); (5) Simpson and Rapport (1970, 1971); (6) Hughes and Gardas (1976); (7) Lis and Sharon (1977), Brown and Hunt (1978); (8) van Heyningen (1974), Wolley and Gommi (1965); (9) Mullin *et al.* (1976a,b); (10) Tate *et al.* (1975), Winand and Kohn (1975); (11) Wolff *et al.* (1974), Lee *et al.* (1977); (12) Lee *et al.* (1976); (13) Cuatrecasas (1973a); (14) Remold and David (1971), Leu *et al.* (1972); (15) Remold (1973); (16) Fox *et al.* (1974); (17) Poste *et al.* (1979b); (18) Higgins *et al.* (1976, 1980b), Liu *et al.* (1978, 1980, 1982); (19) Remold and David (1971); (20) Leu *et al.* (1972); (21) Poste *et al.* (1979a); (22) Rocklin (1976); (23) Poste *et al.* (1979a); (24) Yamamoto and Tokunaga (1981); (25) Fox *et al.* (1974); (26) Amsden *et al.* (1978); (27) Rocklin (1976); Klempner and Rocklin (1982); (28) Baba *et al.* (1979); (29) Bar-Shavit *et al.* (1979); Constantopoulos and Najjar (1973); Nair *et al.* (1978); (30) Sawada *et al.* (1976, 1977); (31) Kobayashi *et al.* (1978); (32) Weitzen *et al.* (1983b); (33) Besancon and Ankel (1976); Krishnamurti *et al.* (1982), Vengris *et al.* (1976); (34) Greene *et al.* (1981); (35) Fleisher *et al.* (1981); (36) Tomaska and Parish (1981, 1982).

Table 4. Carbohydrate Recognition by Lectins

Source	Inhibited by	References[a]
Plants	Various mono- and oligosaccahrides	1
E. coli	α-D-Man	2
Chlamydomonas	α-D-Man	3
Acanthamoebae	Me-α-D-Man, Man, Fru, yeast mannan	4
Slime molds	Lactose, D-Gal	5
Electric eel, chick skeletal muscle and embryonic skeletal muscle cultures, neuroblastoma cells	β-D-Gal	6
Calf heart and lung	β-D-Gal	7
Chick pectoral muscle, chick embryo thigh muscle	Lactose	8
Chicken liver, pancreas, intestine	Lactose	9
Chicken liver, embryonic chick pectoral muscle	Heparin, GalNAc	10
Bovine liver cells	GlcNAc, di-*N*-acetylchitobiose, tri-*N*-acetylchitotriose	11
BHK cells	L-Fuc, GalNAc, thio-di-galactoside	12
Platelets	GalNH$_2$, GlcNH$_2$, ManNH$_2$	13
Oyster	GalNAc, GlcNAc, Gal	14
Mollusk	Blood group substances	15
Snails	Blood group A	16
Lobster	NeuNAc, mucins, GalNAc, KDO	17
Crabs	NeuNAc, M and N blood group substances, CEA	18
Shark serum	Fructosan, rye grass levan, α-(1-3) dextran	19
Human serum	Pneumococcus C polysaccharide	20
Eel serum	Blood group H	21
Rat thymocytes and splenic lymphocytes	Gal, GalNAc, Man, asialoglycoproteins	22
Murine lymphocytes	Fetuin, bovine submaxillary mucin	23
Activated human lymphoid cells	Fetuin, thyroglobulin, *N*-linked fetuin glycopeptides	24

[a] (1) Reviews: Sharon and Lis (1972), Lis and Sharon (1973, 1977), Goldstein and Hayes (1978), Brown and Hunt (1978); (2) Eshdat *et al.* (1978); (3) Weise (1974); (4) Brown *et al.* (1975); (5) Rosen *et al.* (1973, 1974, 1979), Springer *et al.* (1980), Simpson *et al.* (1975); (6) Teichberg *et al.* (1975); (7) de Waard *et al.* (1976); (8) Nowak *et al.* (1977); (9) Beyer *et al.* (1979); (10) Ceri *et al.* (1981); (11) Bowles and Kauss (1976); (12) Dysart and Edwards (1977); (13) Gartner *et al.* (1978); (14) Tripp (1974a); (15) Tripp (1974b); (16) Prokop (1974); (17) Hall *et al.* (1972), Hall and Rowlands (1974), Rostam-Abadi and Pistole (1982); (18) Cohen *et al.* (1974); (19) Sigel (1974); (20) Gotschlich and Edelman (1967); (21) Watkins and Morgan (1952); (22) Kieda *et al.* (1978, 1979); (23) Decker and Marchalonis (1979); (24) Apgar and Cresswell (1982).

Table 5. Surface Glycosyltransferases

Source	Specificity	References[a]
Chick neural retinal cells	UDP-Gal	1
	UDP-GalNAc	2
Chick ventral retinal cells, ventral tectal cells	UDP-Gal: G_{M2}	3
Gastrulating chick embryos	UDP-Gal, UDP-GlcNAc, GDP-Fuc, CMP-NeuNAc	4
Platelets	UDP-Glc: collagen	5
	CMP-NeuNAc: glycoprotein	6
Balb/c 3T3 cells, 3T12 cells	UDP-Gal	7
SV40 3T3 cells	UDP-Gal	8
Virally transformed 3T3 cells	CMP-NeuNAc: asialoglycopeptides	9
Malignant rat dermal fibroblasts	UDP-Gal, CMP-NeuNAc	10
Chlamydomonas	UDP-Gal, UDP-Glc, UDP-GlcNAc, CMP-NeuNAc, GMP-Man, GMP-Fuc	11
Mouse sperm	CMP-NeuNAc: asialoglycoprotein	12
	UDP-Gal	13
	UDP-Gal: poly-*N*-acetyllactosamine	14
Embryonal carcinoma cells	UDP-Gal: poly-*N*-acetyllactosamine	15
Neonatal rat thymocytes and splenic lymphocytes	UDP-Gal	16
Rat splenic lymphocytes	CMP-NeuNAc: glycoprotein	17
Mouse thymic and splenic lymphocytes	UDP-Gal	18
CTL	UDP-Gal	19

[a] (1) Roth *et al.* (1971); (2) McDonough and Lilien (1977); (3) Marchase (1977), Marchase *et al.* (1977), Pierce (1982); (4) Shur (1977a,b); (5) Jamieson *et al.* (1971); (6) Bosmann (1972a); (7) Roth and White (1972); (8) Patt *et al.* (1976); (9) Bosmann (1972b); (10) Lloyd and Cook (1974); (11) McLean and Bosmann (1975); (12) Durr *et al.* (1977); (13) Shur and Bennett (1979); (14) Shur and Hall (1982a,b); (15) Shur (1982b); (16) LaMont *et al.* (1974), Verbert *et al.* (1976); (17) Verbert *et al.* (1977), Hoflack *et al.* (1979); (18) Baker *et al.* (1980); (19) Kurt *et al.* (1981).

Shur, 1982a). A list of reported surface glycosyltransferases is presented in Table 5. In most cases, the carbohydrate receptor for the transferase in the implied recognition is unknown. Complementary reaction between ganglioside G_{M2} (see Fig. 1 for the glycolipid structures) and a cell surface UPD-Gal:G_{M2} galactosyltransferase (G_{M1} synthetase) has been proposed in developing chick neural retina (Marchase, 1977; Marchase *et al.*, 1977).

A galactosyltransferase isoenzyme detected in the serum of patients with colonic, pancreatic, and gastric carcinoma (Podolsky and Weiser, 1975) and on the surface of virally transformed BHK cell tumors in hamsters (Podolsky *et al.*, 1977) appeared to have a high affinity for a gly-

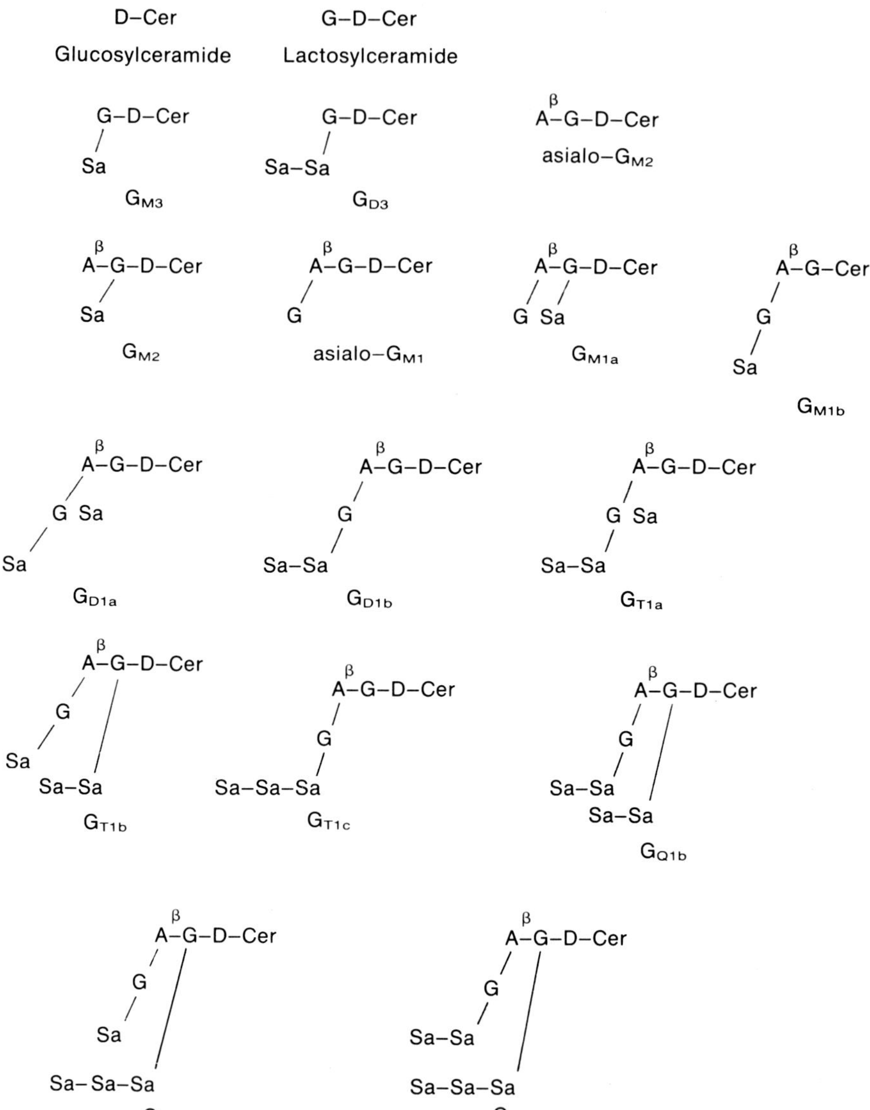

Figure 1. Glycolipids important in biological recognition. Throughout the text, carbohydrate structures will be abbreviated according to the convention of Carver and Grey (1981). Linkages are indicated for hexopyranosides by a line from the position of the substituted hydroxyl of the sugar residue:

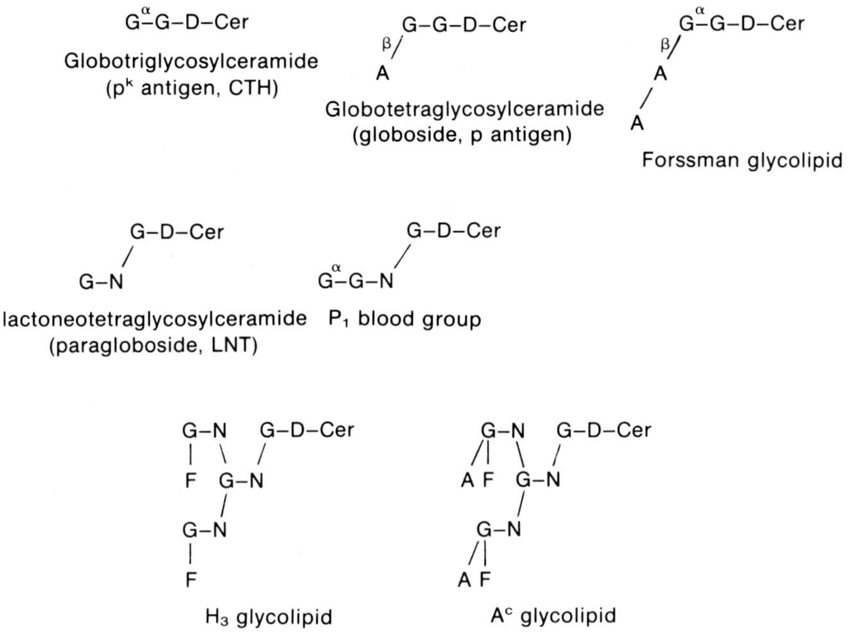

$\overset{\alpha}{\text{G}}$–G–D–Cer

Globotriglycosylceramide
(p^k antigen, CTH)

G–G–D–Cer
β/
A

Globotetraglycosylceramide
(globoside, p antigen)

$\overset{\alpha}{\text{G}}$–G–D–Cer
β/
A
/
A

Forssman glycolipid

G–D–Cer
/
G–N

lactoneotetraglycosylceramide
(paragloboside, LNT)

G–D–Cer
/
$\overset{\alpha}{\text{G}}$–G–N

P₁ blood group

G–N   G–D–Cer
| \ /
F   G–N
    /
G–N
|
F

H₃ glycolipid

G–N   G–D–Cer
/| \ /
A F   G–N
      /
G–N
/|
A F

A^c glycolipid

Figure 1. (*Continued*)

copeptide acceptor purified from human malignant effusions (Podolsky *et al.*, 1978). Addition of this glycopeptide to the media of cells growing in culture caused a significant inhibition of attachment and growth of transformed cells. *In vivo* studies showed that the acceptor inhibited development and progression of tumors in hamsters inoculated with polyoma virus-transformed BHK cells. The structural characterization of this acceptor has not been reported.

Sialic acids can only be substituted at the 4, 8, and 9 positions:

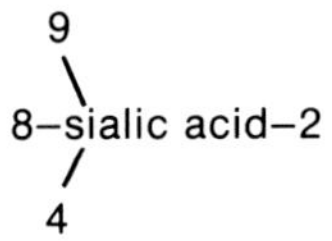

9
\
8–sialic acid–2
/
4

Abbreviations are: D (dextrose) = β-D-glucose (Glc); G = β-D-galactose (Gal); M = α-D-mannose (Man); F = α-L-fucose (Fuc); N = *N*-acetyl-β-D-glucosamine (GlcNAc); A = *N*-acetyl-α-D-galactosamine (GalNAc); S = sialic acids; Sa = *N*-acetyl-α-D-neuraminic acid (NeuNAc); Sg = *N*-glycolyl-α-D-neuraminic acid (NeuNGl); Cer = ceramide.

Table 6. Cell Surface Glycosidases

Source	Activity	References[a]
Bovine brain synaptosome plasma membrane	Ganglioside sialidase	1
Platelets	Glycoprotein sialidase	2
Rat liver plasma membranes	Ganglioside sialidase	3
Virally transformed hamster cell lines	Ganglioside sialidase	4
HSV-transformed cells	Sialidase	5
3T3 cells	Sialidase	6
Human skin fibroblasts	N-acetyl-β-D-glucosaminidase, N-acetyl-α-D-glucosaminidase, α-D-mannosidase, β-D-glucuronidase	7
NIL cells	α-D-Mannosidase	8

[a] (1) Schengrund and Rosenburg (1970); (2) Bosmann (1972a); (3) Schengrund *et al.* (1972); Visser and Emmelot (1973); (4) Schengrund *et al.* (1973); (5) Schengrund *et al.* (1976); (6) Yogeeswaran and Hakomori (1975); (7) von Figura and Voss (1979); (8) Rauvala *et al.* (1981).

There is evidence for a complementary interaction between cell surface glycosyltransferases and aggregation factors involved in embryonic chick cells. An N-acetyl-β-D-galactosaminyl transferase and a glycoprotein acceptor are reportedly involved in reaggregation of neural retinal cells, and a similar interaction between an α-D-mannosyl transferase and a surface glycoprotein in cerebral lobe cells has been studied (McDonough and Lilien, 1977; Rutz and Lilien, 1979).

A UDP-Gal transferase activity was increased in murine *T*-locus mutant sperm with increased fertilization frequency (Shur and Bennett, 1979). A mouse sperm surface UDP-Gal transferase that acts on poly-N-acetyllactosamine structures with terminal N-acetylglucosamine residues was recently implicated in both sperm capacitation (Shur and Hall, 1982a) and binding to the egg zona pellucida during fertilization (Shur and Hall, 1982b). A similar surface galactosyltransferase has also recently been implicated in the binding of poly-N-acetyllactosamine glycoconjugates by embryonal carcinoma cells (Shur, 1982b).

Cell surface glycosidases may also be involved in biological recognition. Cell surface sialidase activities have been reported in bovine brain, platelets, rat liver plasma membranes, and on virally transformed and normal cells (see Table 6). A role for platelet surface neuraminidase in cellular adhesion has been suggested by Bosmann (1972a). A biological role for surface mannosidase in cellular adhesive interactions has also been proposed (Rauvala *et al.*, 1981).

Table 7. Carbohydrate-Mediated Uptake Systems

Source	Specificity	References[a]
Sandhoff disease fibroblasts	Periodate-sensitive marker	1
Generalized gangliosidosis skin fibroblasts	α-Man, β-Man, mannose-containing testicular glycoproteins	2
Fibroblasts	Mannosyl-phosphate	3
Mammalian hepatocytes	Terminal Gal on asialoglycoproteins	4
	Terminal gal IgG immune complexes	5
Avian hepatocytes	Terminal GlcNAc on agalactoglycoproteins	6
Mammalian hepatocytes	G—N— on human lactoferrin / F	7
Mammalian reticuloendothelial cells	Terminal Man, GlcNAc, L-Fuc	8
	Terminal Man on IgM immune complexes	9

[a] (1) Hickman *et al.* (1974); (2) Hieber *et al.* (1976); (3) Kaplan *et al.* (1977), Sly (1977), Ullrich *et al.* (1978); (4) Morell *et al.* (1971), Goldwasser *et al.* (1974), Winkelhake and Nicolson (1976); (5) Thornburg *et al.* (1980); (6) Lunney and Ashwell (1976), Kawasaki and Ashwell (1977); (7) Prieels *et al.* (1978); (8) Schlesinger *et al.* (1976, 1978a,b), Achord *et al.* (1977, 1978), Brown *et al.* (1978), Stahl *et al.* (1978), Stahl and Gordon (1982); (9) Day *et al.* (1980).

Receptors for carbohydrate-mediated uptake have been implicated in recognition systems (reviews: Ashwell and Morell, 1974; Neufeld *et al.*, 1977; Neufeld and Ashwell, 1980). Hickman and Neufeld (1972), studying fibroblasts from the inclusion (I cell) lysosomal storage disease, demonstrated that the I cells' deficiency could be corrected by adding lysosomal hydrolases from the medium of normal human skin fibroblasts. They suggested that the uptake of lysosomal hydrolases requires specific recognition and that the I cell mutation interferes with this process by altering the recognition sites on the hydrolases. Work by several groups (see Table 7) indicated that a mannosyl phosphate residue on lysosomal enzymes was recognized by the lysosomal uptake system.

Another carbohydrate-mediated uptake system is the receptor for glycoproteins on hepatocytes (review: Ashwell and Morell, 1974). Removal of terminal sialic acid from several serum glycoproteins resulted in their rapid clearance after intravenous (i.v.) injection into rats (Morell *et al.*, 1971). This initial finding has been extended by several groups to include the asialoglycoprotein uptake system and a fucosyl-*N*-acetyllactosamine-dependent uptake system in mammalian liver and an agalactoglycoprotein uptake system in avian hepatocytes (see Table 7).

The immune response is eminently involved in recognition phenomena, and there are several examples of carbohydrate-binding proteins involved in aspects of immune recognition. In addition to antibodies that bind complex carbohydrate antigens (Table 1), there is evidence for carbohydrate-binding complement components (Table 2) and for carbohydrate recognition by soluble immune mediators (Table 3), cell surface lectins (Table 4), cell surface enzymes (Table 5), and carbohydrate-mediated uptake systems (Table 7). These recognition phenomena will be explored in Section 2.

Glycoconjugates on the cell surface may reflect the state of differentiation of cells (see Table 8). These changes may be preferentially associated with some tissues and organs such as Forssman antigen (Stern *et al.*, 1978), galactosylceramide, a G_Q ganglioside in the brain (Joffe *et al.*, 1963; Eisenbarth *et al.*, 1979), and brain glycoproteins containing disialosyl linkages (Finne *et al.*, 1977). Differences also occur between fetal and adult cells of the same lineage such as glycosphingolipid levels (Karol *et al.*, 1980) and changes from i to I blood group structures (Fukuda *et al.*, 1980) in erythroid cells. In addition, alterations in structures between cells from the same lineage in varying stages of differentiation have been reported, such as ganglioside content in contiguous intestinal crypt and villous mucosa cells (Glickman and Bouhours, 1976). Changes in glycoconjugate structures have also been associated with differentiation in early stages of mouse embryogenesis (review: Jacob, 1979) and primitive teratocarcinoma cells in culture (review: Martin *et al.*, 1980).

The immune system is comprised of numerous cellular compartments, which can be further subdivided, based on the stage of differentiation of the cells in each compartment. A number of complex carbohydrate differentiation antigens have been studied within the immune system and will be discussed in Section 3.

The implications from numerous studies on the basis of cellular interactions are that some of these specific interactions may depend on complementary interactions between carbohydrate-binding proteins and complex carbohydrate structures on the cell surface. Carbohydrate sequences have been implicated as receptors for parasites, as specific receptors involved in mating reactions, and in specific cellular reaggregation and cellular interactions during embryogenesis (see Table 9). In some cases, surface lectins, glycosyltransferases or glycosidases, and still others, carbohydrate-mediated uptake systems have been implicated in cellular interactions.

Active cooperation or suppression either via soluble mediators, or by direct cellular interaction has been demonstrated in numerous immune reactions. There is preliminary evidence to suggest that in some cases,

Table 8. Carbohydrate Differentiation Antigens

Cell type	Structure	References[a]
Early stages of mouse embryogenesis and primitive teratocarcinoma cells	Forssman antigen	1
	Globotriglycosylceramide and globotetraglycosylceramide	2
	Lectin receptors	3
	Stage-specific embryonic antigen	4
	F9 antigen	5
	T-locus antigens	6
Brain	Galactosylceramide	7
	Disialosyl linkages in glycoproteins	8
Brain, retina, spinal cord, dorsal root ganglia	G_Q ganglioside	9
Fetal to adult erythrocytes	Glycosphingolipid levels	10
	i to I blood group structures	11
Intestinal crypt and villus mucosal cells	Ganglioside content	12
Thymus-derived lymphocytes	Thy-1 glycoprotein (T25)	13
	Thy-1 glycolipids	14
	NeuNAc on Thy-1	15
Thymocyte subset	PNA receptor	16
	SBA receptor	17
Mature T cells	Asialo-G_{M1}	18
CTL	*Vicia villosa* lectin receptor	19
Mouse CFU-S and GvH-reactive T cells	PNA and SBA receptors	20
Human pluripotent hemopoietic stem cell and GvH-reactive T cells	SBA receptors	21
B and T cells	Sialic acid content	22
	SBA receptors	23
	Glycolipids	24
	Glycosylation of T200	25
NK cells	Asialo-G_{M1}	26
Stimulated murine macrophages	*Griffonia simplicifolia* lectin receptor	27
Human monocytes	PNA receptors	28

[a] (1) Stern *et al.* (1978), Willison and Stern (1978); (2) Willison *et al.* (1982); (3) Gachelin *et al.* (1976); (4) Gooi *et al.* (1981); (5) Artzt *et al.* (1973), Jacob (1979), Martin *et al.* (1980); (6) Artzt *et al.* (1974), Marticorena *et al.* (1978); (7) Joffe *et al.* (1963); (8) Finne *et al.* (1977); (9) Eisenbarth *et al.* (1979); (10) Karol *et al.* (1980); (11) Fukuda *et al.* (1980); (12) Glickman and Bouhours (1976); (13) Trowbridge and Hyman (1975); (14) Milewicz *et al.* (1976), Stein-Douglas *et al.* (1976), Stein *et al.* (1978), Wang *et al.* (1978), Kato *et al.* (1979); (15) Johnson *et al.* (1976), Hoessli *et al.* (1980); (16) Reisner *et al.* (1976, 1979), Kornfeld (1978); (17) Reisner and Sharon (1978); (18) Stein *et al.* (1978); (19) Kimura *et al.* (1979a,b); (20) Reisner *et al.* (1978); (21) Y. Reisner (personal communication); (22) Despont *et al.* (1975); (23) Reisner *et al.* (1978); (24) Schwarting and Marcus (1979), Rosenfelder *et al.* (1979, 1980); (25) Morishima *et al.* (1982); (26) Schwarting and Summers (1980); Schwarting and Gajewski (1981); Young *et al.* (1980), Kasai *et al.* (1980), Durdik *et al.* (1980), Habu *et al.* (1981); (27) Maddox *et al.* (1982); (28) O'Keefe and Ashman (1982).

Table 9. Carbohydrate-Recognition in Cellular Interactions

Interaction	Glycoconjugate	References[a]
Phage–bacteria	Cell wall carbohydrates	1
Myxovirus, paramyxovirus, adenovirus, and polyoma virus–cell	Sialic acid receptors	2
Reovirus–cell	GlcNAc receptors	3
Encephalitis virus–cell	α-D-Man receptors	4
Enterovirus, encephalomyelitis virus, dermovirus, levite neurovirus, encephalomyocarditis virus, coxsackie, herpes simplex virus, Shope fibroma virus–cell	Various carbohydrates	5
Sendai virus, rubella virus–cell	Glycolipid	6
Mycoplasma–cell	Carbohydrate receptors	7
Mycoplasma gallisepticum–erythrocyte	Sialic acid receptors	8
Enteric bacteria–cell	D-Man receptors	9
Yeast strain-specific mating reaction	Mannan factor	10
E. coli K12 mating	Periodate-sensitive structures	11
Fertilization of the brown algae *Fucus serratus*	Fuc receptors	12
Mammalian fertilization	Lectin receptors	13
Actinomyces viscosus and *A. naeslundii*–streptococci adherence in human dental plaque; *A. viscosus*-agglutination of neuraminidase-treated erythrocytes	Lactose-like receptors	14
Reaggregation of sea urchin (*Strongylocentrotus purpuratus*) embryo cells	Gal, GalNAc receptors	15
Reaggregation of marine sponges	Proteoglycans	16
	GlcUA receptors	17
Aggregation of HeLa cells	Hyaluronic acid	18
Aggregation of chicken embryonic neural retinal cells	GalNAc	19
Aggregation of chicken embryonic cerebral lobe cells	Man	20
Adhesion of undifferentiated teratocarcinoma cells	Man receptors	21
Adhesion of differentiated (PYS) teratocarcinoma cells	Mucin-like receptors	21
Adhesion of BHK and HeLa cells	Mucoprotein-like receptors	22
SV40 3T3 cell adhesion	Gal receptors	23
Chicken hepatocytes	GlcNAc receptors	24
Rat hepatocytes	Gal receptors	25
NIL, BHK, HeLa, and MDCK cell adhesion	Glycolipids	26
Bacteria–lymphocyte	Various carbohydrate receptors	27
Bacteria–macrophate and bacteria–PMNL	Man receptors	28

Table 9. (Continued)

Interaction	Glycoconjugate	References[a]
Peritoneal macrophage–bacteria	Cell wall carbohydrate receptors	29
Human T cell–SRBC	*N*-Linked glycopeptides	30
Macrophage adhesion	G_{M2}, G_{D1a}	31
Aggregation of glycosidase-digested peritoneal exudate cells	Fuc, GalNAc receptors	32
Natural cytotoxicity	Man	33
Spontaneous cell-mediated cytotoxicity	Ribose, gentibiose, GalNAc, cellobiose, α-lactose	34
NK cytolysis	Fetuin, $GalNH_2$, $GlcNH_2$, Ara, Xyl, Gal, Man; methyl-α-D-mannopyranoside, GlcNAc	35
Human monocyte–target cell	Sialic acid	36
	Various sugars	37
Monocyte–T cell	Various sugars	38
Lymphocyte homing to nodes	Surface sugars	39
Lymphocyte–HEV	Fuc, Man receptors	40
Stem cell homing to spleen	Surface carbohydrates	41
DTH reactivity	Surface carbohydrates	42
MLR	Surface Ia carbohydrate determinants	43
H-2-restricted cytolysis of virus-infected cells	Surface H-2 carbohydrate determinants	44

[a] (1) Wright and Kangasaki (1971); (2) Gottschalk (1966), Levinson *et al.* (1969), Suttajit and Winzler (1971), Howe and Lee (1972); (3) Gelb and Lerner (1965); (4) Homma (1968); (5) Lerner *et al.* (1966), Balazs and Jacobson (1966), Howe and Lee (1972); (6) Haywood (1974), Shortridge *et al.* (1972); (7) Thomas (1969); (8) Gesner and Thomas (1966); (9) Duguid and Old (1980); (10) Crandall *et al.* (1974), Yen and Ballou (1974); (11) Sneath and Lederberg (1960); (12) Bolwell *et al.* (1980); (13) Oikawa *et al.* (1973); (14) McIntire *et al.* (1978), Cisar *et al.* (1979); (15) Asao and Oppenheimer (1979); (16) Cauldwell *et al.* (1973), Henkart *et al.* (1973); (17) Turner and Burger (1973); (18) Kuhns (1974); (19) Lilien *et al.* (1978); (20) Lilien *et al.* (1978); (21) Grabel *et al.* (1979); (22) Allen and Minnikin (1975); (23) Chipowsky *et al.* (1973); (24) Schnaar *et al.* (1978); (25) Weigel *et al.* (1979); (26) Yogeeswaran *et al.* (1974), Huang (1978); (27) Teodorescu *et al.* (1977), Rasanen (1981); (28) Bar-Shavit *et al.* (1977); (29) Freimer *et al.* (1978); (30) Boldt and Armstrong (1976); (31) Riedl *et al.* (1982); (32) Gesner (1966); (33) Stutman *et al.* (1980); (34) MacDermott *et al.* (1980); (35) Ades *et al.* (1981), Weitzen *et al.* (1983b); (36) Czop *et al.* (1978); (37) Muchmore and Blaese (1980); (38) Muchmore and Blaese (1980b), Muchmore *et al.* (1980); (39) Gesner and Ginsburg (1964), Gesner and Woodruff (1969), Berney and Gesner (1970); (40) Stoolman and Rosen (1983); (41) Tonelli and Meints (1978); (42) Kaufmann *et al.* (1981); (43) Hart (1982); (44) Black *et al.* 1981.

intercellular interactions in the immune system occur via complex carbohydrate recognition. This possibility will be explored in Section 4.

There have been several hypotheses involving the importance of carbohydrate recognition in the immune system and the evolutionary origin of these recognition systems. In Section 5 these will be considered, and a hypothesis that considers alterations in cell surface glycoconjugates as the basis for immune surveillance will be forwarded.

2. CARBOHYDRATE-BINDING PROTEINS IN THE IMMUNE SYSTEM

2.1. Antibodies That Recognize Carbohydrate Determinants

The best characterized example of carbohydrate recognition in immunity is the interaction between antibodies and blood group substances (reviews: Morgan, 1970; Horowitz, 1978). The immunochemistry and biology of the blood groups have been intensely studied (Szulman, 1977; Hakomori, 1981; Marcus *et al.*, 1981). People are separated into blood types on the basis of the agglutination of their erythrocytes by antisera. Individuals who lack a particular blood group structure can produce antibodies that react with that antigen and these antibodies are used for classification of that blood type. These reactivities depend, for a number of the blood groups, on carbohydrate structures on the erythrocyte glycoproteins and glycolipids (Watkins, 1966; Hakomori, 1981). Genetic and enzymatic studies have demonstrated that these structures are determined by the presence or absence of individual glycosyltransferases (review: Grollman *et al.*, 1970). The addition or deletion of a single monosaccharide in a blood group structure can change the antigenicity and therefore the blood type.

Studies of blood group substances have established many of the precepts used to demonstrate the participation of complex carbohydrate determinants in biological reactions (review: Morgan, 1970). One of these precepts is hapten inhibiton. Small molecules that fit in a part of the antibody-combining site (haptens) will inhibit serological reactions. Thus, mono-, di-, and oligosaccharides were found to inhibit hemagglutination of erythrocytes or precipitation of blood group substances. Haptens that inhibit at lower concentrations are though to be more like the portion of the antigen (the epitope) that combines with the antibody. A second precept is the loss of recognition after enzymatic removal of specific sugars. Digestion with exoglycosidases decreased reactivity of blood group substances and implicated the released sugar as an important part of the epitope. When impure exoglycosidases were used, addition of a specific inhibitor (e.g., L-fucose for and α-L-fucosidase) blocked the digestion. A third principle is that specific chemical degradations can release fragments that retain inhibitory activity. Chemical degradations and structural characterization of the fragments obtained with partial acid hydrolysis, alkaline degradation, and hydrazinolysis led to structural information that gave rise to the blood group structures (Lloyd and Kabat, 1968). The fourth lesson from studies with the blood groups is that a purified glycosyltransferase can convert a precursor substance into the specific antigenic determinant.

Blood group I/i, ABH, and Lewis (Le) determinants are pictured in Fig. 2. The i determinant contains a linear-repeating poly-N-acetyllactosamine structure, while the I determinant contains a branched structure. The ABH and Le structures are formed on I/i structures. I/i determinants appear on oligosaccharides, glycolipids, and glycoproteins. The oligosaccharide lacto-N-neotetraose is i-reactive, whereas the oligosaccharide lacto-N-neohexose is I-reactive. Glycoproteins carry both O-linked (Oates *et al.*, 1974; Lloyd and Kabat, 1968; Slomiany and Meyer, 1973) and N-linked (Mizoguchi *et al.*, 1982) I/i determinants.

Blood group I/i antibodies (review: Marsh, 1961) are less common than ABO antibodies and are often associated with disease states. Patients with reticulum cell sarcoma, reticulosis, and hemolytic anemia often produce anti-I/i antibodies. The I structure was first characterized on an ovarian cyst fluid glycoprotein that lacked A, B, H, Le[a], and Le[b] specificities (Vicari and Kabat, 1969). This structure, termed "inactive substance," cross-reacted with antipneumococcal type XIV polysaccharide antiserum. The I blood group backbone can terminate in both type 1 and type 2 chains. The type 1 chain is examplified by the lacto-N-tetraose oligosaccharide structure, and the type 2 chain is present in the terminal structure of lacto-N-neotetraose. Combinations of these give rise to structures with the lacto-N-hexose (type 1 and type 2) or the lacto-N-neohexose (two type 2) terminal sequences.

The presence of the *H* gene is required to add an α-L-fucose residue to the 2 position of the terminal β-linked galactose residue in both type 1 and type 2 chains. The product is the H determinant and gives rise to the O blood group. The addition of this linkage to a type 1 chain gives rise to the lacto-N-fucopentaose I structure. This H transferase is subject to further genetic restriction, in that a gene *Se/se* controls the expression of the H determinant in secretions but not on erythrocytes (see Horowitz, 1978). The H determinant is the substrate for the A and B blood group enzymes, and without it, no ABH blood group antigens are expressed, either on erythrocytes or in secretions.

A second fucosyltransferase that is the product of the *Le* gene adds an α-L-fucose residue to the 4 position of the penultimate N-acetylglucosamine residue. This enzyme can only act on type 1 chains, for the 4 position is occupied by galactose in type 2 chains. The product is the Le[a] structure (lacto-N-fucopentaose II terminal sequence) on erythrocytes from an *hh* individual or on secretions from an *se/se* individual, and an Le[b] structure (lacto-N-difucohexose I terminal sequence) on erythrocytes

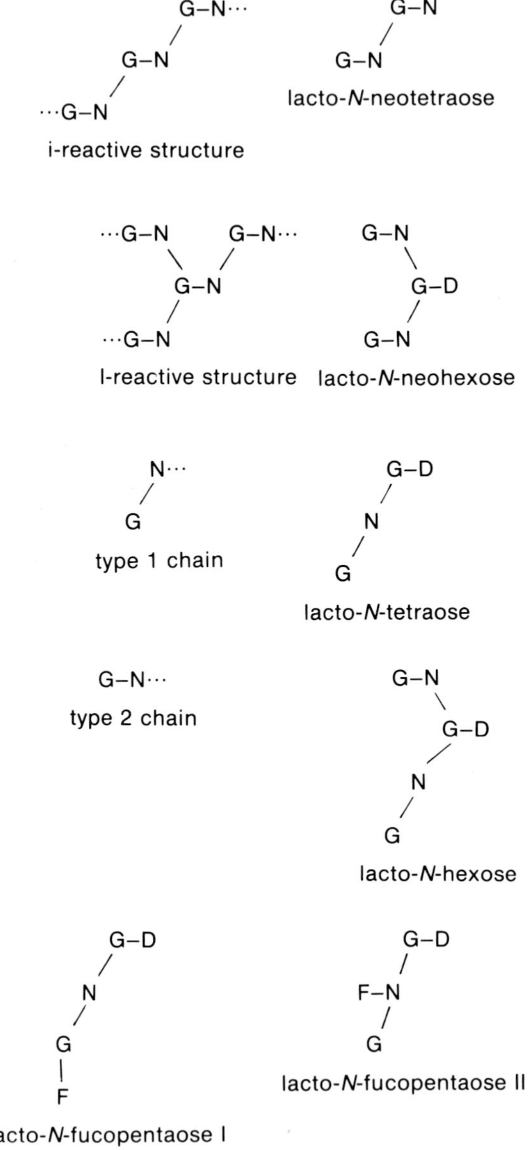

Figure 2. Blood group structures.

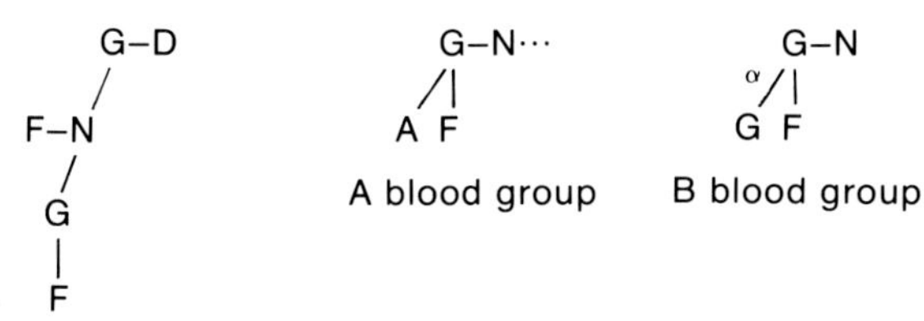

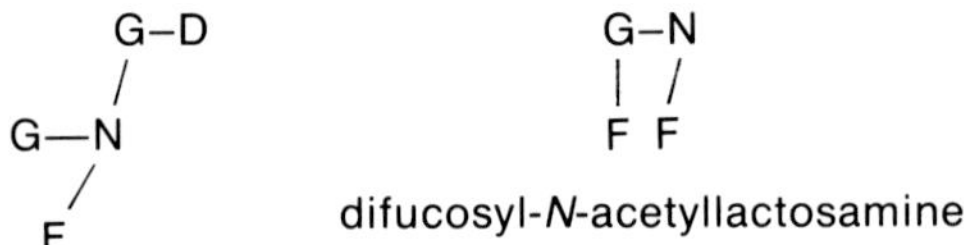

lacto-*N*-difucohexose I

A blood group

B blood group

difucosyl-*N*-acetyllactosamine

lacto-*N*-fucopentaose III

lacto-*N*-difucotetraose

2'-fucosyllactose

```
        CHO CHO CHO
         |   |   |
NH2–Ser–Ser–Thr–Thr–Gly–Val–Ala–Met–COOH    M determinant


        CHO CHO CHO
         |   |   |
NH2–Leu–Ser–Thr–Thr–Glu–Val–Ala–Met–COOH    N determinant


       CHO CHO CHO CHO
        |   |   |   |
NH2–Ser–Ser–Thr–Thr–Gly–Val–Ala–Met–COOH    alternate M structure
```

Figure 2. (*Continued*)

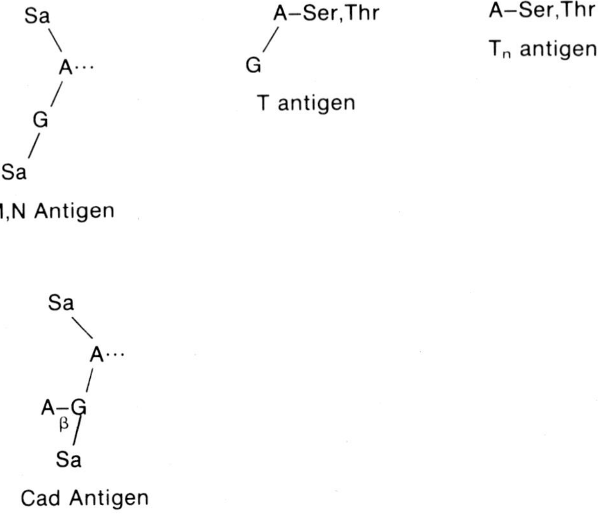

Figure 2. (*Continued*)

from an individual with an *H* gene and on secretions from individuals with *H* and *Se* genes.

The A blood group determinant contains an *N*-acetyl-α-D-galactosamine residue linked to the 3 position of the terminal D-galactose of the H determinant. There are two A enzymes, A_1 and A_2. The A_1 *N*-acetylgalactosaminyltransferase will act on both type 1 and type 2 chains, but the A_2 enzyme will only act on type 2 chains (Horowitz, 1978). The B blood group determinant contains an α-D-galactose linked to the 3 position of the D-galactose of the H determinant.

Four Le structures in secretions have been defined on the basis of *Le/le* and *Se/se* genes: Le^a (*Le, se/se*), Le^b (*Le, Se*), Le^c (*le/le, se/se*), and Le^d (*le/le, Se*). The combinations of A_1, A_2, B, H, Le, and Se gene products give rise to numerous specifications on the secreted products and on erythrocytes. Ten major Lewis antigens (LEW 1–10, Oriol *et al.*, 1980) have been defined. The finding of an additional α-L-fucosyltransferase that adds fucose to the 3 position of the penultimate *N*-acetylglucosamine (Horowitz, 1978) to give a structure with the lacto-*N*-fucopentaose III terminal sequence, predicts additional determinants in blood group chains, as does the finding of blood group-reactive glycolipids (Marcus *et al.*, 1981; Hakomori, 1981) and glycoproteins (Mizuochi *et al.*, 1979; Mizoguchi *et al.*, 1982) with terminal sialic acid residues.

Polyclonal antiserum has been prepared against a synthetic Lewis a antigen (Lemieux *et al.*, 1975). Recently, monoclonal antibodies reactive with blood group structures have been produced (see Table 1). A monoclonal antibody (H11) reactive with type 2 chains containing the H determinant was identified (Knowles *et al.*, 1982) and found to react with difucosyl type 2 structure, difucosyl-*N*-acetyllactosamine, lactodifucotetraose, and 2'-fucosyllactose.

In addition to the ABH, Le, and I/i antigens, there are at least two other blood group systems in which the determinants contain complex carbohydrates: the MNSPr and the P blood groups. The M, N, S, and Pr antigens are present on glycophorin glycoproteins on erythrocytes (reviews: Springer *et al.*, 1976; Anstee, 1981). Although amino acid sequence changes in glycophorin A form the basis of the genetic difference between M and N (Blumenford and Adamany, 1978; Sadler *et al.*, 1979; Prohaskar *et al.*, 1981), the changes lead to differences in antibody reactivity to the heavily glycosylated *N*-terminal peptide region. The M and N antigenic determinants are destroyed by neuraminidase digestion. Springer and coworkers have suggested that the *N*-terminal serine present only on the M determinant is also glycosylated, resulting in a difference in carbohydrate content between M and N determinants (G. F. Springer *et al.*, 1980). A new addition to this family of determinants is the Cad specificity which contains a β-linked GalNAc residue at the 4 position of the Gal residue (Blanchard *et al.*, 1983).

Removal of sialic acid from glycophorin exposes the cryptic Thompson–Freidenreich (T antigen) structure that is reactive with antibodies found in the serum of normal individuals (Springer *et al.*, 1976). Treatment of T antigen with β-galactosidase destroys anti-T reactivity and exposes T_n antigen. Removal of α-linked *N*-acetylgalactosamine destroys reactivity with anti-T_n and removes the last sugar residue linked to serine or threonine. The T and T_n structures are exposed in some disease states (Anstee, 1981) and on some tumors (see below).

The Ss antigens are present on glycophorin B, and although amino acid changes have also been suggested as the genetic basis for the Ss difference, an α-linked *N*-acetylgalactosamine residue at position 25 has also been implicated in the antigenic determinant (Dahr *et al.*, 1980). Another series of antigens on glycophorins A and B are the Pr antigens. At least some of the Pr specificities reside in the *O*-linked disialo-tetrasaccharides (Anstee, 1981).

P blood group antigens are present on glycolipids (Marcus *et al.*, 1981). Individuals of the p blood type lack all P antigens and have the precursors lactosylceramide and lactoneotetraglycosylceramide (Fig. 3). The P^k gene product converts lactosylceramide into the P^k antigen, globotriglycosylceramide. The P_2 gene product converts globotriglycosylceramide into the P structure, globotetraglycosylceramide. The P_1 gene product converts lactoneotetraglycosylceramide into the P_1 structure. The

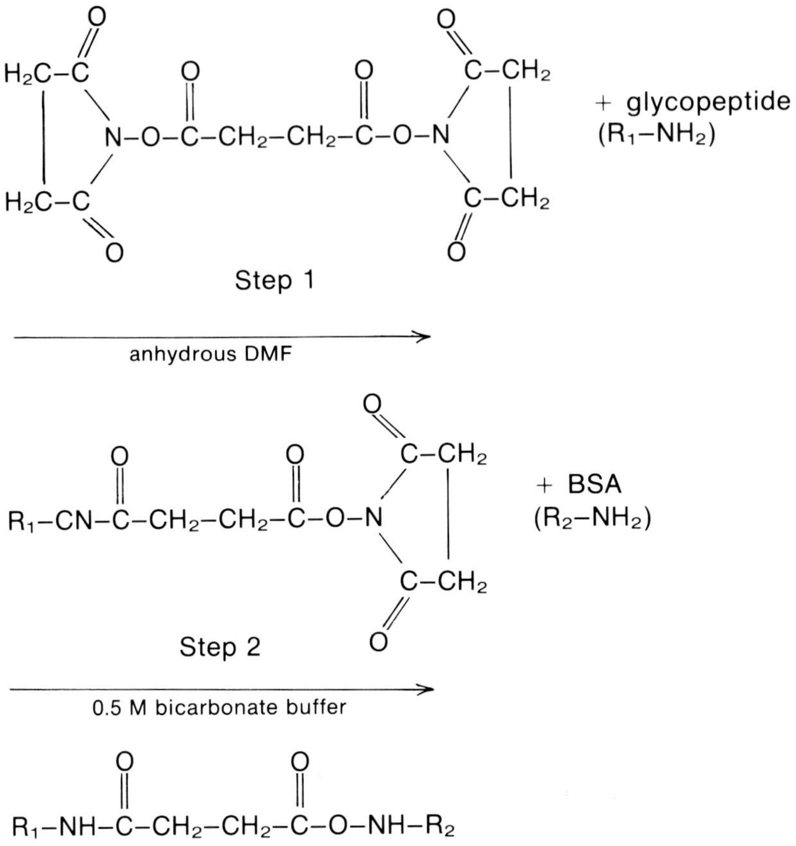

Step 1

anhydrous DMF

Step 2

0.5 M bicarbonate buffer

$$R_1-NH-C(O)-CH_2-CH_2-C(O)-O-NH-R_2$$

Figure 3. Synthesis of neoglycoproteins. Neoglycoproteins are synthesized in a two-step procedure, using a symmetrical bifunctional cross-linking reagent, the di-N-hydroxysuccinimidyl ester of succinic acid (DHS) described by Hill *et al.* (1979). In the first step, dry glycopeptide is reacted with excess diester in anhydrous DMF, and the product is isolated by precipitation with acetone. In the second step, the activated glycopeptide is reacted with BSA in bicarbonate buffer.

majority of individuals (P_1 phenotype) have P and P_1 antigens on their erythrocytes and lack P-reactive antibodies. About 25% of the population contains only P antigens (P_2 phenotype) and produces antibodies to P_1 antigen. The very rare p-phenotype individuals produce antibodies that react with P_1, P, and P^k antigens. Marcus *et al.* (1981) found increased levels of sialosyllactosylceramide (ganglioside G_{M1}) and sialosyllacto-neotetraglycosylceramide in cells of p individuals, presumably due to alternative glycosylation pathways of P blood group precursors. As with other blood group systems, the P blood group antigens are important in transfusion reactions and they may be responsible for aborted fetuses with different P blood types (reviews: Giblett, 1977; Marcus *et al.*, 1981).

The functions of blood group antigens are poorly understood. ABH blood group antigen expression on cultured cells appears to be related to cell cycle in several species (Dawson and Franks, 1967; Kuhns and Bramson, 1968). Lymphocytes of several mouse strains become blood group B positive after phytohemagglutinin (PHA) stimulation (Thomas, 1971). It was proposed that the cell surface display of blood group B determinant was prerequisite for cell division in mouse P815Y tumor cells. To date, no cell surface function related to cell division has been proposed for blood group antigens.

Gershowitz and Neal (1970) examined the possible selection pressures that maintain the blood group polymorphisms in the gene pool. Some diseases are statistically associated with particular blood groups. The authors suggest that epidemics might cause selection due to the availability of antibodies to certain blood group determinants on pathogenic organisms. They also suggest that blood group polymorphisms may be preserved due to other glycosyltransferase-related selection pressures. One possible disease-related function of blood group determinants is that they may serve as viral receptors. M and N blood group glycoproteins are potent inhibitors of hemagglutination by influenza viruses (Springer, 1970). It has also been suggested that glycosyltransferases may be an important factor in the expression of individuality through competition by two parental enzymes for the same acceptor that may lead to unique carbohydrate structures (Grollman *et al.*, 1970). These authors suggested that this individuality might be involved in cellular recognition and graft rejection. Evidence to support this proposal is considered in Section 4.

Springer (1970) presented evidence that blood group-reactive antibodies are produced in response to microbes that contain "blood group" structures. Of 282 strains of intestinal flora tested, 137 showed some ABH specificity. Germfree animals failed to produce blood group antibodies but did so in response to microorganisms or purified blood group substances. In man, some blood group antibodies may arise in response to intestinal flora, to dietary antigens, and to viral and bacterial immunization. Influenza virus preparations produced in chicken eggs contain blood group A and Forssman antigens and when used for immunization, give rise to antibodies to these antigens in the immunized host.

Alterations in carbohydrate structures have been associated with tumors, and a number of these changes have been defined immunochemically. Some of the major changes are related to alterations in blood group structures (Watanabe and Hakomori, 1976; Young and Hakomori, 1978; Cooper and Haesler, 1978; Cooper *et al.*, 1980). Human blood group ABH glycolipids on erythrocytes are constructed from lactoneotetraglycosylceramide (LNT). Antibody to the incomplete blood group glycolipid structure lactotriglycosylceramide, the precursor to LNT, was reactive with some colon tumor cells (Hakomori *et al.*, 1977b). An unusual blood group glycolipid that inhibited anti-A reactions was isolated from a human adenocarcinoma (Hakomori *et al.*, 1967). The authors proposed that glycolipids of this nature arise by an aberration in the synthesis of glycolipids possessing blood group activity (Watanabe and Hokomori, 1976; Hakomori *et al.*, 1977b). They suggest that a genetic or epigenetic program for synthesizing blood group determinants and their carrier carbohydrate chains develops stepwise during ontogenesis and that the program is blocked or modified in the process of oncogenesis. Evidence for this was found in the H_3 and A^c blood group glycolipids that are present in adult erythrocytes in high levels but low in newborn and absent in fetal erythrocytes. These glycolipids may be found in an incomplete form in tumors.

Incomplete blood group structures in human colon carcinoma cells result from lower glycosyltransferase activities (Kim *et al.*, 1978). The loss of A and B transferases results in increased terminal β-linked galactose residues as evidenced by *Ricinus communis* agglutinin binding and by labeling with tritiated sodium borohydride after incubation with galactose oxidase.

Exposed galactose residues represent a major alteration found in numerous tumors (Springer *et al.*, 1982). Antibodies that recognize neuraminidase-treated tumor cells have been detected in serum from unsensitized guinea pigs and C3H mice (Hughes *et al.*, 1973). Antibodies that reacted with autologous neuraminidase-treated human PBL and cultured human lymphoid cells were detected in normal human serum (Rogentine and Plocinik, 1974). Carbohydrate hapten inhibition studies showed that these antibodies were inhibited by oligosaccharides containing terminal β-D-galactose.

Springer *et al.* (1976) studied tumor antigens similar to incomplete M and N blood group structures. In human breast cancers, T antigen was present in malignant but not in benign structures. T antigen was also found in two human colon carcinoma-derived cell lines and in human gastric carcinomas. Antibody to the T-antigen structure was present in sera of all adults tested, possibly due to bacterial immunity (see Springer *et al.*, 1978). Anti-T titration of serum was depressed in 20% of breast carcinoma patients and in 36% of lung and colon carcinoma patients. Carcinoma patients also demonstrated delayed-type hypersensitivity (DTH) and positive macrophage migration-inhibition factor (MIF) assays to T antigen, whereas normal individuals did not. Recently, Kamenov *et al.* (1983) have demonstrated that T antigen-reactive natural antibodies may play a role in rejection of murine leukemias. Other carbohydrate structures associated with tumors include the Hanganutziu–Deicher (H-D) antigen, which contains an N-glycolylneuraminic acid determinant (Nishimaki *et al.*, 1979; Naiki *et al.*, 1981), the Forssman antigen (Hakomori *et al.*, 1977a,b), and, possibly, carcinoembryonic antigen (Pompecki *et al.*, 1981).

Several investigators have produced monoclonal antibodies with restricted tumor reactivities and established that the antigenic determinants involved are carbohydrate in nature. A number of these antibodies are listed in Table 1.

Genetic control over several aspects of the immune system is encoded in the major histocompatibility (*MHC*) locus (*H-2* region in the mouse, *HL-A* region in man). Products of the *H-2* locus (review: Klein, 1975) play major roles in the control of virus susceptibility, immune responses, and cell recognition and interactions (*Ia* genes). In the mouse, the major histocompatibility antigens in allograft rejection are H-2K and H-2D antigens. A third polymorphic histocompatibility locus, *H-2L*, has recently been defined (Hansen and Sachs, 1978).

Ia genes (review: Benacerraf and Katz, 1975) are linked to the MHC in several species including guinea pigs, mice, rats, and rhesus monkeys. Ia cell surface antigens control cell–cell cooperation in immune responses (reviews: Cook *et al.*, 1980; Cullen *et al.*, 1981b). *Ia*-encoded products are secreted by T cells in the form of antigen-specific T-cell factors that control the B-cell response to antigen (Taussig *et al.*, 1974). Ia-related defects in cell cooperation can be found in both T-cell and B-cell populations, implicating at least two classes of Ia antigens (Taussig *et al.*, 1975). T cells secrete both antigen-specific and -nonspecific helper factors (McKenzie and Parish, 1976) and antigen-specific T-cell suppressor factors (Tada *et al.*, 1976) that carry Ia antigens.

There has been controversy for several years as to the participation of carbohydrates in histocompatibility and immune response antigens. Recently, evidence has accumulated indicating that antibodies recognize both protein and carbohydrate antigens for some of these polymorphic systems.

In man, carbohydrates have been implicated in some anti-HLA specificities (Sanderson *et al.*, 1971). Pronase digestion of HLA antigens yielded glycopeptides with most of the carbohydrate, but little of the amino acids, intact, and these glycopeptides specifically inhibited anti-HLA antiserum. Low-molecular-weight oligosaccharides from type XIV pneumococcal polysaccharide inhibited anti-HLA B_t serum. In addition, greater than 70% of HLA-2 antigenic activity was destroyed in 30 min by 5 mM periodate at pH 5 and 4°C in the dark. Several anti-HLA antisera were inhibited by deesterified lipopolysaccharides from some bacteria (Hirata *et al.*, 1973) and by some glycoproteins and polysaccharides (Mittal *et al.*, 1973).

Experiments with murine H-2 alloantigens indicated that the carbohydrate portion of the glycoproteins might play an important role in the antigenic specificity (review: Klein, 1975). Early studies indicated that the carbohydrate units of glycopeptides from several H-2 and Ia specificities were similar by apparent M_r (3200–3300), sugar compositions, and incorporation of radiolabeled monosaccharides (Muramatsu and Nathenson, 1970; Cullen and Nathenson, 1974; Freed and Nathenson, 1977). Recent studies using lectin affinity columns have demonstrated that the Ia glycoprotein oligosaccharides are heterogeneous in structure and that they contain high-mannose structures as well as complex triantennary N-linked oligosaccharides (Cowan *et al.*, 1982a,b).

Recent experiments have implicated carbohydrate determinants in some H-2 specificities (O'Neill and Parish, 1981). Using monoclonal antibodies specific for the H-2K^k specificity, two antigenically distinct subsets of H-2K^k molecules on the surface of lymphocytes were detected. Conventional anti-H-2K^k alloantiserum bound both subsets. The authors proposed that variant forms of H-2 molecules that arise by posttranslational modifications such as glycosylation could lead to changes in the tertiary structure of the H-2 glycoprotein. The reactivities of two monoclonal antibodies recognizing H-2K^k specificities were insensitive to Pronase digestion but sensitive to glycosidases and to sugar inhibitions (O'Neill *et al.*, 1981). Evidence has also been presented for carbohydrate H-2L specificities (Sia and Parish, 1981; see Section 4).

There is substantial evidence for carbohydrate-defined antigens controlled by the *I* region (review: Parish and McKenzie, 1981). Carbohydrate Ia-like antigens have been defined in man, mice, rats, guinea pigs, sheep, and cattle. Evidence for murine Ia specificities that are oligosaccharide in nature was first presented by McKenzie *et al.* (1977). In addition to appearance on B cells, macrophages, epidermal cells, sperm cells, and T cells, Ia antigenic material was found in mouse serum and other tissue fluids, especially after antigenic stimulation. Inhibition of complement-mediated lysis was used to test the alloserum, and inhibition of rosette-forming cells using antibody-coupled erythrocytes was used to test the xenogeneic antisera. Hapten inhibition studies using both allo- and xenogeneic anti-Ia serum indicated carbohydrate specificities for Ia.1, Ia.3, Ia.7, and Ia.15 (Table 3). There is also evidence that the Ia determinants present on antigen-specific helper factors (Howie *et al.*, 1979), suppressor factors (Liew *et al.*, 1980), and transfer factors (Basten *et al.*, 1980) are carbohydrate in nature.

Purification of the carbohydrate Ia antigens indicated that Ia antigens exist in two forms—as cell surface glycoproteins and as glycolipids found both in serum and on spleen cells (Higgins and Parish, 1980). A surprising finding, which supported the presence of the oligosaccharide class of Ia determinants on spleen cells, was reported by Parish *et al.* (1981). They found that neuraminidase digestion of murine splenocytes destroyed old Ia antigens and revealed new Ia specificities not normally expressed by the splenocytes. Neuraminidase digestion unmasked I-A^k-like specificities on A.TH (I-A^s) spleen cells and I-A^s-like antigens on A.TL (I-A^k) spleen cells. The cellular distribution of neuraminidase-exposed antigens resembled that of conventional Ia antigens. In addition to the Ia.3 and .15 specificities exposed, new antigens were involved that did not correlate with any previously described Ia antigens. Sugar inhibition studies demonstrated that the neuraminidase-exposed antigens were carbohydrate in nature. Galactose was an effective inhibitor in these studies. The proportion of α- and β-linked galactose residues associated with the new antigens depended on the target cell used and the anti-Ia serum tested. Furthermore, glycolipid extracts from lymphoid cells were shown to contain the neuraminidase-exposed antigens.

Many of the conflicting reports as to the protein or carbohydrate nature of antigenic determinants of MHC-region gene products in the past are due to the use of antiserum that recognized multiple specificities, some protein in nature and some carbohydrate. This problem has been overcome by the use of monoclonal antibodies that recognize a single specificity. Monoclonal anti-Ia antibodies were tested for their reactivity with glycoprotein and glycolipid classes of Ia molecules (Higgins *et al.*, 1980a). Ten monoclonal antibodies were tested for the ability to precipitate Ia-like glycoproteins from the cell surface, for inhibition of reactivity by sugars and by purified glycoprotein-Ia or glycolipid-Ia, and for neuraminidase sensitivity of the antigen. It was concluded that three of the antibodies recognized carbohydrate Ia specificities (Ia.2, .9, and .17) and seven recognized the glycoprotein class of Ia antigens. Monoclonal antibodies recognizing glycoprotein-defined Ia.2 and .17 specificities were also characterized. These results imply that some Ia specificities, as defined by genetic testing, can occur as both carbohydrate-defined and glycoprotein-defined determinants. Each monoclonal antibody recognized either protein or carbohydrate Ia determinants, but neither reacted with both glycolipid and glycoprotein classes of Ia molecules, implying that the carbohydrate structures on glycolipid Ia determinants are not displayed on the glycoprotein Ia molecules. Sugar inhibition studies indicated distinct carbohydrate specificities for several Ia specificities (Table 1). In a recent report, 8 of 20 anti-Ia monoclonals were found to react with carbohydrate determinants, while the other 12 reacted with protein determinants (Parish *et al.*, 1982).

Low-molecular-weight Ia-like antigens from human serum were also isolated and characterized as glycolipids (Sandrin *et al.*, 1981a). Their antigenicity was destroyed by periodate treatment and digestion with neuraminidase and mixed glycosidases. Sugar inhibition studies by these authors indicated that, for two individuals, different sugars were involved in the Ia-like antigenic determinants. It is interesting that HLA-inhibiting molecules have been reported in human serum from individuals of the appropriate specificities (Van Rood *et al.*, 1970). These results may be complicated by the fact that antibodies to blood group antigens have confusing effects in HLA-Dr testing (Curtoni *et al.*, 1980; Oriol *et al.*, 1980; Park *et al.*, 1980). Antisera used for histocompatibility testing often contain anti-blood group antibodies that give synergistic effects with anti-Dr antibodies in the cytotoxicity of Dr^+ cells of the appropriate blood group, but the antisera fail to lyse Dr^+ cells of the wrong blood group.

A variety of stimuli *in vivo* enhance or suppress serum carbohydrate Ia antigen levels (Parish and McKenzie, 1981). In the mouse, serum Ia antigen levels are stimulated by mitogens and T-independent antigens (Parish and McKenzie, 1977) and are dramatically altered by infections with various agents (Parish *et al.*, 1979). Human serum carbohydrate Ia-like antigen levels were reportedly suppressed in all cancers tested. Murine serum Ia levels rose during allogeneic tumor rejection and dropped rapidly to undetectable levels in mice with progressive syngeneic tumors (Sandrin *et al.*, 1981b). Normal human subjects and patients with diseases other than cancers had high levels of human serum Ia antigens, but 10 of 15 cancer patients had no detectable serum Ia, and the other 5 had severely depressed levels.

If the *I* region codes for carbohydrate structures, then the simplest interpretation is that it codes for glycosyltransferases (Parish and McKenzie, 1981). This would imply the presence of two distinct classes of *Ia*-coded antigens—one protein in nature (glycosyltransferases) and the other carbohydrate in nature (glycolipids). Several models for immune recognition via *MHC*-controlled glycosyltransferases have been proposed (Parish, 1977; Rothenberg, 1978; Parish *et al.*, 1981; Gorczynski, 1984).

The *T* complex is tightly linked to the *H-2* complex (review: Klein and Hammerberg, 1977) and appears to be involved in the contol of differentiation and intercellular interactions during early embryogenesis (Artzt and Bennett, 1975). The embryonic T/t antigens have several similarities to the adult MHC (H-2) antigens (Artzt and Bennett, 1975; Vitetta *et al.*, 1975), and reciprocal expression of T/t and H-2 antigens is observed during differentiation (Stern *et al.*, 1975). The similarities among *T/t*-locus, H-2, and TL antigens, their structural homologies with immunoglobulin, and their roles in intercellular interactions led Vitetta *et al.* (1975) to propose a common origin in a primitive gene concerned with cellular recognition.

Testicular cells express the *T*-locus antigens and have been used to produce antisera. By immunizing wild-type mice with cells mutant at the

T locus, Cheng and Bennett (1980) were able to produce antiserum that reacted with heterozygous testicular cells. After cross-adsorption with heterozygous testicular cells from mice bearing other *T*-locus mutations, each antiserum reacted with a single specificity. They used glycosidase digestion and hapten inhibition to study the nature of the antigenic determinants. Sugar residues were found to be involved in at least six different T/t antigenic determinants (Table 1). These studies indicate that at least some of the *T*-locus genes code for glycosyltransferases or regulators of glycosyltransferases that modify oligosaccharide structures and impart specificity to the T/t antigens by alteration of their terminal sugar residues.

In addition to blood group and MHC reactivities, antibodies with carbohydrate specificities have been detected in normal serum from several species, from patients with certain diseases, and in hyperimmune serum after immunization with glycoconjugates (Table 1). Several murine and human monoclonal antibodies also react with known carbohydrate specificities.

Normal sera from various species were found to contain antibodies that reacted with cell surface carbohydrates (Sela *et al.*, 1975). These antibodies agglutinated erythrocytes of several species, bound to cells from various mouse tissues, and could be isolated by affinity chromatography on fetuin–Sepharose. The binding was inhibited by glycoproteins, glycopeptides, and some lipopolysaccharides. Binding of antibodies to mouse splenocytes was partially inhibited by sialic acid, galactose, *N*-acetylglucosamine, and mannose. The carbohydrate-specific antibodies from chicken and rabbit serum were weakly mitogenic for mouse splenic lymphocytes.

Patients with cold agglutinin syndrome produce monoclonal antibodies in large amounts, and several of those studied recognize carbohydrate antigens (review: Feizi, 1980). An interesting antibody from a patient with Waldenstromn's macroglobulinemia was reported by Tsai *et al.* (1977). The monoclonal antibody reacted with *N*-acetylneuraminic acid residues. This antibody was a cryoglobulin (it precipitated in the cold), and treatment of the antibody with neuraminidase did not alter its agglutination activity, but the asialoantibody was no longer a cryoglobulin. This result implied that the antibody precipitated in the cold due to intermolecular immune binding of *N*-acetylneuraminosyl residues.

"Neoglycoproteins" have been prepared for use as antigens by several groups (Table 2). Immunization with these covalent conjugates of defined carbohydrates and protein has given rise to antibodies reactive with the carbohydrate determinants. Recently, we have prepared immunogens for the production of monoclonal antibodies to carbohydrate antigens by coupling glycopeptides from natural sources to proteins using the divalent cross-linking reagent 1,4-(di-*N*-hydroxysuccinimidyl) succinate (Reading and Hutchins, unpublished). The preparation of neoglyco-proteins by this procedure is outlined in Figure 3. As mice are highly tolerant to many of the carbohydrate antigens, we are using *in vitro* immunization (review: Reading, 1982b) to produce the hybridomas.

The majority of IgG antibodies that react with carbohydrate antigens are restricted to a single rare subclass of IgG in several species. Mouse antidextran responses are restricted to the IgG3 subclass, and rat anti-polysaccharide responses are limited to IgG2c (Perlmutter *et al.*, 1978). In the human, antibody responses to dextran, levan, and teichoic acid are restricted to IgG2 (Yount *et al.*, 1968). Perlmutter *et al.* (1978) noted that even though serum IgG3 subclass levels were low, IgG3 was the major subclass of immunoglobulin on murine splenocytes, and that IgG2 was the major subclass on human peripheral blood lymphocytes. The mouse IgG3, rat IgG2c, and human IgG2 subclasses appear to be antigenically related and are probably the most primitive subclasses. It may be that the immunoglobulins evolved from more primitive carbohydrate-binding hemagglutinins (Section 5).

The glycosylation pattern of immunoglobulins can affect their binding specificities. The sialic acid content of IgA heavy-chain glycopeptides correlates with antibody-binding specificity for dextran-specific hybridoma antibodies (Matsuuchi *et al.*, 1981a). IgA antibodies from Balb/c mice specific for $\alpha(1\text{-}3)$ dextran had sialic acid on virtually all of their glycopeptides, whereas IgA antibodies from the same strain specific for $\alpha(1\text{-}6)$ dextran had sialic acid on only a small subset of their glycopeptides. In C57BL/6 mice, however, IgA antibodies specific for $\alpha(1\text{-}6)$ dextran had sialic acid on most of their glycopeptides. The authors suggested that the carbohydrate moiety on the immunoglobulin molecule may influence the antigen–antibody interaction. A variant IgA anti-$\alpha(1\text{-}3)$ dextran murine myeloma protein with increased sialic acid content had decreased reactivity with polymeric dextran (Matsuuchi *et al.*, 1981b). The parental myeloma protein, when produced in hybrid cells, also showed increased sialic acid content and decreased reactivity with polymeric dextran. These studies demonstrated that the carbohydrate content of an immunoglobulin can influence its binding specificity.

2.2. Carbohydrate-Binding Complement Components

The complement (C') system is a series of enzymes that are activated in a cascade fashion by antigen–antibody complexes and by microorganisms. C' activation results in lysis of cells and stimulation of several other immune reactions. There is considerable evidence that the C' system can be activated, in the absence of specific antibodies, by complex carbohydrate structures. There are two pathways of C' activation. The first is termed the classical pathway, which requires the participation of all nine C′ components C1–C9. The second pathway is termed the alternate or properdin system, requiring the participation of components C3 and C5–C9 and acting in the absence of specific antibody. The alternative pathway is activated by a number of bacterial and viral products (Table 2). Activation of the classical pathway in the absence of specific antibody, possibly by direct interaction of glycoconjugates with the C1q component of the C′ system, has also been reported (Table 2).

C3b, a cleavage product of component C3, attaches to target particles during C′ fixation by both classical and alternate pathways. Chemical kinetic studies provided evidence for an ester linkage between an activated carboxyl function on C3b and hydroxyl groups of glycoconjugates on the target particles (Law *et al.*, 1979).

Sialic acid residues on the surface of target particles appear to regulate C′ reactions. Activators such as rabbit erythrocytes, which are deficient in sialic acid, allow unrestricted formation of C3 convertase and activate C′ via the alternate or properdin pathway. Sheep erythrocytes (SRBC), which contain high levels of sialic acid, are nonactivators of the alternate pathway. Removal of cell-surface sialic acid with neuraminidase (Pangburn and Muller-Eberhard, 1978) or periodate modification of sialic acid residues (Fearon, 1978) converted SRBC from a nonactivator to an activator of the alternate pathway.

Sialic acid residues in the type III capsular polysaccharide of group B *Streptococcus* affect the activation of C′ by the alternate pathway (Edwards *et al.*, 1982). The native polysaccharide containing sialic acid residues does not activate the alternate pathway, but either removal of the sialic acid with neuraminidase or reduction of the carboxylate groups of the sialic acid residues to hydroxymethyl groups results in C′ activation by the alternate C′ pathway.

Sindbis virus activation of the alternate C′ pathway is dependent on the sialic acid content of the virions (Hirsch *et al.*, 1981). When Sindbis virus is grown in BHK cells, the virions have a normal content of sialic acid and do not activate the alternate pathway. Sindbis virus grown in mosquito cells, which are deficient in sialic acid, and Sindbis virus grown in BHK cells that contain sialic acid, but from which the sialic acid had been removed by neuraminidase, activated the alternate pathway much more efficiently. Studies *in vivo* showed that virus deficient in sialic acid was cleared from the circulation of mice much more efficiently than virus rich in sialic acid. Animals rendered deficient in C′ failed to clear the virions. Thus, both the activation *in vitro* and the clearance of virions *in vivo* are significantly affected by the sialic acid content of the virions.

The inverse relationship between sialic acid content (measured by the thiobarbituric acid assay) and alternate-pathway activation was also studied using mouse erythrocytes from 21 inbred strains (Nydegger *et al.*, 1978). In addition, a correlation was noted between alternate-pathway activation and susceptibility to Friend leukemia virus at locus *Fv-2*, which determines susceptibility or resistance to development of leukemia. Resistant strains (5/5) had erythrocytes that were good activators with low sialic acid content, and 6/6 susceptible strains had erythrocytes with low activating capacity and high sialic acid content.

Varki and Kornfeld (1980) reported that the measurement of sialic acid content by Nydegger *et al.* (1978) was probably in error, as the thiobarbituric acid method they used failed to quantitate sialic acids containing *O*-acetyl groups. A good correlation was found between the degree of *O*-acetylation of the erythrocyte sialic acid residues and the susceptibility to C′ lysis, whereas there was no correlation between total erythrocyte sialic acids and C′ sensitivity. The major *O*-acetylated species in all of the murine strains is *N*, 9-*O*-diacetylneuraminic acid. The authors proposed that the genetic trait responsible for the differences in susceptibility of murine erythrocytes to lysis by human C′ might be the degree of *O*-acetylation of the sialic acid residues. From this study, it seems that *O*-acetylation of the C7–C9 chain of sialic acid affects C′ activation in a comparable manner to mild periodate oxidation.

The difference between activating particles with low sialic acid and nonactivating particles with high sialic acid levels was determined to be related to competition between B and β1H for cell-bound C3b (Kazatchkine *et al.*, 1979). With C3b bound to SRBC, which are nonactivators, the affinity of β1H was 5-fold greater than that of B. With C3b bound to neuraminidase-treated SRBC, which are activators, the affinity of β1H was about 15-fold lower, but the affinity of B was unchanged. The authors suggest that this represents a primitive biochemical basis for differentiating cells and causing phagocytosis or cytolysis.

A C′-dependent bactericidal factor, which is not immunoglobulin in nature, has recently been discovered in nonimmune mouse serum (Ihara *et al.*, 1982). This factor binds to the Ra chemotype of *Salmonella* but not to the S or Re chemotypes and is inhibited by *N*-acetylglucosamine.

Evidence has recently been presented to indicate that asparagine-linked oligosaccharides on immunoglobulin molecules may function in C′ activation. Nose and Wigzell (1983) reported that IgG$_{2b}$ monoclonal antibodies made in the presence of tunicamycin, an inhibitor of N-linked glycosylation, bound antigen as well as the normal monoclonal antibody, but failed to fix complement as well as the native antibody.

2.3. Carbohydrate Recognition by Soluble Immune Mediators

Recent evidence indicates that several soluble immune mediators (lymphokines, monokines, helper, suppressor, activator, chemotactic, and toxic factors) bind to complex carbohydrate structures on the surface of their target cells (Table 3). Work from several groups has indicated that carbohydrate residues on the surface of macrophages form part of the receptor for macrophage migration-inhibition factor/macrophage-activation factor (MIF/MAF; Table 3). Glycoconjugates containing α-L-fucose have been implicated as MIF receptors on guinea pig, human, and rat macrophages. Support has been presented for both glycoprotein and acidic glycolipid MIF receptors. Evidence for a different MAF receptor on murine macrophages, possibly a glycoprotein containing mannose and sialic acid, was presented by Yamamoto and Tokunaga (1981). Macrophage migration-stimulation factor (MSF) from fetal bovine serum was inhibited by *N*-acetyl-D-galactosamine but not by L-fucose (Fox *et al.*, 1974). Macrophage and neutrophil chemotactic factors were inhibited by rhamnose and fucose, and leukocyte-inhibition factor (LIF) was blocked by *N*-acetylglucosamine (Amsden *et al.*, 1978; Rocklin, 1976; Klempner and Rocklin, 1982).

In addition to the inhibition of lymphokine reactions with monosaccharides *in vitro*, the suppression of these reactions *in vivo* has been reported by Baba *et al.* (1979). L-Fucose and L-rhamnose inhibited the activity of lymphokine-containing supernatant fractions to induce DTH skin reactions or cause reduction in the macrophage content of peritoneal exudates. L-Fucose also inhibited DTH and the peritoneal macrophage disappearance reaction induced by antigen in immunized guinea pigs.

Tuftsin, a tetrapeptide cleaved by polymorphonuclear leukocytes (PMNL) from leucokinin [which stimulates phagocytosis and migration of PMNL and macrophages (Constantopoulos and Najjar, 1972; Nishioka *et al.*, 1972)], binds to *N*-acetyl-D-neuraminic acid residues on the PMNL (Constantopoulos and Najjar, 1973; Nair *et al.*, 1978) and macrophages (Bar-Shavit *et al.*, 1979).

Lymphotoxin (LT), a soluble toxic factor that is released by antigen- and mitogen-stimulated lymphocytes, also binds to cell surface carbohydrates. LT from guinea pig is inhibited by a glycopeptide from porcine thyroglobulin (Sawada *et al.*, 1976), and its binding to target cells is inhibited by sugars containing β-D-galactosyl or *N*-acetyl-D-galactosaminyl residues (Sawada *et al.*, 1977). Kobayashi *et al.* (1978) isolated and partially characterized a glycopeptide from the surface of LT-sensitive L cells, which inhibited guinea pig LT activity on these cells. α-D-Galactosidase treatment of the glycopeptide significantly diminished its inhibitory activity against LT. Soluble LT molecules released from human cells are related in their specificity to the natural killer (NK) activities of the same cells (Weitzen *et al.*, 1983a). Antibodies to human LT and to human immunoglobulin block lysis of NK-sensitive cells by NK cells. Both NK- and LT-mediated lysis are inhibited by methyl-α-D-mannopyranoside and by *N*-acetyl-D-glucosamine (Weitzen *et al.*, 1983b). These authors suggest that LT may function on the surface of NK cells to effect lysis of targets.

Interferon (IF) also binds to glycoconjugates on the surface of target cells. Bovine brain gangliosides are capable of inhibiting IF binding, and when immobilized, they are able to remove IF from solution (Krishnamurti *et al.*, 1982). Several sialoglycoproteins, however, failed to interfere with IF–Sepharose activity (Besancon and Ankel, 1976). Cholera toxin inhibits IF action, implying that they may have the same receptors (Friedman and Kohn, 1976). The antiviral and antigrowth activities of murine type I (fibroblast) IF, but not type II (immune) IF, are inhibited by gangliosides. Type I IF binds to ganglioside affinity columns and can be eluted with sialyllactose. Murine type II IF passes through such columns unretarded (Ankel *et al.*, 1980). Both human fibroblast and leukocyte interferon are inhibited by gangliosides (Vengris *et al.*, 1976).

Soluble suppressor factors that inhibit either cellular or humoral immune reactions also appear to bind to cell surface carbohydrates on their target cells. Con A-activated human peripheral blood mononuclear cells elaborate soluble immune suppressor supernatant (SISS) activities containing at least two distinct suppressor factors. SISS-T, which inhibits mitogen- and antigen-stimulated T-cell proliferation, shows loss of activity in the presence of *N*-acetyl-D-glucosamine and the activity is retained on *N*-acetyl-D-glucosamine affinity columns. SISS-T binds to the surface glycoprotein receptors recognized by the lectins wheat germ agglutinin and *Agaricus bisporus* lectin, which produce similar inhibition of mitogen- or antigen-induced lymphocyte proliferation (Greene *et al.*, 1981). SISS-B, which inhibits polyclonal B-cell immunoglobulin production, is inhibited by L-rhamnose (Fleisher *et al.*, 1981). The authors were unable to test L-fucose inhibition, as they reported that this monosaccharide alone suppressed pokeweed mitogen-induced immunoglobulin production.

There is also evidence for carbohydrate-mediated recognition in the induction of murine secondary IgG responses. *N*-Acetylgalactosamine at 5 mg/ml, but none of eight other monosaccharides tested, blocked specific secondary IgG responses to 2,4-dinitrophenol (DNP)–hemocyanin and DNP–flagellin (Tomaska and Parish, 1981). *N*-Acetylgalactosamine had no effect on the primary antibody response to DNP–flagellin, on mixed lymphocyte reactions (MLR), or on the generation of cytotoxic T cells. *N*-Acetylgalactosamine blocked the action of isolated helper factor *in vitro*. In these experiments, D-mannose showed inhibition of primary and secondary antibody responses, MLRs, and generation of cytotoxic T cells. The authors suggested that D-mannose inhibited differentiation. In a second report, these authors (Tomaska and Parish, 1982) reported Ia-related differences in inhibition of secondary IgG responses by L-fucose,

D-galactose, *N*-acetylgalactosamine, and *N*-acetylmannosamine. They suggested that the interaction between B lymphocytes and T-cell-replacing factors can involve recognition of *I*-region-controlled carbohydrate structures.

Together, these studies indicate that cell surface glycoconjugates may serve as receptors for MIF, LIF, LT, IF, neutrophil chemotactic factor, macrophage chemotactic factor, T-cell helper factors, SISS-T, and SISS-B. No data are yet available for other soluble immune mediators such as interleukins I and II, B-cell growth factor, B-cell differentiation factor, or antigen-specific T-cell suppressor factors. From the above studies it is probable that several soluble immune mediators make their initial contact with target cells by binding to distinct complex carbohydrate sites on the cell surface.

2.4. Lectins in the Immune System

Agglutinins are widespread in invertebrates (review: Yeaton, 1981a), and an immunological role has been suggested for the agglutinins from several species (reviews: Cohen, 1974; Yeaton, 1981b). Hemagglutinins have been described in the oyster *Crassostrea virginica* (Acton *et al.*, 1969). The agglutination of red blood cells of several species by oyster hemagglutinins was inhibited by *N*-acetylgalactosamine, *N*-acetylglucosamine, or galactose (Tripp, 1974a). Rabbit red cells treated with hemolymph were engulfed more efficiently by oyster phagocytes *in vitro*. Hemolymph proteins from mollusks that react with blood group substances are thought to sensitize erythrocytes to engulfment by hemocytes (review: Tripp, 1974b). Agglutinins from invertebrates are found both in hemolymph and associated with the sexual or reproductive system (Prokop, 1974). These agglutinins have been termed "protectins" to imply their immunological functions. Snails of the genus *Helix* have blood group A-specific agglutinins in their albumin glands. Agglutinins are also found in snail eggs and the eggs of many fishes.

Lobster hemolymph contains a sheep erythrocyte-hemagglutinating activity that is decreased in the presence of *N*-acetylneuraminic acid or *N*-acetylneuraminic acid-containing mucin (Hall *et al.*, 1972). A second lobster agglutinin, which was inhibitable by *N*-acetylgalactosamine, agglutinated mouse, horse, and hamster erythrocytes (Hall and Rowlands, 1974). Lobster hemocytes were found to engulf erythrocytes *in vitro* only if the appropriate agglutinins were present, implying a role in surveillance against foreign bodies.

An *N*-acetylneuraminic acid-inhibitable lectin was also studied in the hemolymph of the horseshoe crab (*Limulus polyphemus*) and the coconut crab (*Birgus latro*) (Cohen *et al.*, 1974). The agglutinins bound to M and

N blood group antigens on erythrocytes and were inhibited by carcinoembryonic antigen (CEA). Recent evidence (Rostam-Abadi and Pistole, 1982) indicates that this lectin also recognizes 2-keto-3-deoxyoctanoate, a component of bacterial lipopolysaccharide (LPS). A carbohydrate-binding protein in crayfish hemocyte lysate binds to fungal β(1-3) glucans and induces a clotting reaction (Soderhall, 1981). A similar clotting reaction is seen in *L. polyphemus* amebocyte lysate after contact with LPS.

Agglutinins are also thought to play a role in opsonization of target cells prior to phagocytosis by hemocytes in earthworms, and may play a role in cellular immunity in allograft rejection (Cooper *et al.*, 1974). Increased levels of these agglutinins are observed after injecting the earthworms with glycoconjugates (Wojdani *et al.*, 1982).

Serum carbohydrate-binding proteins distinct from immunoglobulin have been reported in the shark (Sigel, 1974) and in man and other mammals (Gotschlich and Edelman, 1967). The function of these proteins is unknown, but the authors suggest an immune role.

Lectins have also been found in murine lymphocyte membranes (Kieda *et al.*, 1978, 1979; Decker and Marchalonis, 1979). The activity studied by Kieda *et al.* (1978) was present on thymocytes and spleen cells. This lectin agglutinated trypsinized rabbit erythrocytes and neuraminidase-treated rat erythrocytes. Neuraminidase digestion of the erythrocytes followed by galactose oxidase treatment reduced the agglutination almost completely, implicating the importance of terminal galactose receptors. The agglutination was inhibited by D-galactose, *N*-acetylgalactosamine, and methyl-α-D-mannopyranoside. Asialoglycoproteins were very active inhibitors. Evidence was presented for the presence of at least two lectin specificities on the lymphocytes, one for terminal galactose and one for terminal mannose. These observations were further studied by direct binding of fluorescent neoglycoproteins to the cells (Kieda *et al.*, 1979). Bovine serum albumin (BSA) was substituted with glycosides and labeled with fluorescein. Thymocytes and splenocytes were stained with the neoglycoproteins containing galactose, mannose, or *N*-acetylgalactosamine. In each case, different proportions of the cells were labeled, indicating the presence of different sets of lectins on different cell populations. Lectinlike molecules were also identified on murine lymphocytes after radioiodination and purification by affinity chromatography on fetuin–Sepharose or on bovine submaxillary mucin–Sepharose (Decker and Marchalonis, 1979). The binding was inhibited by free fetuin.

Surface lectins on activated human lymphoid cells were the subject of a recent study by Apgar and Cresswell (1982). These authors measured the binding of membrane glycoprotein micelles to cells. The 25- to 100-nm micelles contained *Lens culinaris* hemagglutinin-affinity-purified membrane glycoproteins from human lymphoblastoid cell lines. They were bound by B and T lymphoblastoid cell line cells and peripheral blood lymphocytes activated either with Con A or in an MLR. The glycoproteins fetuin and porcine thyroglobulin inhibited the binding, and asparagine-linked glycopeptides from fetuin were more active on a weight basis than fetuin. Transferrin, ovalbumin, bovine submaxillary mucin, orosomucoid, and O-linked glycopeptides from fetuin were ineffective at blocking the binding. The authors measured the binding of [^{125}I]fetuin to lymphoid cells and found that this correlated with the glycoprotein micelle binding. They suggested that the cell surface lectin to cell surface glycoprotein interactions might be important in regulation of lymphocyte traffic or the maintenance of lymphoid organ structure. The expression of the lectin after lymphocyte activation suggests a potential role in immune responses.

We have observed the binding of fluorescein- and rhodamine-conjugated asialoglycopeptides to lymphoid and monocyte populations in murine and human hemopoietic tissues (Hutchins and Reading, unpublished). This binding was inhibitable by high concentrations of lactose, suggesting the importance of terminal galactose residues in the binding.

2.5. Glycosyltransferases in the Immune System

Glycosyltransferases have also been reported on lymphoid cells. A cell surface galactosyltransferase activity was assayed on thymocytes and splenocytes of neonatal rats (LaMont *et al.*, 1974). The surface galactosyltransferase activity increased after mitogen stimulation of thymocytes compared to resting T cells. Patt *et al.* (1976) concluded the galactosyltransferase activity of mouse spleen cells was due to internal glycosylation after external hydrolysis of UDP-Gal followed by uptake of the free sugar. This finding was countered by Verbert *et al.* (1976) who demonstrated the existence of ectogalactosyltransferase activity on the lymphocyte surface. They excluded the possibility of misleading results due to precursor hydrolysis and intercellular utilization of free galactose by using a non-phagocytosable exogenous acceptor and by demonstrating transfer of galactose to endogenous cell surface receptors. The cells galactosylated in this way acquired new agglutinating properties with soybean agglutinin, which argued for the external position of the incorporated galactosyl residues. The same group (Verbert *et al.*, 1977) demonstrated the presence of an ectosialyltransferase on rat spleen lymphocytes that was capable of transferring *N*-acetylneuraminic acid to its own membrane and to exogenous acceptors. They (Hoflack *et al.*, 1979) later reported that the reactivity of the ectosialyltransferase could not be related to the presence of a small percentage of broken cells. They found that an inhibitory effect was observed with broken cells when the assay was conducted with a mixture of intact and homogenized lymphocytes. The intracellular inhibitory factor was purified and characterized as CMP-NeuNAc. The endogenous CMP-NeuNAc leads to isotopic dilution of the exogenous, labeled CMP-NeuNAc and appears to inhibit the reaction.

Baker *et al.* (1980) measured galactosyltransferase activity on the surface of thymic and splenic lymphocytes. After immunizing the mice, thymic lymphocytes, but not splenic lymphocytes, showed a significant increase in this activity. The authors suggested that this increase might be due to some intercellular reaction of the immune response. A cell surface galactosyltransferase activity on cytotoxic T lymphocytes (CTL) was correlated with mixed-lymphocyte culture (MLC)-generated secondary CTL cytotoxic activity (Kurt *et al.*, 1981).

2.6. Carbohydrate-Mediated Uptake in the Immune System

Periodate oxidation affected the clearance of purified rat liver lysosomal enzymes when they were reinjected i.v. into rats (Stahl *et al.*, 1976a). Lysosomal β-D-glucuronidase, N-acetyl-β-D-glucosaminidase, α-L-fucosidase, and α-D-mannosidase were cleared rapidly from the circulation before periodate oxidation, but not after. The authors suggested that a carbohydrate recognition marker on these hydrolases was responsible for their rapid clearance. Although the lysosomal and rat preputial gland N-acetyl-β-D-glucosaminidase was cleared rapidly, normal rat serum β-D-glucuronidase and epididymal N-acetyl-β-D-glucosaminidase were cleared slowly from the circulation (Stahl *et al.*, 1976b). Purified lysosomal enzymes were found to compete for the clearance of other purified lysosomal enzymes. This implied that the same receptor was responsible for the uptake of several enzymes. This receptor was different from the fibroblast receptor for lysosomal hydrolase uptake, as rat liver β-D-glucuronidase and rat preputial gland N-acetyl-β-D-glucosaminidase were cleared rapidly *in vivo* but were not efficiently taken up by the fibroblasts.

Schlesinger *et al.* (1976) found that the lysosomal hydrolases β-D-glucuronidase and N-acetyl-β-D-glucosaminidase cleared from the circulation also were taken up by the liver. The enzymes localized in the liver lysosomes and were still enzymatically active. Baynes and Wold (1976) reported that RNase B, but not A, C, or D, was rapidly taken up by the liver. The four forms of RNase differ only in their glycosylation: A being unglycosylated, B containing a high-mannose-type oligosaccharide, and C and D containing complex oligosaccharides, which differ in their degree of sialylation. The half-lives of RNase A, B, C, and D were respectively 577, 15, 862, and 941 min. Digestion with α-D-mannosidase resulted in prolonged circulation for RNase B, shifting the half-life in the circulation from 15 to 733 min. Achord *et al.* (1977) demonstrated that the rapid clearance of asialoagalactoorosomucoid, which contains terminal N-acetylglucosaminyl residues, was inhibited by yeast mannans and by mannose. These results indicated that this receptor recognized both N-acetylglucosamine and mannose.

Schlesinger *et al.* (1978a) reported that β-D-glucuronidase, N-acetyl-β-D-glucosaminidase, ribonuclease B, and asialoagalactoorosomucoid were all cleared by the receptor for terminal N-acetylglucosamine/mannose and that the receptor was on nonparenchymal cells of the liver. RNase B was taken up by reticuloendothelial cells in the spleen and bone marrow as well as by the nonparenchymal cells of the liver (Brown *et al.*, 1978). The same receptor was found on alveolar macrophages from several mammalian species (Schlesinger *et al.*, 1978b; Stahl *et al.*, 1978) and on primary *in vitro* cultures and continuous cell lines of murine macrophages (Stahl and Gordon, 1982). Asialoglycoproteins failed to inhibit the uptake, providing evidence that this was a different receptor from the liver terminal galactose receptor.

β-D-Glucuronidase uptake into liver nonparenchymal cells was blocked by L-fucose as well as mannose and N-acetylglucosamine (Achord *et al.*, 1978). The receptor for mannose/N-acetylglucosamine was purified from rabbit (Kawasaki *et al.*, 1978) and rat (Mizuno *et al.*, 1981; Townsend and Stahl, 1981) livers by affinity chromatography. The receptor bound L-fucose as well as mannose and N-acetylglucosamine but not galactose or mannose-6-phosphate. The binding preference for the receptor on rabbit alveolar macrophages was demonstrated to be similar (Shepherd *et al.*, 1981).

There is recent evidence (Thornburg *et al.*, 1980) that the asialoglycoprotein uptake system may function to clear immune complexes from the circulation. Rat IgG anti-BSA–BSA complexes with half-lives of 4 min, were cleared from the circulation of nonimmune rats and localized in the parenchymal cells of the liver. Coinjection of asialofetuin or galactose inhibited this clearance. The authors suggest that upon antigen binding, IgG molecules undergo an antigen-induced conformational change that results in the exposure of terminal galactosyl groups on an oligosaccharide chain, resulting in recognition by the asialoglycoprotein uptake system and clearance of immune complexes.

The uptake system on reticuloendothelial cells has also been implicated in the uptake of soluble immune complexes (Day *et al.*, 1980). BSA–IgM complexes are rapidly cleared from the circulation of nonimmune rats, and the clearance is inhibited by ovalbumin and mannan but not by asialofetuin. The complexes localized in hepatic nonparenchymal cells and in other organs rich in reticuloendothelial cells. Digestion of the immune complexes with α-D-mannosidase abolished the rapid clearance. An antigen-induced conformational change in the IgM molecule, exposing a high-mannose-type oligosaccharide, was proposed as the basis of the up-take. This hypothesis was tested *in vitro* by binding to Con A before and after complexing IgM with BSA. The IgM was tightly bound to the Con A column only after binding to BSA.

Recently, Nose and Wigzell (1983) prepared monoclonal IgG_{2b} antibodies devoid of asparagine-linked carbohydrate chains by treatment of the hybridoma cells with tunicamycin. They found that antigen-antibody complexes produced with the normal IgG_{2b} were cleared rapidly from the circulation of mice, but complexes produced with the carbohydrate deficient antibodies were not rapidly eliminated.

3. DIFFERENTIATION ANTIGENS IN THE IMMUNE SYSTEM

Glycoconjugates can also serve as markers of immune cellular differentiation. Antigenicity of the thymus-dependent (T cell) differentiation alloantigen Thy-1 may depend on both glycoprotein and glycolipids in T-cell membranes. Thy-1-positive ($Thy-1^+$) and Thy-1-negative ($Thy-1^-$) variants were characterized as to the nature of the genetic defect by Trowbridge and Hyman (1975). They found that variants that produced no Thy-1 did produce immunologically cross-reacting Lactoperoxidase (LPO)-^{125}I-labeled material that migrated in SDS–PAGE as a 25,000-dalton component (T25). Based on their studies, they proposed that the alleles of Thy-1 are determined by a difference in sugar sequence in the outer region of a carbohydrate moiety of T25 and that Thy-1.1 and Thy-1.2 genes code for glycosyltransferases that add the appropriate sugar.

Milewicz *et al.* (1976) demonstrated that treatment of C3H mouse bone marrow cells with thymic factor (TF) or with *Vibrio cholera* neuraminidase (VCN) exposed Thy-1.2 and G_{M1} antigens. Anti-Thy-1.2 was not cytotoxic for VCN-treated AKR (Thy-1.1) marrow cells. Pretreatment of TF-treated marrow cells with cholera toxin or choleragenoid, both of which bind G_{M1}, abrogated the cytotoxicity of anti-Thy-1.2 or anti-G_{M1} antisera in the presence of C'. Pretreatment of $Thy-1.2^+$ C3H thymocytes with cholera toxin or choleragenoid greatly reduced the cytotoxicity of anti-Thy-1.2 antiserum.

Stein-Douglas *et al.* (1976) reported that although rabbit antiserum to G_{M1} reacts with murine thymocytes and T cells, this reactivity is independent of Thy-1 type, and it appears to be associated with cross-reacting antibodies, probably to asialo-G_{M1}, G_{D1b}, or both, rather than antibodies to G_{M1} (for structures of the glycolipids, see Fig. 1). They (Stein *et al.*, 1978) reported that absorption of antiserum to G_{M1} on asialo-G_{M1} columns rendered the antiserum anti-G_{M1} specific and demonstrated that G_{M1} and Thy-1.2 antigen cap independently on C3H thymocytes, which indicates that G_{M1} is not Thy-1.2 antigen. Protease treatment of lymphocytes revealed receptors for anti-G_{M1} on most cells. These studies indicate that T and B cells differ in their accessibility of G_{M1} to antibody, and not necessarily in their G_{M1} content. They reported that purified antibodies to asialo-G_{M1} react with mature T cells in all strains of mice tested, but in contrast to anti-G_{M1}, these antibodies did not react with most thymocytes or with Pronase-treated B cells.

Wang *et al.* (1978) reported that Thy-1 antigenicity was associated with glycolipids of brain and thymocytes. They immunized spleen cells with brain or thymocyte glycolipids from AKR (Thy-1.1) or C3H (Thy-1.2) mice and generated antibody-forming cells that were specific for Thy-1.1 or Thy-1.2, respectively. The response was measured using a modified plaque-forming cell (PFC) assay using thymocyte target cells in agarose. A brain G_{M1} ganglioside preparation purified by column and thin-layer chromatography (TLC) from AKR or C3H mice contained Thy-1.1- or Thy-1.2-active glycolipids, respectively. The Thy-1.1- or Thy-1.2-active glycolipids could be separated from G_{M1} by a third TLC system and were estimated to be only a small percentage of the total brain G_{M1} preparations. Thy-1-active glycolipids isolated from thymocytes had similar TLC mobility. They suggested that Thy-1 antigenicity lies in the carbohydrate structures that are conjugated to either lipid or protein carriers, because the observed Thy-1 specificity was associated with both glycolipids and glycoproteins. These results were extended (Kato *et al.*, 1979) using two-dimensional TLC-purified material. They reported that neuraminidase digestion or mild acid treatment of Thy-1 glycolipids abrogated the PFC response. The neuraminidase sensitivity of Thy-1-active glycoconjugates was further demonstrated on Thy-1-active shed materials, Thy-1-active glycolipids, and Thy-1-active glycoprotein, using either the immune response assay for Thy-1 allotypes or a radioimmunoassay (RIA) for Thy-1.2. Neuraminidase treatment of the shed complexes, glycolipids, or glycoprotein from C3H mice resulting in the loss of Thy-1.2 activity and digestion of all three forms from AKR mice resulted in loss of Thy-1.1 activity. An unusual finding was that both Thy-1.1-active shed complexes and Thy-1.1 glycolipid exhibited Thy-1.2 activity after digestion with neuraminidase, and this conversion was established using three sources of neuraminidase and both assays for Thy-1.2 activity.

Johnson *et al.* (1976) demonstrated the importance of sialic acid in the Thy-1.2 antigen S-49 murine lymphoblastoid cells. They reported that cytotoxic inhibitory activity was lost after neuraminidase treatment of the Thy-1.2 alloantigen and that sialic acid also inhibited the cytotoxicity of AKR anti-C3H Thy-1.2 serum and C' for S-49 cells. Trypsin also increased the inhibitory activity, indicating that the Thy-1.2 determinant was associated with a glycoprotein.

In addition to the Thy-1 alloantigens, which are markers of all T cells, there are glycoconjugate markers of T-cell subsets. Sialic acid content and sialyltransferases are reportedly indicators of T-cell maturation (Despont *et al.*, 1975). Thymocytes were reported to contain the lowest amount of sialic acid, whereas hydrocortisone-resistant thymocytes, T cells, and B cells contained higher amounts, and macrophages contained the highest. The sialyltransferase activities of T cells of different stages paralleled their sialic acid content.

Painter and White (1976) demonstrated that incubation of mouse thymocytes with ConA causes a twofold increase in surface sialyltransferase. Hoessli *et al.* (1980) reported that lymph node Thy-1 glycoprotein antigens are more acidic than thymus Thy-1 antigens. The charge difference was abolished if the cells were treated with neuraminidase.

Attachment of sialic acid residues to the peanut agglutinin (PNA) receptor may be an important step in the maturation of murine thymocytes (Reisner *et al.*, 1976). PNA binds T antigen and reacts with *N*-acetyllactosamine. The major immature thymocyte subpopulation could be separated from the immunocompetent minor subpopulation by agglutination with PNA. The nonagglutinated cells were the hydrocortisone-resistant thymocytes, induced graft-versus-host reaction (GvHR), responded with mitogenesis to stimulation by PHA and Con A, and displayed low Thy-1 and high H-2 antigen levels. The agglutinated fraction responded only to Con A, and had high Thy-1 and low H-2 antigen levels. The nonagglutinated fraction bound almost no PNA, but after neuraminidase digestion, it bound the same amount as the agglutinated fraction.

Kornfeld (1978) determined terminal oligosaccharide structures of glycopeptides from calf thymocyte plasma membranes. One class of glycopeptides which was *O*-glycosidically linked and had the T-antigen structure, which is a high-affinity receptor for PNA (Reisner *et al.*, 1976; Lotan *et al.*, 1975), was present as a major component of surface glycoproteins.

PNA also binds to a subpopulation of peripheral lymphocytes that are T cells or null cells (London *et al.*, 1978). In contrast to the PNA$^+$ thymocytes that were corticosteroid and radiation sensitive, the 5% of the spleen cells that were PNA$^+$ were corticosteroid and radiation resistant.

Roelants *et al.* (1979) studied PNA reactivity of T cells during development. PNA$^+$ and PNA$^-$ T lymphocytes were found simultaneously in the liver at day 10 of gestation, in the thymus at day 11, and in the spleen at day 18. Prethymocytes in nude mice comprised both PNA$^+$ and PNA$^-$ cells. The authors found no evidence for maturation from PNA$^+$ to PNA$^-$ cells and suggested that cortical and medullary thymocytes develop independently.

Evidence for maturation of PNA^- cells from PNA^+ cells has been presented by other groups. Irle *et al.* (1978) found that PNA^+ thymocytes were unresponsive to PHA, to MLR, or to GvHR, and were unable to provide helper function for a primary *in vitro* response to SRBC. In culture, PNA^+ cells proliferated in response to Con A and a 24-hr supernatant medium from Con A-stimulated lymph node cultures. The blasts formed displayed mitogen-induced cytotoxicity and gave rise to PNA^-, medium-sized lymphocytes that were poor in Thy-1, rich in H-2, PHA responsive, and immunologically competent. In addition to mitogens and stimulated lymph node culture supernatant medium, thymic epithelial culture supernatant medium (TES) induces maturation from PNA^+ to PNA^- thymocytes (Kruisbeek and Astaldi, 1979). TES or interleukin II (IL-2) stimulates PNA^+ thymocytes to produce CTL responses.

Con A stimulates PNA^- but not PNA^+ cells to produce IL-2 (Kruisbeek *et al.*, 1980). Wagner *et al.* (1980) demonstrated that PNA^+ thymocytes were Lyt-123$^+$, whereas 90% of the PNA^- thymocytes were Lyt-1$^+$, about 10% being Lyt-123$^+$. In the presence of IL-2, PNA^+, Lyt-123$^+$ cells mounted cytotoxic responses. PNA^-, Lyt-123$^+$ cells represented a highly reactive pool of primary CTL precursors. PNA^-, Lyt-1$^+$ helper cells made the PNA^- fraction independent of IL-2. Conlon *et al.* (1982) reported that PNA^+ murine thymocytes were unresponsive to mitogen and interleukin I (IL-1) but were stimulated by mitogen in the presence of IL-2. They also reported that the PNA^- cells were mitogen responsive without IL-1 or IL-2. Reisner *et al.* (1979) found that 60–80% of human thymocytes bind PNA. PNA^+ thymocytes responded poorly to MLR and PHA stimulation. Acute leukemic blast cells were found to be PNA^+, while chronic leukemic peripheral blood leukocytes were PNA^-.

Receptors for the *N*-acetyl-D-galactosamine-inhibitable soybean agglutinin (SBA) also change with T-cell maturation (Reisner and Sharon, 1978). Immature thymocytes are PNA^+, SBA^+, mature thymocytes are PNA^-, SBA^+, and most splenic T cells are PNA^-, SBA^-. The receptors on the more mature cells are masked by the addition of sialic acid, for neuraminidase treatment exposes PNA and SBA receptors on mature T cells.

Murine hemopoietic stem cells defined by colony formation units in the spleen of irradiated recipients (CFU-S) are PNA^+ and SBA^+ (Reisner *et al.*, 1978). Agglutination of spleen cells by these lectins was used to obtain a population of $CFU-S^+$, $GvHR^-$ cells that were used to reconstitute lethally irradiated allogeneic (F_1 hybrid) mice without complications of GvHR. The authors also reported that agglutination with SBA could be used to separate T and B cells. In man, stem cells appear to be SBA^-. Recently, a combination of SBA agglutination and SRBC rosetting has been used to remove human GvHR-reactive cells in human allogeneic bone marrow transplantation (Y. Reisner, personal communication).

Glycolipid patterns of lymphoid cells also change as a function of differentiation. Schwarting and Marcus (1979) reported that 50–80% of Ig^+ cells reacted with antibodies to globoside, paragloboside, lactosylceramide, trihexosylceramide, G_{M1}, and asialo-G_{M1} (for structures, see Fig. 1), whereas 90% of T cells were unreactive with all six antibodies. Chronic lymphocytic leukemic (CLL) B lymphocytes from 7 of 8 patients were unreactive with all six antibodies. Rosenfelder *et al.* (1979) found characteristic differences in glycolipids between mitogen-stimulated B and T cells using biosynthetic radiolabeling with $[^{14}C]Gal$. Rosenfelder *et al.* (1980) reported that ganglioside G_{M3} may be associated with murine plasma cells. They obtained strikingly different ganglioside patterns of T and B lymphoblastoid tumor cell lines, particularly differences between levels of G_{M3} and G_{M1}.

Mitogenic assays can only yield limited information about lectin receptors, as cells that respond may do so directly, through lectin binding to their surface receptors, or indirectly, through stimulation by other cell populations that were triggered by the lectins. Cells that do not respond may bind lectin but not be triggered (see Hellstrom *et al.*, 1976). With these limitations in mind, some information may be obtained from studies with lectin stimulation. Using 5-bromo-2'-deoxyuridine and light inactivation of human lymphocytes stimulated by mitogens, followed by restimulation with different mitogens, Touraine *et al.* (1976) reported that Con A and PHA appear to stimulate the same subset of T cells. PWM was found to stimulate two cell populations, one was Con A and PHA reactive and the other was not. Lipsick *et al.* (1980) reported that two lectins from chicken tissues that were inhibitable by lactose and asialofetuin and purpureum, the lectin from *Dictyostelium purpureum*, were mitogenic for a population of nonadherent, Thy-1$^-$ mouse spleen cells and for spleen cells from nude (athymic) mice. The responding cells were thought to be B lymphocytes.

Another interesting glycoconjugate difference between B and T cells was found in the Leu 200 series of high-molecular-weight glycoproteins (Morishima *et al.*, 1982). These authors found that a monoclonal antibody against Leu 200 precipitated proteins with bands in SDS–PAGE at 190, 205, 215, and 230K from B- and T-cell lines. The T-cell lines generally had the 190 and 205K bands, while the B-cell lines had 190, 205, and 215K bands. The 230K band was only observed with one T- and one B-cell line. Neuraminidase treatment resulted in an increased mobility of the 215 and 230K bands but not the 190 and 205K bands. Oligosaccharides from all cell lines showed a shift in molecular weight after neuraminidase digestion. The authors concluded that the 215 and 230K bands have more sialic acid than the 190 and 205K bands and that the 215 and 230K bands may have larger or more highly branched carbohydrate chains.

A glycoprotein differentiation antigen, M_r 145K (T-145), on CTL has also been described (Kimura and Wigzell, 1977). After activation by alloantigens or Con A, Lyt-1$^-$2$^+$ blast T cells are T-145$^+$, but Lyt-1$^+$2$^-$ blasts and normal Lyt-1$^-$2$^+$ T cells are T-145$^-$. T-145$^+$ cells were directly correlated with CTL. *Vicia villosa* lectin absorbents depleted CTL from alloactivated populations of T cells via the lectin-specific interaction with T-145 (Kimura *et al.*, 1979a). This procedure did not affect NK or antibody-dependent cell-mediated cytotoxic cells (ADCC or K cells) (Kimura *et al.*, 1979b). The best inhibitor of *V. villosa* lectin was determined to be a disaccharide containing an *N*-acetylgalactosamine residue α-linked to galactose (Kaladas *et al.*, 1981). Although this lectin appeared to have a unique reactivity with CTL when used for cell affinity chromatography, less specific reactivity was obtained using fluorescence-activated cell sorting (FACS) (MacDonald *et al.*, 1981).

A ganglioside has also been reported as a differentiation antigen associated with NK cells. Schwarting and Summers (1980) analyzed glycosphingolipids (GSL) from C57BL/6 thymocytes and splenic T cells by high-pressure liquid chromatography (HPLC). Glucosylceramide and lactosylceramide were the major GSL of thymocytes from 1 to 30 weeks of age. Asialo-G_{M1} was only found in trace amounts. In splenic T cells, asialo-G_{M1} increased with age such that at 10 weeks, the splenic T cells had 10 to 20 times the concentration found in thymocytes. C57BL/6 *bg/ bg* (beige) mice, which lack NK activity, had low levels of asialo-G_{M1} in splenic T cells. Schwarting and Gajewski (1981) studied the neutral GSL and gangliosides from thymocytes and splenic T lymphocytes from normal and beige mice. In addition to the alterations in asialo-G_{M1} noted above, they found that a terminally sialylated ganglioside, which was a derivative of asialo-G_{M1}, was sialidase sensitive, and was deficient in thymic and splenic T cells of beige mice. Other deletions were noted in the ganglioside components of beige mice thymus and spleen cells. The authors suggested that glycolipid defects may cause lack of proper maturation of NK cells. C' and antibody to asialo-G_{M1}, but not to G_{M1}, globoside, or asialo-G_{M2}, abolished NK activity (Young *et al.*, 1980). Cytotoxic alloimmune T cells were not affected by anti-asialo G_{M1}, implying that asialo-G_{M1} display may be characteristic of NK cells. Anti-asialo-G_{M1}, rabbit antiserum, and C' eliminated NK activity of spleen cells of several strains of mice (Kasai *et al.*, 1980). Asialo-G_{M1} absorbed the anti-NK activity. Absorption of anti-brain-associated T-cell antigen with asialo-G_{M1} also diminished its anti-NK activity without affecting its anti-T activity. Durdik *et al.* (1980) reported that activated NK cells display a higher level of Thy-1 than resting cells. Either anti-Thy-1 or anti-asialo-G_{M1} and C′, but not anti-globoside or anti-asialo-G_{M2}, eliminated NK activity.

Anti-asialo-G_{M1} was also effective *in vivo* against NK cells (Habu *et al.*, 1981). Intravenous injection of anti-asialo-G_{M1} abolished NK activity against the YAC-1 lymphoma cell line in spleen cells from nude mice, as well as from control mice. In anti-asialo-G_{M1} treated nude mice, the incidence of tumor take and growth was enhanced with subcutaneously transplanted tumors. Asialo-G_{M1} has also been reported on leukemic cells of patients with acute lymphoblastic leukemia (Nakahara *et al.*, 1980).

In contrast to lymphoid cells, little is known about glycoconjugate differentiation markers in the macrophage–monocyte lineages. Stimulated macrophages express a new glycoprotein receptor reactive with one of the *Griffonia simplicifolia* isolectins specific for α-linked galactose residues (Maddox *et al.*, 1982). This lectin distinguishes stimulated macrophages from resident populations. The reactive macrophages, separated by adherence to lectin–Sepharose, mediate cellular cytotoxicity to tumor cells, whereas the nonadherent cells do not.

Lectins were also useful for separating murine granulocyte–macrophage colony-forming cells (CFC) by either agglutination or FACS (Nicola *et al.*, 1980). CFC were PNA⁺, SBA⁺, *Helix pomatia* agglutinin (HPA)⁺, and PWM⁺. PNA was also useful in defining human normal and leukemic cells of the monocyte lineage (O'Keefe and Ashmen, 1982). They found that monocytes, monoblasts, and a population of cells that were probably precursors of monoblasts were PNA⁺, but the classical myeloid blast cells were PNA⁻.

Another line of evidence for the importance of glycoconjugates in the differentiation of myeloid cells is the finding that tunicamycin, an inhibitor of asparagine-linked glycoprotein glycosylation, induces differentiation in both human (HL-60) and murine (M1) myeloid leukemia cell lines in culture (Nakayasu *et al.*, 1980). The authors suggest that glycosylation of cellular proteins has an important role in maintaining these myeloid leukemia cells in an undifferentiated state in culture.

Differences in the glycosylation of the murine B-cell and spleen-derived adherent cell Ia glycoproteins of the same Ia haplotype have been reported (Cullen *et al.*, 1981a). I-A and I-E molecules from purified B cells and splenic adherent cells from various haplotypes revealed a consistent difference in the isoelectric focusing (IEF) patterns of Ia chains. Digestion with neuraminidase, kinetics of labeling, and subcellular distribution indicated that the extra acid IEF bands in B-cell Ia molecules represented a more heavily sialylated form of α chains not present in adherent cells. The authors suggested that the selective differential glycosylation of B-cell and adherent cell Ia antigens could have some relation to cell-type-specific recognition of those cells. Further Ia glycosylation differences were reported by Cowan *et al.* (1982b). Murine I-A^k chains from whole spleens displayed three electrophoretic bands: α_1, α_2, and α_3. Only α_1 and α_2 were labeled with [^{3}H]fucose. Both α_1 and α_2 chains contained sialic acid, but α_3 did not. The authors suggested that α_3 may contain a high-mannose carbohydrate structure.

4. CARBOHYDRATE-MEDIATED RECOGNITION IN CELLULAR INTERACTIONS

Glycoconjugates play a role in bacterial adherence to human lymphocyte subpopulations (Teodorescu *et al.*, 1977). B cells were bound by the majority of the strains tested, and T cells were separated into four unique subsets on the basis of bacterial binding. The adherence of bacteria to lymphocytes appears to be dependent on their binding to carbohydrates on the lymphocyte surface (Rasanen, 1981). Some monosaccharides inhibited bacterial binding as did treatment of the lymphocyte with lectins, enzymes, or sodium metaperiodate.

Bacteria also bind to mouse peritoneal macrophages and to human PMNL via a mannose-inhibitable mechanism (Bar-Shavit *et al.*, 1977). The mannose-binding component appeared to be on the surface of *E. coli*, and *S. typhi.* Evidence was presented for a distinct mechanism for bacterial adherence to mouse peritoneal exudate macrophages based on carbohydrate-binding components on the macrophages that recognize bacterial cell wall carbohydrate components (Freimer *et al.*, 1978).

Human T lymphocytes also bind to the surface of SRBC (Bentwich *et al.*, 1973). This binding is enhanced after treatment of either the lymphocytes or the SRBC with neuraminidase (Galili and Schlesinger, 1974). More stable rosettes are formed by thymocytes and blast cells from untreated patients with acute lymphocytic leukemia (ALL) (Borella and Sen, 1975). Specific glycopeptides inhibit human lymphocyte–SRBC rosettes (Boldt and Armstrong, 1976). Trypsin treatment of SRBC reduced rosette formation, and the glycopeptides released inhibited rosette formation by untrypsinized erythrocytes. This finding suggests that rosetting occurs through the binding of glycoconjugates on SRBC by a receptor on human T cells. Inhibition was also observed with asparagine-linked model glycopeptides rich in sialic acid, galactose, *N*-acetylglucosamine, and mannose. Fetuin glycopeptides, human transferrin glycopeptide, and human erythrocyte glycopeptide inhibited rosetting. Stepwise degradation of fetuin glycopeptide established that exposed galactosyl residues were important determinants of inhibitory activity. Neuraminidase treatment of SRBC increased rosetting, presumably by exposing galactosyl residues on glycoproteins. Human T lymphocytes activated in MLR form more stable SRBC rosettes than nonactivated lymphocytes (Galili and Schlesinger, 1976). A similar receptor seems to be involved in the "natural attachment" of T cells to other cells of the same species (Galili *et al.*, 1978). This developmentally regulated receptor is present on thymocytes but can only be demonstrated on peripheral T cells after removal of sialic acid from either the T-cell membrane or that of the target cell. The authors suggested that natural attachment may contribute to the avidity of cellular interactions that occur after specific immune attachment.

Rosette formation between thymocytes and autologous erythrocytes is mediated by receptors on thymocytes that primarily recognize self H-2L molecules on erythrocytes. On the basis of sugar inhibition studies and the sensitivity of the receptors and acceptors to protease and glycosidase treatments, it appears that a protein receptor on thymocytes recognizes the carbohydrate portion of a glycoprotein on erythrocytes (Sia and Parish, 1981). The thymocyte receptor appears to recognize terminal galactose, mannose, and sialic acid residues on a branched-chain carbohydrate structure on erythrocytes, with mouse strains of different H-2 haplotype expressing carbohydrate structures that differ in the linkage of these three terminal sugars. The authors suggested that because H-2L are carbohydrate H-2 antigens, *H-2* regions may code for glycosyltransferases and that the thymocyte receptors for erythrocytes are these glycosyltransferases.

The peripheral blood leukocytes from a high proportion of chronic lymphocytic leukemia (CLL) patients form rosettes with mouse red blood cell (MRBC) (Catovsky *et al.*, 1976). Null, pre-B, and B-cell-type ALL cells do not form MRBC rosettes. Rosette formation is inhibited by anti-immunoglobulin heavy chains and by aggregated IgG, suggesting a topological relationship with immunoglobulin, Fc receptors, or both. Normal B cells rosette with MRBC, and a non-B, non-T lymphocyte population from normal individuals rosettes with MRBC only after neuraminidase treatment (Gupta *et al.*, 1976). The MRBC receptors for rosette formation have not been characterized.

Rosetting has also been used to investigate glycolipid-binding receptors on the surface of rat alveolar and peritoneal macrophages (Riedl *et al.*, 1982). The macrophages fail to rosette with untreated SRBC, but form rosettes after treatment of the SRBC with G_{M2}, G_{D1a}, or a mixture of bovine gangliosides. Pretreatment of the SRBC with G_{M1} did not induce rosette formation. The authors suggest that the macrophage recognition sites for gangliosides might be important in cellular interactions and in immune surveillance.

Gesner (1966) was the first to suggest that, within the immune system, saccharide structures at the mammalian cell surface act as "recognition sites" that participate in mediating normal activities of cells. He found that rat peritoneal exudate cells (PEC) agglutinated after glycosidase digestion. This agglutination was inhibited by L-fucose and N-acetyl-D-galactosamine. He suggested that lysosomal glycosidases released from damaged tissue might alter cell surface structures for adhesion and stimulate local reactions.

Lymphocytes and monocytes are capable of several forms of target cell recognition and destruction (review: Bennett *et al.*, 1980). Specific T-cell receptors for antigen are present on antigen-primed CTL, and antibody serves as a recognition factor for ADCC (K) cells. NK cells (Herberman and Holden, 1978) can also bind to target cells, and interestingly, beige mice that lack NK activity were also deficient in ADCC activity (Roder, 1980). NK target structures have been isolated from murine and human lymphoma and leukemic cells (Roder *et al.*, 1979).

Recent work by Nose and Wigzell (1983) suggests that carbohydrates may function in cellular recognition of antibodies in ADCC killing. They found that IgG2b antibodies deficient in asparagine-linked oligosaccharides bound antigen as well, but failed to act in ADCC assays as well as the completely glycosylated monoclonal antibody. The same carbohydrate-deficient antibodies failed to be bound by the Fc receptor (receptor for the heavy immunoglobulin chain fragment) on mouse macrophages in comparison to the complete antibodies.

Evidence for glycoconjugate recognition of target cells in natural cytotoxicity has been reported by several groups. Addition of D-mannose to the murine natural cytotoxic assay blocked cytotoxicity of the METH A fibrosarcoma target cells (Stutman *et al.*, 1980). All of the sugars tested inhibited NK lysis of the lymphoma YAC-1 target. None of the sugars tested affected CTL killing.

Using Chang liver cells as targets, spontaneous cell-mediated cytotoxicity (CMC), but not ADCC, was inhibited by D-ribose, D-gentibiose, N-acetyl-D-galactosamine, D-cellobiose, and α-lactose (MacDermott *et al.*, 1980).

Ades *et al.* (1981) demonstrated dose-dependent inhibition of NK cytolysis of target cells using a panel of monosaccharides. They found that fetuin, galactosamine, glucosamine, arabinose, xylose, galactose, and mannose, but not mucin, inhibited NK assays. Pretreatment of effector cells with glucosamine, galactosamine, arabinose, or galactose inhibited subsequent NK activity. Treatment of effector cells with a crude glycosidase mixture also resulted in loss of NK activity.

Recent studies with the specificity of NK cells indicate that they may recognize a high-molecular-weight glycoprotein, but evidence has not been presented for the recognition of carbohydrate versus protein determinants. Cytotoxicity, but not the initial binding to target cells, was inhibited by sugars (R. Herberman, personal communication). Other workers have demonstrated that glycopeptides isolated from K562 leukemic cell plasma membranes can inhibit NK cytotoxicity (Decker, *et al.*, 1984; Taryan, *et al.*, 1984).

Glycolipid expression on target cells has also been compared with NK susceptibility (Young *et al.*, 1981). Although display of asialo-G_{M2} on L5178Y cells is associated with susceptibility to NK attack, there is no correlation between asialo-G_{M2} level and NK sensitivity. The authors added that asialo-G_{M2} is not found in a variety of other oncogenic cell lines that are susceptible to NK attack.

Monocyte–macrophage recognition of target cells may also be dependent on carbohydrate structure. Human monocyte–target cell interaction is modulated by membrane sialic acid (Czop *et al.*, 1978). In the absence of exogenous proteins, monocytes ingest a variety of natural particulate activators of the human alternate complement pathway. SRBC that do not activate human C' via the alternate pathway or initiate a direct monocyte phagocytic response can be modified to exhibit both functions by the deletion (with neuraminidase) or alteration (with periodate oxidation followed by borohydride reduction) of membrane sialic acid residues. The authors suggest that the capacity of the nonimmune host to respond to desialylated particles by initiating the monocyte ingestive process represents a primitive biochemical basis for differentiation of nonself from self.

Inoculation of mice with neuraminidase-treated tumor cells results in macrophage activation *in vivo*, and these macrophages are able to kill the same tumor cells *in vitro* (Alley and Snodgrass, 1978). Macrophages are able to distinguish "old" RBC from "young" RBC in syngeneic mice, and this may be related to the loss of surface sialic acid residues (Leibovich and Knyszynski, 1980). Erythrocyte survival in the circulation is decreased after digestion with neuraminidase (Aminoff *et al.*, 1976), but because the desialylated erythrocytes are cleared by the liver (Durocher *et al.*, 1975), the clearance may be due to the asialoglycoprotein receptor on hepatocytes (Ashwell and Morell, 1974). Phagocytosis was not inhibited by several monosaccharides or sugar phosphates.

Cytotoxic human monocytes appear to recognize xenogeneic targets through sugar-specific receptors on the monocyte cell surface (Muchmore and Blaese, 1980). Monocyte killing of chicken RBC was inhibited by cellobiose, gentibiose, lactose, maltose, mannose, melibiose, and trehalose but not by 15 other sugars. Killing of rat RBC, chicken RBC, and horse RBC was inhibited to different extents by arabinogalactan, cellobiose, or mannose. None of the sugars inhibited PHA-induced proliferation of lymphocytes, and PHA-induced cytolysis of chicken RBC was not inhibited by cellobiose.

The same authors (Muchmore and Blaese, 1980; Muchmore *et al.*, 1980) reported that monocyte–T cell interactions are, at least in part, mediated by sugar-specific cellular receptors. Tetanus toxoid-induced secondary antigen-induced proliferation of human peripheral blood mononuclear cells is dependent on monocyte–T cell interactions, and is inhibited by L-altrose, L-fucose, gentibiose, *N*-acetylgalactosamine, D-mannose, L-rhamnose, and sorbose, but not by 23 other sugars. Ia (Dr$_W$)-specific antisera totally block the binding of radioactive sugars to monocytes of the appropriate Dr$_W$ type.

One of the most complicated and intriguing intercellular interactions in the immune response is the trafficking, homing, and recirculation of lymphocytes (review: Ford and Gowans, 1969), which depends in part on organ- and tissue-specific receptors (Fichtelius, 1969a). Chicken primitive lymphocytes home to epithelial primordium of thymus and of bursa Fabricii early in histogenesis (Fichtelius, 1969b). There is also a selective immigration of young blood lymphocytes to the gut epithelium of adult rats.

Transfused rat thoracic duct lymphocytes (TDL) treated with glycosidases from *Clostridium perfringens* showed altered organ localization (Gesner and Ginsburg, 1964). Enzyme digestion decreased localization of ^{32}P-labeled cells in the spleen and lymph nodes. Increased radioactivity was found in the liver. The effect of the enzyme mixture was specifically inhibited by L-fucose and *N*-acetyl-D-galactosamine but not by the other sugars tested. Treated cells suspended in syngeneic rat serum were loosely aggregated after 30 min at 37°C. This was not observed in control samples. Neuraminidase-treated rat TDL or cells treated with 0.1 mM periodate, when injected into rats, demonstrated an increased distribution to the liver, and decreased localization in the lymph nodes and spleen (Gesner and Woodruff, 1969). Trypsin-treated cells remained in the circulation and did not accumulate in lymph nodes or liver but did localize in the white pulp of the spleen (Gesner *et al.*, 1969).

Neuraminidase treatment, but not digestion with trypsin, altered the circulation patterns of rat thymocytes (Berney and Gesner, 1970). Treatment of mouse T lymphocytes or lymph node cells with lectins markedly altered the distribution of the cells after i.v. injection (Gillette *et al.*, 1973; Freitas and de Sousa, 1975). These experiments are difficult to interpret, for the lectin-coated cells could aggregate each other, bind other cells and glycoproteins in the circulation, or cause their adherence to organ capillary beds. In addition, lectins can cause capping and shedding of receptors from the cell surface and steric hindrance of other cell surface components. Treatment of lymphocytes with phospholipases also altered their localization *in vivo* (Freitas and de Sousa, 1976), but this may be related to changes in membrane fluidity and cell deformability, rather than cellular recognition.

Intravenous inoculation of Newcastle disease virus (NDV) altered the character of the recirculating pool of small lymphocytes. Woodruff and Woodruff (1972) suggested that the lymphocytopenia observed in mice and rats during NDV infection may be due to the modification of the surface of TDL, preventing their interaction with endothelial cells and thereby impairing lymphocyte homing.

In perfused organs *in vitro*, neuraminidase-treated lymphocytes were found to migrate into tissues in increased numbers compared to untreated cells (Ford *et al.*, 1976). These authors suggested that removal of sialic acid only results in increased *in vivo* localization of cells in the liver, which removes them from the circulation. This is most likely due to the terminal galactose-binding protein described by Ashwell and Morell (1974).

Using glutaraldehyde-fixed sections of lymph nodes, Anderson and Anderson (1976) and Woodruff *et al.* (1977) observed the adherence of rat TDL to the endothelium of high-endothelial venules (HEV). They found that the determinants on the TDL were trypsin sensitive but that sialic acid was not essential for this interaction. TDL–HEV binding was inhibited by trypsin-treated TDL supernatant fractions, suggesting that the enzyme had released a polypeptide with HEV-binding activity. The binding of lymphocytes to HEV was conserved in allogeneic combinations (Butcher *et al.*, 1979a), and with other species the binding was directly related to the evolutionary distance between the species tested and the mouse (Butcher *et al.*, 1979b).

Recent studies with lymphocyte–HEV binding *in vitro* have implicated lymphocyte binding to glycoconjugates on the endothelial cells of HEV (Stoolman and Rosen, 1983). L-Fucose, D-mannose, and the L-fucose-rich sulfated polysaccharide fucoidin specifically inhibited the attachment. L-Fucose showed stereoselective inhibition at 10–37 mM and fucoidin produced 50% inhibition at 10^{-8} M.

Using adoptive transfer of peritoneal exudate T lymphocytes from mice immunized with SRBC, Kaufmann *et al.* (1981) demonstrated that DTH caused by these cells was reduced if their surface carbohydrate structures were modified. Neuraminidase digestion of the T cells leads to a transient reduction of DTH in syngeneic recipients. Neuraminidase digestion followed by galactose oxidase markedly reduces DTH. Incubation with periodate causes permanent loss of DTH-transferring capacity. The effects of galactose oxidase following neuraminidase digestion and of periodate oxidation are reversed by borohydride reduction. Decreased DTH reactions parallel reduced migration of cells into sites of antigen deposition. Trapping of cells in the liver can account for the loss of neuraminidase-treated cells and the periodate-oxidized cells but not for the loss of neuraminidase–galactose oxidase-treated cells, as these cells do not localize in the liver in increased numbers. The effect is not solely related to the asialoglycoprotein uptake system in the liver, for galactose oxidase treatment of the asialoglycoproteins inhibits their binding, and borohydride reduction reverses this inhibition.

There is recent evidence implicating the asparagine-linked oligosaccharides on Ia and H-2 glycoproteins in cellular recognition by thymic lymphocytes. MLRs require recognition by T cells of foreign Ia molecules on the stimulator cells. Tunicamycin treatment of murine lymphocytes specifically inhibits *N*-glycosylation of proteins and blocks the stimulation of MLRs (Hart, 1982). This effect may be directly due to lack of glycosylation, for the Ia molecules appear to be inserted into the membrane without oligosaccharide chains. Specifically immune CTL will lyse virus-infected cells only if they recognize self H-2 glycoproteins on the target cells. Tunicamycin treatment of murine lymphocytes also blocks *H-2*-restricted cytolysis of virus-infected cells (Black *et al.*, 1981). Again, H-2 molecules appear to be inserted into the membrane without glycosylation.

A recognition role for the carbohydrate chains on Ia antigens has been proposed by Cowing and Chapdelaine (1983). They found that allogeneic macrophages stimulate T cells in an MLR, but allogeneic B cells do not. Removal of sialic acid from the B cells leads to allogeneic stimulation of MLRs. This indicates that T cells may discriminate between Ia antigens on B lymphocytes and macrophages on the basis of differences in the carbohydrate chains.

5. CONCLUSIONS AND FUTURE DIRECTIONS

Evidence has been presented to indicate the fundamental importance of carbohydrate structure recognition in the immune system. Carbohydrate structures appear to be recognized by antibodies, C′ components, lymphokines and monokines, surface lectins, surface enzymes, and carbohydrate-mediated uptake systems. Carbohydrate structures appear to change during differentiation of cells in the immune system. Several cellular immune interactions may depend on carbohydrate recognition. The evidence for specific recognition of complex carbohydrate structures in some immune interactions is tenuous at best. The structural bases for most of these recognition events remain unknown. In addition, little is known about the nature of the cell surface carbohydrate-binding proteins. In the future, efforts must continue toward understanding carbohydrate structure and function in the immune system.

Several theories have been advanced as to the evolution of carbohydrate-binding proteins that appear to function in immune recognition.

Parish (1977) suggested that invertebrates recognize nonself via glycosyltransferase recognition factors and that this recognition leads to phagocytosis. Rothenberg (1978) proposed that the *MHC*-encoded recognition units are ancestrally related to the invertebrate hemagglutinins and were modified glycosyltransferases. Muchmore and Blaese (1980) proposed that the recognition systems involved in T-cell, B-cell cooperation, T-macrophage cooperation, and macrophage–target cell recognition are also evolutionarily related to the invertebrate hemagglutinins. Parish *et al.* (1981) proposed that the *I*-region gene products are glycolipid glycosyltransferases, which function in both synthesis and recognition of self Ia glycolipids. Gorczynski (1984) has proposed that these carbohydrate recognition units are involved in carbohydrate antigen recognition and presentation by macrophages. It is possible that antibodies, MHC antigens, the C' system, the clotting system, cell surface lectins, cell surface glycosyltransferases, and organ homing systems are all evolutionarily related to the primitive carbohydrate-binding proteins of invertebrates.

I would like to conclude with a thought on the nature of immune surveillance. If the evidence that implicates carbohydrate recognition by antibodies, NK, ADCC, and macrophage surveillance over neoplastic cells is correct, this must indicate a fundamental relationship between cancer and altered carbohydrate structure. As multicellular organisms evolved, cooperation among the cells of the organism was required for successful competition. In addition to surveillance over foreign invaders, it became necessary to search for altered "self." If a particular cell became altered in its regulation, it would have to be eliminated in order for the host to survive. Primitive immune surveillance systems, possibly in the form of carbohydrate-binding agglutinins (primitive immunoglobulins) and wandering phagocytic cells (primitive reticuloendothelial cells), arose and were most likely maintained because of their benefit to the host's survival potential. At the same time, cells that became altered in their regulation, but remained transparent to the primitive immune surveillance systems, would threaten the host's survival potential. It seems reasonable to propose that mutations that allowed organisms to control expression of surface recognition structures in tight regulation with growth and differentiation controls would have a selective advantage. Cells with this control could present "doomsday antigens" on their surfaces, signaling for their elimination but leading to the survival of the host. If these antigens were protein, invaders such as a virus that could shut down host protein synthesis could block the elaboration of the "doomsday antigens." Any invaders that required glycosylation of their own proteins would allow carbohydrate signals to reach the cell surface. Glycosylation, which uses existing transferases, would be immune to the immediate effects of shutdown of host DNA, RNA, and protein synthesis. I propose that the regulation of glycosylation in cells became tightly coupled to differentiation and growth control, and that the immune surveillance system evolved to detect and eliminate cells presenting "doomsday antigens."

Several clues point to a fundamental relationship between cell surface carbohydrate structure and cancer. Although the available space does not permit review of the extensive literature, there are many studies on carbohydrate alterations in virus-infected, virus-transformed, and tumor cells (reviews: Lai and Duesberg, 1972; Hakomori, 1973, 1975; Brady and Fishman, 1974; Glick, 1974; Steiner *et al.*, 1975; Smets *et al.*, 1975; Richardson *et al.*, 1975; Kurth, 1976; Emmelot *et al.*, 1976; Atkinson and Hakimi, 1980). Another important point is that a number of monoclonal antibodies that are tumor reactive recognize altered carbohydrate structure (see Section 2). Lectins are also able to recognize alterations in tumor cells and have been used to detect altered surface components in murine (Reading *et al.*, 1980a,b) and human tumors (Reading, 1982a). In addition, there are hints that several of the important mechanisms of immune surveillance may be probing surface carbohydrate structures (see Section 4).

This hypothesis, if correct, could have a great impact on the understanding of biological and therapeutic aspects of cancer. A more precise understanding of alterations in cellular regulation in tumor cells might lead to more specific metabolic therapies. Diagnosis might be improved by using carbohydrate-binding antibodies and lectins, and these same reagents might be useful for some therapeutic aspects of cancer management.

Although numerous carbohydrate alterations have been studied in tumor cells, the basis of the structural differences and the control of these alterations in glycosylation are generally unknown. In the future, intensive studies of the structure, enzymology, and altered regulation of glycosylation in normal, virus-infected, and tumor cells may lead to a better understanding of host defense systems against altered "self" in cancer and other diseases.

REFERENCES

Achord, D. T., Brot, F. E., and Sly, W. S., 1977, Inhibition of the rat clearance system for agalacto-orosomucoid by yeast mannans and by mannose, *Biochem. Biophys. Res. Commun.* **77**:409–415.

Achord, D. T., Brot, F. E., Bell, C. E., and Sly, W. S., 1978, Human β-glucuronidase: *In vivo* clearance and *in vitro* uptake by a glycoprotein recognition system on reticuloendothelial cells, *Cell* **15**:269–278.

Acton, R. T., Bennett, J. C., Evans, E. E., and Schrohenloher, R. E., 1969, Physical and chemical characterization of an oyster hemagglutinin, *J. Biol. Chem.* **214**:4128–4135.

Ades, E. W., Hinson, A., and Decker, J. M., 1981, Effector cell sensitivity to sugar moieties. I. Inhibition of human natural killer cell activity by monosaccharides, *Immunobiology* **160**:248–258.

Allen, A., and Minnikin, S. M., 1975, The binding of the mucoprotein from gastric mucus to cells in tissue culture and the inhibition of cell adhesion, *J. Cell Sci.* **17**:617–631.

Alley, C. D., and Snodgrass, M. J., 1978, Effect of inoculation with neuraminidase-treated tumor cells on macrophage cytotoxicity in vitro, *Cancer Res.* **38**:2332–2338.

Alving, C. R., and Richards, R. L., 1977, Immune reactivities of antibodies against glycolipids. II. Comparative properties, using liposomes, of purified antibodies against mono, di, and trihexosyl ceramide haptens, *Immunochemistry* **14**:383–389.

Aminoff, D., Bell, W. C., Fulton, I., and Ingebrigtsen, N., 1976, Effect of sialidase on the viability of erythrocytes in circulation, *Am. J. Hematol.* **1**:419–432.

Amsden, A., Ewan, V., Yoshida, T., and Cohen, S., 1978, Studies on cellular receptors for lymphokines. I. Interactions of chemotactic factors with monosaccharides, *J. Immunol.* **120**:542–549.

Anderson, A. O., and Anderson, N. D., 1976, Lymphocyte emigration from high endothelial venules in rat lymph nodes, *Immunology* **31**:731–748.

Ankel, H., Krishnamurti, C., Besancon, F., Stefanos, S., and Falcoff, E., 1980, Mouse fibroblast (type I) and immune (type II) interferons: Pronounced differences in affinity for gangliosides and in antiviral and antigrowth effects on mouse leukemia L-1210R cells, *Proc. Natl. Acad. Sci. USA* **77**:2528–2532.

Anstee, D. J., 1981, The blood group MNSs-active sialoglycoproteins, *Semin. Hematol.* **18**:13–31.

Apgar, J. R., and Cresswell, P., 1982, Expression of cell surface lectins on activated lymphoid cells, *Eur. J. Immunol.* **12**:570–576.

Artzt, K., and Bennett, D., 1975, Analogies between embryonic (T/t) antigens and adult major histocompatibility (H-2) antigens, *Nature (London)* **256**:545–547.

Artzt, K., Dubois, P., Bennett, D., Condamine, H., Babinet, C., and Jacob, F., 1973, Surface antigens common to mouse cleavage embryos and primitive teratocarcinoma cells in culture, *Proc. Natl. Acad. Sci. USA* **70**:2988–2992.

Artzt, K., Bennett, D., and Jacob, F., 1974, Primitive teratocarcinoma cells express a differentiation antigen specified by a gene at the T-locus in the mouse, *Proc. Natl. Acad. Sci. USA* **71**:811–814.

Asao, M. I., and Oppenheimer, S. B., 1979, Inhibition of cell aggregation by specific carbohydrates, *Exp. Cell Res.* **120**:101–110.

Ashall, F., Bramwell, M. E., and Harris, H., 1982, A new marker for human cancer cells. I. The Ca antigen and the CA1 antibody, *Lancet* **2**:1–6.

Ashwell, G., and Morell, A. G., 1974, The role of surface carbohydrates in the hepatic recognition and transport of circulating glycoproteins, *Adv. Enzymol.* **41**:99–128.

Atkinson, P. H., and Hakimi, J., 1980, Alterations in glycoproteins of the cell surface, in: *The Biochemistry of Glycoproteins and Proteoglycans* (W. J. Lennarz, ed.), pp. 191–239, Plenum Press, New York.

Baba, T., Yoshida, T., and Cohen, S., 1979, Suppression of cell-mediated immune reactions by monosaccharides, *J. Immunol.* **122**:838–841.

Baker, A. P., Smith, W. J., and Holden, D. A., 1980, Development of an immunological response and changes in the activity of an ectogalactosyltransferase, *Cell. Immunol.* **51**:186–191.

Balazs, E. A., and Jacobson, B., 1966, Interaction of amino sugars and amino sugar-containing macromolecules with viruses, cells, and tissues, in: *The Amino Sugars,* Vol. IIB (E. A. Balazs and R. W. Jeanloz, eds.), pp. 361–395, Academic Press, New York.

Barondes, S. H., 1980, Endogenous cell-surface lectins: Evidence that they are cell adhesion molecules, in: *The Cell Surface: Mediator of Developmental Processes* (N. K. Wessels and S. Subtelny, eds.), pp. 349–363, Academic Press, New York.

Barondes, S. H., 1981, Lectins: Their multiple endogenous cellular functions, *Annu. Rev. Biochem.* **50:**207–231.

Bar-Shavit, Z., Ofek, I., Goldman, R., Mirelman, D., and Sharon, N., 1977, Mannose residues on phagocytes and receptors for the attachment of *Escherichia coli* and *Salmonella typhi, Biochem. Biophys. Res. Commun.* **78:**455–460.

Bar-Shavit, Z., Stabinsky, Y., Fridkin, M., and Goldman, R., 1979, Tuftsin–macrophage interaction: Specific binding and augmentation of phagocytosis, *J. Cell Physiol.* **100:**5–62.

Basten, A., Croft, S., Parish, C. R., and McKenzie, I. F. C., 1980, Transfer of cell mediated immunity with cell free leukocyte extracts. III. Demonstration of IA antigens in the specific component, *Cell. Immunol.* **56:**440–451.

Basu, M., Basu, S., Shanabruch, W. G., Moskal, J. R., and Evans, C. H., 1976, Lectin and cholera toxin binding to guinea pig tumor (104c1) cell surfaces before and after glycosphingolipid incorporation, *Biochem. Biophys. Res. Commun.* **71:**385–392.

Baynes, J. W., and Wold, F., 1976, Effect of glycosylation on the *in vivo* circulating half-life of ribonuclease, *J. Biol. Chem.* **251:**6016–6024.

Benacerraf, B., and Katz, D. H., 1975, The nature and function of histocompatibility-linked immune response genes, in: *Immunogenetics and Immunodeficiency* (B. Benacerraf, ed.), pp. 117–177, University Park Press, Baltimore.

Bennett, M., Kumar, V., Levy, E., and Rodday, P., 1980, Genetic resistance to tumors: Roles of marrow-dependent and -independent cells, in: *Genetic Control of Natural Resistance to Infection and Malignancy* (E. Skamene, P. A. L. Kongshavn, and M. Landy, eds.), pp. 431–443, Academic Press, New York.

Bentwich, Z., Douglas, S. D., Skutelsky, E., and Kunkel, H. G., 1973, Sheep red cell binding to human lymphocytes treated with neuraminidase: Enhancement of T cell binding and identification of a subpopulation of B cells, *J. Exp. Med.* **137:**1532–1537.

Berger, E. G., Buddecke, E., Kamerling, J. P., Kobata, A., Paulson, J. C., and Viegenthart, J. F. G., 1982, Structure, biosynthesis, and functions of glycoprotein glycans, *Experientia* **38:**1129–1162.

Berney, S. N., and Gesner, B. M., 1970, The circulatory behaviour of normal and enzyme altered thymocytes in rats, *Immunology* **18:**681–691.

Besancon, F., and Ankel, H., 1976, Specificity and reversibility of interferon ganglioside interaction, *Nature (London)* **259:**576–578.

Beyer, E. C., Tokuasu, K. T., and Barondes, S. H., 1979, Localization of an endogenous lectin in chicken liver, intestine, and pancreas, *J. Cell Biol.* **82:**565–571.

Black, P. L., Vitetta, E. S., Forman, J., Kang, C.-Y., May, R. D., and Uhr, J. W., 1981, Role of glycosylation in the H-2-restricted cytolysis of virus-infected cells, *Eur. J. Immunol.* **11:**48–55.

Blanchard, D., Cartron, J.-P., Fournet, B., Montrevil, J., van Halbeek, H., and Vliegenthart, J. F. G., 1983, Primary structure of the oligosaccharide determinant of blood group Cad specificity, *J. Biol. Chem.* **268:**7691–7695.

Bloch, R., Maccecchini, M. L., Jumblatt, R., Buttrick, P., and Burger, M. M., 1977, Sugar-specific antibodies reactive towards cell-surface carbohydrates, *Eur. J. Biochem.* **80:**261–266.

Blumenford, O. O., and Adamany, A. M., 1978, Structural polymorphism within the amino-terminal region of MM, NN, and MN glycoproteins (glycophorins) of the human erythrocyte membrane, *Proc. Natl. Acad. Sci. USA* **75:**2727–2731.

Boldt, D. H., and Armstrong, J. P., 1976, Rosette formation between human lymphocytes and sheep erythrocytes: Inhibition of rosette formation by specific glycopeptides, *J. Clin. Invest.* **57:**1068–1078.

Boldt, D. H., MacDermott, R. P., and Jorolan, E. P., 1975, Interaction of plant lectins with purified human lymphocyte populations: Binding characteristics and kinetics of proliferation, *J. Immunol.* **114:**1532–1536.

Bolwell, G. P., Callow, J. A., and Evans, L. V., 1980, Fertilization in brown algae. III. Preliminary characterization of putative gamete receptors from eggs and sperm of *Fucus serratus, J. Cell Sci* **43:**209–224.

Borella, L., and Sen, L., 1975, E receptors on blasts from untreated acute lymphocytic leukemia (ALL): Comparison of temperature dependence of E rosettes formed by normal and leukemic lymphoid cells, *J. Immunol.* **114:**187–190.

Bortolussi, R., Ferrieri, P., Bjorksten, B., and Quie, P. G., 1979, Capsular K1 polysaccharide of *Escherichia coli*: Relationship to virulence in newborn rats and resistance to phagocytosis, *Infect. Immun.* **25:**293–298.

Bosmann, H. B., 1972a, Platelet adhesiveness and aggregation. II. Surface sialic acid, glycoprotein: N-acetylneuraminic acid transferase and neuraminidase of human blood platelets, *Biochim. Biophys. Acta* **279:**456–474.

Bosmann, H. B., 1972b, Sialyl transferase activity in normal and RNA- and DNA-virus transformed cells utilizing desialyzed, trypsinized cell plasma membrane external surface glycoproteins as exogeneous acceptors, *Biochem. Biophys. Res. Commun.* **49:**1256–1262.

Bowles, D. J., and Kauss, H., 1976, Isolation of a lectin from liver plasma membrane and its binding to cellular membrane receptors *in vitro, FEBS Lett.* **66:**16–19.

Brady, R. O., and Fishman, P. H., 1974, Biosynthesis of glycolipids in virus-transformed cells, *Biochim. Biophys. Acta* **355:**121–148.

Brown, J. C., and Colling, R. G., 1982, Properties of cold agglutinin and group carbohydrate-specific antibodies isolated from group C streptococcal antisera, *Mol. Immunol.* **19:**457–465.

Brown, J. C., and Hunt, R. C., 1978, Lectins, *Int. Rev. Cytol.* **52:**277–349.

Brown, R. C., Bass, H., and Coombs, J. P., 1975, Carbohydrate binding proteins involved in phagocytosis by *Acanthamoeba, Nature (London)* **254:**434–435.

Brown, T. L., Henderson, L. A., Thorpe, S. R., and Baynes, J. W., 1978, The effect of α-mannose-terminal oligosaccharides on the survival of glycoproteins in the circulation, *Arch. Biochem. Biophys.* **188:**418–428.

Butcher, E. C., Scollay, R. G., and Weissman, I. L., 1979a, Lymphocyte adherence to high endothelial venules: Characterization of a modified *in vitro* assay, and examination of the binding of syngeneic and allogeneic lymphocyte populations, *J. Immunol.* **123:**1996–2003.

Butcher, E. C., Scollay, R., and Weissman, I., 1979b, Evidence of continuous evolutionary changes in structures mediating adherence of lymphocytes to specialized venules, *Nature (London)* **280:**496–498.

Carver, J. P., and Grey, A. A., 1981, Determination of glycopeptide primary structure by 360-MHz proton magnetic resonance spectroscopy, *Biochemistry* **20:**6607–6616.

Catovsky, D., Cherchi, M., Okos, A., Hedge, U., and Galton, A. G., 1976, Mouse red-cell rosettes in B-lymphoproliferative disorders, *Br. J. Haematol.* **33:**173–177.

Cauldwell, C. B., Henkart, P., and Humphreys, T., 1973, Physical properties of sponge aggregation factor: A unique proteoglycan complex, *Biochemistry* **12:**3051–3055.

Ceri, H., Kobiler, D., and Barondes, S. H., 1981, Heparin-inhibitable lectin: Purification from chicken liver and embryonic chicken muscle, *J. Biol. Chem.* **256:**390–394.

Cheng, C. C., and Bennett, D., 1980, Nature of the antigenic determinants of T locus antigens, *Cell* **19:**537–543.

Childs, R. A., Kapadia, A., and Feizi, T., 1980, Expression of blood group I and i active carbohydrate sequences on cultured human and animal cell lines assessed by radioimmunoassays with monoclonal cold agglutinins, *Eur. J. Immunol.* **10:**379–384.

Chipowsky, S., Lee, C., and Roseman, S., 1973, Adhesion of cultured fibroblasts to insoluble analogues of cell-surface carbohydrates, *Proc. Natl. Acad. Sci. USA* **70:**2309–2312.

Cisar, J. O., Kolenbrander, P. E., and McIntire, F. C., 1979, Specificity of coaggregation reaction between human oral streptococci and strains of *Actinomyces viscosus* or *Actinomyces naeslundii, Infect. Immun.* **24:**742–752.

Cohen, E. (ed.), 1974, *Biomedical Perspectives of Agglutinins of Invertebrate and Plant Origins, Ann. N.Y. Acad. Sci.* **234.**

Cohen, E., Rozenberg, M., and Massaro, E. J., 1974, Agglutinins of *Limulus polyphemus* (horseshoe crab) and *Birgus latro* (coconut crab), *Ann. N.Y. Acad. Sci.* **234:**28–33.

Conlon, P. J., Henney, C. S., and Gillis, S., 1982, Cytokine-dependent thymocyte responses: Characterization of IL 1 and IL 2 target subpopulations and mechanism of action, *J. Immunol.* **128:**797–801.

Constantopoulos, A., and Najjar, V. A., 1972, Tuftsin, a natural and general phagocytosis stimulating peptide affecting macrophages and polymorphonuclear granulocytes, *Cytobios* **6:**97–100.

Constantopoulos, A., and Najjar, V. A., 1973, The requirement for membrane sialic acid in the stimulation of phagocytosis by the natural tetrapeptide, tuftsin, *J. Biol. Chem.* **248:**3819–3822.

Cook, R. G., Vitetta, E. S., Uhr, J. W., and Capra, J. D., 1980, The I region of the murine major histocompatibility complex: Genetics and structure, in: *Membranes, Receptors and the Immune Response* (E. P. Cohen and H. Kohler, eds.), pp. 95–105; Liss, New York.

Cooper, E. L., Lemmi, C. A. E., and Moore, T. C., 1974, Agglutinins and cellular immunity in earthworms, *Ann. N.Y. Acad. Sci.* **234:**34–50.

Cooper, H. S., and Haesler, W. E., 1978, Blood group substances as tumor antigens in the distal colon, *Am. J. Clin. Pathol.* **69:**594–598.

Cooper, H. S., Coc, J., and Patchefsky, A. S., 1980, Immunohistologic study of blood group substances in polyps of the distal colon, *Am. J. Clin. Pathol.* **73:**345–350.

Cooper, N. R., and Morrison, D. C., 1978, Binding and activation of the first component of human complement by the lipid A region of lipopolysaccharides, *J. Immunol.* **120:**1862–1868.

Cowan, E. P., Cummings, R. D., Schwartz, B. D., and Cullen, S. E., 1982a, Analysis of murine Ia antigen glycosylation by lectin affinity chromatography, *J. Biol. Chem.* **257:**11241–11248.

Cowan, E. P., Schwartz, B. D., and Cullen, S. E., 1982b, Murine I-A^k α-chain subspecies with glycosylation differences, *J. Immunol.* **128:**2019–2025.

Cowing, C., and Chapdelaine, J. M., 1982, T cells discriminate between Ia antigens expressed on allogeneic accessory cells and B cells: A potential function for carbohydrate side chains on Ia molecules, *Proc. Natl. Acad. Sci. U.S.A.* **80:**6000–6004.

Crandall, M., Lawrence, L. M., and Saunders, R. M., 1974, Molecular complementarity of yeast glycoprotein mating factors, *Proc. Natl. Acad. Sci. USA* **71:**26–29.

Critchley, D. R., and Vicker, M. G., 1977, Glycolipids as membrane receptors important in growth regulation and cell–cell interactions, in: *Dynamic Aspects of Cell Surface Organization* (G. Poste and G. L. Nicolson, eds.), pp. 307–370, Elsevier/North-Holland, Amsterdam.

Cuatrecasas, P., 1973a, Interaction of *Vibrio cholerae* enterotoxin with cell membranes, *Biochemistry* **12:**3547–3558.

Cuatrecasas, P., 1973b, Gangliosides and membrane receptors for cholera toxin, *Biochemistry* **12:**3558–3566.

Cullen, S. E., and Nathenson, S. G., 1974, Further characterization of Ia (immune response region associated) antigen molecules, in: *The Immune System: Genes, Receptors, Signals* (E. E. Sercarz, A. R. Williamson, and C. F. Fox, eds.), pp. 191–200, Academic Press, New York.

Cullen, S. E., Kindle, C. S., Shreffler, D. C., and Cowing, C., 1981a, Differential glycosylation of murine B cell and spleen adherent cell Ia antigens. *J. Immunol.* **127**:1478–1484.

Cullen, S. E., Rose, S. M., and Kindle, C. S., 1981b, Ia antigens: Molecular components in immune regulation?, in: *Current Trends in Histocompatibility* (R. A. Reisfeld and S. Ferrone, eds.), pp. 391–413, Plenum Press, New York.

Curtoni, E. S., Borelli, I., Cornaglia, B. M., Olivetti, E., and Peyretti, F., 1980, Antibodies for other blood systems present in workshop DR sera and interaction with anti-DR antibodies, in: *Histocompatibility Testing* (P. I. Teraski, ed.) pp. 900–902, UCLA Tissue Typing Laboratory, Los Angeles.

Czop, J. K., Fearon, D. T., and Austen, K. F., 1978, Membrane sialic acid on target particles modulates their phagocytosis by a trypsin-sensitive mechanism on human monocytes, *Proc. Natl. Acad. Sci. USA* **75**:3831–3835.

Dahr, W., Gielen, W., and Beyreuther, K., 1980, Structure of the Ss blood group antigens. I. Isolation of Ss-active glycopeptides and differentiation of the antigens by modification of methionine, *Z. Physiol. Chem.* **361**:145–152.

Dawson, A., and Franks, D., 1967, Factors affecting the expression of blood group antigen A in cultured cells. *Exp. Cell Res.* **47**:377–385.

Day, J. F., Thornburg, R. W., Thorpe, S. R., and Baynes, J. W., 1980, Carbohydrate-mediated clearance of antibody antigen complexes from the circulation, *J. Biol. Chem.* **255**:2360–2365.

Dazzo, F. B., 1980, Lectins and their saccharide receptors as determinants of specificity in the *Rhizobium*–legume symbiosis, in: *The Cell Surface: Mediator of Developmental Processes* (S. Subtelny and N. K. Wessels, eds.), pp. 277–304, Academic Press, New York.

Decker, J. M., and Marchalonis, J. J., 1979, Lectin-like molecules on murine T and B lymphocytes, *Fed. Proc. Abstr.* **38**:934.

Decker, J. M., Hinson, A., and Ades, E. W., 1984, Inhibition of human NK cell cytotoxicity against K562 cells with glycopeptides from K562 plasma membranes, (submitted).

Despont, J. P., Abel, C. A., and Grey, H. M., 1975, Sialic acids and sialyltransferases in murine lymphoid cells: Indicators of T cell maturation, *Cell. Immunol.* **17**:487–494.

de Waard, A., Hickman, S., and Kornfeld, S., 1976, Isolation and properties of β-galactoside binding lectins of calf heart and lung, *J. Biol. Chem.* **251**:7581–7587.

Dillner-Centerlind, M.-L., Axelsson, B., Hammarstrom, S., Hellstrom, U., and Perlmann, P., 1980, Interaction of lectins with human T lymphocytes: Mitogenic properties, inhibitory effects, binding to the cell membrane and to isolated surface glycopeptides, *Eur. J. Immunol.* **10**:434–442.

Dippold, W. G., Lloyd, K. O., Li, L. T. C., Ikeda, H., Oettgen, H. F., and Old L. J., 1980, Cell surface antigens of human malignant melanoma: Definition of six antigenic systems with mouse monoclonal antibodies, *Proc. Natl. Acad. Sci. USA* **77**:6114–6118.

Duguid, J. P., and Old, D. C., 1980, Adhesive properties of *Enterobacteriaceae*, in: *Bacterial Adherence* (E. H. Beachey, ed.), pp. 186–217, Chapman & Hall, London.

Durdik, J. M., Beck, B. N., and Henney, C. S., 1980, Asialo G_{M1} and Thy 1 as cell surface markers of murine NK cells, in: *Natural Cell-Mediated Immunity against Tumors* (R. E. Herberman, ed.), pp. 37–46, Academic Press, New York.

Durocher, J. R., Payne, R. C., and Conrad, M. E., 1975, Role of sialic acid in erythrocyte survival, *Blood* **45**:11–20.

Durr, R., Shur, B., and Roth, S., 1977, Sperm-associated sialyltransferase activity, *Nature* (*London*) **265:**547–548.

Dysart, J., and Edwards, J. G., 1977, A membrane-bound haemagglutinin from cultured hamster fibroblasts (BHK 21 cells), *FEBS Lett.* **75:**96–100.

Edwards, M. S., Kasper, D. L., Jennings, H. J., Baker, C. J., and Nicholson-Weller, A., 1982, Capsular sialic acid prevents activation of the alternative complement pathway by type III, group B streptococci, *J. Immunol.* **128:**1278–1283.

Eisenbarth, G. S., Walsh, F. S., and Nirenberg, M., 1979, Monoclonal antibody to a plasma membrane antigen of neurons, *Proc. Natl. Acad. Sci. USA* **76:**4913–4917.

Emmelot, P., van Beek, W. P., and Smets, L. A., 1976, Cell surface carbohydrate and cell transformation: A general change signifying tumorigenicity, in: *Membrane Alterations as Basis of Liver Injury* (H. Popper, L. Bianchi, and W. Reulter, eds.), pp. 179–195, University Park Press, Baltimore.

Eshdat, T., Ofek, I., Yashouv-Gan, Y., Sharon, N., and Mirelman, D., 1978, Isolation of a mannose-specific lectin from *Escherichia coli* and its role in the adherence of the bacteria to epithelial cells, *Biochem. Biophys. Res. Commun.* **85:**1551–1559.

Etzler, M. E., 1974, Use of plant agglutinins in characterization of glycoprotein and glycolipids from mammalian cells, *Ann, N.Y. Acad. Sci.* **234:**260–275.

Fearon, D. T., 1978, Regulation by membrane sialic acid of β1H-dependent decay-dissociation of amplification C3 convertase of the alternative complement pathway, *Proc. Natl. Acad. Sci. USA* **75:**1971–1975.

Feizi, T., 1980, The monoclonal antibodies of cold agglutinin syndrome, *Med. Biol.* **58:**123–127.

Fichtelius, K.-E., 1969a, Introduction: Organ- and tissue-specific cell receptors, in: *Cellular Recognition* (R. T. Smith and R. A. Good, eds.), p. 69, Appleton–Century–Crofts, New York.

Fichtelius, K.-E., 1969b, Homing of lymphocytes to the gut epithelium, in: *Cellular Recognition* (R. T. Smith and R. A. Good, eds.), pp. 71–78, Appleton–Century–Crofts, New York.

Finne, J., Krusius, T., and Rauvala, H., 1977, Occurrence of disialosyl groups in glycoproteins, *Biochem. Biophys. Res. Commun.* **74:**405–410.

Fleisher, T. A., Greene, W. C., Blaese, R. M., and Waldmann, T. A., 1981, Soluble suppressor supernatants elaborated by concanavalin A-activated human mononuclear cells. II. Characterization of a soluble suppressor of B cell immunoglobulin production, *J. Immunol.* **126:**1192–1197.

Ford, W. L., and Gowans, J. L., 1969, The traffic of lymphocytes, *Semin. Hematol.* **6:**67–83.

Ford, W. L., Sedgley, M., Sparshott, S. M., and Smith, M. E., 1976, The migration of lymphocytes across specialized vascular endothelium. II. The contrasting consequences of treating lymphocytes with trypsin or neuraminidase, *Cell Tissue Kinet.* **9:**351–361.

Fox, R. A., Gregory, D. S., and Feldman, J. D., 1974, Macrophage receptors for migration inhibitory factor (MIF), migration stimulatory factor (MSF), and agglutinating factor, *J. Immunol.* **112:**1867–1872.

Freed, J. H., and Nathenson, S. G., 1977, Similarity of the carbohydrate structures of H-2 and Ia glycoproteins, *J. Immunol.* **119:**477–482.

Fredman, P., Richert, N. D., Magnani, J. L., Willingham, M. C., Pastan, I., and Ginsburg, V., 1983, A monoclonal antibody that precipitates the glycoprotein receptor for epidermal growth factor is directed against the human blood group H type 1 antigen, *J. Biol. Chem.* **258:**11206–11210.

Freimer, N. B., Ogmundsdottir, H. M., Blackwell, C. C., Sutherland, I. W., Graham, L., and Weir, D. M., 1978, The role of cell wall carbohydrates in binding of microorganisms to mouse peritoneal exudate macrophages, *Acta Pathol. Microbiol. Immunol. Scand.* **86:**53–57.

Freitas, A. A., and de Sousa, M., 1975, Control mechanisms of lymphocyte traffic: Modification of the traffic of ^{51}Cr-labeled mouse lymph node cells by treatment with plant lectins in intact and splenectomized hosts, *Eur. J. Immunol.* **5:**831–838.

Freitas, A. A., and de Sousa, M., 1976, Control mechanism of lymphocyte traffic: Altered migration of ^{51}Cr-labeled mouse lymph node cells pretreated *in vitro* with phospholipases, *Eur. J. Immunol.* **6:**703–711.

Friedman, R. M., and Kohn, L. D., 1976, Cholera toxin inhibits interferon action, *Biochem. Biophys. Res. Commun.* **70:**1078–1084.

Fukuda,, M., Fukuka, M. N., Papayannopoulou, T., and Hakomori, S.-I., 1980, Membrane differentiation in human erythroid cells: Unique profiles of cell surface glycoproteins expressed in erythroblasts *in vitro* from three ontogenic stages, *Proc. Natl. Acad. Sci. USA* **77:**3474–3478.

Gachelin, G., Buc-Caron, M.-H., Lis, H., and Sharon, N., 1976, Saccharides on teratocarcinoma cell plasma membranes: Their investigation with radioactively labelled lectins, *Biochim. Biophys. Acta* **436:**825–832.

Galili, U., and Schlesinger, M., 1974, The formation of stable E rosettes after neuraminidase treatment of either human peripheral blood lymphocytes or of sheep red blood cells, *J. Immunol.* **112:**1628–1634.

Galili, U., and Schlesinger, M., 1976, The formation of stable E-rosettes by human T lymphocytes activated in mixed lymphocyte reactions, *J. Immunol.* **117:**730–735.

Galili, U., Galili, N., Vanky, F., and Klein, E., 1978, Natural species-restricted attachment of human and murine T lymphocytes to various cells, *Proc. Natl. Acad. Sci. USA* **75:**2396–2400.

Gartner, T. K., Williams, D. C., and Minion, F. C., 1978, Thrombin-induced platelet aggregation is mediated by a platelet plasma membrane-bound lectin, *Science* **200:**1281–1283.

Gelb, L. D., and Lerner, A. M., 1965, Reovirus hemagglutination: Inhibition by N-acetyl-D-glucosamine, *Science* **147:**404–405.

Gershowitz, H., and Neal, J. V., 1970, The blood group polymorphisms: Why are they there?, in: *Blood and Tissue Antigens* (D. Aminoff, ed.), pp. 33–49, Academic Press, New York.

Gesner, B. M., 1966, Cell surface sugars as sites of cellular reactions, *Ann. N.Y. Acad. Sci.* **129:**758–766.

Gesner, B. M., and Ginsburg, V., 1964, Effect of glycosidases on the fate of transfused lymphocytes, *Proc. Natl. Acad. Sci. USA* **52:**750–755.

Gesner, B., and Thomas, L., 1966, Sialic acid binding sites: Role in hemagglutination by *Mycoplasma gallisepticum, Science* **151:**590–591.

Gesner, B. M., and Woodruff, J. J., 1969, Factors affecting the distribution of lymphocytes, in: *Cellular Recognition* (R. T. Smith and R. A. Good, eds.), pp. 79–90, Appleton–Century–Crofts, New York.

Gesner, B. M., Woodruff, J. J., and McCluskey, R. T., 1969, An autoradiographic study of the effect of neuraminidase or trypsin on transfused lymphocytes, *Am. J. Pathol.* **57:**215–230.

Gewurz, H., Shin, H. S., and Mergenhagen, S. E., 1968, Interactions of the complement system with endotoxic lipopolysaccharide: consumption of each of the six terminal complement components. *J. Exp. Med.* **128:**1049–1057.

Giblett, E. R., 1977, Some perspectives on blood group genetics and immunology, in: *Perspectives on Blood Group Immunology and Genetics* (J. F. Mohn, R. W. Plunkett, R. K. Cunningham, and R. M. Lambert, eds.), pp. 437–448, Karger, Basel.

Gill, D. M., and King, C. A., 1975, The mechanism of action of cholera toxin in pigeon erythrocyte lysates, *J. Biol. Chem.* **250:**6424–6432.

Gillette, R. W., McKenzie, G. O., and Swanson, M. H., 1973, Effect of concanavalin A on the homing of labeled T lymphocytes, *J. Immunol.* **111:**1902–1905.

Ginsburg, V., and Neufeld, E. F., 1969, Complex heterosaccharides of animals, *Annu. Rev. Biochem.* **38:**371–388.

Glaudemans, C. P. J., Das, M. K., and Vrana, M., 1978, Homogeneous murine immunoglobulins with anti carbohydrate specificity, *Methods Enzymol.* **50:**316–323.

Glick, M. C., 1974, Chemical components of surface membranes related to biological properties, in: *Biology and Chemistry of Eukaryotic Cell Surfaces*, Vol. 7 (E. Y. C. Lee and E. E. Smith, eds.), pp. 213–240, Academic Press, New York.

Glickman, R. M., And Bouhours, J. F., 1976, Characterization, distribution and biosynthesis of the major ganglioside of rat intestinal mucosa, *Biochim. Biophys. Acta* **424:**17–25.

Goldman, R., 1974, Effect of concanavalin A on phagocytosis by macrophages, *FEBS Lett.* **46:**209–213.

Goldstein, I. J., and Hayes, C. E., 1978, The lectins: Carbohydrate-binding proteins of plants and animals, *Adv. Carbohydr. Chem. Biochem.* **35:**127–128.

Goldstein, I. J., Hughes, R. C., Monsigny, M., Osawa, T., and Sharon N., 1980, What should be called a lectin?, *Nature (London)* **285:**66.

Goldwasser, E., Kung, C. K.-H., and Eliason, J., 1974, On the mechanism of erythropoietin-induced differentiation, *J. Biol. Chem.* **249:**4202–4206.

Gooi, H. C., Feizi, T., Kapadia, A., Knowles, B. B., Solter, D., and Evans, M. J., 1981, Stage-specific embryonic antigen involves alpha 1-3 fucosylated type 2 blood group chains, *Nature (London)* **292:**156–158.

Gooi, H. C., Thorpe, S. J., Hounsell, E. F., Rumpold, H., Kraft, D., Foster, O., and Feizi, T., 1983, Marker of peripheral blood granulocytes and monocytes of man recognized by two monoclonal antibodies VEP8 and VEP9 involves the trisaccharide 3-fucosyl-N-acetyllactosamine, *Eur. J. Immunol.* **13:**306–312.

Gorczynski, R. M., 1984, Macrophages, self–non-self discrimination and cell surface carbohydrate receptors in the immune system (submitted).

Gotschlich, E. C., and Edelman, G. M., 1967, Binding properties and specificity of C-reactive protein, *Proc. Natl. Acad. Sci. USA* **57:**706–712.

Gottschalk, A., 1966, Interactions between glycoproteins and viruses, in: *The Amino Sugars*, Vol. IIB (E. A. Balazs and R. W. Jeanloz, eds.), pp. 337–359, Academic Press, New York.

Grabel, L. B., Rosen, S. D., and Martin, G. R., 1979, Teratocarcinoma stem cells have a cell surface carbohydrate-binding component implicated in cell–cell adhesion, *Cell* **17:**477–484.

Greene, W. C., Fleisher, T. A., and Waldmann, T. A., 1981, Soluble suppressor supernatants elaborated by concanavalin A-activated human mononuclear cells. I. Characterization of a soluble suppressor of T cell proliferation, *J. Immunol.* **126:**1185–1191.

Gregson, N. A., and Hammer, C. T., 1982, Some immunological properties of antisera raised against the trisialoganglioside GT$_{1b}$, *Mol. Immunol.* **19:**543–550.

Grollman, E. F., Kobata, A., and Ginsburg, V., 1970, Enzymatic basis of blood types in man, *Ann. N.Y. Acad. Sci.* **169:**153–160.

Grollman, E. F., Lee, G., Ramos, S., Lazo, P. S., Kaback, R., Friedman, R. M., and Kohn, L. D., 1978, Relationships of the structure and function of the interferon receptor to hormone receptors and establishment of the antiviral state, *Cancer Res.* **38:**4172–4185.

Gupta, S., Good, R. A., and Siegal, F. P., 1976, Rosette-formation with mouse erythrocytes. II. A marker for human B and non-T lymphocytes, *Clin. Exp. Immunol.* **25:**319–327.

Habu, S., Fukui, H., Shimamura, K., Kasai, M., Nagai, Y., Okumura, K. O., and Tamaoki, N., 1981, *In vivo* effects of anti-asialo G_{M1}. I. Reduction of NK activity and enhancement of transplanted tumor growth in nude mice, *J. Immunol.* **127**:34–37.

Hakomori, S.-I., 1973, Glycolipids of tumor cell membrane, *Adv. Cancer Res.* **18**:265–312.

Hakomori, S.-I., 1975, Structures and organization of cell surface, glycolipids dependency on cell growth and malignant transformation, *Biochim. Biophys. Acta* **417**:55–89.

Hakomori, S.-I., 1981, Blood group ABH and Ii antigens of human erythrocytes: Chemistry, polymorphism, and their developmental change, *Semin. Hematol.* **18**:39–62.

Hakomori, S.-I., Koscielak, J., Bloch, K. J., and Jeanloz, R. W., 1967, Immunologic relationship between blood group substances and a fucose-containing glycolipid of human adenocarcinoma, *J. Immunol.* **98**:31–38.

Hakomori, S.-I., Wang, S. M., and Young, W. W., Jr., 1977a, Isoantigenic expression of Forssman glycolipid in human gastric and colonic mucosa: Its possible identity with "A-like antigen" in human cancer, *Proc. Natl. Acad. Sci. USA* **74**:3032–3027.

Hakomori, S.-I., Watanabe, K., and Laine, R. A., 1977b, Glycosphingolipids with blood group A, H, and I activity: Their status in group A_1 and A_2 erythrocytes and their changes associated with ontogeny and oncogeny, in: *Human Blood Groups* (J. F. Mohn, R. W. Plunkett, R. K. Cunningham, and R. M. Lambert, eds.), pp. 150–163, Karger, Basel.

Hall, J. L., and Rowlands, D. T., Jr., 1974, Heterogeneity of lobster agglutinins. II. Specificity of agglutinin–erythrocyte binding, *Biochemistry* **13**:828–832.

Hall, J. L., Rowlands, D. T., Jr., and Nilson, U. R., 1972, Complement-unlike hemolytic activity in lobster hemolymph, *J. Immunol.* **109**:816–823.

Hankins, C. N., and Shannon, L. M., 1978, The physical and enzymatic properties of a phytohemagglutinin from mung beans, *J. Biol. Chem.* **253**:7791–7797.

Hansen, T. H., and Sachs, D. H., 1978, Isolation and antigenic characterization of the product of a third polymorphic H-2 locus, H-2L, *J. Immunol.* **121**:1469–1472.

Hart, G. W., 1982, The role of asparagine-linked oligosaccharides in cellular recognition by thymic lymphocytes, *J. Biol. Chem.* **257**:151–158.

Haywood, A. M., 1974, Characteristics of Sendai virus receptors in a model membrane, *J. Mol. Biol.* **83**:427–436.

Heath, E. C., 1971, Complex polysaccharides, *Annu. Rev. Biochem.* **40**:29–56.

Hellstrom, U., Dillner, M.-L., Hammarstrom, S., and Perlmann, P., 1976, The interaction of mitogenic and nonmitogenic lectins with T lymphocytes: Association of cellular receptor sites, *Scand. J. Immunol.* **5**:45–54.

Henkart, P., Humphreys, S., and Humphreys, T., 1973, Characterization of sponge aggregation factor: A unique proteoglycan complex, *Biochemistry* **12**:3045–3050.

Herberman, R. B., and Holden, H. T., 1978, Natural cell-mediated immunity, *Adv. Cancer Res.* **27**:305–377.

Hickman, S., and Neufeld, E. F., 1972, A hypothesis for I-cell disease: Defective hydrolases that do not enter lysosome, *Biochem. Biophys. Res. Commun.* **49**:992–999.

Hickman, S., Shapiro, L. J., and Neufeld, E. F., 1974, A recognition marker required for uptake of lysosomal enzyme by cultured fibroblasts, *Biochem. Biophys. Res. Commun.* **57**:55–61.

Hieber, V., Distler, J., Myerowitz, R., Schmickel, R. D., and Jourdian, G. W., 1976, The role of glycosidically bound mannose in the assimilation of β-galactosidase by generalized gangliosidosis fibroblasts, *Biochem. Biophys. Res. Commun.* **73**:710–717.

Higgins, T. J., and Parish, C. R., 1980, Extraction of the carbohydrate-defined class of Ia antigens from murine spleen cells and serum, *Mol. Immunol.* **17**:1065–1073.

Higgins, T. J., Sabatino, A., Remold, H., and David, J., 1976, Enhancement of migration inhibitory macrophages (Mφ) with Mφ glycolipids (GL), *Fed. Proc.* Abstr. **35**:389.

Higgins, T. J., Parish, C. R., Hogarth, P. M., McKenzie, I. F. C., and Hammerling, G. J., 1980a, Demonstration of carbohydrate- and protein-determined Ia antigens by monoclonal antibodies, *Immunogenetics* **11**:467–482.

Higgins, T. J., Liu, D. Y., Remold, H. G., and David, J. R., 1980b, Further characterization of the putative glycolipid receptor for MIF: Role of fucose associated with an acidic glycolipid, *Biochem. Biophys. Res. Commun.* **93**:1259–1265.

Hill, M., Bechet, J. J., and d'Albis, A., 1979, Disuccinimidyl esters as bifunctional crosslinking reagents for proteins: Assays with myosin, *FEBS Lett.* **102**:282–286.

Hirano, T., Hashimoto, H., Shiokawa, Y., Iwamori, M., Nagai, Y., Kasai, M., Ochiai, Y., and Okumura, K. O., 1980, Antiglycolipid autoantibody detected in the sera from systemic lupus erythematosus patients, *J. Clin. Invest.* **66**:1437–1440.

Hirata, A. A., McIntire, F. C., Terasaki, P. I., and Mittal, K. K., 1973, Cross-reactions between human transplantation antigens and bacterial lipopolysaccharides, *Transplantation* **15**:441–445.

Hirsch, H. E., and Parks, M. E., 1976, Serological reactions against glycolipid-sensitized liposomes in multiple sclerosis, *Nature (London)* **264**:785–787.

Hirsch, R. L., Winkelstein, J. A., and Griffin, D. E., 1980, The role of complement in viral infections: Activation of the classical and alternative complement pathway by sindbis virus, *J. Immunol.* **124**:2507–2510.

Hirsch, R. L., Griffin, D. E., and Winkelstein, J. A., 1981, Host modification of sindbis virus sialic acid content influences alternative complement pathway activation and virus clearance, *J. Immunol.* **127**:1740–1743.

Hoessli, D., Bron, C., and Pink, R. L., 1980, T-lymphocyte differentiation is accompanied by increase in sialic acid content of Thy-1 antigen, *Nature (London)* **283**:576–578.

Hoflack, B., Cacan, R., Montreuil, J., and Verbert, A., 1979, Detection of ectosialyltransferase activity using whole cells: Correction of misleading results due to the release of intracellular CMP-N-acetylneuraminic acid, *Biochim. Biophys. Acta.* **568**:348–356.

Holmgren, J., Mansson, J.-E., and Svennerholm, L., 1974, Tissue receptor for cholera exotoxin: structural requirements of G_{M1} ganglioside in toxin binding and inactivation. *Med. Biol.* **52**:229–233.

Homma, R., 1968, Reaction of Japanese encephalitis virus with mannan, *Acta Virol.* **12**:385–396.

Horowitz, M. I., 1978, Immunological aspects, in: *The Glycoconjugates,* Vol. II (M. I. Horowitz and W. Pigman, eds.), pp. 387–436, Academic Press, New York.

Howe, C., and Lee, L. T., 1972, Virus–erythrocyte interactions, *Adv. Virus Res.* **17**:1–50.

Howie, S., Parish, C. R., David, C. S., McKenzie, I. F. C., Maurer, P. H., and Feldman, M., 1979, Serological analysis of antigen-specific helper factors specific for poly-L(Tyr, Glu)-poly-DLAla–poly-LLys((T,G)-A–L) and $LGlu^{60}$-$LAla^{30}$-$LTyr^{10}$(GAT), *Eur. J. Immunol.* **9**:501–506.

Huang, R. T. C., 1978, Cell adhesion mediated by glycolipids, *Nature (London)* **276**:624–626.

Hughes, R., and Gardas, A., 1976, Phenotypic reversion of ricin-resistant hamster fibroblasts to a sensitive state after coating with glycolipid receptors, *Nature (London)* **264**:63–66.

Hughes, R. C., Palmer, P. D., and Sanford, B. H., 1973, Factors involved in the cytotoxicity of normal guinea pig serum for cells of murine tumor TA3 sublines treated with neuraminidase, *J. Immunol.* **111**:1071–1080.

Ihara, I., Harada, Y., Ihara, S., and Kawakami, M., 1982, A new complement-dependent bactericidal factor found in nonimmune mouse sera: Specific binding to polysaccharide of Ra chemotype *Salmonella, J. Immunol.* **128**:1256–1260.

Irle, C., Piguet, P.-F., and Vassalli, P., 1978, In vitro maturation of immature thymocytes into immunocompetent T cells in the absence of direct thymic influence, *J. Exp. Med.* **148**:32–45.

Itoh, N., and Yamashina, I., 1975, Interaction of antimannan with glycopeptides, *Biochem. Biophys. Res. Commun.* **67**:840–845.

Jacob, F., 1979, Cell surface and early stages of mouse embryogenesis, *Curr. Top. Dev. Biol.* **13**:117–137.

Jamieson, G. A., Urban, C. L., and Barber, A. J., 1971, Enzymatic basis for platelet: collagen adhesion as the primary step in haemostasis, *Nature New Biol.* **234**:5–7.

Ji, T. H., and Nicolson, G. L., 1974, Lectin binding and perturbation of the outer surface of the cell membrane induces a transmembrane organizational alteration at the inner surface, *Proc. Natl. Acad. Sci. USA* **71**:2212–2216.

Joffe, S., Rapport, M. M., and Graf, L., 1963, Identification of an organ specific lipid hapten in brain, *Nature (London)* **197**:60–62.

Johannsen, R., Sedlacek, H. H., and Schmidtberger, R., 1979, Characteristics of cytotoxic antibodies against neuraminidase-treated lymphocytes in man, *J. Natl. Cancer Inst.* **62**:733–742.

Johnson, B. J., Kucich, U. N., and Maurelli, A. T., 1976, Studies on the antigenic determinants of the Thy-1.2 alloantigen as expressed by the murine lymphoblastoid line S-49.1 TB 2.3, *J. Immunol.* **116**:1669–1672.

Jolley, M. E., Glaudemans, C. P. J., Rudikoff, S., and Potter, M., 1974, Structural requirements for the binding of derivatives of D-galactose to two homogeneous murine immunoglobulins, *Biochemistry* **13**:3179–3184.

Kaladas, P. M., Kabat, E. A., Kimura, A., and Ersson, B., 1981, The specificity of the combining site of the lectin from *Vicia villosa* seeds which reacts with cytotoxic T-lymphoblasts, *Mol. Immunol.* **18**:969–977.

Kamenov, B., Kieran, M. W., Leigh, J. B., Greenberg, A. H., and Longnecker, B. M., 1983, A new model for leukemia-lymphoma metastasis. I. Differential growth and rejection of murine lymphoid-leukemia cell lines in the bone marrow, *Proceedings, Symposium on Metastasis and Invasion,* Houston.

Kano, K., and Milgrom, F., 1970, Antigens shared by human tissues and erythrocytes, in: *Histocompatibility Testing* (P. I. Terasaki, ed.), pp. 443–452, Williams & Wilkins, Baltimore.

Kaplan, A., Achord, D. T., and Sly, W. S., 1977, Phosphohexosyl components of a lysosomal enzyme are recognized by pinocytosis receptors on human fibroblasts, *Proc. Natl. Acad. Sci. USA* **74**:2026–2030.

Karol, R. A., Kundu, S. K., Suzuki, A., and Marcus, D. M., 1980, Immunological reactivity and concentration of glycosphingolipids in adult and umbilical cord erythrocytes, *Blood Transfus. Immunohaematol.* **23**:589–598.

Karsenti, E., and Arvrameas, S., 1973, The use of concanavalin A in the study of the dynamics of lymphocyte membrane glycans, *FEBS Lett.* **32**:238–242.

Kasai, M., Iwamori, M., Nagai, Y., Okumura, K., and Tada, T., 1980, A glycolipid on the surface of mouse natural killer cells, *Eur. J. Immunol.* **10**:175–180.

Kasper, D. L., Winkelhake, J. L., Zollinger, W. D., Brandt, B. L., and Artenstein, M. S., 1973, Immunochemical similarity between polysaccharide antigens of *Escherichia coli* O7:K1(L):NM and group B *Neisseria meningitidis*, *J. Immunol.* **110**:262–268.

Kato, K. P., Wang, T. J., and Esselman, W. J., 1979, Radiolabeling and isolation of Thy-1 active glycolipids from murine brain and lymphoma cell lines, *J. Immunol.* **123**:1977–1984.

Kaufmann, S. H. E., Schauer, R., and Hahn, H., 1981, Carbohydrate surface constituents of T cells mediating delayed-type hypersensitivity that control entry into sites of antigen deposition, *Immunobiology* **160**:184–195.

Kawasaki, T., and Ashwell, G., 1977, Isolation and characterization of an avian hepatic binding protein specific for N-acetylglucosamine-terminated glycoproteins, *J. Biol. Chem.* **252**:6536–6543.

Kawasaki, T., Etoh, R., and Yamashina, I., 1978, Isolation and characterization of a mannan-binding protein from rabbit liver, *Biochem. Biophys. Res. Commun.* **81:**1018–1024.

Kazatchkine, M. D., Fearon, D. T., and Austen, K. F., 1979, Human alternative complement pathway: Membrane-associated sialic acid regulates the competition between B and β1H for cell-bound C3b, *J. Immunol.* **122:**75–81.

Kieda, C. M. T., Bowles, D. J., Ravid, A., and Sharon, N., 1978, Lectins in lymphocyte membranes, *FEBS Lett.* **94:**391–396.

Kieda, C. M. T., Roche, A.-C., Delmotte, F., and Monsigny, M., 1979, Lymphocyte membrane lectins: Direct visualization by the use of fluoresceinyl–glycosylated cytochemical markers, *FEBS Lett.* **99:**329–332.

Kim, Y. S., Whitehead, J. S., Siddiqui, B., and Tsao, D., 1978, Glycoconjugate alterations in malignant and inflammatory disease of the colon, in: *Glycoproteins and Glycolipids in Disease Processes* (E. F. Walborg, Jr., ed.), pp. 295–310, American Chemical Society, Washington, D.C.

Kimura, A. K., and Wigzell, H., 1977, Cell surface glycoproteins of murine cytotoxic T lymphocytes. I. T 145, a new cell surface glycoprotein selectively expressed on Ly1⁻ 2⁺ cytotoxic T lymphocytes, *J. Exp. Med.* **147:**1418–1434.

Kimura, A. K., Wigzell, H., Holmquist, G., Ersson, B., and Carlsson, P., 1979a, Selective affinity fractionation of murine cytotoxic T lymphocytes (CTL), *J. Exp. Med.* **149:**473–484.

Kimura, A., Orn, A., Holmquist, G., Wigzell, H., and Ersson, B., 1979b, Unique lectin-binding characteristics of cytotoxic T lymphocytes allowing their distinction from natural killer cells and "K" cells, *Eur. J. Immunol.* **9:**575–578.

Klein, J., 1975, *Biology of the Mouse Histocompatibility-2 Complex,* Springer-Verlag, Berlin.

Klein, J., and Hammerberg, C., 1977, The control of differentiation by the T complex, *Immunol. Rev.* **33:**70–104.

Klempner, M. S., and Rocklin, R. E., 1982, Specific binding of leukocyte inhibitory factor to neutrophil plasma membranes, *J. Immunol.* **128:**2040–2043.

Knowles, R. W., Bai, Y., Daniels, G. L., and Watkins, W., 1982, Monoclonal antitype 2 H: An antibody detecting a precursor of the A and B blood group antigens, *J. Immunogenet.* **9:**69–76.

Kobayashi, Y., Sawada, J.-I., and Osawa, T., 1978, Isolation and characterization of an inhibitory glycopeptide against guinea pig lymphotoxin from the surface of L cells, *Immunochemistry* **15:**61–66.

Koprowski, H., Herlyn, M., Steplewski, Z., and Sears, H. F., 1981, Specific antigen in serum of patients with colon carcinoma, *Science* **212:**53–54.

Kornfeld, R., 1978, Structure of the oligosaccharides of three glycopeptides from calf thymocyte plasma membranes, *Biochemistry* **17:**1415–1423.

Kornfeld, S., and Kornfeld, R., 1978, Use of lectins in the study of mammalian glycoproteins, in: *The Glycoconjugates,* Vol. II (M. I. Horowitz and W. Pigman, eds.), pp. 434–449, Academic Press, New York.

Koskimies, S., 1980, Human lymphoblastoid cell line producing specific antibody against Rh-antigen D, *Scand. J. Immunol.* **11:**73–77.

Krishnamurti, C., Besancon, F., Justesen, J., Poulsen, K., and Ankel, H., 1982, Inhibition of mouse fibroblast interferon by gangliosides, *Eur. J. Biochem.* **124:**1–6.

Kristiansen, T., 1974, Studies on blood group substances. V. Blood group substance A coupled to agarose as an immunosorbent, *Biochim. Biophys. Acta* **263:**567–574.

Kruisbeek, A. M., and Astaldi, G. C. B., 1979, Distinct effects of thymic epithelial culture supernatants on T cell properties of mouse thymocytes separated by the use of peanut agglutinin, *J. Immunol.* **123:**984–991.

Kruisbeek, A. M., Zijlstra, J. J., and Krose, T. M., 1980, Distinct effects of T cell growth factors and thymic epithelial factors on the generation of cytotoxic T lymphocytes by thymocyte subpopulations, *J. Immunol.* **125:**995–1002.

Kuhns, W. J., 1974, Sponge aggregation: A model for studies on cell–cell interactions, *Ann. N.Y. Acad. Sci.* **234:**58–74.

Kuhns, W. J., and Bramson, S., 1968, Variable behavior of blood group H on Hela cell populations synchronized with thymidine, *Nature (London)* **219:**938–939.

Kundu, S. K., Marcus, D. M., and Veh, R. W., 1980, Preparation and properties of antibodies to G_{D3} and G_{M1} gangliosides, *J. Neurochem.* **34:**184–188.

Kurt, K. A., Shur, B. D., and Lindquist, R. R., 1981, Cytolytic T lymphocyte galactosyltransferase activity, *Fed. Proc.* **40:**1150.

Kurth, R., 1976, Surface alterations in cells infected by avian leukosis virus, *Biomembranes* **8:**167–233.

Lai, M. M. C., and Duesberg, P. H., 1972, Differences between the envelope glycoproteins and glycopeptides of avian tumor viruses released from transformed and nontransformed cells, *Virology* **50:**359–372.

LaMont, J. T., Perrotto, J. L., Weiser, M. M., and Isselbacher, K. J., 1974, Cell surface galactosyltransferase and lectin agglutination of thymus and spleen lymphocytes, *Proc. Natl. Acad. Sci. USA* **71:**3726–3730.

Law, S. K., and Levine, R. P., 1977, Interaction between the third complement protein and cell surface macromolecules, *Proc. Natl. Acad. Sci. USA* **74:**2701–2705.

Law, S. K., Lichtenberg, N. A., and Levine, R. P., 1979, Evidence for an ester linkage between the labile binding site of C3b and receptive surfaces, *J. Immunol.* **123:**1388–1394.

Layton, J. E., 1980, Anti-carbohydrate activity of T cell-reactive chicken anti-mouse immunoglobulin antibodies, *J. Immunol.* **125:**1993–1997.

Lee, G., Aloj, S. M., Brady, R. O., and Kohn, L. D., 1976, The structure and function of glycoprotein hormone receptors: Ganglioside interactions with human chorionic gonadotropin, *Biochem. Biophys. Res. Commun.* **73:**370–377.

Lee, G., Aloj, S. M., and Kohn, L. D., 1977, The structure and function of glycoprotein hormone receptors: Ganglioside interactions with luteinizing hormone, *Biochem. Biophys. Res. Commun.* **77:**434–441.

Leibovich, S. J., and Knyszynski, A., 1980, *In vitro* recognition of "old red" blood cells by macrophages from syngeneic mice: Characteristics of the macrophage–red blood cell interaction, *J. Reticuloendothel. Soc.* **27:**411–419.

Lemieux, R. V., Bundle, D. R., and Baker, D. A., 1975, The properties of a "synthetic" antigen related to the human blood-group Lewis a, *J. Am. Chem. Soc.* **97:**4076–4083.

Lerner, A. M., Bailey, E. J., and Tillotson, J. R., 1966, Enterovirus hemagglutination: Inhibition by several enzymes and sugars, *J. Immunol.* **95:**1111–1115.

Leu, R. W., Eddleston, A. L. W. F., Hadden, J. W., and Good, R. A., 1972, Mechanism of action of migration inhibitory factor (MIF). I. Evidence for a receptor for MIF present on the peritoneal macrophage but not on the alveolar macrophage, *J. Exp. Med.* **136:**589–603.

Levinson, B., Pepper, D., and Belyavin, G., 1969, Substituted sialic acid prosthetic groups as determinants of viral hemagglutination, *J. Virol.* **3:**477–483.

Liew, F. Y., Sia, D. Y., Parish, C. R., and McKenzie, I. F. C., 1980, *MHC*-coded determinants on antigen specific suppressor factor for delayed-type hypersensitivity and surface phenotype of cells producing the factor, *Eur. J. Immunol.* **10:**305–309.

Lilien, J., Hermolin, J., and Lipke, P., 1978, Molecular interactions in specific cell adhesion, in: *Specificity of Embryological Interactions*, Vol. IV (D. R. Garrod, ed.), pp. 132–155, Chapman & Hall, London.

Lipsick, J. S., Beyer, E. C., Barondes, S. H., and Kaplan, N. O., 1980, Lectins from chicken tissues are mitogenic for Thy-1 negative murine spleen cells, *Biochem. Biophys. Res. Commun.* **97:**56–61.

Lis, H., and Sharon, N., 1973, The biochemistry of plant lectins (phytohemagglutinins), *Annu. Rev. Biochem.* **42:**541–565.

Lis, H., and Sharon, N., 1977, Lectins: Their chemistry and applications to immunology, in: *The Antigens,* Vol. IV (M. Sela, ed.), Academic Press, New York, pp. 429–529.

Liu, D. Y., Higgins, T. J., Petschek, K. D., Remold, H. G., and David, J. R., 1978, Fucose and sialic acid are required on the macrophage (Mϕ) for its response to migration inhibitory factor (MIF) and on Mϕ glycolipids (GSL) for their ability to enhance the Mϕ response to MIF, *Fed. Proc.* Abstr. **37:**1400.

Liu, D. Y., Petschek, K. D., Remold, H. G., and David, J. R., 1980, Role of sialic acid in the macrophage glycolipid receptor or MIF, *J. Immunol.* **124:**2042–2047.

Liu, D. Y., Petschek, K. D., Remold, H. G., and David, J. R., 1982, Isolation of a guinea pig macrophage glycolipid with the properties of the putative migration inhibitory factor receptor, *J. Biol. Chem.* **257:**159–162.

Lloyd, C. W., and Cook, G. M. W., 1974, On the mechanism of the increased aggregation by neuraminidase of 16C malignant rat dermal fibroblasts *in vitro*, *J. Cell Sci.* **15:**575–590.

Lloyd, K. O., and Kabat, E. A., 1968, Immunochemical studies on blood groups. XLI. Proposed structures for the carbohydrate portions of blood group A, B, H, Lewis[a], and Lewis[b] substances, *Proc. Natl. Acad. Sci. USA* **61:**1470–1477.

London, J., Berrih, S., and Bach, J.-F., 1978, Peanut agglutinin. I. A new tool for studying T lymphocyte subpopulations, *J. Immunol.* **121:**438–443.

Loos, M., Bitter-Suermann, D., and Dierich, M., 1974, Interaction of the first (C1), the second (C2) and the fourth (C4) component of complement with different preparations of bacterial lipopolysaccharides and with lipid A, *J. Immunol.* **112:**935–940.

Lotan, R., and Nicolson, G. L., 1978, Membrane glycoproteins: Dynamics and affinity isolation, in: *Glycoproteins and Glycolipids in Disease Processes* (E. F. Walborg, Jr., ed.), pp. 256–271, American Chemical Society, Washington, D.C.

Lotan, R., and Nicolson, G. L., 1979, Purification of cell membrane glycoproteins by lectin affinity chromatography, *Biochim. Biophys. Acta* **559:**329–376.

Lotan, R., Skutelsky, E., Danon, D., and Sharon, N., 1975, The purification, composition, and specificity of the anti-T lectin from peanut (*Arachis hypogaea*), *J. Biol. Chem.* **250:**8518–8523.

Lunney, J., and Ashwell, G., 1976, Hepatic receptor of avian origin capable of binding specifically modified glycoproteins, *Proc. Natl. Acad. Sci. USA* **73:**341–343.

MacDermott, R. P., Kienker, L. J., and Muchmore, A. W., 1980, Inhibition of spontaneous but not antibody dependent cell mediated cytotoxicity by simple sugars, *Fed. Proc.* **39:**4893.

MacDonald, H. R., Mach, J. P., Schreyer, M., Zaech, P., and Cerottini, J. C., 1981, Flow cytometric analysis of the binding of *Vicia villosa* lectin to T lymphoblasts: Lack of correlation with cytolytic function, *J. Immunol.* **126:**883–886.

McDonough, J., and Lilien, J., 1977, The turnover of a tissue specific cell surface ligand which inhibits lectin induced capping, *J. Supramol. Struct.* **7:**409–418.

McIntire, F. C., Vatter, A. E., Baros, J., and Arnold, J., 1978, Mechanism of coaggregation between *Actinomyces viscosus* T14V and *Streptococcus sanguis* 34, *Infect. Immun.* **21:**978–988.

McKenzie, I. F. C., and Parish, C. R., 1976, Secretion of Ia antigens by a subpopulation of T cells which are Ly-1$^+$, Ly-2$^-$, and Ia$^-$, *J. Exp. Med.* **144:**847–851.

McKenzie, I. F. C., Clarke, A., and Parish, C. R., 1977, Ia antigenic specificities are oligosaccharide in nature: Hapten-inhibition studies, *J. Exp. Med.* **145:**1039–1053.

McLean, R. J., and Bosmann, H. B., 1975, Cell–cell interactions: Enhancement of glycosyl transferase ectoenzyme systems during *Chlamydomonas* gametic contact, *Proc. Natl. Acad. Sci. USA* **72:**310–313.

Maddox, D. E., Shibata, S., and Goldstein, I. J., 1982, Stimulated macrophages express a new glycoprotein receptor reactive with *Griffonia simplicifolia* I-B$_4$ isolectin, *Proc. Natl. Acad. Sci. USA* **79:**166–170.

Magnani, J. L., Brockhaus, M., Smith, D. F., Steplewski, Z., and Koprowski, H., 1981, A monosialoganglioside is a monoclonal antibody-defined antigen of colon carcinoma, *Science* **212:**55–56.

Magnani, J. L., Nilsson, B., Brockhaus, M., Zopf, D., Steplewski, Z., Koprowski, H., and Ginsburg, V., 1982, A monoclonal antibody-defined antigen associated with gastrointestinal cancer is a ganglioside containing sialylated lacto-N-fucopentaose II, *J. Biol. Chem.* **257:**14365–14369.

Marchase, R. B., 1977, Biochemical investigations of retinotectal adhesive specificity, *J. Cell Biol.* **75:**237–257.

Marchase, R. B., Pierce, M., and Roth, S., 1977, Complementarity between the ganglioside GM$_2$ and the enzyme GM$_1$ synthetase is a possible recognition mechanism in the chick retino-tectal projection, *J. Supramol. Struct. Suppl.* **1:**32.

Marcus, D. M., Kundu, S. K., and Suzuki, A., 1981, The P blood group system: Recent progress in immunochemistry and genetics, *Semin. Hematol.* **18:**63–71.

Marcus, R. L., Shin, H. S., and Mayer, M. M., 1971, An alternate complement pathway: C-3 cleaving activity, not due to $\overline{C4, 2a}$, on endotoxic lipopolysaccharide after treatment with guinea pig serum; relation to properdin, *Proc. Natl. Acad. Sci. USA* **68:**1351–1354.

Marsh, W. L., 1961, Anti-i: A cold antibody defining the Ii relationship in human red cells, *Br. J. Haematol.* **7:**200–209.

Marticorena, P., Artzt, K., and Bennett, D., 1978, Relationship of F9 antigen and genes of the T/t complex, *Immunogenetics* **7:**337–347.

Martin, G. R., Grabel, L. B., and Rosen, S. D., 1980, Use of teratocarcinoma cells as a model systems for studying the cell surface during early mammalian development. in: *The Cell Surface: Mediator of Developmental Processes* (S. Subtelny and N. K. Wessells, eds.), pp. 325–348, Academic Press, New York.

Matsuuchi, L., Wims, L. A., and Morrison, S. L., 1981a, A variant of the dextran-binding mouse plasmacytoma J558 with altered glycosylation of its heavy chain and decreased reactivity with polymeric dextran, *Biochemistry* **20:**4827–4835.

Matsuuchi, L., Sharon, J., and Morrison, S. L., 1981b, An analysis of heavy chain glycopeptides of hybridoma antibodies: Correlation between antibody specificity and sialic acid content, *J. Immunol.* **127:**2188–2190.

Milewicz, C., Miller, H. C., and Esselman, W. J., 1976, Membrane expression of Thy-1.2 and G$_{M1}$ ganglioside on differentiating T lymphocytes, *J. Immunol.* **117:**1774–1780.

Mittal, K.-K., Terasaki, P. I., Springer, G. F., Desai, P. R., McIntire, F. C., and Hirata, A. A., 1973, Inhibition of anti-HL-A alloantisera by glycoproteins, polysaccharides, and lipopolysaccharides from diverse sources, *Transplant. Proc.* **5:**499–506.

Mizoguchi, A., Mizuochi, T., and Kobata, A., 1982, Structures of the carbohydrate moieties of secretory component purified from human milk, *J. Biol. Chem.* **257:**9612–9621.

Mizuno, Y., Kozutsumi, Y., Kawasaki, T., and Yamashina, I., 1981, Isolation and characterization of a mannan-binding protein from rat liver, *J. Biol. Chem.* **256:**4247–4252.

Mizuochi, T., Yamashita, K., Fukikawa, K., Kisiel, W., and Kobata, A., 1979, The carbohydrate of bovine prothrombin. Occurrence of Gal β 1-3 GlcNAc grouping in asparagine-linked sugar chains. *J. Biol. Chem.* **254:**6419–6425.

Momoi, M., Kennet, R. H., and Glick, M. C., 1980, A surface glycoprotein as a human neuroblastoma antigen detected by monoclonal antibodies, *Prog. Cancer Res. Ther.* **12:**177–181.

Morgan, W. T. J., 1970, Molecular aspects of human blood-group specificity, *Ann. N.Y. Acad. Sci.* **169:**118–133.

Morishima, Y., Ogata, S.-I., Collins, N. H., Dupont, B., and Lloyd, K. O., 1982, Carbohydrate differences in human high molecular weight antigens of B- and T-cell lines, *Immunogenetics* **15:**529–535.

Morita, A., Tsao, D., and Kim, Y. S., 1980, Identification of cholera toxin binding glycoproteins in rat intestinal microvillus membranes, *J. Biol. Chem.* **255:**2549–2553.

Morell, A. G., Gregoriadis, G., Scheinberg, I. H., Hickman, J., and Ashwell, G., 1971, The role of sialic acid in determining the survival of glycoproteins in circulation, *J. Biol. Chem.* **246:**1461–1467.

Morrison, D. C., and Kline, L. F., 1977, Activation of the classical and properdin pathways of complement by bacterial lipopolysaccharides (LPS), *J. Immunol.* **118:**362–368.

Muchmore, A. V., and Blaese, R. M., 1980, Evidence that monocyte mediated cellular recognition phenomena are mediated by receptors with specificity for simple oligosaccharides, in: *Macrophage Recognition in Immunity* (E. R. Unanue and A. S. Rosenthal, eds.), pp. 505–517, Academic Press, New York.

Muchmore, A. V., Decker, J. M., and Blaese, R. M., 1980, Evidence that specific oligosaccharides block early events necessary for the expression of antigen-specific proliferation by human lymphocytes, *J. Immunol.* **125:**1306–1311.

Mullin, B. R., Fishman, P. H., Lee, G., Aloj, S. M., Ledley, F. D., Winand, R. J., Kohn, L. D., and Brady, R. O., 1976a, Thyrotropin–ganglioside interactions and their relationship to the structure and function of thyrotropin receptors, *Proc. Natl. Acad. Sci. USA* **73:**842–846.

Mullin, B. R., Aloj, S. M., Fishman, P. H., Lee, G., Kohn, L. D., and Brady, R. O., 1976b, Cholera toxin interaction with thyrotropin receptors on thyroid plasma membranes, *Proc. Natl. Acad. Sci. USA* **73:**1679–1683.

Muramatsu, T., and Nathenson, S. G., 1970, Studies on the carbohydrate portion of membrane-located mouse H-2 alloantigens, *Biochemistry* **9:**4875–4883.

Naiki, M., and Marcus, D. M., 1977, Binding of N-acetylgalactosamine-containing compounds by a human IgM paraprotein, *J. Immunol.* **119:**537–539.

Naiki, M., Marcus, D. M., and Ledeen, R., 1974, Properties of antisera to ganglioside G_{M1} and asialo-G_{M1}, *J. Immunol.* **113:**84–93.

Naiki, M., Ikuta, K., Fujii, Y., and Kato, S., 1981, Is N-glycolylneuraminic acid a tumor associated antigenic determinant in avian as well as human?, in: *Glycoconjugates* (T. Yamakawa, T. Osawa, and S. Handa, eds.), pp. 185–186, Japan Societies Scientific Press, Tokyo.

Nair, R. M. G., Ponce, B., and Fudenberg, H. H., 1978, Interactions of radiolabeled tuftsin with human neutrophils, *Immunochemistry* **15:**901–907.

Nakahara, K., Ohashi, T., Oda, T., Hirano, T., Kasai, M., Okumura, K., and Tada, T., 1980, Asialo GM_1 as a cell-surface marker detected in acute lymphoblastic leukemia, *N. Engl. J. Med.* **302:**674–677.

Nakayasu, M., Terada, M., Tamura, G., and Sugimura, T., 1980, Induction of differentiation of human and murine myeloid leukemia cells in culture by tunicamycin, *Proc. Natl. Acad. Sci. USA* **77:**409–413.

Neufeld, E. F., and Ashwell, G., 1980, Carbohydrate recognition systems for receptor-mediated pinocytosis, in: *The Biochemistry of Glycoproteins and Proteoglycans* (W. J. Lennarz, ed.), pp. 241–266, Plenum Press, New York.

Neufeld, E. F., Sando, G. N., Garvin, A. J., and Rome, L. H., 1977, The transport of lysosomal enzymes, *J. Supramol. Struct.* **6:**95–101.

Nicola, N. A., Burgess, A. W., Staber, F. G., Johnson, G. R., Metcalf, D., and Battye, F. L., 1980, Differential expression of lectin receptors during hemopoietic differentiation:

Enrichment for granulocyte–macrophage progenitor cells, *J. Cell. Physiol.* **103:**217–237.

Nicolson, G. L., 1974, The interactions of lectins with animal cell surfaces, *Int. Rev. Cytol.* **39:**89–190.

Nashimaki, T., Kano, K., and Milgrom, F., 1979, Hanganutzui–Deicher antigen and antibody in pathologic sera and tissues, *J. Immunol.* **122:**2314–2318.

Nishioka, K., Constantopoulos, A., Satoh, P. S., and Najjar, V. A., 1972, The characteristics, isolation and synthesis of the phagocytosis stimulating peptide tuftsin, *Biochem. Biophys. Res. Commun.* **47:**172–179.

Nose, M., and Wigzell, H., 1983, Biological significance of carbohydrate chains on monoclonal antibodies, *Proc. Natl. Acad. Sci. USA* **80:**6632–6636.

Nowak, T. P., Kobiler, D., Roel, L. E., and Barondes, S. H., 1977, Developmentally regulated lectin from embryonic chick pectoral muscle, *J. Biol. Chem.* **252:**6026–6030.

Nowinski, R., Berglund, C., Lane, J., Lostrum, M., Bernstein, I., Young, W., Hakomori, S.-I., Hill, L., and Cooney, M., 1980, Human monoclonal antibody against Forssman antigen, *Science* **210:**537–539.

Nudelman, E., Hakomori, S.-I., Knowles, B. B., Solter, D., Rowinski, R. C., Tam, M. R., and Young, W. W., Jr., 1980, Monoclonal antibody directed to the stage-specific embryonic antigen(SSEA-1) reacts with a branched glycosphingolipid similar in structure to I antigen, *Biochem. Biophys. Res. Commun.* **97:**443–451.

Nudelman, E., Hakomori, S.-I., Kannagi, R., Levery, S., and Yeh, M.-Y., 1982, Characterization of a human melanoma-associated ganglioside antigen defined by a monoclonal antibody, 4.2, *J. Biol. Chem.* **257:**12752–12756.

Nydegger, U. E., Fearon, D. T., and Austen, K. F., 1978, Autosomal locus regulates inverse relationship between sialic acid content and capacity of mouse erythrocytes to activate human alternative complement pathway, *Proc. Natl. Acad. Sci. USA* **75:**6078–6082.

Oates, M. D. G., Rosbottom, A. C., and Schrager, J., 1974, Further investigations into the structure of human gastric mucin: The structural configuration of the oligosaccharide chains, *Carbohydr. Res.* **34:**115–137.

Oikawa, T., Yanagimachi, R., and Nicolson, G. L., 1973, Wheat germ agglutinin blocks mammalian fertilization, *Nature (London)* **241:**256–259.

O'Keefe, D., and Ashman, L., 1982, Peanut agglutinin: A marker for normal and leukemic cells of the monocyte lineage, *Clin. Exp. Immunol.* **48:**329–338.

O'Neill, H. C., and Parish, C. R., 1981, Monoclonal antibody detection of two classes of H-2K^k molecules, *Mol. Immunol.* **18:**713–722.

O'Neill, H. C., Parish, C. R., and Higgins, T. J., 1981, Monoclonal antibody detection of carbohydrate-defined and protein-defined H-2K^k antigens, *Mol. Immunol.* **18:**663–675.

Oriol, R., Baur, M. P., Danilovs, J., and Mayr, W., 1980, Combined ABH–Lewis–secretor antigens, in: *Histocompatibility Testing* (P. I. Terasaki, ed.), pp. 585–589, UCLA Tissue Typing Laboratory, Los Angeles.

Painter, R. G., and White, A., 1976, Effect of concanavalin A on expression of cell surface sialyltransferase activity of mouse thymocytes, *Proc. Natl. Acad. Sci. USA* **73:**837–841.

Pangburn, M. K., and Muller-Eberhard, H. J., 1978, Complement C3 convertase: Cell surface restriction of β1H control and generation of restriction of neuraminidase-treated cells, *Proc. Natl. Acad. Sci. USA* **75:**2416–2420.

Parish, C. R., 1977, Simple model for self–non-self discrimination in invertebrates, *Nature (London)* **267:**711–713.

Parish, C. R., and McKenzie, I. F. C., 1977, Mitogens and T-independent antigens stimulate T lymphocytes to secrete Ia antigens, *Cell. Immunol.* **33:**134–144.

Parish, C. R., and McKenzie, I. F. C., 1981, Carbohydrate-defined antigens controlled by the I region, in: *Current Trends in Histocompatibility* (R. A. Reisfeld and S. Ferrone, eds.), pp. 231–263, Plenum Press, New York.

Parish, C. R., Freeman, R. R., McKenzie, I. F. C., Cheers, C., and Cole, G. A., 1979, Ia antigens in serum during different murine infections, *Infect. Immun.* **26:**422–426.

Parish, C. R., Higgins, T. J., and McKenzie, I. F. C., 1981, Lymphocytes express Ia antigens of foreign haplotype following treatment with neuraminidase, *Immunogenetics* **12:**1–20.

Parish, C. R., Higgins, T. J., and McKenzie, I. F. C., 1982, Carbohydrate Ia antigens in mouse and man, in: *Ia Antigens* (S. Ferrone and C. S. David, eds.), pp. 143–160, CRC Press, Boca Raton, Fla.

Park, M. S., Oriol, R., Terasaki, P. I., and Nakata, S., 1980, Lewis-related specificities of the 8th workshop, in: *Histocompatibility Testing* (P. I. Terasaki, ed.), pp. 929–930, UCLA Tissue Typing Laboratory, Los Angeles.

Patt, L. M., Endres, R. O., Lucas, D. O., and Grimes, W. J., 1976, Ectogalactosyltransferase studies in fibroblasts and concanavalin A-stimulated lymphocytes, *J. Cell Biol.* **68:**799–802.

Perlmutter, R. M., Hansburg, D., Briles, D. E., Nicolotti, R. A., and Davie, J. M., 1978, Subclass restriction of murine anti-carbohydrate antibodies, *J. Immunol.* **121:**566–572.

Pierce, M., 1982, Quantification of ganglioside GM_1 synthetase activity on intact chick neural retinal cells, *J. Cell Biol.* **93:**76–81.

Pierce, M., Turley, E. A., and Roth, S., 1981, Cell surface glycosyltransferase activities, *Int. Rev. Cytol.* **65:**1–47.

Pierce, N. F., 1973, Differential inhibitory effects of cholera toxoids and ganglioside on the enterotoxins of *Vibrio cholerae* and *Escherichia coli, J. Exp. Med.* **137:**1009–1023.

Podolsky, D. K., and Weiser, M. M., 1975, Galactosyltransferase activities in human sera: Detection of a cancer-associated isoenzyme, *Biochem. Biophys. Res. Commun.* **65:**545–551.

Podolsky, D. K., Weiser, M. M., Westwood, J. C., and Gammon, M., 1977, Cancer-associated serum galactosyltransferase activity, *J. Biol. Chem.* **252:**1807–1813.

Podolsky, D. K., Weiser, M. M., and Isselbacher, K. J., 1978, Inhibition of growth of transformed cells and tumors by an endogenous acceptor of galactosyltransferase, *Proc. Natl. Acad. Sci. USA* **75:**4426–4430.

Pompecki, R., Shively, J. E., and Todd, C. W., 1981, Demonstration of elevated anti Lewis antibodies in sera of cancer patients using a carcinoembryonic antigen–polyethylene glycol immunoassay, *Cancer Res.* **41:**3023–3027.

Poste, G., Kirsh, R., and Filder, I. J., 1979a, Cell surface receptors for lymphokines, I. The possible role of glycolipids as receptors for macrophage migration inhibitory factor (MIF) and macrophage activation factor (MAF), *Cell. Immunol.* **44:**71–88.

Poste, G., Allen, H., and Matta, K. L., 1979b, Cell surface receptors for lymphokines. II. Studies on the carbohydrate composition of the MIF receptor on macrophages using synthetic saccharides and plant lectins, *Cell. Immunol.* **44:**89–98.

Potter, M., 1977, Antigen-binding myeloma proteins of mice, *Adv. Immunol.* **25:**141–211.

Prieels, J. P., Pizzo, S. V., Glasgow, L. R., Paulson, J. C., and Hill, R. L., 1978, Hepatic receptor that specifically binds oligosaccharides containing fucosyl α-1-3 N-acetylglucosamine linkages, *Proc. Natl. Acad. Sci. USA* **75:**2215–2219.

Prohaska, R., Koerner, T. A. W., Jr., Armitage, I. M., and Furthmayr, H., 1981, Chemical and carbon-13 nuclear magnetic resonance studies of the blood group M and N active sialoglycopeptides from human glycophorin A, *J. Biol. Chem.* **256:**5781–5791.

Prokop, O., 1974, Protectins: Past, present problems, and perspectives, *Ann. N.Y. Acad. Sci.* **234:**228–231.

Pukel, C. S., Lloyd, K. O., Travassos, L. R., Dippold, W. G., Oettgen, H. F., and Old, L. J., 1982, G_{D3}, a prominent ganglioside of human melanoma: Detection and characterization by mouse monoclonal antibody, *J. Exp. Med.* **155:**1133–1147.

Rasanen, L., 1981, Adherence of bacterial to human lymphocyte subpopulations and the role of monosaccharides in bacterial binding, *Cell. Immunol.* **58**:19–28.

Rauvala, H., and Finne, J., 1979, Structural similarity of the terminal carbohydrate sequences of glycoproteins and glycolipids, *FEBS Lett.* **97**:1–8.

Rauvala, H., Carter, W. G., and Hakomori, S.-I., 1981, Studies on cell adhesion and recognition. II. The occurrence of α-mannosidase at the fibroblast cell surface and its possible role in cell recognition, *J. Cell Biol.* **88**:149–159.

Reading, C. L., 1982a, Analysis of tumor cell lectin receptors using a HPLC-enzyme assay, *J. Cell. Biochem. Suppl.* **6**:72.

Reading, C. L., 1982b, Theory and methods for *in vitro* immunization and monoclonal antibody production, *J. Immunol. Methods* **53**:261–291.

Reading, C. L., Belloni, P. N., and Nicolson, G. L., 1980a, Selection and in vivo properties of lectin-attachment variants of malignant lymphosarcoma cell lines, *J. Natl. Cancer Inst.* **64**:1241–1249.

Reading, C. L., Brunson, K. W., Torrianni, M., and Nicolson, G. L., 1980b, Malignancies of murine metastatic lymphosarcoma cell lines and clones correlate with decreased cell surface display of RNA tumor virus envelope glycoprotein gp70, *Proc. Natl. Acad. Sci. USA* **77**:5943–5947.

Reichert, C. M., and Goldstein, I. J., 1979, The immunochemistry of antibodies sharing concanavalin A's anti-mannosyl binding specificity, *J. Immunol.* **122**:1138–1145.

Reisner, Y., and Sharon, N., 1978, Lectin receptors as markers for lymphocyte subpopulations in mouse and man, in: *Molecular Mechanisms of Biological Recognition* (M. Balaban, ed.), pp. 95–106, Elsevier, Amsterdam.

Reisner, Y., Linker-Israeli, M., and Sharon, N., 1976, Separation of mouse thymocytes into two subpopulations by the use of peanut agglutinin, *Cell. Immunol.* **25**:129–134.

Reisner, Y., Itzicovitch, L., Meshorer, A., and Sharon, N., 1978, Hemopoietic stem cell transplantation using mouse bone marrow and spleen cells fractionated by lectins, *Proc. Natl. Acad. Sci. USA* **75**:2933–2936.

Reisner, Y., Biniaminov, M., Rosenthal, E., Sharon, N., and Ramot, B. O., 1979, Interaction of peanut agglutinin with normal human lymphocytes and with leukemic cells, *Proc. Natl. Acad. Sci. USA* **76**:447–451.

Remold, H. G., 1973, Requirement for α-L-fucose on the macrophage membrane receptor for MIF, *J. Exp. Med.* **138**:1065–1076.

Remold, H. G., and David, J. R., 1971, Further studies on migration inhibitory factor (MIF): Evidence for its glycoprotein nature, *J. Immunol.* **107**:1090–1098.

Richardson, C. L., Baker, S. R., Morre, D. J., and Keenan, T. W., 1975, Glycosphingolipid synthesis and tumorigenesis: A role for the Golgi apparatus in the origin of specific receptor molecules of the mammalian cell surface, *Biochim. Biophys. Acta* **417**:175–184.

Riedl, M., Forster, O., Rumpold, H., and Bernheimer, H., 1982, A ganglioside-dependent cellular binding mechanism in rat macrophages, *J. Immunol.* **128**:1205–1210.

Rocklin, R. E., 1976, Role of monosaccharides in the interaction of two lymphocyte mediators with their target cells, *J. Immunol.* **116**:816–820.

Roder, J. C., 1980, Different genes regulate tumor cell recognition and cytolysis by NK cells in the mouse, in: *Genetic Control of Natural Resistance to Infection and Malignancy* (E. Skamene, P. A. L. Kongshavn, and M. Landy, eds.), pp. 405–412, Academic Press, New York.

Roder, J. C., Rosen, A., Fenyo, E. M., and Troy, F. A., 1979, Target–effector interaction in the natural killer cell system: Isolation of target structures, *Proc. Natl. Acad. Sci. USA* **76**:1405–1409.

Roelants, G. E., London, J., Mayor-Withey, K. S., and Serrano, B., 1979, Peanut agglutinin. II. Characterization of the Thy-1, Tla and Ig phenotype of peanut agglutinin-positive cells in adult, embryonic and nude mice using double immunofluorescence, *Eur. J. Immunol.* **9**:139–145.

Rogentine, G. N., Jr., and Plocinik, B. A., 1974, Carbohydrate inhibition studies of the naturally occurring human antibody to neuraminidase-treated human lymphocytes, *J. Immunol.* **113**:848–858.

Roseman, S., 1970, The synthesis of complex carbohydrates by multiglycosyltransferase systems and their potential function in intercellular adhesion, *Chem. Phys. Lipids* **5**:270–297.

Rosen, S. D., Kafka, J. A., Simpson, D. L., and Barondes, S. H., 1973, Developmentally regulated, carbohydrate-binding protein in *Dictyostelium discoideum, Proc. Natl. Acad. Sci. USA* **70**:2554–2557.

Rosen, S. D., Simpson, D. L., Rose, J. E., and Barondes, S. H., 1974, Carbohydrate-binding protein from *Polysphondylium pallidum* implicated in intercellular adhesion, *Nature (London)* **252**:128, 149–151.

Rosen, S. D., Kaur, J., Clark, D. L., Pardos, B. T., and Frazier, W. A., 1979, Purification and characterization of multiple species (isolectins) of a slime mold lectin implicated in intercellular adhesion, *J. Biol. Chem.* **254**:9408–9415.

Rosenfelder, G., Van Eijk, R. V. W., and Muhlradt, P. F., 1979, Metabolic carbohydrate-labelling of glycolipids from mouse splenocytes, *Eur. J. Biochem.* **97**:229–237.

Rosenfelder, G., Herbst, H., and Braun, D. G., 1980, Glycolipids as markers of murine T and B lymphoblastoid tumour cell lines, *FEBS Lett.* **114**:213–218.

Rostam-Abadi, H., and Pistole, T. G., 1982, Lipopolysaccharide-binding lectin from the horseshoe crab, *Limulus polyphemus,* with specificity for 2-keto-3-deoxyoctonate (KDO), *Dev. Comp. Immunol.* **6**:209–218.

Roth, S., and White, D., 1972, Intercellular contact and cell-surface galactosyl transferase activity, *Proc. Natl. Acad. Sci. USA* **69**:485–489.

Roth, S., McGuire, E. J., and Roseman, S., 1971, Evidence for cell-surface glycosyltransferases: Their potential role in cellular recognition, *J. Cell Biol.* **51**:536–547.

Rothenberg, B. E., 1978, The self recognition concept: An active function for the molecules of the major histocompatibility complex based on the complementary interaction of protein and carbohydrate, *Dev. Comp. Immunol.* **2**:23–37.

Rovis, L., Kabat, E. A., and Potter, M., 1972, Immunological studies on a mouse myeloma protein having specific affinity for 2-acetamido-2-deoxy-D-mannose, *Carbohydr. Res.* **23**:223–227.

Rutz, R., and Lilien, J., 1979, Functional characterization of an adhesion component from the embryonic chick neural retina, *J. Cell Sci.* **36**:323–342.

Sadler, J. E., Paulson, J. C., and Hill, R. L., 1979, The role of sialic acid in the expression of human MN blood group antigens, *J. Biol. Chem.* **254**:2112–2119.

Sanderson, A. R., Cresswell, P., and Welsh, K. I., 1971, Involvement of carbohydrate in the immunochemical determinant area of HLA substances, *Nature New Biol.* **230**:8–12.

Sandrin, M. S., McKenzie, I. F. C., Higgins, T. J., and Parish, C. R., 1981a, Isolation and characterization of low molecular weight Ia-like antigens from normal human serum, *Mol. Immunol.* **18**:513–519.

Sandrin, M. S., Henning, M. M., Vaughan, H. A., McKenzie, I. F. C., and Parish, C. R., 1981b, Serum Ia levels during tumor growth in mice and humans, *J. Natl. Cancer Inst.* **66**:279–283.

Sarkar, S., and Menge, A. C., 1977, Cell antigens recognized by rabbit antibodies specific for oligomannosyl determinants, *J. Supramol. Struct.* **6**:617–632.

Sawada, J. i., Shioiri-Nakano, K., and Osawa, T., 1976, Cytotoxic activity of purified guinea pig lymphotoxin against various cell lines, *Jpn. J. Exp. Med.* **46**:263–267.

Sawada, J.-i., Kobayashi, Y., and Osawa, T., 1977, The effect of pulse treatment of target cells with guinea pig lymphotoxin and the nature of its binding to target cells, *Jpn. J. Exp. Med.* **47**:93–98.

Schengrund, C. L., and Rosenberg, A., 1970, Intracellular location and properties of bovine brain sialidase, *J. Biol. Chem.* **245**:6196–6200.

Schengrund, C. L., Jensen, D. S., and Rosenberg, A., 1972, Localization of sialidase in the plasma membrane of rat liver cells, *J. Biol. Chem.* **247**:2742–2746.

Schengrund, C. L., Lausch, R. N., and Rosenberg, A., 1973, Sialidase activity in transformed cells, *J. Biol. Chem.* **248**:4424–4428.

Schengrund, C. L., Rosenberg, A., and Repman, M. A., 1976, Ecto-ganglioside-sialidase activity of herpes simplex virus-transformed hamster embryo fibroblasts, *J. Cell Biol.* **70**:555–561.

Schlesinger, P. H., Rodman, J. S., Frey, M., Lang, S., and Stahl, P., 1976, Clearance of lysosomal hydrolases following intravenous infusion. The role of liver in the clearance of β-glucuronidase and N-acetyl-β-D-glucosaminidase, *Arch. Biochem. Biophys.* **177**:606–614.

Schlesinger, P. H., Doebber, T. W., Mandrell, B. F., White, R., DeSchryver, C., Rodman, J. S., Miller, M. J., and Stahl, P., 1978a, Plasma clearance of glycoproteins with terminal mannose and N-acetylglucosamine by liver nonparenchymal cells, *Biochem. J.* **176**:103–109.

Schlesinger, P. H., Rodman, J. S., Miller, J., Enders, G. H., and Stahl, P., 1978b, Mannose glucose specific receptor on alveolar macrophages, *Fed. Proc.* **37**:1655.

Schnaar, R. L., Weigel, P. H., Kuhlenschmidt, M. S., Chuan Lee, Y., and Roseman, S., 1978, Adhesion of chicken hepatocytes to polyacrylamide gels derivatized with N-acetylglucosamine, *J. Biol. Chem.* **253**:7940–7951.

Schwarting, G. A., and Gajewski, A., 1981, Glycolipids of murine lymphocyte subpopulations: A defect in the levels of sialidase-sensitive sialosylated asialo G_{M1} in beige mouse lymphocytes, *J. Immunol.* **126**:2403–2407.

Schwarting, G. A., and Marcus, D. M., 1979, Cell surface glycosphingolipids of normal and leukemic human lymphocytes, *Clin. Immunol. Immunopathol.* **14**:121–129.

Schwarting, G. A., and Summers, A., 1980, Gangliotetraosylceramide is a T cell differentiation antigen associated with natural cell-mediated cytotoxicity, *J. Immunol.* **124**:1691–1694.

Sela, B.-A., Wang, J. L., and Edelman, G. M., 1975, Antibodies reactive with cell surface carbohydrates, *Proc. Natl. Acad. Sci. USA* **72**:1127–1131.

Sharon, N., Lis, H., 1972, Lectins: Cell-agglutinating and sugar-specific proteins, *Science* **177**:949–959.

Sharon, N., and Lis, H., 1980, Glycoproteins, in: *The Proteins,* Vol. V (H. Neurath and R. L. Hill, eds.), pp. 1–144, Academic Press, New York.

Shepherd, V. L., Lee, Y. C., Schlesinger, P. H., and Stahl, P. D., 1981, L-Fucose-terminated glycoconjugates are recognized by pinocytosis receptors on macrophages, *Proc. Natl. Acad. Sci. USA* **78**:1019–1022.

Shortridge, K. F., Biddle, F., and Pepper, D. S., 1972, Rubella virus nonspecific hemagglutination inhibitor: Evidence for the role of glycolipid bound to low density (β) lipoprotein, *Clin. Chim. Acta* **42**:285–294.

Shur, B. D., 1977a, Cell-surface glycosyltransferases in gastrulating chick embryos. I. Temporally and spatially specific patterns of four endogenous glycosyltransferase activities, *Dev. Biol.* **58**:23–39.

Shur, B. D., 1977b, Cell-surface glycosyltransferases in gastrulating chick embryos. II. Biochemical evidence for a surface localization of endogenous glycosyltransferase activities, *Dev. Biol.* **58**:40–55.

Shur, B. D., 1982a, Cell surface glycosyltransferase activities during fertilization and early embryogenesis, in: *The Glycoconjugates,* Vol. III, Part A (M. I. Horowitz, ed.), pp. 146–186, Academic Press, New York.

Shur, B. D., 1982b, Evidence that galactosyltransferase is a surface receptor for poly(N)-acetyllactosamine glycoconjugates on embryonal carcinoma cells, *J. Biol. Chem.* **257:**6871–6878.

Shur, B. D., and Bennett, D., 1979, A specific defect in galactosyltransferase regulation on sperm bearing mutant alleles of the T/t Locus, *Dev. Biol.* **71:**243–259.

Shur, B. D., and Hall, N. G., 1982a, Sperm surface galactosyltransferase activities during in vitro capacitation, *J. Cell Biol.* **95:**567–573.

Shur, B. D., and Hall, N. G., 1982b, A role for mouse sperm surface galactosyltransferase in sperm binding to the egg zona pellucida, *J. Cell Biol.* **95:**574–579.

Shur, B. D., and Roth, S., 1975, Cell surface glycosyltransferases, *Biochim. Biophys. Acta* **415:**473–512.

Sia, D., and Parish, C. R., 1981, Anti-self receptors, *Immunogenetics* **12:**587–599.

Sigel, M. M., 1974, Primitive immunoglobulins and other proteins with binding functions in the shark, *Ann. N.Y. Acad. Sci.* **234:**198–215.

Simpson, D. L., Rosen, S. D., and Barondes, S. H., 1975, Pallidin: Purification and characterization of a carbohydrate-binding protein from *Polysphondium palladium* implicated in intercellular adhesion, *Biochim. Biophys. Acta* **412:**109–119.

Simpson, D. L., Thorne, D. R., and Loh, H. H., 1978, Lectins: Endogenous carbohydrate-binding proteins from vertebrate tissues: Functional role in recognition processes?, *Life Sci.* **22:**727–748.

Simpson, L. L., and Rapport, M. M., 1970, Ganglioside inactivation of botulinum toxin, *J. Neurochem.* **18:**1341–1343.

Simpson, L. L., and Rapport, M. M., 1971, The binding of botulinum toxin to membrane lipids: Sphingolipids, steroids and fatty acids, *J. Neurochem.* **18:**1751–1759.

Slomiany, B. L., and Meyer, K., 1973, Oligosaccharides produced by acetolysis of blood group active (A + H) sulfated glycoproteins from hog gastric mucin, *J. Biol. Chem.* **248:**2290–2295.

Sly, W. S., 1977, Receptor-mediated uptake and clearance of lysosomal enzymes, *J. Supramol. Struct. Suppl.* **1:**36.

Smets, L. A., van Beek, W. P., Collard, J. G., Temmink, H., van Gils, B., and Emmelot, P., 1975, Comparative evaluation of plasma membrane alterations associated with neoplasia, in: *Cellular Membranes and Tumor Cell Behavior,* pp. 269–286, 28th Annu. Symp. Fundamental Cancer Research, Williams & Wilkins, Baltimore.

Smith, D. F., and Ginsburg, V., 1980, Antibodies against sialyloligosaccharides coupled to protein, *J. Biol. Chem.* **255:**55–59.

Sneath, P. H. A., and Lederberg, J., 1960, Inhibition by periodate of mating in *Escherichia coli* K-12, *Proc. Natl. Acad. Sci. USA* **47:**86–90.

Soderhall, K., 1981, Fungal cell wall β-1,3-glucans induce clotting and phenoloxidase attachment to foreign surfaces of crayfish hemocyte lysate, *Dev. Comp. Immunol.* **5:**565–573.

Solter, D., and Knowles, B. B., 1978, Monoclonal antibody defining a stage-specific mouse embryonic antigen (SSEA-1), *Proc. Natl. Acad. Sci. USA* **75:**5565–5569.

Springer, G. F., 1970, Importance of blood-group substances in interactions between man and microbes, *Ann. N.Y. Acad. Sci.* **169:**134–152.

Springer, G. F., Desai, P. R., Yang, H. J., Schachter, H., and Narasimhan, S., 1976, Interrelations of blood group M and precursor specificities and their significance in human carcinoma, in: *Human Blood Groups* (J. F. Mohn, R. W., Plunkett, R. K. Cunningham, and R. M. Lambert, eds.), pp. 179–187, Karger, Basel.

Springer, G. F., Desai, P. R., and Murthy, M. S., 1978, Human carcinoma-associated precursors of the blood group MN antigens, in: *Glycoproteins and Glycolipids in Disease Processes* (E. F. Walborg, Jr., ed.), pp. 311–325, American Chemical Society, Washington, D. C.

Springer, W. R., Haywood, P. L., and Barondes, S. H., 1980, Endogenous cell surface lectin in *Dictyostelium*: Quantitation, elution by sugar, and elicitation by divalent immunoglobulin, *J. Cell Biol.* **87:**682–690.

Springer, G. F., Yang, H. J., Mbawa, E., and Grohlich, D., 1980, Highly active aminoterminal sialoglycopenta-, hexa- and heptapeptides from human erythrolyte NM antigens, *Naturwissenchaften* **67:**473–474.

Springer, G. F., Murthy, M. S., Fry, W. A., Tegmeyer, H., and Svanlon, E. F., 1982, Patient's immune response to breast and lung carcinoma-associated Thomsen–Friedenreich (T) specificity, *Klin. Wochenschr.* **60:**121–131.

Staerk, J., Ronneberger, H. J., Wiegandt, H., and Ziegler, W., 1974, Interaction of ganglioside G_{Gtet} and its derivatives with choleragen, *Eur. J. Biochem.* **48:**103–110.

Stahl, P., and Gordon, S., 1982, Expression of a mannosyl-fucosyl receptor for endocytosis on cultured primary macrophages and their hybrids, *J. Cell Biol.* **93:**49–56.

Stahl, P., Six, H., Rodman, J. S., Schlesinger, P., Tulsiani, D. R. P., and Touster, O., 1976a, Evidence for specific recognition sites mediating clearance of lysosomal enzymes *in vivo*, *Proc. Natl. Acad. Sci. USA* **73:**4045–4049.

Stahl, P., Rodman, J. S., and Schlesinger, P., 1976b, Clearance of lysosomal hydrolases following intravenous infusion, *Arch. Biochem. Biophys.* **117:**594–605.

Stahl, P. D., Rodman, J. S., Miller, M. J., and Schlesinger, P. H., 1978, Evidence for receptor-mediated binding of glycoproteins, glycoconjugates, and lysosomal glycosidases by alveolar macrophages, *Proc. Natl. Acad. Sci. USA* **75:**1399–1403.

Stein, K. E., Schwarting, G. A., and Marcus, D. M., 1978, Glycolipid markers of murine lymphocyte subpopulations, *J. Immunol.* **120:**676–679.

Stein-Douglas, K. E., Schwarting, G. A., Naiki, M., and Marcus, D. M., 1976, Gangliosides as markers for murine lymphocyte subpopulations, *J. Exp. Med.* **143:**822–832.

Steiner, S., Gacto, M., and Steiner, M. R., 1975, Fucolipids in normal and transformed cells, in: *Cellular Membranes and Tumor Cell Behavior*, pp. 309–324, 28th Annu. Symp. Fundamental Cancer Research, Williams & Wilkins, Baltimore.

Steinitz, M., Seppala, I., Eichmann, K., and Klein, G., 1979, Establishment of a human lymphoblastoid cell line with specific antibody production against group A streptococcal carbohydrate, *Immunobiology* **156:**41–47.

Stern, P. L., Martin, G. R., and Evans, M. J., 1975, Cell surface antigens of clonal teratocarcinoma cells at various stages of differentiation, *Cell* **6:**455–465.

Stern, P. L., Willison, K. R., Lennox, E., Galfre, G., Milstein, C., Secher, D., Ziegler, A., and Spencer, T., 1978, Monoclonal antibodies as probes for differentiation and tumor-associated antigens: A Forssman specificity on teratocarcinoma stem cells, *Cell* **14:**775–783.

Stoolman, L. M., and Rosen, S. D., 1983, Possible role for cell surface carbohydrate-binding molecules in lymphocyte recirculation, *J. Cell Biol.* **96:**722–729.

Strombeck, D. R., and Harrold, D., 1974, Binding of cholera toxin to mucins and inhibition by gastric mucin, *Infect. Immun.* **10:**1266–1272.

Stutman, O., Dien, P., Wisun, R. E., and Lattime, E. C., 1980, Natural cytotoxic cells against solid tumors in mice: Blocking of cytotoxicity by D-mannose, *Proc. Natl. Acad. Sci. USA* **77:**2895–2898.

Suttajit, M., and Winzler, R. J., 1971, Effect of modification of N-acetylneuraminic acid on the binding of glycoproteins to influenza virus and on susceptibility to cleavage by neuraminidase, *J. Biol. Chem.* **246:**3398–3404.

Sweeley, C. C., Fung, Y.-K., Macher, B. A., Moskal, J. R., and Nunez, H. A., 1978, Structure and metabolism of glycolipids, in: *Glycoproteins and Glycolipids in Disease Processes* (E. F. Walborg, Jr., ed.), pp. 47–85, American Chemical Society, Washington, D.C.

Szulman, A. E., 1977, The ABH and Lewis antigens of human tissues during prenatal and postnatal life, in: *Human Blood Groups* (J. F. Mohn, R. W. Plunkett, R. K., Cunningham, and R. M. Lambert, eds.), pp. 426–436, Karger, Basel.

Tada, T., Tanguchi, M., and David, C. S., 1976, Properties of the antigen-specific suppressive T-cell factor in the regulation of antibody response of the mouse. IV. Special subregion assignment of the gene(s) that codes for the suppressive T-cell factor in the H-2 histocompatibility complex, *J. Exp. Med.* **144:**713–725.

Targan, S., Decker, J., and Ades, E. W., 1984, Mechanism of inhibition of natural killing by a glycopeptide isolated from the K562 plasma membrane, (submitted).

Tate, R. L., Holmes, J. A., and Kohn, L. D., 1975, Characteristics of solubilized thyrotropin receptor from bovine thyroid plasma membranes, *J. Biol. Chem.* **250:**6527–6533.

Taussig, M. J., Mozes, E., and Isaac, R., 1974, Antigen-specific thymus cell factors in the genetic control of the immune response to poly-(tyrosyl,glutamyl)-poly-D,L-alanyl–polylysyl, *J. Exp. Med.* **140:**310–312.

Taussig, M. J., Munro, A. J., Campbell, R., David, C. S., and Staines, N. A., 1975, Antigen-specific T-cell factor in cell cooperation. Mapping within the I region of the H-2 complex and ability to cooperate across allogeneic barriers, *J. Exp. Med.* **142:**694–700.

Teichberg, V. I., Silman, I., Beitsch, D. D., and Resheff, G., 1975, A β-D-galactoside binding protein from electric organ tissue of *Electrophorus electricus, Proc. Natl. Acad. Sci. USA* **72:**1383–1387.

Teodorescu, M., Mayer, E. P., and Dray, S., 1977, Identification of five human lymphocyte subpopulations by their differential binding of various strains of bacteria, *Cell. Immunol.* **29:**353–362.

Thomas, D. B., 1971, Cyclic expression of blood group determinants in murine cells and their relationship to growth control, *Nature (London)* **233:**317–321.

Thomas, L., 1969, Relationships between mycoplasmas and mammalian cells, in: *Cellular Recognition* (R. T. Smith and R. A. Good, eds.), pp. 139–144, Appleton–Century–Crofts, New York.

Thornburg, R. W., Day, J. F., Baynes, J. W., and Thorpe, S. R., 1980, Carbohydrate-mediated clearance of immune complexes from the circulation, *J. Biol. Chem.* **255:**6820–6825.

Tomaska, L. D., and Parish, C. R., 1981, Inhibition of secondary IgG responses by N-acetyl-D-galactosamine, *Eur. J. Immunol.* **11:**181–186.

Tomaska, L. D., and Parish, C. R., 1982, Inhibition of secondary IgG responses by monosaccharides: Evidence for I-region control, *J. Immunogenet.* **9:**63–68.

Tonelli, Q., and Meints, R. H., 1978, Sialic acid: A specific role in hematopoietic spleen colony formation, *J. Supramol. Struct.* **8:**67–78.

Touraine, J. L., Touraine, F., Hadden, J. W., Hadden, E. M., and Good, R. A., 1976, 5-Bromodeoxyuridine–light inactivation of human lymphocytes stimulated by mitogens and allogeneic cells: Evidence for distinct T-lymphocyte subsets, *Immunology* **52:**105–117.

Townsend, R., and Stahl, P., 1981, Isolation and characterization of a mannose N-acetylglucosamine fucose-binding protein from rat liver, *Biochem. J.* **194:**209–214.

Trenkner, E., and Sarkar, S., 1977, Microbial carbohydrate specific antibodies distinguish between different stages of differentiating mouse cerebellum, *J. Supramol. Struct.* **6:**465–472.

Tripp, M. R., 1974a, Oyster hemolymph proteins, *Ann. N.Y. Acad. Sci.* **234:**18–22.

Tripp, M. R., 1974b, Molluscan immunity, *Ann. N.Y. Acad. Sci.* **234**:23–27.

Trowbridge, I, S., and Hyman, R., 1975, Thy-1 variants of mouse lymphomas: Biochemical characterization of the genetic defect, *Cell* **6**:279–287.

Tsai, C.-M., Zopf, D. A., Wistar, R., Jr., and Ginsburg, V., 1976, A human cold agglutinin which binds lacto-N-*neo*tetraose, *J. Immunol.* **117**:717–721.

Tsai, C.-M., Zopf, D. A., Yu, R. K., Wistar, R., Jr., and Ginsburg, V., 1977, A Waldenstrom macroglobulin that is both a cold agglutinin and a cryoglobulin because it binds N-acetylneuraminosyl residues, *Proc. Natl. Acad. Sci. USA* **74**:4591–4594.

Turner, R. S., and Burger, M. M., 1973, Involvement of a carbohydrate group in the active site for surface guided reassociation of animal cells, *Nature (London)* **244**:509–510.

Uchida, T., and Nagai, Y., 1980, Affinity chromatographic purification of anti-glycolipid antibodies and their application to the membrane studies. *J. Biochem.* **87**:1829–1841.

Uemura, K.-I., Yuzawa-Watanabe, M., Kitazawa, N., and Taketomi, T., 1980, Liposome agglutination and liposome membrane immune-damage assays for the characterization of antibodies to glycosphingolipids, *J. Biochem. Tokyo* **87**:1641–1648.

Ullrich, K., Mersmann, G., Weber, E., and von Figura, K., 1978, Evidence for lysosomal enzyme recognition by human fibroblasts via a phosphorylated carbohydrate moiety, *Biochem. J.* **170**:643–650.

van Dijk, W. C., Verbrugh, H. A., van der Tol, M. E., Peters, R., and Verhoef, J., 1979, Role of *Escherichia coli* K capsular antigens during complement activation, C3 fixation, and opsonization, *Infect. Immun.* **25**:603–609.

van Heyningen, W. E., 1974, Gangliosides as membrane receptors for tetanus toxin, cholera toxin and serotonin, *Nature (London)* **249**:415–417.

van Heyningen, W. E., Carpenter, C. C. J., Pierce, N. F., and Greenough, W. B., III, 1971, Deactivation of cholera toxin by ganglioside, *J. Infect. Dis.* **124**:415–418.

Van Rood, J. J., Van Leeuwen, A., Koch, C. T., and Frederiks, E., 1970, HL-A inhibiting activity in serum, in: *Histocompatibility Testing* (P. I. Terasaki, ed.), pp. 483–485, Williams & Wilkins, Baltimore.

Varki, A., and Kornfeld, S., 1980, An autosomal dominant gene regulates the extent of 9-O-acetylation of murine erythrocyte sialic acids, *J. Exp. Med.* **152**:532–544.

Vengris, V. E., Reynolds, F. H., Jr., Hollenberg, M. D., and Pitha, P. M., 1976, Interferon action: Role of membrane gangliosides, *Virology* **72**:486–493.

Verbert, A., Cacan, R., and Montreuil, J., 1976, Ectogalactosyltransferase, *Eur. J. Biochem.* **70**:49–53.

Verbert, A., Cacan, R., Debeire, P., and Montreuil, J., 1977, Peculiar behavior of ecto-sialytransferase toward exogenous acceptors, *FEBS Lett.* **74**:234–238.

Vicari, G., and Kabat, E. A., 1969, Immunochemical studies on blood groups. XLII. Isolation and characterization from ovarian cyst fluid of a blood group substance lacking A, B, H, Le[a] and Le[b] specificity, *J. Immunol.* **102**:821–825.

Vicari, G., Sher, A., Cohn, M., and Kabat, E. A., 1970, Immunochemical studies on a mouse myeloma protein with specificity for certain β-linked terminal residues of N-acetyl-D-glucosamine, *Immunochemistry* **7**:829–838.

Visser, A., and Emmelot, P., 1973, Studies on plasma membranes. XX. Sialidase in hepatic membranes, *J. Membr. Biol.* **14**:73–84.

Vitetta, E. S., Artzt, K., Bennett, D., Boyse, E. A., and Jacob, F., 1975, Structural similarities between a product of the T/t-locus isolated from sperm and teratoma cells, and H-2 antigens isolated from splenocytes, *Proc. Natl. Acad. Sci. USA* **72**:3215–3219.

Voac, D., Sacks, S., Alderson, T., Takei, F., Lennox, E., Jarvis, J., Milstein, C., and Darnborough, J., 1980, Monoclonal anti-A from a hybrid-myeloma: Evaluation as a blood grouping reagent, *Vox Sang.* **39**:134–140.

von Figura, K., and Voss, B., 1979, Cell surface-associated lysosomal enzymes in cultured human skin fibroblasts, *Exp. Cell Res.* **121**:267–276.

Wagner, H., Hardt, C., Bartlett, R., Rollinghoff, M., and Pfizenmaier, K., 1980, Intrathymic differentiation of cytotoxic T lymphocyte (CTL) precursors. I. The CTL immunocompetence of peanut agglutinin-positive (cortical) and negative (medullary) Lyt 123 thymocytes, *J. Immunol.* **125:**2532–2538.

Walborg, E. F., Jr., 1978, Current concepts of glycoprotein structure, in: *Glycoproteins and Glycolipids in Disease Processes* (E. F. Walborg, Jr., ed.), pp. 5–20, American Chemical Society, Washington, D.C.

Wang, T. J., Freimuth, W. W., Miller, H. C., and Esselman, W. J., 1978, Thy-1 antigenicity is associated with glycolipids of brain and thymocytes, *J. Immunol.* **121:**1361–1365.

Watanabe, K., and Hakomori, S.-I., 1976, Status of blood group carbohydrate chains in ontogenesis and in oncogenesis, *J. Exp. Med.* **144:**644–653.

Watkins, W. M., and Morgan, W. T. J., 1952, Neutralization of the anti-H agglutinin in eel serum by simple sugars, *Nature (London)* **169:**825–826.

Watkins, W. M., 1966, Blood-group specific substances, in: *The Glycoproteins: Their Composition, Structure and Function* (A. Gottschalk, ed.), pp. 425–515, Elsevier, Amsterdam.

Weigel, P. H., Schnaar, R. L., Kuhlenschmidt, M. S., Schmell, E., Lee, R. T., Lee, Y. C., and Roseman, S., 1979, Adhesion of hepatocytes to immobilized sugars: A threshold phenomenon, *J. Biol. Chem.* **254:**10830–10838.

Weise, L., 1974, Nature of sex specific glycoprotein agglutinins in *Chlamydomonas, Ann. N.Y. Acad. Sci.* **234:**383–395.

Weitzen, M. L., Yamamoto, R. S., and Granger, G. A., 1983a, Identification of human lymphocyte-derived lymphotoxins with binding and cell lytic activity on NK sensitive cell lines in vitro, *Cell. Immunol.* **77:**30–41.

Weitzen, M. L., Inninis, E., Yamamoto, R. S., and Granger, G. A., 1983b, Inhibition of human NK induced cell-lysis and soluble cell-lytic molecules with anti-human LT antisera and various saccharides, *Cell. Immunol.* **77:**42–51.

Willison, K. R., and Stern, P. L., 1978, Expression of a Forssman Antigenic specificity in the preimplantation mouse embryo, *Cell* **14:**785–793.

Willison, K. R., Karol, R. A., Suzuki, A., Kundu, S. K., and Marcus, D. M., 1982, Neutral glycolipid antigens as developmental markers of mouse teratocarcinoma and early embryos: An immunologic and chemical analysis, *J. Immunol.* **129:**603–609.

Winand, R. J., and Kohn, L. D., 1975, Thyrotropin effects on thyroid cells in culture, *J. Biol. Chem.* **250:**6534–6540.

Winkelhake, J. L., and Kasper, D. L., 1972, Affinity chromatography of antimeningococcal antiserum, *J. Immunol.* **109:**824–833.

Winkelhake, J. L., and Nicolson, G. L., 1976, Effects of exoglycosidase treatments on autochthonous antibody survival time in the circulation, *J. Biol. Chem.* **251:**1074–1080.

Winkelstein, J. A., and Tomasz, A., 1977, Activation of the alternative pathway by pneumococcal cell walls, *J. Immunol.* **118:**451–454.

Winkelstein, J. A., Bochini, J. A., Jr., and Schiffman, G., 1976, The role of the capsular polysaccharide in the activation of the alternative pathway by the pneumococcus, *J. Immunol.* **116:**367–370.

Winkelstein, J. A., Abramovitz, A. S., and Tomasz, A., 1980, Activation of C3 via the alternative complement pathway results in fixation of C3b to the pneumococcal cell wall, *J. Immunol.* **124:**2502–2506.

Wojdani, A., Stein, E. A., Lemmi, C. A., and Cooper, E. L., 1982, Agglutinins and proteins in the earthworm, *Lumbricus terrestris,* before and after injection of erythrocytes, carbohydrates, and other materials, *Dev. Comp. Immunol.* **6:**613–624.

Wolff, J., Winand, R. J., and Kohn, L. D., 1974, The contribution of subunits of thyroid stimulating hormone to the binding and biological activity of thyrotropin, *Proc. Natl. Acad. Sci. USA* **71:**3460–3464.

Wolley, D. W., and Gommi, B. W., 1965, Serotonin receptors. VII. Activities of various pure gangliosides as the receptors, *Proc. Natl. Acad. Sci. USA* **53**:959–963.

Woodruff, J. J., and Woodruff, J. F., 1972, Virus-induced alterations of lymphoid tissues. III. Fate of radiolabeled thoracic duct lymphocytes in rats inoculated with Newcastle disease virus, *Cell. Immunol.* **5**:307–317.

Woodruff, J. J., Katz, I. M., Lucas, L. E., and Stamper, H. B., Jr., 1977, An *in vitro* model of lymphocyte homing. II. Membrane and cytoplasmic events involved in lymphocyte adherence to specialized high-endothelial venules of lymph nodes, *J. Immunol.* **119**:1603–1610.

Wright, A., and Kanegasaki, S., 1971, Molecular aspects of lipopolysaccharides, *Physiol. Rev.* **51**:748–784.

Yamaga, K. M., Kubo, R. T., and Etlinger, H. M., 1978, Studies on the question of conventional immunoglobulin on thymocytes from primitive vertebrates. I. Presence of anti-carbohydrate antibodies in rabbit anti-trout Ig sera, *J. Immunol.* **120**:2068–2073.

Yamamoto, S., and Tokunaga, T., 1981, D-Mannose as a component of the macrophage surface receptor for macrophage-activating factor (MAF) in mice, *Cell. Immunol.* **61**:319–331.

Yeaton, R. W., 1981a, Invertebrate lectins. I. Occurrence, *Dev. Comp. Immunol.* **5**:391–402.

Yeaton, R. W., 1981b, Invertebrate lectins. II. Diversity of specificity, biological synthesis and function in recognition, *Dev. Comp. Immunol.* **5**:535–545.

Yeh, M. Y., Hellstrom, I., and Hellstrom, K. E., 1981, Clonal variation in expression of a human melanoma antigen defined by a monoclonal antibody, *J. Immunol.* **126**:1312–1317.

Yen, P. H., and Ballou, C. E., 1974, Partial characterization of the sexual agglutination factor from *Hansenula wengei* Y-2340 type 5 cells, *Biochemistry* **13**:2428–2437.

Yogeeswaran, G., and Hakomori, S.-I., 1975, Cell contact-dependent ganglioside changes in mouse 3T3 fibroblasts and a suppressed sialidase activity on cell contact, *Biochemistry* **14**:2151–2156.

Yogeeswaran, G., Laine, R. A., and Hakomori, S.-I., 1974, Mechanism of cell contact-dependent glycolipid synthesis: Further studies with glycolipid glass complex, *Biochem. Biophys. Res. Commun.* **59**:591–599.

Young, W. W., Jr., and Hakomori, S.-I., 1978, Status of blood group carbohydrate chains in human tumors, in: *Glycoproteins and Glycolipids in Disease Processes* (E. F. Walborg, Jr., ed.), pp. 357–371, American Chemical Society, Washington, D.C.

Young, W. W., Jr., Hakomori, S.-I., and Levine, P., 1979, Characterization of anti-Forssman (anti-Fs) antibodies in human sera: Their specificity and possible changes in patients with cancer, *J. Immunol.* **123**:92–96.

Young, W. W., Jr., Hakomori, S.-I., Durdik, J. M., and Henney, C. S., 1980, Identification of ganglio-N-tetraosylceramide as a new cell surface marker for murine natural killer (NK) cells, *J. Immunol.* **124**:199–201.

Young, W. W., Jr., Durdik, J. M., Urdal, D., Hakomori, S.-I., and Henney, C. S., 1981, Glycolipid expression in lymphoma cell variants: Chemical quantity, immunologic reactivity, and correlations with susceptibility to NK cells, *J. Immunol.* **126**:1–6.

Yount, W. J., Dorner, M. M., Kunkel, H. G., and Kabat, E. A., 1968, Studies on human antibodies. VI. Selective variations in subgroup composition and genetic markers, *J. Exp. Med.* **127**:633–646.

Zopf, D. A., Ginsburg, A., and Ginsburg, V., 1975, Goat antibody directed against human Le[b] blood group hapten, lacto-N-difucohexaose I, *J. Immunol.* **115:**1525–1529.
Zopf, D. A., Tsai, C.-M., and Ginsburg, V., 1978, Antibodies against oligosaccharides coupled to proteins: Characterization of carbohydrate specificity by radioimmune assay, *Arch. Biochem. Biophys.* **185:**61–71.

6

Nervous System Glycoproteins

Molecular Properties and Possible Functions

Paul T. Kelly

1. INTRODUCTION

One of the goals of the cellular neurobiologist is to acquire information on the structure and function of membrane components at a molecular level. Within this area of study, one of the major interests has focused on membrane glycoproteins. Plasma membranes that have most attracted the interests of neuroscientists are membranes that are structurally or functionally distinct to the nervous system, e.g., membranes that make up axons, dendrites, growing neurites, synaptic vesicles, myelin, and synaptic junctions. The brain contains a vast diversity of cell types, both neuronal and nonneuronal; cellular entities that differ on the basis of anatomical location and neurophysiological properties. Thus, studies during the past two decades have concentrated on the identification and characterization of membrane glycoproteins that may be useful in distinguishing, or are unique to, specific classes of neurons, and more recently, different types of synapses. Advances in the areas of biochemistry, cellular and subcellular fractionation, cytochemistry, and immunology have made possible new approaches that have lead to a better understanding of the molecular and functional properties of membrane glycoproteins in the nervous system. One of the ultimate goals of the neurobiologist is to dissect the synapse into its constituent molecules and to elucidate the

Paul T. Kelly • Division of Biology, Kansas State University, Manhattan, Kansas 66506.

structural and functional contribution of each, and then in a synthetic approach, to determine intermolecular relationships that will provide a comprehensive understanding of how the synapse works. The development of procedures to isolate subcellular fractions highly enriched in subsynaptic structures such as synaptic junctions and postsynaptic densities has made possible the beginning of this difficult and complex task. The ability to adapt and grow cellular elements of nervous tissues under defined culture conditions has made possible the study of specific physiological and cell–cell interaction properties of individual neurons. The ability to culture embryonic neurons now allows one to study the behavior and molecular properties of neurons as they differentiate and form synaptic connections.

This chapter was written with an emphasis on membrane glycoproteins that make up various subcellular compartments of synapses in the nervous system. A number of reviews on nervous system glycoproteins (Gombos and Zanetta, 1978; Matus, 1978; Mahler, 1979; Margolis and Margolis, 1979; Cotman and Kelly, 1980; Gurd, 1982; Margolis and Margolis, 1983), glycolipids (Morgan *et al.*, 1977), and glycosaminoglycans (Margolis and Margolis, 1979, 1983; Brunngraber, 1972), as well as monographs on the macromolecular architecture of synapses in the central nervous system (Cotman and Kelly, 1980; Gurd, 1982; Matus, 1978) have recently appeared. Much of this previous information has not been recounted in the present chapter. I have tried instead to review new findings from the last four years, some of which have not yet appeared in press.

2. STRUCTURE AND COMPOSITION OF GLYCOPROTEIN OLIGOSACCHARIDES IN THE NERVOUS SYSTEM

The composition and structure of carbohydrates in brain glycoproteins have been covered in detail in recent monographs and chapters (Margolis and Margolis, 1979, 1983). What follows is a brief overview of the principal features of the carbohydrate moieties of nervous system glycoproteins. Glycosaminoglycans (mucopolysaccharides) will not be discussed (see Margolis and Margolis, 1979).

A majority of the carbohydrate (85–90%) in nervous system glycoproteins is linked to the amide nitrogen of protein asparagine residues through N-acetylglucosamine (GlcNAc) (Margolis and Margolis, 1983). Many different forms of these N-glycosidically linked sugars are present in brain. Glycopeptides prepared from protease digests of lipid-free protein extracts from brain have been analyzed by gel-filtration and affinity chromatography on Con A–Sepharose (Finne and Krusius, 1982; Di Benedetta *et al.*, 1969; Brunngraber *et al.*, 1973). Glycopeptide fractions prepared by these methods differ in molecular size and molar ratios of carbohydrates. To date, these fractions are not homogeneous and are unsuitable for detailed structural analysis (Margolis and Margolis, 1979). The "neutral," high-mannose glycopeptides fractionated by these methods possess a high affinity for Con A and appear to contain only mannose and GlcNAc. However, most of the N-glycosidically linked oligosaccharides of brain glycoproteins are "complex" and are characterized as (1) containing galactose, fucose, and sialic acid in addition to mannose and glucosamine, and (2) being more extensively branched when compared to the neutral, high-mannose glycopeptides.

Much of the fucose in complex oligosaccharides is linked peripheral to GlcNAc residues and only 29% is linked to the proximal GlcNAc, which is N-glycosidically linked to asparagine (Krusius and Finne, 1977, 1978). Approximately 25% of the GlcNAc is terminal while greater than half is O-substituted at the 4-position. Over 50% of the galactose is 3-O-substituted, usually by sialic acid, while the remainder is primarily terminal (Krusius and Finne, 1977). Most of the sialic acid in brain glycoproteins is at the terminal, nonreducing position and about 10% is 8-O-substituted by another sialic acid (Finne et $al.$, 1977). The core pentasaccharide of brain glycoproteins most often contains a branch point at a disubstituted mannose residue. Additional mannose residues, peripheral to the branch point, are ususally mono- and sometimes disubstituted (Krusius and Finne, 1977).

Approximately 15% of the carbohydrate in brain glycoproteins is O-glycosidically linked, through GalNAc, to the hydroxyl groups of serine and threonine (Margolis and Margolis, 1983). These oligosaccharides contain a core disaccharide that consists of galactose and GalNAc. Sialic acid is commonly found substituted at the C-3 of galactose and/or the C-6 of GalNAc. The resulting disialylated tetrasaccharide makes up about 50% of all O-glycosidically linked carbohydrates in brain glycoproteins (Margolis and Margolis, 1979). Other sugars (e.g., rhamnose, xylose, and mannosamine), which have been detected in small quantities in brain glycopeptide fractions (Brunngraber and Brown, 1964; Brunngraber et $al.$, 1970), are not established components of animal glycoproteins (Margolis and Margolis, 1979). Further studies are necessary to determine whether or not these unusual sugars are true constituents of nervous system glycoproteins.

The principal carbohydrate residues present in synaptic junction (SJ) and postsynaptic density (PSD) glycoproteins are mannose (43 and 130 nmoles carbohydrate/mg protein, respectively), galactose (44 and 31 nmoles/mg protein), glucosamine (45 and 46 nmoles/mg protein), fucose (32 and 17 nmoles/mg protein), sialic acid (19 nmoles/mg protein in SJs; not determined for PSDs), and galactosamine (20 and 13 nmoles/mg protein) (Churchill *et al.*, 1976). When comparisons are made among synaptic plasma membrane (SPM), SJ, and PSD fractions, galactose and mannose are more concentrated in SJs and PSDs, respectively (Churchill *et al.*, 1976). Slightly differing values for the carbohydrate content of individual synaptic fractions have been reported by numerous laboratories (Breckenridge *et al.*, 1972; Vincendon *et al.*, 1973; Margolis *et al.*, 1975; Krusius *et al.*, 1978; Mahler, 1979).

Sulfated sugars have been found in brain glycoproteins and have been identified as galactose 6-sulfate and GlcNAc 6-sulfate (Margolis and Margolis, 1970; Allen *et al.*, 1976). Brain sulfated glycoproteins are distinct from glycosaminoglycans in that the former contain higher levels of mannose and fucose when compared to keratan sulfate (Margolis and Margolis, 1979). In brain, sulfated glycoproteins are not confined to any one subcellular fraction. For example, in CNS and PNS myelin, both major glycoproteins are sulfated (Matthieu *et al.*, 1975a,b). Sulfated glycoproteins are present in SPM, which are distinct from those found in myelin or glycosaminoglycans (Simpson *et al.*, 1976).

3. GLYCOPROTEINS OF SYNAPTIC MEMBRANES AND SYNAPTIC JUNCTIONS

Membrane glycoproteins, especially those that extend into the extracellular environment, play important roles in cell–cell recognition and intercellular adhesion (Barondes, 1976; Hausman and Moscona, 1976; Thiery *et al.*, 1977). Within the nervous system, glycoproteins have been shown to mediate adhesion between specific cellular elements (see Section 5). Macromolecules have been isolated and characterized from nervous system tissues that display lectinlike properties and participate in intercellular recognition and adhesion (Barondes *et al.*, 1974; Barondes and Rosen, 1976; Simpson *et al.*, 1977, 1978; Kobiler *et al.*, 1978). The results from such studies have led many to postulate that glycoproteins extending from the surface of neurons are in strategic locations to be involved in developmental recognition events that direct the formation of specific synaptic connections between neurons and, secondarily, to maintain the adhesive stability of synapses once they have been formed. Because of these important functional implications associated with synaptic glycoproteins, researchers have attempted to identify and characterize glycocomponents present on the external surface of synaptic membranes.

The synapse is the basic structural–functional subunit of the nervous system (Fig. 1). It is the point of close physical contact and site of physiological communication between neurons. The vast majority of synapses in the CNS are axodendritic and possess a very characteristic morphol-

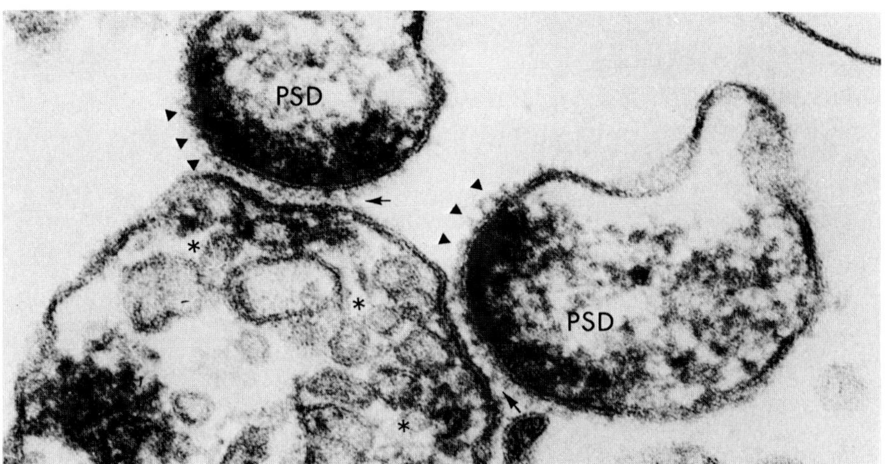

Figure 1. Electron micrograph of two asymmetric synapses from a synaptosomal fraction prepared from adult rat forebrain. The presynaptic terminal contains numerous synaptic vesicles (asterisks); arrows point to synaptic clefts containing electron-dense material. The postsynaptic membrane overlying the postsynaptic density (PSD) displays numerous bristles or knobs (arrowheads) that extend outward from the neuron's surface; these knobs may correspond to the postsynaptic membrane glycoproteins that bind Con A.

ogy, the major feature of which is the presence of prominent PSD ($\sim$ 300–500 Å thick) on the cytoplasmic surface of the postsynaptic membrane (PSM) (Gray, 1959; Bloom, 1970). This SJ complex is characteristic of the "asymmetric," type I synapse [Gray's (1959) designation]. In the cerebral cortex and cerebellum of the mammalian brain, approximately 80% of all synapses are of this type (Colonnier, 1968). A subcellular fraction highly enriched in asymmetric SJs has proven valuable in elucidating many of the functional workings of synapses (Fiszer and De Robertis, 1967; Davis and Bloom, 1973; Cotman and Taylor, 1972; Kelly and Cotman, 1977, 1978, 1981; Blomberg *et al.*, 1977; Kelly *et al.*, 1979; Rostas *et al.*, 1979; Gurd, 1980; Carlin *et al.*, 1980; Kelly and Montgomery, 1982). SJ fractions (Fig. 2) consist of 15–20% intact SJs and 55% PSM specializations [this is the postsynaptic junctional membrane with attached PSD; presynaptic structures are absent (Churchill *et al.*, 1976)]. Approximately two-thirds of the junctional structures in SJ fractions contain morphologically intact regions of postsynaptic plasma membrane (Rostas *et al.*, 1979; Kelly and Cotman, 1981; Kelly and Montgomery, 1982). SJs possessing similar molecular and morphological properties can be prepared from a variety of species, including human (Rostas *et al.*, 1979; Nieto-Sampedro *et al.*, 1982). SJs have also been isolated from discrete brain regions and, except for cerebellum, display molecular and morphological

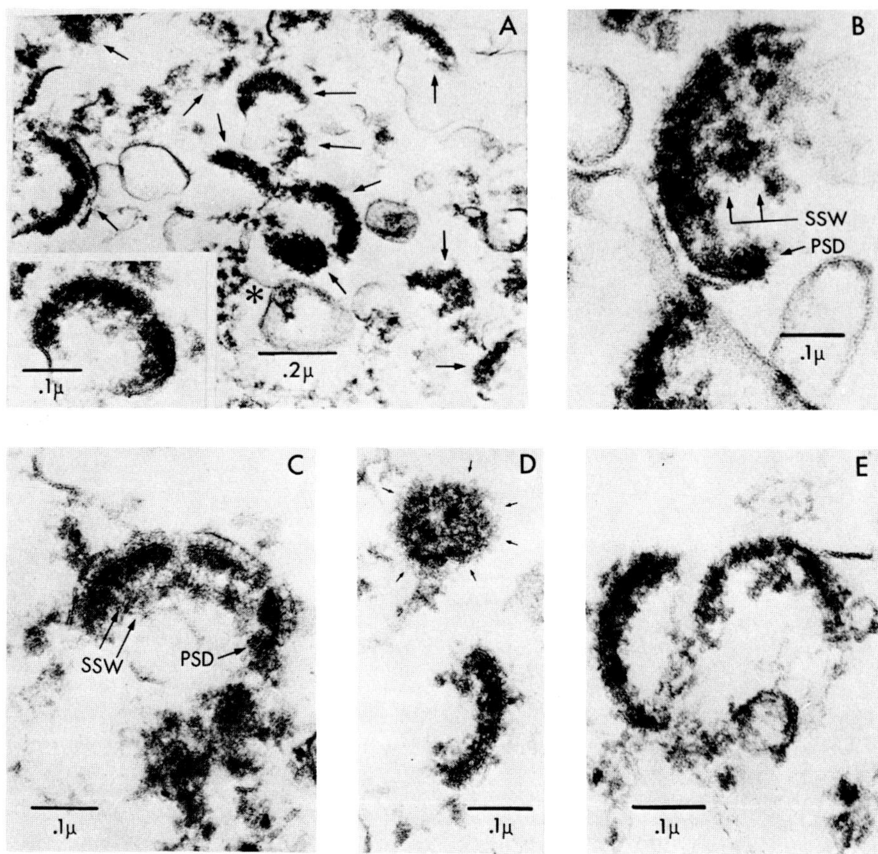

Figure 2. Electron micrographs of synaptic junction (SJ) fractions from rat forebrain (A) and cerebellum (B–E). Arrows in (A) designate postsynaptic membrane specializations (postsynaptic membrane with underlying PSD, no presynaptic structures present). Subsynaptic web (SSW) and PSDs are shown in planes parallel to the postsynaptic membrane [asterisk (A); structure circumscribed by arrows in (D)]. (C) shows an intact SJ with pre- and postsynaptic membranes, cleft material, SSW, and PSD.

properties that are highly conserved (Kelly and Cotman, 1977; Rostas *et al.*, 1979; Carlin *et al.*, 1980; Groswald *et al.*, 1983).

Early cytochemical studies demonstrated that the synaptic cleft (the intercellular space at the point of synaptic contact) contains high levels of carbohydrates (Rambourg and Leblond, 1967; Bondareff, 1967; Bondareff and Sjöstrand, 1969; Pfenninger, 1973). Cytochemistry, however, does not reveal the precise molecular nature of carbohydrate-containing components at synapses. During the 1970s, plant lectins were used as molecular probes to elucidate the subcellular location and molecular identity of synaptic glycoproteins. Histochemical studies using ferritin conjugates of Con A and *Ricinus communis* agglutinin (RCA) have shown that the external surface of the PSM at the synaptic cleft is enriched in receptors for these lectins (Matus *et al.*, 1973; Bittiger and Schnebli, 1974; Cotman and Taylor, 1974; Kelly *et al.*, 1976). The apparent density of lectin receptors localized to the PSM at the synapse is 6- to 10-fold greater when compared to receptors on adjacent, extrajunctional membranes or receptors on the presynaptic membrane (Fig. 3) (Kelly *et al.*, 1976). Lectin receptors on the PSM either comprise, or are in very close proximity to, the "bristle," knoblike specializations that extend approximately 90 Å into the cleft (Van der Loos, 1963; Cotman and Taylor, 1972). By ultrastructural appearance, these "bristles" somewhat resemble the fibrous nature of the PSD and may represent a structural, transmembrane extension of the PSD (Matus and Walters, 1975). Lectin receptors in the membrane overlying the PSD do not patch or aggregate in the presence of saturating amounts of multivalent lectins, which distinguishes them from extrajunctional receptors, which readily aggregate in response to lectin binding (arrowheads, Fig. 3A) (Kelly *et al.*, 1976). The restricted mobility of these Con A-binding receptors is an intrinsic property of the PSM, for it is observed following detachment of presynaptic membranes by mechanical sheer force (Fig. 3). These observations suggest that a direct or indirect interaction of membrane components with the molecular scaffolding of the underlying PSD may serve in restricting their ability to laterally diffuse and aggregate in response to multivalent lectins. The distribution and mobility of these postsynaptic lectin receptors is a property common to different neuronal types [cerebellar Purkinje cells, and hippocampal pyramidal and granule cells (Kelly *et al.*, 1976)].

Postsynaptic receptors for Con A are retained in subcellular fractions that are highly enriched in SJ structures (Cotman and Taylor, 1974). Isolated SJs display large amounts of Con A receptors on their PSM with little or no labeling of extrajunctional membranes or the cytoplasmic surface of PSDs (Fig. 3B). The receptors for Con A in SJs were identified by the binding of [^{125}I]-Con A to glycoproteins in SJ fractions that had first been separated in SDS gels (Kelly and Cotman, 1977; Gurd, 1977). SJ glycoproteins in four molecular weight classes account for greater than 80% of the Con A-binding activity in this fraction. Results from different laboratories have assigned these glycoproteins (GPs) to the following molecular weight classes: GP-I (160–180K), GP-II (125–130K), GP-III (115–125K), and GP-IV (95–110K) (see Fig. 3C). GP-III binds approximately 47% of the total [^{125}I]-Con A, GP-IV binds 28%, GP-I binds 21%, and GP-II binds 5–7% (Kelly and Cotman, 1977; Gurd, 1977). GP-III is frequently resolved on gels as a narrowly spaced doublet (Kelly and Mont-

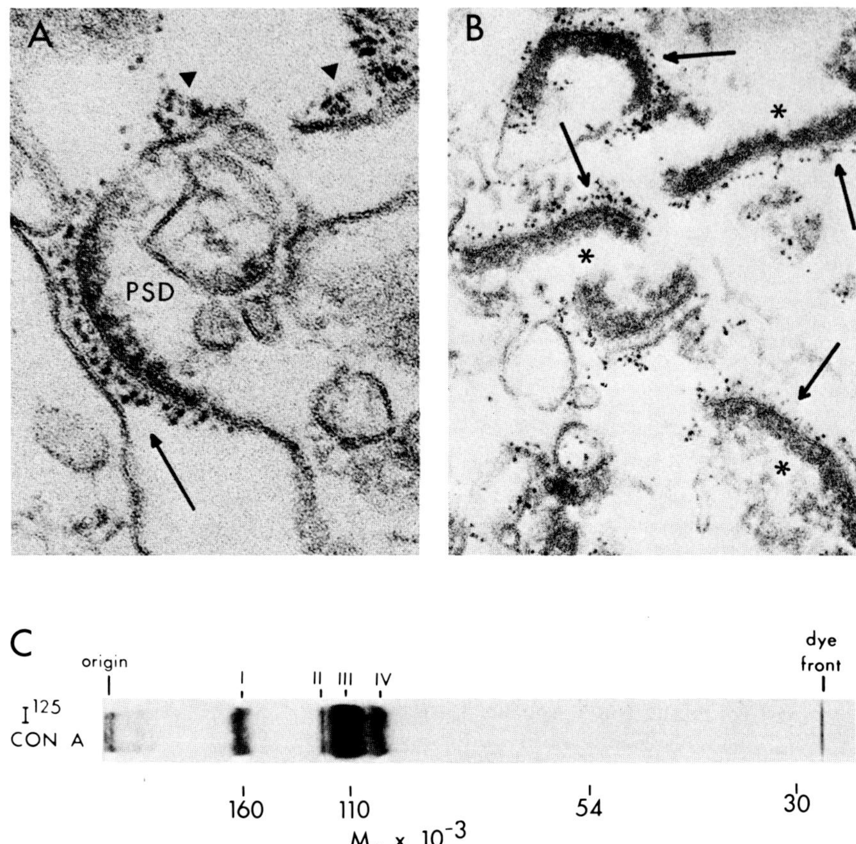

Figure 3. (A) Ferritin–Con A binding to dendritic postsynaptic membrane of hippocampal Purkinje cell at 37°C. Conjugates bound to the junctional surface of the postsynaptic membrane are dense and linearly uniform (arrow), while binding to extrajunctional membranes is often organized in clusters or aggregates (arrowheads). (B) Ferritin–Con A binding to isolated SJs; electron-dense ferritin particles bind to the external surface of the postsynaptic membrane overlying the PSD (arrows); binding is not observed on the cytoplasmic face of the PSD (asterisks). The distribution (linearity and density) of ferritin–Con A binding to synaptic structures in SJ fractions is very similar to that observed for tissue homogenates [compare to (A)]. (C) Autoradiogram showing the pattern of [^{125}I]-Con A-binding glycoproteins in forebrain SJ fractions (55 μg protein). Approximate M_r are indicated along the bottom of the autoradiogram; the four major Con A-binding glycoproteins are labeled I, II, III, and IV.

gomery, 1982; Mena *et al.*, 1981). Estimates on the relative distribution of the different Con A-binding GPs, based on protein staining, indicate that GP-IV comprises 61%, GP-III 24%, and GP-I about 15% of the Con A-positive GPs in SJs (Gurd, 1980). Thus, in terms of the specific relative binding of [^{125}I]-Con A, GP-III has the greatest number of lectin receptors per polypeptide (GP-I has about 20% less and GP-IV has 65% less compared to GP-III). Through the use of additional lectins, Gurd and co-workers (see Gurd, 1982) have shown that the three major molecular weight classes of Con A-binding GPs contain fucose, sialic acid, and *N*-acetyl-D-glucosamine, in addition to mannose. These GPs are sensitive to mild proteolytic digestion (Kelly and Cotman, 1977) and their capacity to bind Con A is greatly reduced (greater than 70–95%) by digestion with endoglycosidase H or α-mannosidase (Gurd, 1980; Gurd and Fu, 1982). Moreover, when intact SJs are treated with endoglycosidase H, greater than 90% of the Con A-binding activity of each GP is removed (Groswald and Kelly, unpublished data). These results support early cytochemical findings that showed that the carbohydrate (lectin-binding) moieties of these postsynaptic GPs extend from the external surface of the membrane and, hence, are capable of physically interacting with proteolytic and glycolytic enzymes. Although the exact structure of the oligosaccharides of these GPs is unknown, their sensitivity to specific endoglycosidases and carbohydrate composition inferred by gel chromatography of pronase-liberated ^{3}H-labeled oligosaccharides, as well as studies on their differential lectin-binding properties, suggests that they are high-mannose, complex asparagine-linked chains (Gurd and Fu, 1982).

Results from different laboratories have shown that the Con A-binding GPs are uniquely associated with the PSM of asymmetric, type I synapses (studies to date have been performed primarily on tissues of forebrain origin). These GPs are not detected in subcellular fractions enriched in myelin, microsomes, mitochondria, synaptic vesicles, extrajunctional synaptic membranes, cytoplasm, and axolemma (Gurd, 1980; Kelly and Montgomery, 1982; Mena *et al.*, 1981). The unequivocal localization of these presumptive, synapse-specific components in neural tissues awaits immunohistochemical studies at the EM level that use antibodies against individual GPs.

The Con A-binding receptors in SJ fractions were the first GPs to be identified in brain that extend from the external surface of the postsynaptic junctional membrane. Similar molecular weight classes of Con A-binding GPs have been identified in SJ fractions prepared from forebrain and cerebral cortex, as well as forebrains from a number of species [rat, rabbit, chicken, steer, human, mouse, dog, lizard (Rostas *et al.*, 1979; Nieto-Sampedro *et al.*, 1982)]. Moreover, discrete brain regions obtained by microdissection and which contain predominantly one neuronal type [hippocampal CAl and dentate gyrus molecular layer, and caudate nucleus (Rostas *et al.*, 1979)] yield SJ fractions with similar patterns of Con A-binding GPs. Thus, on the basis of Con A binding, these GPs do not appear to be preferentially distributed among different synaptic types. The possibility of synapse-specific differences in the carbohydrate sequences of individual oligosaccharides that may distinguish one GP from another awaits further investigation.

The developmental appearance (by biochemical analysis) of the four prominent molecular weight classes of Con A-binding GPs in SJs closely parallels the process of synaptogenesis (Kelly and Cotman, 1981; de Silva *et al.*, 1979). Between postnatal days 5 and 90, the amount of Con A binding to the molecular weight regions of GP-I, GP-II, GP-III, and GP-IV increases approximately 50, 53, 49, and 136%, respectively (estimates based on results by Kelly and Cotman, 1981). It is not known whether these increases are a consequence of developmental changes in the number of GPs, the carbohydrate content of their respective oligosaccharides, or both. Developmental studies that have examined the *in vivo* incorporation of labeled fucose and ManNAc have shown that the oligosaccharides of these SJ GPs continue to be modified following the initial formation of synaptic contacts (Fu *et al.*, 1981). Gurd has also shown that the Con A-binding GPs in SJs are substrates for endogeneous neuraminidases (Margolis and Margolis, 1983). Whether or not the sialic acid content of these GPs is modulated during synapse formation remains unknown.

Using ferritin–Con A conjugates we have recently localized the surface distribution of these Con A-binding GPs in rat hippocampal dentate gyrus and CAl regions during the period of brain development that coincides with synaptogenesis (postnatal days 5 through 17). Receptors for Con A are first detected on the surface of the PSM as soon as the earliest morphological signs of an immature PSD are observed (Fig. 4; Kelly, unpublished observations). These studies have also demonstrated that early appearing Con A receptors display restricted lateral mobility, a feature reminiscent of their counterparts in the mature PSM. Additional experiments have been partially successful in demonstrating that immature Con A receptors may possess a more dynamic state of lateral mobility (Kelly, unpublished observations). The conditions that most enhance their mobility are incubations of synapses in drugs that disrupt microtubules (colchicine and nocodazol) and microfilaments (cytochalasin B), together with antibodies against Con A to induce patching or aggregation of previously bound ferritin–Con A (Fig. 4). It appears that this additional, external cross-linking ligand (anti-Con A) is critical for experimentally increasing the lateral mobility of these receptors.

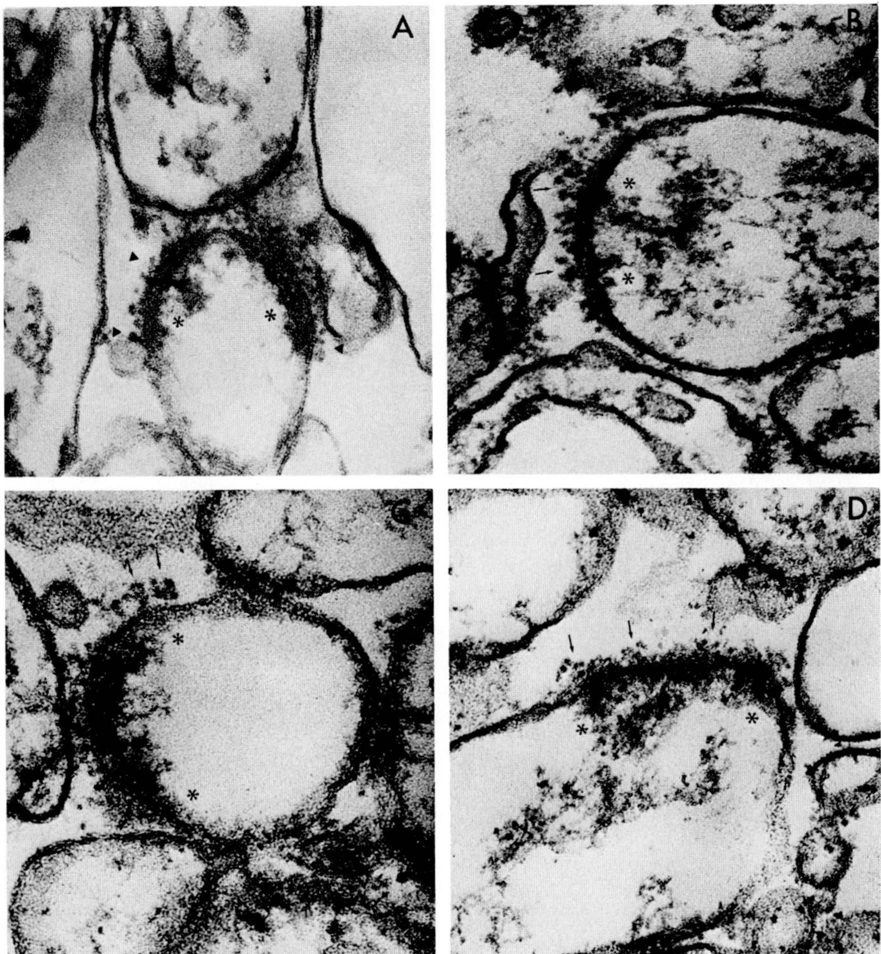

Figure 4. Ferritin–Con A binding to dendritic postsynaptic membranes of granule cells in newborn rat hippocampus. (A) Eight-day tissue homogenates, incubated at 37°C for 15 min (control labeling conditions), pelleted by centrifugation and reincubated as described below; note uniform labeling (arrowheads) of membrane overlying PSD (asterisks) under these conditions. (B–D) Nine-day tissue homogenates, incubated for 15 min at 37°C in the presence of cytochalasin B (10 μg/ml), nocodazole (0.5 μg/ml), and ferritin Con–A. Following incubation at 37°C, samples were washed free of unbound ferritin–Con A by centrifugation at 10,000g, resuspended in buffer containing cytochalasin B and nocodazole plus anti-Con A antibody (1:50 dilution). Samples were then incubated 15 min at 37°C, fixed with 4% glutaraldehyde, and processed for EM. In the presence of nocodazole, cytochalasin B, and anti-Con A, lectin receptors on the external surface of the postsynaptic membrane display a high tendency to aggregate into clusters [see arrows in (B–D)].

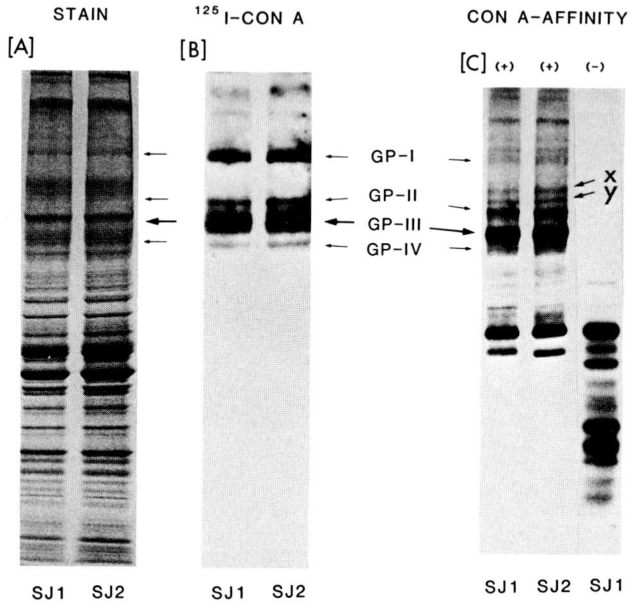
STAIN
[A]
125I-CON A
[B]
CON A-AFFINITY
[C]
(+) (+) (-)
GP-I
GP-II
GP-III
GP-IV
x
y
SJ1 SJ2
SJ1 SJ2
SJ1 SJ2 SJ1

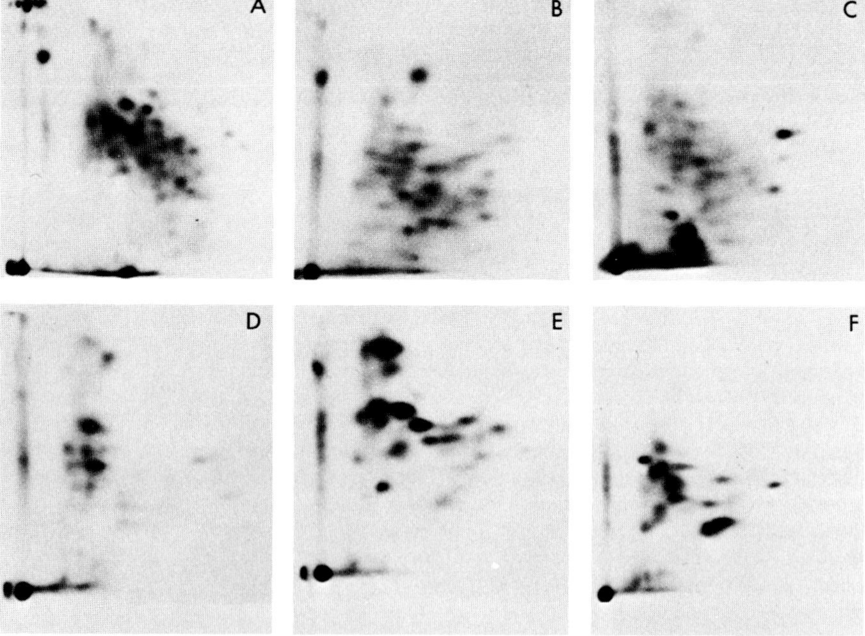
A
B
C
D
E
F

As discussed earlier, SJs isolated from a number of distinct regions in the mammalian forebrain contain four major Con A-binding GPs, each one possessing an apparent molecular weight that is indistinguishable from its counterpart in other brain regions (Kelly and Cotman, 1977; Rostas *et al.*, 1979). Because of their possible involvement in cell–cell recognition and synapse formation, a number of laboratories have attempted to identify specific molecular properties that may distinguish one Con A-binding GP from another. Mena and Cotman (1982) have isolated GP-I, GP-II, GP-III, and GP-IV from SJs by affinity chromatography on Con A–Sepharose. Con A-positive GPs were then separated on SDS gels, their respective bands cut out, iodinated, and processed for two-dimensional peptide mapping by the method described by Elder *et al.* (1977). Comparisons of tryptic fingerprints for each of these GPs indicated that some similarities were apparent between GP-I, GP-II, and GP-III; however, GP-IV was clearly distinct from the other three. The greatest apparent similarities were between GP-II (125–130K) and GP-III (115–125K). However, this apparent relatedness may have been unavoidable because (1) the very close molecular weights of GP-II and GP-III make it difficult to clearly separate them on one-dimensional gels, and (2) GP-II is a relatively minor Con A-binding component compared to GP-III and sometimes runs in the high-molecular-weight shoulder of the GP-III band. In addition, the method of peptide mapping of proteins in individual gel slices used by Mena and Cotman (1982) is sensitive to artifacts when applied to proteins that are minor staining components, such as GP-II. This is a consequence of nonprotein-related peptide spots that result from interfering substances in the acrylamide and/or gel slice. Such autoradiographic spots are generated when a mock gel slice or a slice containing

---

Figure 5. (Upper) Coomassie blue-stained gel (A) of rat SJs (50 μg) and the autoradiogram (B) of this same gel showing the pattern of [^{125}I]-Con A-binding glycoproteins (GPs) from two separate preparations of SJs (1 and 2); arrows indicate GP-I, -II, -III, and -IV. SJ fractions were then solubilized in 2% SDS, iodinated by the chloramine-T method, and fractionated on a Con A–Sepharose affinity column in the presence of 0.1% SDS (w/v) and 2 mM dithiothreitol. Affinity chromatography resulted in Con A-positive (+) (glycoproteins retarded and subsequently eluted with 0.2 M α-mannoside) and Con A-negative fractions (−) (material unretarded by Con A–Sepharose). (C) The autoradiographic patterns of these fractions; radioactive bands that correspond in molecular weight to their counterparts in panel (B) are designated with arrows. (Lower) Autoradiograms of tryptic/chymotryptic ^{125}I-labeled peptides generated from affinity-purified Con A-binding glycoproteins of SJs shown in (C) of upper panel. (A) GP-I, (B) GP-X [145K glycoprotein shown in (C) of upper panel; arrow], (C) GP-Y [140K glycoprotein shown in (C) of upper panel; arrow], (D) GP-II, (E) GP-III, and (F) GP-IV. Approximately 5000 cpm of each sample was subjected to two-dimensional electrophoresis (left to right) followed by chromatography (bottom to top); autoradiographic exposures ranged from 5 to 7 days.

very little protein is radioiodinated and processed for peptide mapping (Elder *et al.*, 1977; Elder, personal communication; Kelly, unpublished observations). Thus, when mapping a number of different, but minor, protein components from the same gel, one must be able to subtract those autoradiographic spots that are protein-independent from the genuine ^{125}I-labeled peptides that are specific to the protein being examined.

Experiments from this laboratory and those done in collaboration with James Gurd have also examined the major Con A-binding GPs in SJs for possible molecular similarities. These studies have been done with a modification of the technique initially described by Elder *et al.* (1977). The principal difference is that the Con A-binding GPs are first labeled in solution with either [^{125}I]- or [^{32}P]-ATP prior to affinity chromatography, gel electrophoresis, and peptide mapping. Using this modification, all Con A-binding GPs are simultaneously labeled, in solution, and under identical conditions (Fig. 5). Using this procedure, the labeling of minor protein components in individual gel slices and the possible introduction of nonprotein-related autoradiographic spots are avoided. Our results differ from those of Mena and Cotman (1982) in that the tryptic or tryptic/chymotryptic peptide finderprints of GP-I, GP-II, GP-III, and GP-IV appear very dissimilar (Fig. 5) (Gurd *et al.*, 1983; Kelly and Hay, in preparation). These results suggest that individual Con A-binding GPs do not share a common polypeptide backbone. However, subtle differences in the amino acid or carbohydrate sequence of a single GP could exist between different synaptic types or between brain regions. Not until larger and completely homogeneous quantities of these Con A-binding GPs are obtained will more detailed studies on specific molecular differences and similarities be possible.

In collaboration with James Gurd, we have examined the endogenous phosphorylation of Con A-binding GPs in SJs (Gurd *et al.*, 1983). These experiments show that these GPs are phosphorylated *in vitro* by cAMP-dependent kinases. The incorporation of ^{32}P into GP-III displays the greatest increase in the presence of cAMP (198% of control values), although the labeling of all four GPs is stimulated by cAMP. cAMP selectively stimulated the phosphorylation of only two of the six phosphopeptides associated with GP-III; all peptides are phosphorylated to some degree in the absence of cAMP. The ability of intrinsic cAMP-dependent kinases, which are presumably located on the cytoplasmic surface of the PSM, to phosphorylate these GPs (whose oligosaccharide moieties extend into the synaptic cleft) provides further evidence for their transmembrane orientation. We have recently observed in SJ fractions that GP-I is phosphorylated by an endogenous Ca^{2+}/calmodulin-dependent protein kinase II (Kelly *et al.*, 1983, 1984). The observation that GP-I and GP-III are retained in isolated PSDs (Fig. 6) adds further support to the notion that these GPs span the PSM and are anchored to the cytoskeletal scaffolding of the PSD (Kelly and Cotman, 1977; Gurd *et al.*, 1983; Groswald *et al.*, 1982; 1983). These GPs appear to be the first transmembrane components that have been characterized and localized to the PSM of CNS asymmetric synapses. Such a molecular orientation appears to ideally suit these GPs for a role in the regulation of postsynaptic function(s).

Lectin mapping studies showed that the external surface orientation, lateral mobility, and relative density of receptors for Con A on postsynaptic junctional membranes are indistinguishable between forebrain and cerebellum type I synapses (Kelly *et al.*, 1976). Con A-binding GPs have recently been identified in SPM and SJ fractions isolated from rodent cerebellum (Groswald *et al.*, 1982; 1983). The Con A-binding GPs that characterize SJ fractions from forebrain and are markers for the PSM of asymmetric, type I synapses (Gurd, 1980; Kelly and Montgomery, 1982; Mean *et al.*, 1981), are absent, or undetectable in cerebellum SJ fractions (Fig. 6). Although the numbers and approximate M_r of Con A-binding GPs in cerebellum and forebrain SJ fractions are somewhat analogous, their similarities end here. In cerebellum SJ fractions, five prominent Con A-binding GPs are resolved on SDS gels. The major glycoprotein (on the basis of protein staining and Con A binding) has an apparent M_r of 240K and will be referred to as GP-A (Fig. 6). Cerebellum SJs also contain GPs of 190, 180, 128, and 113K. When cerebellar GPs are compared on the basis of relative [^{125}I]-Con A binding, GP-A > GP-128K > GP-190K = GP-180K > GP-113K. Comparisons made between SJs from cerebral cortex and cerebellum indicate that GP-III and GP-A bind roughly equivalent amounts of [^{125}I]-Con A per mg total SJ protein. GP-A appears to be a cerebellum-specific GP; it is undetectable in soluble fractions from cerebellum as well as plasma membrane purified from cerebral cortex, spinal cord, liver, kidney, and lens. GP-A is almost totally resistant to the detergent extraction used in the purification of cerebellum SJ fractions. These studies suggest that the GP composition of the PSM in cerebellum and forebrain SJ fractions is quite different.

We have obtained some evidence concerning the molecular basis for the observed differences in GPs between forebrain and cerebellum SJ fractions. The major Con A-binding GP, GP-A (240K), in cerebellum SJ fractions is undetectable in forebrain subcellular fractions (Groswald and Kelly, 1984). In both forebrain and cerebellum SJ fractions, a narrowly spaced protein doublet is present (230 and 235K), which in cerebellum migrates just ahead of GP-A. The peptide fingerprints of either band in this doublet are similar (greater than 90% of the tryptic/chymotryptic peptides of each comigrate with its counterpart following two-dimensional separations), and the doublet in forebrain SJs is indistinguishable from the cerebellum doublet. This doublet is indistinguishable from fodrin [a

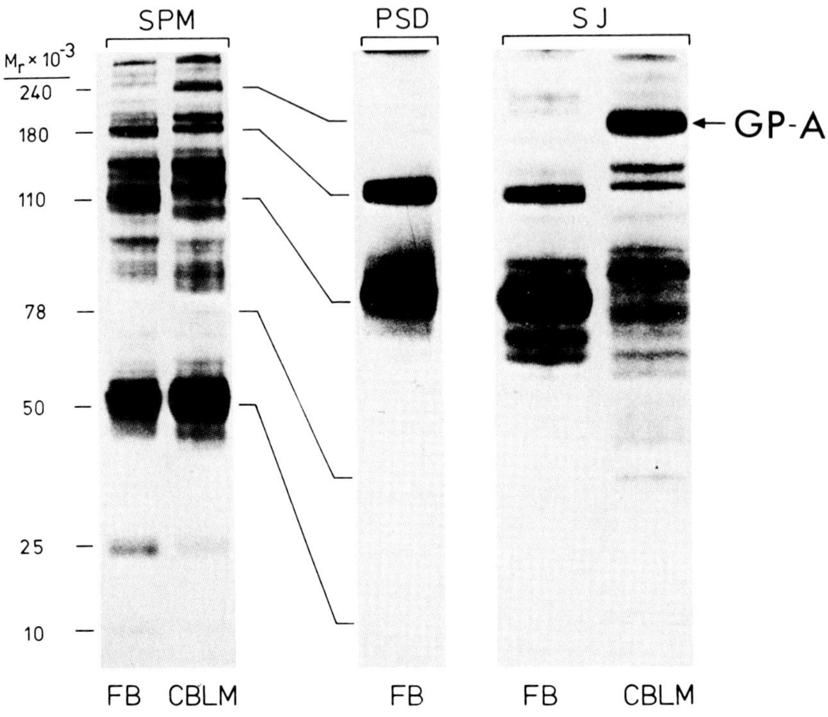

Figure 6. Autoradiograms of [^{125}I]-Con A-binding glycoproteins in synaptic plasma membrane (SPM) and synaptic junction (SJ) fractions from forebrain (FB) and cerebellum (CBLM), and a postsynaptic density (PSD) fraction from FB. Approximate M_r are indicated at the left; the cerebellum-specific glycoprotein (GP-A) is indicated at the right. Electrophoresis was carried out in one-dimensional SDS gels. Each sample contained approximately 55 μg protein. Autoradiograms were exposed for 24 hr, except for the PSD fraction, which was exposed for 96 hr.

protein doublet of 230 and 235K (Levine and Willard, 1981)]; an antifodrin antibody (kindly provided by Mark Willard and co-workers) reacts with the 235K polypeptide of this doublet in both forebrain and cerebellum SJs. More interesting, however, is the finding that the fingerprint of cerebellum GP-A is very similar to the 235K polypeptide of the SJ-fodrin doublet (greater than 50% of the tryptic/chymotryptic peptides of each comigrate following two-dimensional peptide fingerprinting). This represents an extremely high degree of structural relatedness, as unrelated proteins share less than 3% of their respective tryptic/chymotryptic peptides. Neither purified fodrin or SJ-fodrin binds appreciable amounts of Con A. These results suggest that differences in related macromolecules between forebrain and cerebellum SJs may reside in the oligosaccharide rather than the polypeptide region of these GPs. This could result from neuroanatomical differences in the activity and/or specificity of glycosyltransferases or other carbohydrate-modifying enzymes. Goodrum *et al.* (1979) have shown that SPM fractions, and the junctional and non-junctional subfractions derived from SPMs, contain GP galactosyltransferase activity. The nonjunctional fraction (Triton X-100-soluble SPM fraction) contained the highest specific activity in the presence of exogenous acceptor, suggesting an enrichment of the enzyme in this fraction. In contrast, the SJ fraction has the greatest endogenous activity in the absence of added acceptor. These results suggest that there is a relative enrichment of endogenous acceptors for galactosyltransferases in SJ compared to SPM fractions.

The amount of GP-A in cerebellum SJs is dependent on whether or not the parent SPM fraction has bound Ca^{2+} (Groswald and Kelly, 1984). If Ca^{2+}-free solutions are used during the isolation of SJs from cerebellum, greater than 80% of the GP-A normally found in cerebellum SJs is extracted during the Triton X-100 treatment of SPMs. Cerebellum SPMs isolated in the absence of Ca^{2+}, treated with 100 μM Ca^{2+} prior to extraction in 0.25% Triton X-100, yield SJ fractions with normal amounts of GP-A. SPMs isolated in the absence of Ca^{2+}, then treated with 100 μM Ca^{2+} and washed (to remove unbound Ca^{2+}), yield SJ fractions with normal amounts of GP-A. The ability of SPM-bound Ca^{2+} to protect GP-A from being solubilized by Triton may be mediated by calmodulin, for isolated SJ (Kelly *et al.*, 1983) and PSD fractions (Grab *et al.*, 1979) contain large amounts of calmodulin ($\sim$ 5–10 μg calmodulin/mg SJ protein). GP-A is related to fodrin and fodrin is known to interact with calmodulin as it can be affinity-purified using immobilized calmodulin (Glenney *et al.*, 1982).

The major Con A-binding GPs in the PSM of forebrain SJs are phosphorylated in a cAMP-stimulated manner (see above; Gurd *et al.*, 1983). Using cerebellum SJs and the same *in vitro* phosphorylation paradigm, we have observed the Ca^{2+}/calmodulin-stimulated phosphorylation of a protein with a molecular weight similar to that of GP-A. In cerebellum SJ fractions, the endogenous phosphorylation of this 240K protein is stimulated approximately 1000% by the addition of Ca^{2+} (0.5 mM) and calmodulin (1 μg/70 μg SJ protein) (Kelly and Hay, unpublished results). In these experiments, a Ca^{2+}/calmodulin-stimulated phosphorylation of fodrin is not observed. The possibility that GP-A may be a substrate for a Ca^{2+}/calmodulin-dependent protein kinase, and the observation that the solubility of GP-A in Triton X-100 is greatly affected by whether or not SPM fractions have been pretreated with Ca^{2+}, together suggest a functional coupling between this physiologically important ion and this cerebellum SJ glycoprotein.

Our results support the notion that GP-A is an integral PSM GP in SJs, with its oligosaccharide portion extending into the synaptic cleft. The nearly quantitative recovery of GP-A in SJs from the parent SPM fraction may be due to its being an integral membrane GP as well as its possible anchorage to a membrane-associated cytoskeleton. Membrane GPs have been shown to interact with intracellular cytoskeletal proteins e.g., spectrin, tubulin, and actin–myosin (Bourguignon and Singer, 1977; Schlessinger et al., 1977; Branton et al., 1981; Prives et al., 1982). The PSM of asymmetric, type I synapses has a specialized cytoskeletal structure, the PSD (Cotman and Kelly, 1980). PSDs contain tubulin and actin (Blomberg et al., 1977; Kelly and Cotman, 1978), and SJ fractions contain myosin (Beach et al., 1981), a Ca^{2+}-modulated Mg^{2+}-ATPase (Cotman and Kelly, 1980), and α-actinin (Kelly and Bright, unpublished results). α-Actinin is an actin-binding protein whose proposed function is to cross-link and possibly anchor actin filaments to the plasma membrane in nonmuscle cells and thin filaments to the Z disk in skeletal muscle (Stromer and Goll, 1972; Mooseker and Tilney, 1975; Mooseker, 1976). Like α-actinin, fodrin appears to be a bona fide actin-binding protein (Levine and Willard, 1981), and GP-A is structurally related to fodrin. Whether or not GP-A is an actin-binding protein remains unknown. The proposed anchorage role of actin-binding proteins like fodrin and α-actinin, and possibly GP-A, would appear to ideally suit them for restricting the lateral mobility of synaptic glycoproteins and receptors by anchoring them to the PSD's cytoskeletal scaffolding. The anchoring of the PSD's cytoskeletal system to membrane components such as the Con A-binding GPs, could constitute a mechanism that modulates and restricts the lateral mobility of these postsynaptic components (Cotman and Kelly, 1980). Therefore, if GP-A does in fact bind actin at the synapse, its proposed function would be to bind actin filaments in the PSD and thus anchor the PSD to the cytoplasmic surface of the PSM. Recent immunofluorescence studies have shown that fodrin is localized to the cortical cytoplasm (resembing a cell lining) in neurons, nonneuronal tissues, cultured fibroblasts, and mouse 3T3 cells (Levine and Willard, 1981). Its intracellular location, therefore, prohibits a direct involvement in cell–cell interaction. Circumstantial evidence suggests that GP-A's carbohydrate portion extends into the synaptic cleft (Kelly et al., 1976) and could, therefore, interact with molecules located on the neighboring presynaptic membrane. Such an orientation suggests a possible role in cell–cell interaction for GP-A. GP-A may represent a molecular extension (connection) from the PSD (where GP-A's postulated actin-binding region is anchored), through the PSM and into the synaptic cleft. Within the cleft, we postulate that GP-A's oligosaccharide region may be involved in interacting with the presynaptic membrane of an adjacent neuron. A molecular extension of this nature represents a simple linkage mechanism that could underlie cell–cell adhesion at the synapse.

4. CELL SURFACE GLYCOCOMPONENTS OF CULTURED NEURONS: DISTRIBUTION AND DYNAMICS OF LECTIN RECEPTORS AT THE GROWTH CONE

Pfenninger and Maylie-Pfenninger (1981a,b) have used ferritin–lectin conjugates to examine the distribution of glycoconjugates on the surface of sprouting neurons in cultures derived from superior cervical ganglion (SCG) and spinal cord (SC). The densities of lectin receptors on the surface of cell bodies, neurite shafts, and growth cones were determined. In these studies, three general types of lectin-binding patterns were observed: (1) roughly equal density of lectin receptors covering the entire neuron (e.g., SCG neurons labeled with RCA I), (2) a gradual increase of lectin receptors in a gradient along the cell body–growth cone axis (e.g., SC neurons labeled with RCA II), and (3) a gradual decrease of lectin receptors in a gradient along the same perikaryal–growth cone axis (e.g., SCG and SC neurons labeled with wheat germ agglutinin). Moreover, the density of receptors for each lectin was different between these two types of neurons. Control experiments were carried out to guard against artifacts in lectin mapping patterns that could have resulted from nonspecific adsorption of media-derived glycocomponents to neuronal membranes as well as changes in the density and mobility of receptors that could have resulted from tissue dissociation prior to plating neurons in culture.

Although these studies have not identified the molecular nature of cell-surface glycocomponents responsible for lectin binding (e.g., glycolipids or GPs), they show that the membrane of the growth cone is biochemically different from that of the perikaryon, as well as the adjoining neurite shaft (Pfenninger and Maylie-Pfenninger, 1981a,b). In addition, these studies show that the gradient of lectin receptors along the cell body–growth cone axis is inversely related to the density profile of intramembranous particles (IMPs; the putative freeze-fracture counterparts of protein or GP clusters spanning the membrane). In fact, the density of lectin-binding sites determined in these studies was far greater than the densities reported for IMPs in the growth cone membrane (Pfenninger, 1978).

In the framework of brain development, the growth cone represents the growing neuron's migratory organelle and most distal cellular element in the pathway leading to the location of an appropriate synaptic partner (target cell). If one assumes that such a specialized developmental organelle requires distinct molecular properties, then the studies discussed above support the notion that the growth cone's surface is different from the rest of the neuron. Those glycocomponents that are present in high concentrations on the growth cone's surface may be candidates for regulating cell–cell interactions along the axon's migratory path, and may ultimately be involved in signaling the identity of the correct target cell on which to form a synapse.

In a companion study, Pfenninger and Maylie-Pfenninger (1981b) examined the relative rates of turnover/appearance and clearance of glycoconjugates (lectin receptors) along the surface of growing neurons. To study the dynamics of membrane components during axonal growth, single-label pulse-chase experiments with ferritin–lectin conjugates and free lectins were carried out. Sympathetic neurons labeled in culture with ferritin–wheat germ agglutinin or ferritin–RCA displayed a high concentration of lectin receptors near the distal, growth cone end of the neurite. On the other hand, when living neurons are first labeled with lectin conjugates and chased in lectin-free medium for 3–20 min, membrane surfaces devoid of lectin readily appear on growth cones, their filopodia, and plasma membranes that overlie vesicle clusters. These lectin-free areas can subsequently be shown to contain lectin receptors when pulse-chased neurons are fixed with aldehyde and relabeled with newly added lectin conjugates. In a second group of experiments, neurons were first labeled with free lectin at saturating concentrations, chased in lectin-free medium, fixed with glutaraldehyde, and finally relabeled with ferritin conjugates of the same lectin. These experiments resulted in heavy labeling of the growth cone's surface at filopodia and directly above intracellular regions containing vesicle clusters. These results support the notion that lectin receptors (cell surface glycoconjugates) are preferentially inserted into plasma membrane compartments of the growth cone and that these events parallel the process of membrane expansion accompanying neurite growth. In this context, the estimated rate of membrane expansion is approximately 0.5 $\mu m^2/min$, and the extent of lectin labeling on growth cones following a 10- to 15-min double-label pulse-chase experiment is a few to several square micrometers (Pfenninger and Maylie-Pfenninger, 1981b). These results suggest that the appearance of new lectin receptors is a phenomenon primarily localized to the growth cone. Such an interpretation is in close agreement with the process of distal neuritic growth proposed earlier by Hughes (1953) and Bray (1970, 1973). The precise mechanism(s) by which membrane expansion and insertion of new membrane glycocomponents occur, however, remains unknown. It has been proposed that the clusters of small vesicles immediately beneath the growth cone's plasma membrane may represent a pool of new plasmalemmal components that have been preassembled and transported to sites of active membrane expansion (Pfenninger and Maylie-Pfenninger, 1981b). Pfenninger *et al.* (1982) have recently developed subcellular fractionation techniques in which fractions enriched in growth cone structures can be isolated from fetal rat brain (day 17 of gestation). Studies of this nature should ultimately lead to the identification and characterization of specific molecular components that are responsible for the growth cone's specialized functions during synaptogenesis.

5. CELL SURFACE GLYCOPROTEINS: INVOLVEMENT IN CELL– CELL RECOGNITION AND ADHESION

At a cellular level, the development of the mammalian CNS is an impressive achievement. During the earliest stages of development, the organization of embryonic cells in the primitive nervous system involves cell–cell recognition and this appears to require the formation of selective and specific histogenetic adhesions between brain cells. Once neural and nonneural cells have begun the phenomenon of cell sorting, the ensuing process encompasses the precise manner in which neurons differentiate, send out neuritic processes, and form synaptic connections. In vertebrate nervous systems, the precision with which neuronal circuitry and synaptic contacts are formed is highly conserved between animals of the same species and to a large degree between different species. The human brain contains an estimated 10^{10} neurons (Jacobson, 1970, p. 69), some of which may have up to 80,000 synaptic contacts on its surface (Cotman and Kelly, 1980). With this degree of complexity, together with the enormous numbers of cellular entities, synaptic partners could be formed during development in roughly $10^{3\text{ billion}}$ different combinations (Cotman and Banker, 1974). At least 10^{12} bits of information would be required to specify such a multitude of synaptic patterns (estimate from Bremerman, 1963; see Cotman and Banker, 1974); however, estimates of the amount of information in human DNA are only 10^5 bits. Thus, developmental neurobiologists have concentrated their studies on elucidating more generalized rules and mechanisms that may control the processes by which synaptic circuitry is established.

Many of the mechanisms that have been proposed to control neuronal differentiation and synapse formation invoke the involvement of chemoaffinity gradients and cell surface recognition macromolecules. Early this century, Ramón y Cajal (1905, 1908) proposed that the directionality of neurite outgrowth was guided by diffusible substances (neurotropism), and Hamburger (1962) proposed that the chemical makeup of the surrounding substratum may control the direction of fiber growth. Probably the most quoted hypothesis was put forth by Sperry (1963) and postulates that specific chemical affinities or matches are made between growing axons and appropriate target cells on which synapses are to be formed. Other concepts, many of which share certain features, postulate mechanisms involving specific adhesiveness and fasiculation (Weiss, 1947), autonomous organization of axonal tract projections (Gaze, 1970), and selective atrophy of fiber branches (Jacobson, 1970, pp. 159–169).

To date, little information is available concerning the identity of cell surface macromolecules that are directly involved in neuronal recognition at the point of synapse formation. The postsynaptic Con A-binding GPs discussed earlier in this chapter are candidates for possessing such recognition and/or adhesion properties, but their precise function at the synapse remains unknown. There are, however, a few nervous system GPs that have been biochemically and immunologically characterized, and their involvement in cell–cell adhesion has been well documented. One such GP has been named *retina cognin* (r-cognin) and has been studied by Hausman, Moscona, and co-workers (Hausman and Moscona, 1975). It is a retina-specific factor that was first purified from tissue culture media in which embryonic retinal cells were grown. In a bioassay using neural retinal cells, r-cognin significantly accelerates the rate at which cell–cell aggregates are formed. R-cognin has an apparent M_r of 50,000 and is composed of 15% by weight carbohydrate (GlcNAc, mannose, galactose, and sialic acid). R-cognin's biological activity is rapidly destroyed by trypsin while being refractory to treatments that degrade or alter its carbohydrate composition (e.g., periodate oxidation, neuraminidase and hyaluronidase treatments). R-cognin is not a substrate for, nor does it contain galactosyltransferase activity. A GP very similar, if not identical, to r-cognin has been isolated from embryonic retinal membranes by butanol extraction (Hausman and Moscona, 1976). Because of its extraction and partitioning properties in butanol, membrane-associated r-cognin possess hydrophobic characteristics that may allow it to interact with lipids in the membrane bilayer. When assayed for cell-aggregating activity, the amount of r-cognin that can be isolated from retinal membranes greatly decreases after embryonic day 13, a time that coincides with the establishment of mature cellular and morphological organization of the retina (Hausman and Moscona, 1976). Immunohistochemical studies have shown that r-cognin is a cell surface component of embryonic neural retinal cells, and estimates of its cellular content place 10^4–10^5 r-cognin molecules/cell (Hausman and Moscona, 1979). Tissue-specific aggregation factors can be isolated, using similar techniques, from neural retina, spinal cord and cerebrum of the chick (Hausman and Moscona, 1976). Cogninlike factors have also been isolated from chick heart, skeletal muscle, and liver. These cogninlike molecules show no cross-reactivity with anti-r-cognin antibodies and do not interfere with r-cognin in a cognin-mediated cell aggregation assay. Anti-r-cognin antibodies show little, if any, cross-reactivity to ligatin (R. Hausman, personal communication), a 10,000-dalton, peripheral cell surface protein that inhibits the reassociation of dissociated chick retinal cells (Jakoi and Marchase, 1979; Marchase *et al.*, 1981).

Using a polyclonal antibody against r-cognin, freshly trypsinized 10-day retinal cells are refractory to complement-mediated cell lysis until approximately 4 hr in culture, the period of time shown to be necessary for r-cognin to be regenerated on the cell surface. The reestablishment of antibody-dependent cell lysis is blocked by cycloheximide and thus the regeneration of cell surface r-cognin appears dependent on protein synthesis (Hausman and Moscona, 1979). The ability of retinal cells to display anti-r-cognin-dependent lysis is age-dependent. By embryonic day 16, r-cognin cannot be detected on cultured retinal cells by this assay. In addition, the ability of exogenously added r-cognin to confer antibody-dependent lysis to freshly trypsinized cells decreases sharply during development and cannot be detected by embryonic day 14. Although retinal cells displayed the maximum sensitivity to anti-r-cognin-mediated lysis, cells from 10-day embryonic optic tectum and cerebral cortex are also sensitive, although to lesser degrees, and liver and red blood cells are refractory (Hausman and Moscona, 1979). Thus, there may be cognins or cogninlike components that are specific to different neural tissues and share a similar antigenic determinant, but that differ from one another in terms of intercellular adhesion specificity.

Recent immunohistochemical studies have demonstrated that r-cognin displays a patchy distribution on retinal cell surfaces (R. Hausman, personal communication). Moreover, with improved fixation techniques and longer growth periods for retinal cells in culture, it appears that during embryonic development the amount of r-cognin on the cell soma decreases while that on neurites increases (neuritic processes are very sensitive to removal by trypsinization when retinal cells are initially placed in culture). Preliminary studies on the localization of r-cognin at the EM level have shown that it is present at what appear to be immature ribbon synapses in the embryonic retina (early forming synapses are difficult to morphologically identify due to the paucity of synaptic vesicles in presynaptic structures). The immunohistochemical localization of r-cognin at retinal synapses of posthatching chicks has not yet been carried out. Subcellular fractionation studies with synaptosomes isolated from the chick retina have demonstrated the codistribution of anti-r-cognin immunoreactivity and α-bugarotoxin binding (a marker for postsynaptic nicotinic acetylcholine receptors). These experiments suggest that r-cognin may be present in early formed postsynaptic structures (R. Hausman, personal communication).

Thus, r-cognin appears to play an important role in the process of mutual recognition and morphogenetic association of embryonic cells. It is a cell surface, cell-adhesion GP that may or may not play a role in recognition and adhesion events that underlie synapse formation. Future experiments are necessary to test whether or not anti-r-cognin antibodies will block or alter the specificity of synapse formation in tissue culture. Moreover, since cognins specific to different neural tissues appear to share one or more distinct antigenic determinants, their respective biological uniqueness in terms of cell–cell adhesion specificity may be expressed by molecular determinants not yet identified. More detailed biochemical and immunological (e.g., monoclonal antibodies against differing cognin domains) studies are required to answer such questions.

Edelman and co-workers have isolated and characterized a cell adhesion molecule (CAM) that is a GP and appears to be involved in adhesive interactions between neural cells and the formation of nervous system tissues (Thiery *et al.*, 1977). Antibodies raised against chick neural-CAM (N-CAM) have been shown to disrupt (1) neural cell aggregation (Rutishauser *et al.*, 1976; Chuong *et al.*, 1982), (2) fasciculation of fiber outgrowth from cultured spinal ganglia (Rutishauser *et al.*, 1978; Jørgensen *et al.*, 1980), (3) the normal histogenesis of cultured whole neural retina (Buskirk *et al.*, 1980), and (4) neuron–muscle interactions *in vitro* (Grumet *et al.*, 1982). N-CAM was first isolated from culture media in which 10-day embryonic chick retinas were grown (Brackenbury *et al.*, 1977; Thiery *et al.*, 1977). A high-molecular-weight CAM precursor (240K, proCAM) is present on the surface of retinal cells grown in monolayers (Rutishauser *et al.*, 1976). Early studies revealed that retinal or brain cells from embryos of the appropriate age could bind just as well to cells of different tissue types (e.g., retina to brain), as to homologous cell types. Moreover, antibodies to N-CAM successfully block both homologous and heterologous cell–cell interactions between these two tissues (Rutishauser *et al.*, 1976). These results suggest that mechanisms of intercellular adhesion may be the same for most, if not all, brain and retinal cells and indicate that the existence of tissue-specific ligands mediating intercellular events is unlikely (Brackenbury *et al.*, 1977).

CAM has been isolated and purified from 14-day embryonic chick brain by extracting plasma membranes with 0.5% Nonidet P-40 (Hoffman *et al.*, 1982). Thus, CAM appears to be an integral membrane component. In solution, CAM is polydisperse (M_r 0.5×10^6–1.2×10^6), while in SDS it displays an apparent M_r of 200–250K. CAM has an unusual carbohydrate content: 13 moles of sialic acid and 1.4 moles of galactose per 100 moles of amino acid. Digestion of CAM with neuraminidase results in the formation of a 140K degradation product. On a weight basis, approximately 1% of the total membrane protein in embryonic brain is CAM. Neutral proteolytic activity is associated with purified CAM and it has been shown to undergo spontaneous autolytic conversion to a 65K polypeptide and a sialic acid-rich 100K fragment. The 65K polypeptide appears to contain most, if not all, of the antigenic determinants so far described for CAM. The 65K degradation product neutralizes greater than 90% of the adhesion blocking activity of anti-CAM Fab' in an aggregation bioassay (Hoffman *et al.*, 1982). CAM, therefore, appears similar to r-cognin in that its biological activity does not reside in the molecule's carbohydrate-rich domains. A neural cell adhesion molecule has recently been isolated from embryonic and fetal human brain that is similar, if not identical, to N-CAM (McClain and Edelman, 1982).

Jørgensen *et al.* (1980) have shown that the nervous system-specific D2 antigen is immunologically cross-reactive with CAM. D2 is an integral membrane component that appears to mediate adhesion among neurites in cultured rat ganglia. Anti-D2 antibodies (Fab' fragments) inhibit neurite fasciculation in cultured rat sympathetic ganglia. D2 is present on the surface of cell bodies and neurites of primary fetal neurons. In the adult, D2 is found mainly on the external surface of synaptic membranes. Like CAM, a soluble form of D2 is released into culture medium and has an apparent M_r of 140K.

CAM is not nervous system-specific as antigenic determinants of CAM have been shown to be present on embryonic muscle cells from chick (Grumet *et al.*, 1982). Using an *in vitro* cell aggregation system, myoblasts will adhere to nerve cells from embryonic retina. This neuronal–nonneuronal adhesion can be blocked with Fab' fragments prepared from anti-CAM antibodies. Artificial membrane vesicles prepared from lipids and affinity-purified CAM bind to the surface of myoblasts and myotubes and this binding is inhibited by anti-CAM Fab'. These results suggest that the CAM adhesion mechanism is involved in nerve–muscle as well as nerve–nerve interactions (Grumet *et al.*, 1982).

As mentioned earlier, artificial vesicles can be prepared that contain N-CAM and either brain, or synthetic lipids. They contain about 0.6% protein, most of which is present on the vesicle's external surface (Rutishauser *et al.*, 1982). N-CAM vesicles, like anti-N-CAM Fab', cause a marked decrease in the rate of retinal cell aggregation. Purified [^{125}I]-N-CAM displays neither significant binding nor consistent inhibition of retinal cell aggregation in this *in vitro* assay. However, these preparations contain N-CAM in an aggregated form that can be dispersed by brief exposure to acidic pH (pH 3). Following neutralization, acid-dispersed N-CAM binds in significant amounts to retinal cells and its binding is blocked by anti-N-CAM Fab′ (Rutishauser *et al.*, 1982). Acid-dispersed N-CAM alters the rate of retinal cell aggregation in rotary suspension assays; at 200 rpm, N-CAM at 100 μg/ml caused a 40% inhibition of aggregation. At lower speeds (90 rpm), however, the rate of aggregation was stimulated about 30% and may be due to the self-aggregating properties of N-CAM and not a direct stimulation of cell–cell cross-bridging. Thus, purified N-CAM is not a cell-aggregating molecule analogous to r-cognin, the latter of which greatly stimulates the rates at which neural cells aggregate and may in fact possess bifunctional and/or multivalent cell–cell adhesion domains. Neural retinal membranes contain proCAM, an apparent precursor to N-CAM. proCAM appears to be a cell surface glycoprotein that upon proteolytic degradation releases CAM into the culture medium (Rutishauser *et al.*, 1982). Whereas CAM appears to be monospecific in the sense that it can only interact or bind to itself, proCAM may contain an additional molecular domain that could give it bifunctional properties, e.g., (1) one to interact and possibly intercalate proCAM into the membrane's lipid bilayer, and (2) the other to bind CAM domains of proCAMs on adjacent cells to which adhesive bonds are being established. In this context, it would seem that comparisons with r-cognin may be more informative if made with proCAM rather than N-CAM, once the molecular and biological properties of proCAM are better understood.

The precise role of CAM as a ligand in cell adhesion will be better understood once the detailed molecular nature of cell–cell bonds are elucidated. The rapid and specific aggregation of synthetic vesicles containing CAM suggests that inter-CAM bridges of a bimolecular nature can establish a type of membrane–membrane linkage (Rutishauser *et al.*, 1982). Whether or not CAM may also bind to other types of membrane components in addition to itself, and whether or not adhesive interactions between multiple molecules may be more indicative of the natural biological mechanism(s) await further investigation.

The results from studies on CAM support the notion that this cell surface GP is involved directly in the ligating mechanism for the Ca^{2+}-independent adhesion that is specific among nerve cells and their fiber outgrowths (Rutishauser *et al.*, 1978; Jørgensen *et al.*, 1980). In terms of interneuronal recognition and its regulation of synapse formation, it will be important to determine whether or not CAM plays a role in this very important nervous system process. Knowledge of the distinct molecular domains in N-CAM, its possible self-degradation and self-aggregation properties will undoubtedly provide clues to understanding its ultimate biological function(s). The availability of numerous monoclonal antibodies and knowledge of their antigenic determinants (Chuong *et al.*, 1982) are providing important molecular clues for studies on CAM's cell–cell linkage properties, developmental and subcellular expression, and physiological significance.

6. BRAIN CELL SURFACE GLYCOCOMPONENTS INVOLVED IN GROWTH INHIBITION

Glycopeptides have been isolated from mouse and bovine brain cell surfaces that inhibit protein synthesis, and division of cells in culture (Kinders *et al.*, 1979, 1980a,b). These glycopeptides are released from cells by mild proteolysis and can be purified by gel filtration, affinity chromatography with *Ulex europaeus*, isoelectric focusing, and SDS–PAGE. The purified molecules inhibit protein synthesis and division of "normal" cells in culture, whereas they are less or completely ineffective against tumor or transformed cell lines (Kinders and Johnson, 1981, 1982; Kinders *et al.*, 1982). Affinity-purified mouse glycopeptides are effective at a concentration of 2×10^5 molecules per target cell. The protein synthesis and cell growth inhibitory activities copurify and display the same degree of target cell specificity. It is possible, therefore, that a single molecule contains both biological activities. Antibodies raised against mouse and bovine glycopeptides have shown that these molecules share cross-reacting epitopes (Kinders and Johnson, 1982).

These glycopeptides have been shown to react with target cell surfaces, although binding alone is insufficient to convey biological activity. In fact, both "normal" and transformed cell lines bind biologically active molecules (Kinders and Johnson, 1981). By using a mouse fibrosarcoma cell line that is refractory to the glycopeptide's biological activity, it has been demonstrated that G_{M1} sensitizes these tumor cells to the growth inhibitory activity (Kinders *et al.*, 1982). Although recent studies indicate that G_{M1} is not the cell-surface receptor for the inhibitory glycopeptide, the reconstitution of G_{M1} on the cell surface indirectly confers sensitivity to fibrosarcoma cells.

These glycopeptides apparently inhibit protein synthesis at the level of nascent polypeptide elongation (Kinders *et al.*, 1980a). For the inhibitory activity to be manifested, however, glycopeptides must first react with the cell surface, for their direct addition to cell-free protein synthesis reactions does not inhibit polypeptide elongation. In light of this observation and putative role of cell surface binding, it is likely that a second messenger is required to initiate the sequence of events that result in protein synthesis inhibition.

A curious property of both mouse and bovine glycopeptides is related to their inhibitory action on cell growth. Both inhibitors block cell division in a reversible manner and a large majority of inhibited cells appear stalled in G2 (Kinders and Johnson, 1981, 1982). The ability of these molecules to reversibly synchronize cells in G2 may be a valuable tool for studies on cell cycle kinetics. Although the majority of studies carried out with these glycopeptides have focused on their inhibition of protein synthesis, cell growth, and tumorigenesis, it is possible that these substances play important roles in nervous system development. It is conceivable that such surface components physiologically act during brain development to regulate cell division, cell–cell interactions and/or differentiation.

7. GLYCOPROTEINS OF SYNAPTIC VESICLES AND SECRETORY ORGANELLES

Zanetta *et al.* (1981) have recently examined the GP composition of rat brain synaptic vesicles by sequential fractionation on lectin affinity columns (Con A, *Ulex europaeus*, *Ricinus sanguinis*, and wheat germ agglutinin). Approximately 25% of the total protein in synaptic vesicle preparations bound to one or a combination of the four affinity columns. The most abundant GP fraction is *Ulex europaeus*-positive (fucose-rich) and represents about 20% of the total vesicle protein. The Con A-positive fraction (mannose-rich) was the next most prominent (5%), followed by *Ricinus*-positive (3%, D-galactopyranoside-rich) and wheat germ-positive (1%, GlcNAc-rich). In vesicle preparations, GP-bound sugars are present at 40 μg sugar/mg protein (synaptic membranes contain ~ 60–70 μg sugar/mg protein). These studies revealed approximately 60 synaptic vesicle GPs that were identified on the basis of periodic acid–Schiff (PAS) staining of the different lectin-affinity fractions. The major PAS-positive band has a M_r of 48K, and less prominent components are detected that possess the following M_rs: 140, 120, 65, 43, and 23K. In this context, the GP composition of brain synaptic vesicles is much more complex than that reported for chromaffin granules (see below). It has been suggested that this great complexity in GP composition may result from the heterogeneous population of synaptic vesicle types that are present in cerebral cortex tissues; classes of vesicles that may be distinct from one another on the basis of neurotransmitter compartmentation (Zanetta *et al.*, 1981). However, this diversity of GPs may also be due in part to contaminating soluble and plasma membrane impurities that are difficult to remove, or quantitatively account for in synaptic vesicle preparations isolated from brain (Margolis and Margolis, 1983).

We have recently examined the GP content of rat forebrain synaptic vesicles by the binding of [^{125}I]-Con A to SDS gels (P. Kelly, R. K. Margolis, and R. U. Margolis, unpublished observations). In gels that contain equivalent amounts of synaptic membrane and synaptic vesicle protein

(55 μg/lane), only one major Con A-binding GP (40–45K) was observed in synaptic vesicles while SPMs contained approximately 30 GPs that bound greater or equivalent amounts of [^{125}I]-Con A. The prominent Con A-binding GP in synaptic vesicles was not associated with a prominent Coomassie blue staining band and was not detected in significant amounts in synaptic membrane fractions. Synaptic vesicles contained between 9 and 11 minor glycocomponents (i.e., at least five-fold less Con A-binding activity compared to the 40–45K GP); 6 of these appear to have Con A-binding counterparts in purified synaptic membrane fractions and possess apparent M_r of 220, 180, 125, 77, 58, and 26K. Three minor Con A-binding components appear enriched in synaptic vesicle fractions when compared to synaptic membranes (M_r 85, 16, and 12–14K). Due to the fact that most of the Con A-binding GPs in synaptic vesicles also appear present in synaptic membrane fractions, it is difficult to know precisely how pure these vesicle preparations are. Alternatively, one might expect some degree of similarity in molecular composition between these to membrane fractions, for synaptic vesicles are known to fuse and become part of the presynaptic membrane during transmitter release (Heuser and Reese, 1973; Heuser *et al.*, 1979; von Wedel *et al.*, 1981). The problem of identifying membrane components that are specific to either of these two compartments of the synaptic apparatus will require careful and continued attention. Only then will studies be able to focus on specific functions of synaptic vesicle GPs in neurotransmission.

Glycocomponents of neuronal and nonneuronal secretory organelles have recently been reviewed (Giannattasio *et al.*, 1979; Margolis and Margolis, 1983), and studies on the GP composition of prolactin and chromaffin granules have been reported (Zanini *et al.*, 1980; Giannattasio *et al.*, 1980). Purified membranes of chromaffin granules contain approximately 30 protein bands on SDS gels; only 3 of these are major GPs (Winkler and Westhead, 1980). The major membrane GP is dopamine β-hydroxylase (DBH), which is also present in much smaller amounts in the soluble compartment of the chromaffin granule. The protein and carbohydrate regions of the soluble and membrane-bound forms of DBH appear identical except the latter appears to contain a distinct hydrophobic peptide that is absent from the soluble DBH (Slater *et al.*, 1981; Fischer-Colbrie *et al.*, 1982). The two other prominent GPs (GP-1 and GP-3) in granule membranes contain approximately 30% carbohydrate, which is considerably greater than the 5% value reported for DBH (Fischer-Colbrie *et al.*, 1982). In terms of membrane orientation, much of the oligosaccharide regions of granule membrane GPs are present on the inside surface of the vesicle (Huber *et al.*, 1979; Abbs and Phillips, 1980). The inward orientation of GP carbohydrates in granule membranes is in close agreement with many of the current models of membrane assembly and the insertion of newly synthesized proteins destined to become components of the cell surface (Lingappa *et al.*, 1978; Lodish and Rothman, 1979).

8. GLYCOPROTEINS INVOLVED IN NEUROTRANSMISSION

The synapse is the principle site of cell–cell communication in nervous system tissues and houses receptors in the PSM that are in large part responsible for mediating the physiological action of neurotransmitters. Although the pharmacology of a number of putative transmitter receptors in central and peripheral nervous systems is well known, the molecular composition, macromolecular architecture, and precise subcellular localization of only a few are understood in detail. In terms of molecular properties, there is a strong likelihood that most, if not all, transmitter receptors will turn out to be membrane GPs.

To date, the acetylcholine receptor (AChR) is the best understood; much of our knowledge coming from studies on the electric organs of sea rays (*Torpedo californica* and related species; see reviews by Raftery *et al.*, 1980; Conti-Tronconi and Raftery, 1982). The AChR can be affinity-purified with immobilized analogs of ACh or antireceptor toxins (Olsen *et al.*, 1972; Schmidt and Raftery, 1972; Karlsson *et al.*, 1972; Raftery, 1973; Lindstrom *et al.*, 1978; Gonzalez-Ros *et al.*, 1980), and can be reconstituted into vesicles of known lipid composition (Michaelson and Raftery, 1974; Michaelson *et al.*, 1976; Schiebler and Hucho, 1978; Epstein and Racker, 1978; Hazelbauer and Changeux, 1979; Gonzalez-Ros *et al.*, 1980). Reconstituted AChR–vesicle complexes display many of the normal physiological properties of the native receptor (e.g., neurotransmitter recognition, proper inside-out orientation, and ion translocation properties). The AChR has a doughnut-shaped appearance by EM when negatively stained (Cartaud *et al.*, 1973). It is a large complex that protrudes about 5 nm from the external-membrane surface, has a diameter of about 8–9 nm (Karlin, 1980), and comprises a total extracellular surface of about 220 nm^2. The AChR monomer has an M_r of about 250,000 and is composed of two α subunits (M_r 38K), and one each of β (M_r 50K), γ (M_r 57K), and δ subunits (M_r 64K). Each monomer contains two ACh-binding sites and a cation channel whose opening they regulate (Damle and Karlin, 1978; Karlin, 1980; Anholt *et al.*, 1980). Experiments using affinity-labeling analogs have demonstrated that ACh is bound in part by the α subunits (Karlin, 1980), and affinity analogs of local anesthetics label the δ subunit (Saitoh *et al.*, 1980). The different subunits can be distinguished by peptide mapping (Gullick *et al.*, 1981; Lindstrom *et al.*, 1979a; Froehner and Rafto, 1979), amino acid sequence analyses (Raftery *et al.*, 1980), and specific poly- and monoclonal antibodies (Claudio and Raftery, 1977; Lindstrom *et al.*, 1979b; Tzartos *et al.*, 1981; Gullick *et al.*, 1981). The AChR is a GP containing small amounts of galactose, mannose, glucose, and GlcNAc (Vandlen *et al.*, 1979). Peptide mapping studies have identified fragments in each of the different subunits that contain carbohydrates, and in each case only one prominent carbohydrate-containing peptide is generated from each subunit (Gullick *et al.*, 1981). The main immunogenic region of the AChR is in the α subunit and most of the monoclonal antibodies that have been obtained are directed against it (Tzartos *et al.*, 1981). One monoclonal antibody is directed against a region on the α subunit that is oriented toward the extracellular space and that contains both its carbohydrate region and the site that is labeled with the AChR-combining site-specific reagent [4-(*N*-maleimido)benzyl]-tri[^{3}H]methylammonium iodide. This region extends into the extracellular space, appears to be highly conserved among different species, and does not overlap with toxin-binding nor carbohydrate regions of this subunit (Tzartos *et al.*, 1981). Antibodies to this highly immunogenic region have been shown to passively transfer the acute form of experimental autoimmune myasthenia gravis in animals (Tzartos and Lindstrom, 1980). Although the function of this immunogenic determinant(s) is unknown, it has been proposed to play a role in the interaction between the AChR and components of the basement membrane involved in the localization of this receptor to the synaptic junction (Patrick *et al.*, 1973; Burden *et al.*, 1979; Tzartos *et al.*, 1981). It is also possible that the oligosaccharide region of this receptor may determine its proper orientation in the membrane and thus facilitate its anchorage in the PSM, possibly through interactions with underlying cytoskeletal components of the synapse (Prives *et al.*, 1982). In addition, it has been proposed that external regions of membrane GPs, such as the AChR, may be involved in intercellular recognition and adhesion at the synapse (Barondes, 1976).

With regard to the AChR discussed above, less is known about transmitter receptors in the mammalian nervous system. Receptors for ACh are present in brain and have been localized to the PSM (Lentz and Chester, 1977). Studies have demonstrated the partial purification of rat brain nicotinic AChR by affinity chromatography using α-cobratoxin. Brain AChR possesses hydrodynamic and toxin-binding properties similar to its counterpart in the electric organ (McQuarrie *et al.*, 1976, 1978). Rat brain AChR interacts with the lectins Con A, wheat germ agglutinin, and castor bean, demonstrating that the GP nature of this receptor is similar to that of the eel AChR (Salvaterra *et al.*, 1977). Studies have also indicated that the cholera toxin receptor appears to be a ganglioside (G_{M1}; Van Heyningen, 1974). Acetylcholinesterase, the enzyme that degrades ACh and is a component of the basal lamina at the neuromuscular junction

(Weinberg *et al.*, 1981), contains carbohydrate (Wenthold *et al.*, 1974; Stefanovic *et al.*, 1975), and the L-glutamate-binding receptor appears to be a GP on the basis of Con A binding (Michaelis *et al.*, 1974).

9. GLYCOPROTEINS OF CENTRAL AND PERIPHERAL NERVOUS SYSTEM MYELIN

The major GP of CNS myelin is the myelin-associated glycoprotein (MAG; Quarles *et al.*, 1973). It has an apparent M_r of 100,000 and by PAS staining of total myelin, MAG constitutes approximately 1% of total myelin protein (7.2 μg MAG/mg myelin; Quarles *et al.*, 1981). Thus, MAG is a relatively minor component of myelin. A simple procedure for the purification of MAG from myelin has been reported and consists of solubilization in lithium diiodosalicylate followed by phase partitioning in phenol (Quarles and Pasnak, 1977). The extraction recovers about 50% of the initial MAG content in myelin and results in a fraction of MAG that is pure enough for detailed chemical analysis.

In vivo labeling studies, chemical analysis, and lectin affinity chromatography indicate that MAG's carbohydrate content consists of fucose, mannose, galactose, GlcNAc, and sialic acid (Quarles *et al.*, 1973, 1977; Quarles and Everly, 1977; Zanetta *et al.*, 1977). MAG is about one-third carbohydrate by weight: 34% GlcNAc, 23% mannose, 20% galactose, 18% NANA, and 5% fucose (Barbarash *et al.*, 1981). The high sialic acid content of this GP is reflected in the observation that the apparent M_r is decreased about 10% following enzymatic digestion with neuraminidase (Quarles, 1976). Gel filtration studies on labeled glycopeptides prepared from MAG indicate that this GP contains oligosaccharides that are quite heterogeneous in size (Quarles, 1976). MAG is labeled *in vivo* with [^{35}S]sulfate, probably on its sugar residues (Matthieu *et al.*, 1975a; Simpson *et al.*, 1976). The sulfation of MAG, along with its high sialic acid content, are reflected in its acidic isoelectric point (pI 3.8; Quarles, 1979). Recent advances in the isolation, purification, and characterization of anti-MAG antibodies will likely result in a more detailed understanding of the structure and composition of this GP.

MAG is probably not a major structural component of compact, multilamellar myelin. Sternberger *et al.* (1979) have demonstrated immunocytochemically (peroxidase–antiperoxidase staining with an antibody against MAG) that MAG is present in the cytoplasm of CNS oligodendroglia before myelin formation. As myelination proceeds, and sheaths grow in length and thickness, immunocytochemical staining for MAG is only observed in the periaxonal region of myelinated axons (the inside portion of the sheath adjacent to the axon). Surfaces of unmyelinated axons are not stained with this antibody. Barring physical problems related to the ability of the antibody and peroxidase conjugates to penetrate the interior of multilamellar myelin, MAG is undetectable in compacted myelin sheaths. Immunocytochemical results agree with early biochemical findings demonstrating that MAG is highly concentrated in subcellular fractions enriched in single myelin membranes (fraction $W_1 3$; Waehneldt *et al.*, 1977; McIntyre *et al.*, 1978). MAG is present in lower amounts (threefold) in myelin fractions that are characterized by many large, multilamellar myelin sheaths. The periaxonal localization of MAG, together with its paucity in compacted myelin, have led many to suggest that it may function in stabilizing the integrity of adhesion between neurons and oligodendroglia at the myelin–axon interface (Quarles, 1979).

Antibodies against MAG do not react with P_o, the major GP of PNS myelin. Interestingly, however, anti-MAG immunoreactivity is observed in developing Schwann cells of the PNS, and at the periaxonal region of myelinated PNS axons (Sternberger *et al.*, 1979). Figlewicz *et al.* (1982) have recently demonstrated using biochemical (peptide mapping, *in vivo* labeling with [^{3}H]fucose, and gel electrophoresis) and immunological (antigen precipitation) techniques that MAG is also present in PNS myelin. Thus, MAG is not a CNS-specific myelin GP; however, the amount of MAG in PNS myelin (based on total protein weight) is at least an order of magnitude lower than that present in the CNS.

Sato *et al.* (1982) have shown that a putative neutral protease is present in highly purified myelin and that its proteolytic activity appears specific for the degradation of MAG as well as myelin basic protein (BP); other myelin-associated proteins appear unaffected by this presumptive protease. In highly purified myelin, the degradation of MAG by this proteolytic activity results in the formation of a 94K fragment, designated dMAG (MAG derivative). Conversion of MAG to dMAG is much more rapid in human compared to rat brain myelin, with half-times for degradation being approximately 20 min vs. 2 hr, respectively. Sato *et al.* (1982) have presented preliminary results showing that the conversion of MAG to dMAG is more rapid in myelin isolated from multiple sclerosis brains. The inference from these studies, therefore, is that the proteolytic breakdown of MAG (or myelin BP) may represent one of the early events in the etiology of demyelination that eventually results in pathological states such as multiple sclerosis. During brain development, as discussed earlier, the early synthesis of MAG in oligodendroglia prior to myelin formation, together with its final deposition at the periaxonal junction, implicates a role for MAG in the establishment of functional interactions between myelin and axonal membranes. In light of the recent finding of MAG in the periaxonal region of PNS myelin, it appears reasonable to postulate a similar role for this GP in mediating the specificity of myelin–axon interactions in the PNS. Consistent with this notion are reports that in numerous experimental and pathological cases, Schwann cells are known to myelinate CNS axons and, conversely, oligodendroglia can form myelin sheaths around peripheral axons (Aguayo *et al.*, 1979; Spencer, 1979). A direct involvement of MAG in cell–cell interactions seems to require its presence on the cell surface; however, little evidence exists concerning its possible transmembrane orientation or molecular architecture within the myelin membrane.

The major GP of PNS myelin (Everly *et al.*, 1973; Wood and Dawson, 1973), designated P_o, comprises approximately 50–70% of the total protein in PNS myelin (Ishaque *et al.*, 1980). It has an apparent M_r of 30,000 and is insoluble in neutral chloroform/methanol (Quarles, 1979). P_o is present in myelin isolated from peripheral nerves of a number of species and detailed information on its amino acid and carbohydrate composition has been reported. The sugar content on P_o is approximately 5% by weight, which is much lower than the sugar content of MAG in CNS myelin (Ishaque *et al.*, 1980). The carbohydrates in P_o appear to reside in a single, asparagine-linked nonasaccharide that contains GlcNAc (2.6 moles), mannose (2.7 moles), galactose (1 mole), fucose (0.8 mole), and sialic acid (0.8 mole) (Roomi and Eylar, 1978). This oligosaccharide is attached to asparagine through an *N*-glycosidic linkage. P_o is sulfated, possibly on one or more of its sugar residues (Matthieu *et al.*, 1975b).

Purified PNS myelin treated with trypsin results in the conversion of P_o to a 19K fragment that remains in the particulate fraction, contains the complete nonasaccharide, and is enriched in hydrophobic amino acids when compared to P_o (Roomi and Eylar, 1978). These results suggest that a region of this GP, approximately one-third of the distance from its N-terminal end, is cleaved by trypsin and thus extends from the lipid bilayer of the myelin membrane. In addition, the 19K oligosaccharide-containing region is resistant or protected from the action of trypsin and remains particulate following trypsinization. There have been reports on the presence of additional, relatively minor, low-molecular-weight GPs in PNS myelin (Roomi *et al.*, 1978a,b; Singh *et al.*, 1978). However, their identity and/or possible relatedness to P_o await more detailed experimentation.

Due to the large amounts of P_o in purified PNS myelin, it is likely to be a major structural component in the lipid–protein matrix of the myelin membrane. Correlative biochemical and EM studies (Sea and Peterson, 1975), surface labeling via vectoral probes (Peterson and Gruener, 1978), and EM histochemical staining for carbohydrates (Peterson and Pease, 1972; Wood and McLaughlin, 1975) indicate that the oligosaccharide and a portion of the polypeptide region of P_o reside in the intraperiod zone where the external leaflets of Schwann cell membranes are closely apposed and form multilamellar myelin. Detailed information on the possible transmembrane orientation and molecular architecture of P_o in the myelin membrane is presently unknown.

During development, the P_o content of peripheral nerves increases and correlates well with the overall increase of myelin in the PNS (Wiggins *et al.*, 1975; Wood and Engel, 1976). Immunocytochemical studies have shown that P_o is present in newly formed myelin sheaths of peripheral nerves and in the perinuclear regions of Schwann cells that contain numerous Golgi membranes (Trapp *et al.*, 1981). Autoradiographic studies by Gould (1977) indicate that the fucosylation of P_o occurs in the Golgi region from which point the GP is transported to the outermost extent of the myelin sheath. From this location, P_o displays an apparently slow lateral mobility, diffusing into multilamellar regions at a rate 100-fold slower than that observed for membrane lipids.

Numerous studies have shown that the P_o content of PNS myelin decreases more rapidly than other myelin-associated proteins during experimentally induced (nerve crush, section, or ligation) Wallerian degeneration (Wood and Dawson, 1974b; Peterson and Sea, 1975; Luttges *et al.*, 1976). Experimental allergic neuritis (EAN), an *in vivo* state of demyelination, has in one case been produced by immunizing animals with purified P_o (Wood and Dawson, 1974a). Other investigators have been unable to induce EAN with purified P_o (Brostoff *et al.*, 1975; Kitamura *et al.*, 1976), and results suggest that the P_2 basic protein is the principal antigen involved (Brostoff, 1977; Brostoff *et al.*, 1977). Whether or not P_o, or its natural breakdown product(s), is directly involved in the etiology of EAN or is an indirect consequence of this type of demyelination remains unknown.

10. CONCLUDING REMARKS

It is likely that GPs, and their oligosaccharides, play a number of important roles in the multitude of cellular functions of the nervous system. Carbohydrate-containing polypeptides are major components of the limiting plasma membrane of both neuronal and nonneuronal cells. They are not only critical structural components of these membranes but undoubtedly play important roles in a cell's ability to interact with its neighbors as well as to regulate its own internal metabolic workings. GPs involved in the physiological action and degradation of neurotransmitters possess important functional roles in the nervous system. Many glycoconjugates have been identified in nervous system tissues; however, a detailed understanding of the composition and structure of complex oligosaccharides associated with only a few GPs has been determined. In spite of the progress made in these areas, the function of the majority of brain GPs remains unknown. Major obstacles in the study of brain GPs are due to the vast cellular heterogeneity of nervous system tissues as well as difficulties in obtaining large and homogeneous quantities of individual GPs. A number of GPs have been localized to the cell surface but only a few have been shown to be enriched or localized to specific cell types or cellular domains such as synaptic and growth cone membranes, synaptic vesicles, SJs, and periaxonal regions of central and peripheral nervous system myelin. Moreover, only in the last year has it been demonstrated that the GP composition is clearly dissimilar between SJs isolated from different brain regions.

Previous studies have demonstrated the involvement of GPs in cell–cell adhesion. However, the described functions for such cell surface, nervous system-specific GPs have in large part been elucidated or partially understood using *in vitro* systems. Their specific roles within the nervous system, either in a developmental or differentiation setting, await future investigation. In this context, an important goal in understanding nervous system development is the identification of cell surface glycocomponents involved in neuronal recognition and the formation of synaptic contacts. Progress in technological areas as well as the use of monoclonal antibodies as specific molecular probes should provide answers to some of these elusive questions, and the molecules involved, in the coming decade.

ACKNOWLEDGMENTS. Much of the work cited in this review that was performed in the author's laboratory was supported by grants from the National Institutes of Health (NS-15554 and NS-00605), the National Science Foundation (BNS-8106259), and the Sloan Foundation. I wish to thank Paul Montgomery and Dr. Douglas Groswald for their helpful comments during the preparation of this review.

REFERENCES

Abbs, M. T., and Phillips, J. H., 1980, Organization of the proteins of the chromaffin granule membrane, *Biochim. Biophys. Acta* **595**:200–221.

Aguayo, A. J., Brag, G. M., Perkins, C. S., and Duncan, I. D., 1979, Axon-sheath cell interactions in peripheral and central nervous system transplants, *Soc. Neurosci. Symp.* **4**:361–383.

Allen, W. S., Otterbein, C., Varma, R., Varma, R. S., and Wardi, A. H., 1976, Nondialyzable sulfated sialoglycopeptide fractions derived from bovine heifer brain glycoproteins: Isolation, characterization, and carbohydrate–peptide linkage studies, *J. Neurochem.* **26**:879–885.

Anholt, R., Lindstrom, J., and Montal, M., 1980, Functional equivalence of monomeric and dimeric forms of purified acetylcholine receptors from *Torpedo californica* in reconstituted lipid vesicles, *Eur. J. Biochem.* **109**:481–487.

Barbarash, G. R., Figlewicz, D. A., and Quarles, R. H., 1981, Myelin associated glycoprotein: Purification and partial characterization, *Trans. Am. Soc. Neurochem.* **12**:165.

Barondes, S. H. (ed.), 1976, *Neuronal Recognition*, Plenum Press, New York.

Barondes, S. H., and Rosen, S. D., 1976, Cell surface carbohydrate-binding proteins: Role in cell recognition, in: *Neuronal Recognition* (S. H. Barondes, ed.), pp. 332–356, Plenum Press, New York.

Barondes, S. H., Rosen, S. D., Simpson, D. L., and Kafka, J. A., 1974, Agglutinins of formalinized erythrocytes: Changes in activity with development of Dictyostelium discoideum and embryonic chick brain, in: *Dynamics of Degeneration and Growth in Neurons* (F. K. Olson and Y. Zotterman, eds.), pp. 449–454, Pergamon Press, Elmsford, N.Y.

Beach, R., Kelly, P. T., Babitch, J., and Cotman, C. W., 1981, Identification of myosin in isolated synaptic junctions, *Brain Res.* **225**:75–95.

Bittiger, H., and Schnebli, H. P., 1974, Binding of concanavalin A and ricin to synaptic junctions of rat brain, *Nature (London)* **249**:370–371.

Blomberg, F., Cohen, R. S., and Siekevitz, P., 1977, The structure of postsynaptic densities isolated from dog cerebral cortex. II. Characterization and arrangement of some major proteins within the structure, *J. Cell Biol.* **74**:204–225.

Bloom, F. E., 1970, Correlating structure and function of synaptic ultrastructure, in: *The Neurosciences: Second Study Program* (E. D. Schmidt, ed.), p. 729, The Rockefeller University Press, New York.

Bondareff, W., 1967, An intercellular substance in rat cerebral cortex: Submicroscopic distribution of ruthenium red, *Anat. Rec.* **157**:527–536.

Bondareff, W., and Sjöstrand, J., 1969, Cytochemistry of synaptosomes, *Exp. Neurol.* **24**:450–458.

Bourguignon, L. Y. W., and Singer, S. J., 1977, Transmembrane interactions and the mechanism of capping of surface receptors by their specific ligands, *Proc. Natl. Acad. Sci. USA* **74**:5031–5035.

Brackenbury, R., Thiery, J.-P., Rutishauser, U., and Edelman, G. M., 1977, Adhesion among neural cells of the chick embryo, *J. Biol. Chem.* **252**:6835–6840.

Branton, D., Cohen, C. M., and Tyler, J., 1981, Interaction of cytoskeletal proteins on the human erythrocyte membrane, *Cell* **24**:24–32.

Bray, D., 1970, Surface movements during the growth of single explanted neurons, *Proc. Natl. Acad. Sci. USA* **65**:905–910.

Bray, D., 1973, Branching patterns of isolated sympathetic neurons, *J. Cell Biol.* **56**:702–712.

Breckenridge, W. C., Breckenridge, J. E., and Morgan, I. G., 1972, Glycoproteins of the synaptic region, *Adv. Exp. Med. Biol.* **32**:135–153.

Brostoff, S. W., 1977, Immunological responses to myelin and components, in: *Myelin* (P. Morell, ed.), pp. 415–446, Plenum Press, New York.

Brostoff, S. W., Karkhanis, Y. D., Carlo, D. J., Reuter, W., and Eylar, E. H., 1975, Isolation and partial characterization of the major proteins of rabbit sciatic nerve myelin, *Brain Res.* **86**:449–458.

Brostoff, S. W., Levit, S., and Powers, J., 1977, Isolation of a disease inducing peptide from bovine PNS myelin P_2 protein, *Trans. Am. Soc. Neurochem.* **8**:205.

Brunngraber, E. G., 1972, Biochemistry, function and neuropathology of glycoproteins in brain tissue, in: *Function and Structural Proteins of the Nervous System* (A. N. Davison, P. Mandel, and I. G. Morgan, eds.), pp. 109–133, Plenum Press, New York.

Brunngraber, E. G., and Brown, B. D., 1964, Fractionation of brain macromolecules. II. Isolation of protein-linked sialomucopolysaccharides from subcellular, particulate fractions from rat brain, *J. Neurochem.* **11**:449–459.

Brunngraber, E. G., Aro, A., and Brown, B. D., 1970, Differential determination of glucosamine, galactosamine, and mannosamine in glycopeptides derived from brain tissue glycoproteins, *Clin. Chim. Acta* **29:**333–342.

Brunngraber, E. G., Hof, H., Susz, J., Brown, B. D., Aro, A., and Chang, I., 1973, Glycopeptides from rat brain glycoproteins, *Biochim. Biophys. Acta* **304:**781–796.

Burden, S. J., Sargent, P. B., and McMahan, U. T., 1979, Acetylcholine receptors in regenerating muscle accumulate at original synaptic sites in the absence of the nerve, *J. Cell Biol.* **82:**412–425.

Buskirk, D. R., Thiery, J.-P., Rutishauser, U., and Edelman, G. M., 1980, Antibodies to a neural cell adhesion molecule disrupt histogenesis in cultured chick retina, *Nature (London)* **285:**488–489.

Carlin, R. K., Grab, D. J., Cohen, R. S., and Siekevitz, P., 1980, Isolation and characterization of postsynaptic densities from various brain regions: Enrichment of different types of postsynaptic densities, *J. Cell Biol.* **86:**831–843.

Cartaud, J., Benedetti, E. L., Cohen, J. B., Meunier, J.-C., and Changeux, J.-P., 1973, Presence of a lattice structure in membrane fragments rich in nicotinic receptor protein from the electric organ of *Torpedo marmorata, FEBS Lett.* **33:**109–113.

Chuong, C.-M., McClain, D. A., Streit, P., and Edelman, G. M., 1982, Neural cell adhesion molecules in rodent brains isolated by monoclonal antibodies with cross-species reactivity, *Proc. Natl. Acad. Sci. USA* **79:**4234–4238.

Churchill, L., Cotman, C., Banker, G., Kelly, P., and Shannon, L., 1976, Carbohydrate composition of central nervous system synapses: Analysis of isolated synaptic junctional complexes and postsynaptic densities, *Biochim. Biophys. Acta* **448:**57–72.

Claudio, T., and Raftery, M. A., 1977, Immunological comparison of acetylcholine receptors and their subunits from species of electric ray, *Arch. Biochem. Biophys.* **181:**484–489.

Colonnier, M., 1968, Synaptic patterns on different cell types in the different laminae of the cat visual cortex: An electron microscope study, *Brain Res.* **9:**268–287.

Conti-Tronconi, B. M., and Raftery, M. A., 1982, The nicotinic cholinergic receptor: Correlation of molecular structure with functional properties, *Annu. Rev. Biochem.* **51:**491–530.

Cotman, C. W., and Banker, G. A., 1974, The making of a synapse, in: *Reviews of Neuroscience,* Vol. 1 (S. Ehrenpreis and I. J. Kopin, eds.), pp. 1–62, Raven Press, New York.

Cotman, C. W., and Kelly, P. T., 1980, Macromolecular architecture of CNS synapses, in: *The Cell Surface and Neuronal Function* (C. W. Cotman, G. Poste, and G. L. Nicolson, eds.) pp. 505–533, Elsevier/North-Holland, Press, Amsterdam.

Cotman, C. W., and Taylor, D., 1972, Isolation and structural studies on synaptic complexes from rat brain, *J. Cell Biol.* **55:**697–711.

Cotman, C. W., and Taylor, D., 1974, Localization and characterization of concanavalin A receptors in the synaptic cleft, *J. Cell Biol.* **62:**236–242.

Damle, V. N., and Karlin, A., 1978, Affinity labeling of one of two α-neurotoxin binding sites in acetylcholine receptor from *Torpedo californica, Biochemistry* **17:**2039–2045.

Davis, G., and Bloom, F. E., 1973, Isolation of synaptic junctional complexes from rat brain, *Brain Res.* **62:**135–153.

deSilva, N. S., Gurd, J. W., and Schwartz, C., 1979, Developmental alterations of rat brain synaptic membranes: Reaction of glycoproteins with plant lectins, *Brain Res.* **165:**283–293.

Di Benedetta, C., Brunngraber, E. G., Whitney, G., Brown, B. D., and Aro, A., 1969, Compositional patterns of sialofucohexosaminoglycans derived from rat brain glycoproteins, *Arch. Biochem. Biophys.* **131:**404–413.

Elder, J. H., Pickett, R. A., Hampton, J., and Lerner, R. A., 1977, Radioiodination of proteins in single polyacrylamide gel slices, *J. Biol. Chem.* **252:**6510–6515.

Epstein, M., and Racker, E., 1978, Reconstitution of carbamylcholine-dependent sodium ion flux and desensitization of the acetylcholine receptor from *Torpedo californica, J. Biol. Chem.* **253**:6660–6662.

Everly, J. L., Brady, R. O., and Quarles, R. H., 1973, Evidence that the major protein in rat sciatic nerve myelin is a glycoprotein, *J. Neurochem.* **21**:329–334.

Figlewicz, D. A., Quarles, R. H., Johnson, D., Barbarash, G. R., and Sternberger, N. H., 1982, Biochemical demonstration of the myelin-associated glycoprotein in the peripheral nervous system, *J. Neurochem.* **37**:749–758.

Finne, J., and Krusius, T., 1982, Preparation and fractionation of glycopeptides, *Methods Enzymol.* **83**:269–277.

Finne, J., Krusius, T., Rauvala, H., and Hemminki, K., 1977, The disialosyl group of glycoproteins: Occurrences in different tissues and cellular membranes, *Eur. J. Biochem.* **77**:319–323.

Fischer-Colbrie, R., Schachinger, M., Zangerle, R., and Winkler, H., 1982, Dopamine β-hydroxylase and other glycoproteins from the soluble content and the membranes of adrenal chromaffin granules: Isolation and carbohydrate analysis, *J. Neurochem.* **38**:725–732.

Fiszer, S., and De Robertis, E., 1967, Action of Triton X-100 on ultrastructure and membrane-bound enzymes of isolated nerve endings from rat brain, *Brain Res.* **5**:31–44.

Froehner, S. C., and Rafto, S., 1979, Comparison of the subunits of *Torpedo californica* acetylcholine receptor by peptide mapping, *Biochemistry* **18**:301–307.

Fu, S. C., Cruz, T. F., and Gurd, J. W., 1981, Development of synaptic glycoproteins: Effect of postnatal age on the synthesis and concentration of synaptic membrane and synaptic junctional fucosyl and sialyl glycoproteins, *J. Neurochem.* **36**:1338–1351.

Gaze, R. M. (ed.), 1970, *The Formation of Nerve Connections*, Academic Press, New York.

Giannattasio, G., Zanini, A., and Meldolesi, J., 1979, Complex carbohydrates of secretory organelles, in: *Complex Carbohydrates of Nervous Tissue* (R. U. Margolis and R. K. Margolis, eds.), pp. 327–345, Plenum Press, New York.

Giannattasio, G., Zanini, A., Rosa, P., Meldolesi, J., Margolis, R. K., and Margolis, R. U., 1980, Molecular organization of prolactin granules. III. Intracellular transport of sulfated glycosaminoglycans and glycoproteins of the bovine prolactin granule matrix, *J. Cell Biol.* **86**:273–279.

Glenney, J. R., Glenney, P., and Weber, K., 1982, Erythroid spectrin, brain fodrin, and intestinal brush border proteins (TW-260/240) are related molecules containing a common calmodulin-binding subunit bound to a variant cell type-specific subunit, *Proc. Natl. Acad. Sci. USA* **79**:4002–4005.

Gombos, G., and Zanetta, J. P., 1978, Recent methods for the separation and analysis of central nervous system glycoproteins, in: *Research Methods in Neurochemistry*, Vol. 4 (N. Marks and R. Rodnight, eds.), pp. 307–343, Plenum Press, New York.

Gonzales-Ros, J. M., Paraschos, A., and Martinez-Carrion, M., 1980, Reconstitution of functional membrane-bound acetylcholine receptor from isolated *Torpedo californica* receptor protein and electroplax lipids, *Proc. Natl. Acad. Sci. USA* **77**:1796–1800.

Goodrum, J. F., Bosmann, H. B., and Tanaka, R., 1979, Glycoprotein galactosyltransferase activity in synaptic junctional complexes isolated from rat forebrain, *Neurochem. Res.* **4**:331–337.

Gould, R. M., 1977, Incorporation of glycoproteins into peripheral nerve myelin, *J. Cell Biol.* **75**:326–338.

Grab, D. J., Berzins, K., Cohen, R. S., and Siekevitz, P., 1979, Presence of calmodulin in PSDs isolated from canine cerebral cortex, *J. Biol. Chem.* **254**:8690–8696.

Gray, E. G., 1959, Axo-somatic and axo-dendritic synapses of the cerebral cortex: An electron microscope study, *J. Anat.* **93**:420–430.

Groswald, D. E., Montgomery, P. R., and Kelly, P. T., 1982, Type I, asymmetric synapses from the rat cerebellum, *Soc. Neurosci. Abstr.* 116.1

Groswald, D. E., Montgomery, P. R., and Kelly, P. T., 1983, Synaptic junctions isolated from cerebellum and forebrain: comparisons of morphological and molecular properties, *Brain Res.* **278**:63–80.

Groswold, D. E., and Kelly, P. T., 1984, Evidence that a cerebellum-enriched synaptic junction glycoprotein is related to fodrin and resists extractin with Triton in a calcium-dependent manner, *J. Neurochem.* **42**:534–546.

Grumet, M., Rutishauser, U., and Edelman, G. M., 1982, Neural cell adhesion molecule is on embryonic muscle cells and mediates adhesion to nerve cells *in vitro*, *Nature (London)* **295**:693–695.

Gullick, W. J., Tzartos, S., and Lindstrom, J., 1981, Monoclonal antibodies as probes of acetylcholine receptor structure. 1. Peptide mapping, *Biochemistry* **20**:2173–2180.

Gurd, J. W., 1977, Identification of lectin receptors associated with rat brain PSDs, *Brain Res.* **126**:154–159.

Gurd, J. W., 1980, Subcellular distribution and partial characterization of the three major classes of concanavalin A receptors associated with rat brain synaptic junctions, *Can. J. Biochem.* **58**:941–951.

Gurd, J. W., 1982, Molecular characterization of synapses of the central nervous system, in: *Molecular Approaches to Neurobiology* (I. R. Brown, ed.), pp. 99–130, Academic Press, New York.

Gurd, J. W., and Fu, S. C., 1982, Concanavalin A receptors associated with rat brain synaptic junctions are high mannose-type oligosaccharides, *J. Neurochem.* **39**:719–725.

Gurd, J. W., Bisson, N., and Kelly, P. T., 1983, Synaptic junction glycoproteins are phosphorylated by cyclic-AMP dependent protein kinases, *Brain Res.* **269**:287–296.

Gurd, J. W., Gordon-Weeks, P., and Evans, W. H., 1982, Biochemical and morphological comparisons of PSDs prepared from rat, hamster, and monkey brains by phase partitioning, *J. Neurochem.* **39**:1117–1124.

Hamburger, V., 1962, Specificity in neurogenesis, *J. Cell. Comp. Physiol.* **60**(Suppl.):81–92.

Hausman, R. E., and Moscona, A. A., 1975, Purification and characterization of the retina-specific cell-aggregating factor, *Proc. Natl. Acad. Sci. USA* **72**:916–920.

Hausman, R. E., and Moscona, A. A., 1976, Isolation of retina-specific cell-aggregating factor from membranes of embryonic neural retina tissue, *Proc. Natl. Acad. Sci. USA* **73**:3594–3598.

Hausman, R. E., and Moscona, A. A., 1979, Immunologic detection of retina cognin on the surface of embryonic cells, *Exp. Cell Res.* **119**:191–204.

Hazelbauer, G. L., and Changeux, J.-P., 1979, Reconstitution of a chemically excitable membrane, *Proc. Natl. Acad. Sci. USA* **71**:1479–1483.

Heuser, J. E., and Reese, T. S., 1973, Evidence for recycling of synaptic vesicle membrane during transmitter release at the frog neuromuscular junction, *J. Cell Biol.* **57**:315–344.

Heuser, J. E., Reese, T. S., Dennis, M. J., Jan, L., and Evans, L., 1979, Synaptic vesicle exocytosis captured by quick freezing and correlated with quantal transmitter release, *J. Cell Biol.* **81**:275–300.

Hoffman, S., Sorkin, B. C., White, P. C., Brackenbury, R., Mailhammer, R., Rutishauser, U., Cunningham, B. A., and Edelman, G. M., 1982, Chemical characterization of a neural cell adhesion molecule purified from embryonic brain membranes, *J. Biol. Chem.* **257**:7720–7729.

Huber, E., König, P., Schuler, G., Aberer, W., Plattner, H., and Winkler, H., 1979, Characterization and topography of the glycoproteins of adrenal chromaffin granules, *J. Neurochem.* **32**:35–47.

Hughes, A. F., 1953, The growth of embryonic neurites: A study of cultures of chick neural tissue, *J. Anat.* **87**:150–162.

Ishaque, A., Roomi, M. W., Szymanska, I., Kowalski, S., and Eylar, E. H., 1980, The P_o glycoprotein of peripheral nerve myelin, *Can. J. Biochem.* **58**:913–921.

Jacobson, M., 1970, *Developmental Neurobiology*, Holt, Rinehart & Winston, New York.

Jakoi, E. R., and Marchase, R. B., 1979, Ligatin from embryonic chick neural retina, *J. Cell Biol.* **80**:642–650.

Jørgensen, O. S., Delouvee, A., Thiery, J.-P., and Edelman, G. M., 1980, The nervous system specific protein D2 is involved in adhesion among neurites from cultured rat ganglia, *FEBS Lett.* **111**:39–42.

Karlin, A., 1980, Molecular properties of nicotinic acetylcholine receptors, in: *The Cell Surface and Neuronal Function* (C. W. Cotman, G. Poste, and G. L. Nicolson, eds.), pp. 191–260, Elsevier/North-Holland, Amsterdam.

Karlsson, E., Heilbronn, E., and Widlund, L., 1972, Isolation of the nicotinic acetylcholine receptor by biospecific chromatography on insolubilized *Naja naja* neurotoxin, *FEBS Lett.* **28**:107–111.

Kelly, P. T., and Cotman, C. W. 1977, Identification of glycoproteins and proteins at synapses in the central nervous system, *J. Biol. Chem.* **252**:786–793.

Kelly, P. T., and Cotman, C. W., 1978, Characterization of tubulin and actin and identification of a distinct postsynaptic density polypeptide, *J. Cell Biol.* **79**:173–183.

Kelly, P. T., and Cotman, C. W., 1981, Developmental changes in morphology and molecular composition of isolated synaptic junctional structures, *Brain Res.* **206**:251–271.

Kelly, P. T., and Montgomery, P. R., 1982, Subcellular localization of the 52,000 molecular weight major postsynaptic density protein, *Brain Res.* **233**:265–286.

Kelly, P., Cotman, C. W., Gentry, C., and Nicolson, G. L., 1976, Distribution and mobility of lectin receptors on synaptic membranes of identified neurons in the central nervous system, *J. Cell Biol.* **71**:487–496.

Kelly, P. T., Largen, M., and Cotman, C. W., 1979, Cyclic AMP-stimulated protein kinases at brain synaptic junctions, *J. Biol. Chem.* **254**:1564–1575.

Kelly, P. T. McGuinness, T. L., and Greengard, P., 1983, Calcium/calmodulin-dependent phosphorylation in synaptic junctions, *Soc. Neurosci. Abstr.* **296.19**.

Kelly, P. T., McGuinness, T. L., and Greengard, P., 1984, Evidence that the major PSD protein is a component of a calcium/calmodulin-dependent protein kinase, *Proc. Natl. Acad. Sci. U.S.A.* **81**:945–949.

Kinders, R. J., and Johnson, T. C., 1981, Glycopeptides prepared from mouse cerebrum inhibit protein synthesis and cell division in baby hamster kidney cells, but not in their polyoma virus-transformed analogs, *Exp. Cell Res.* **136**:31–41.

Kinders, R. J., and Johnson, T. C., 1982, Isolation of cell-surface glycopeptides from bovine cerebral cortex that inhibit cell growth and protein synthesis in normal but not in transformed cells, *Biochem. J.* **206**:1–9.

Kinders, R., Johnson, T., and Rachmeler, M., 1979, An inhibitor of protein synthesis prepared by protease treatment of mouse cerebral cortex cells, *Life Sci.* **24**:43–50.

Kinders, R. J., Hughes, J. V., and Johnson, T. C., 1980a, Glycopeptides from brain inhibit rates of polypeptide chain elongation, *J. Biol. Chem.* **255**:6368–6372.

Kinders, R. J., Milenkovic, A. G., Nordin, P., and Johnson, T. C., 1980b, Characterization of cell-surface glycopeptides from mouse cerebral cortex that inhibit cell growth and protein synthesis, *Biochem. J.* **190**:605–614.

Kinders, R. J., Rintoul, D. A., and Johnson, T. C., 1982, Ganglioside GM1 sensitizes tumor cells to growth inhibitory glycopeptides, *Biochem. Biophys. Res. Commun.* **107**:663–669.

Kitamura, K., Suzuki, M., and Uyemura, K., 1976, Purification and partial characterization of two glycoproteins in bovine peripheral nerve myelin membrane, *Biochim. Biophys. Acta* **455:**806–816.

Kobiler, D., Beyer, E. C., and Barondes, S. H., 1978, Developmentally regulated lectins from chick muscle, brain and liver have similar chemical and immunological properties, *Dev. Biol.* **64:**265–272.

Krusius, T., and Finne, J., 1977, Structural features of tissue glycoproteins: Fractionation and methylation analysis of glycopeptides derived from rat brain, kidney and liver, *Eur. J. Biochem.* **78:**369–379.

Krusius, T., and Finne, J., 1978, Characterization of a novel sugar sequence from rat brain glycoproteins containing fucose and sialic acid, *Eur. J. Biochem.* **84:**395–403.

Krusius, T., Finne, J., Margolis, R. U., and Margolis, R. K., 1978, Structural features of microsomal, synaptosomal, mitochondrial and soluble glycoproteins of brain, *Biochemistry* **17:**3849–3854.

Lentz, T. L., and Chester, J., 1977, Localization of acetylcholine receptors in central synapses, *J. Cell Biol.* **72:**258–267.

Levine, L., and Willard, M., 1981, Fodrin: Axonally transported polypeptides associated with the internal periphery of many cells, *J. Cell Biol.* **90:**631–643.

Lindstrom, J., Einarson, B., and Merlie, J., 1978, Immunization of rats with polypeptide chains from *Torpedo* acetylcholine receptor causes an autoimmune response to receptors in rat muscle, *Proc. Natl. Acad. Sci. USA* **75:**769–773.

Lindstrom, J., Merlie, J., and Yogeeswaran, G., 1979a, Biochemical properties of acetylcholine receptor subunits from *Torpedo californica, Biochemistry* **18:**4465–4470.

Lindstrom, J., Walter, B., and Einarson, B., 1979b, Immunochemical similarities between subunits of acetylcholine receptors from *Torpedo electrophorus*, and mammalian muscle, *Biochemistry* **18:**4470–4480.

Lingappa, V. R., Katz, F. N., Lodish, H. F., and Blobel, G., 1978, A signal sequence for the insertion of a transmembrane glycoprotein: Similarities in the secretory proteins in primary structure and function, *J. Biol. Chem.* **253:**8667–8670.

Lodish, H. F., and Rothman, J. E., 1979, The assembly of cell membranes, *Sci. Am.* **240:**48–63.

Luttges, M. W., Kelly, P. T., and Gerren, R. A., 1976, Degenerative changes in mouse sciatic nerves: Electrophoretic and electrophysiologic characterizations, *Exp. Neurol.* **50:**706–733.

McClain, D. A., and Edelman, G. M., 1982, A neural cell adhesion molecule from human brain, *Proc. Natl. Acad. Sci. USA* **79:**6380–6384.

McIntyre, R. J., Quarles, R. H., Webster, H. deF., and Brady, R. O., 1978, Isolation and characterization of myelin-related membranes, *J. Neurochem.* **30:**991–1002.

McQuarrie, C., Salvaterra, P. M., De Blas, A., Routes, J., and Mahler, H. R., 1976, Studies on nicotinic acetylcholine receptors in mammalian brain, *J. Biol. Chem.* **251:**6335–6339.

McQuarrie, C., Salvaterra, P. M., and Mahler, H. R., 1978, Studies on nicotinic acetylcholine receptors in mammalian brain, *J. Biol. Chem.* **253:**2743–2747.

Mahler, H. R., 1979, Glycoproteins of the synapse, in: *Complex Carbohyrates of Nervous Tissue* (R. U. Margolis and R. K. Margolis, eds.), pp. 165–184, Plenum Press, New York.

Marchase, R. B., Harges, P., and Jakoi, E. R., 1981, Ligatin from embryonic chick neural retina inhibits retinal cell adhesion, *Dev. Biol.* **86:**250–255.

Margolis, R. K., and Margolis, R. U., 1970, Sulfated glycopeptides from rat brain glycoproteins, *Biochemistry* **9:**4389–4396.

Margolis, R. K., and Margolis, R. U., 1979, Structure and distribution of glycoproteins and glycosaminoglycans, in: *Complex Carbohydrates of Nervous Tissue* (R. U. Margolis and R. K. Margolis, eds.), pp. 45–73, Plenum Press, New York.

Margolis, R. K., and Margolis, R. U., 1983, Glycoproteins and proteoglycans, in: *Handbook of Neurochemistry* (A. Lajtha, ed.), pp. 177–204, Vol. 5, 2nd ed., Plenum Press, New York.

Margolis, R. K., Margolis, R. U., Preti, C., and Lai, D., 1975, Distribution and metabolism of glycoproteins and glycosaminoglycans in subcellular fractions of brain, *Biochemistry* **14**:4797–4804.

Matthieu, J.-M., Quarles, R. H., Poduslo, J. F., and Brady, R. O., 1975a, [^{35}S]Sulfate incorporation into myelin glycoproteins. I. Central nervous system, *Biochim. Biophys. Acta* **392**:159–166.

Matthieu, J.-M., Everly, J. L., Brady, R. O., and Quarles, R. H., 1975b, [^{35}S]Sulfate incorporation into myelin glycoproteins. II. Peripheral nervous system, *Biochim. Biophys. Acta* **392**:167–174.

Matus, A. I., 1978, The chemical synapse: Structure and function, in: *Intercellular Junctions and Synapses* (J. O. Feldman, N. B. Gilula, and J. D. Pitts, eds.), pp. 99–139, Chapman & Hall, London.

Matus, A. I., and Walters, B. B., 1975, Ultrastructure of the synaptic junctional lattice isolated from mammalian brain, *J. Neurocytol.* **4**:369–375.

Matus, A., DePetris, S., and Raff, M. C., 1973, Mobility of concanavalin A receptors in myelin and synaptic membranes, *Nature New Biol.* **244**:278–280.

Mena, E. E., and Cotman, C. W., 1982, Synaptic cleft glycoproteins contain homologous amino acid sequences, *Science* **216**:422–424.

Mena, E. E., Foster, A. C., Fagg, G. E., and Cotman, C. W., 1981, Identification of synapse specific components: Synaptic glycoproteins, proteins, and transmitter binding sites, *J. Neurochem.* **37**:1557–1566.

Michaelson, D. M., and Raftery, M. A., 1974, Purified acetylcholine receptor: Its reconstitution to a chemically excitable membrane, *Proc. Natl. Acad. Sci. USA* **71**:4768–4772.

Michaelis, E. K., Michaelis, M. L., and Boyarsky, L. L., 1974, High-affinity glutamic acid binding to brain synaptic membranes, *Biochim. Biophys. Acta* **367**:338–348.

Michaelson, D. M., Duguid, J. R., Miller, D. L., and Raftery, M. A., 1976, Reconstitution of a purified acetylcholine receptor, *J. Supramol. Struct.* **4**:419–425.

Mooseker, M. S., 1976, Brush border motility: Microvillar contraction in Triton-treated brush borders isolated from intestinal epithelium, *J. Cell Biol.* **71**:417–433.

Mooseker, M. S., and Tilney, L. G., 1975, The organization of an actin filament–membrane comples: Filament polarity and membrane attachment in the microvilli of intestinal epithelial cell, *J. Cell Biol.* **67**:725–743.

Morgan, I. G., Gombos, G., and Tettamanti, G., 1977, Glycoproteins and glycolipids of the nervous system, in: *The Glycoconjugates* (M. I. Horowitz and W. Pigman, eds.), Vol. 1, pp. 351–383, Academic Press, New York.

Nieto-Sampedro, M., Bussineau, C. M., and Cotman, C. W., 1982, Isolation, morphology, and protein and glycoprotein composition of synaptic junctional fractions from the brain of lower vertebrates: Antigen PSD-95 as a junctional marker, *J. Neurosci.* **2**:722–734.

Olsen, R. W., Meunier, J.-C., and Changeux, J.-P., 1972, Progress in the purification of the cholinergic receptor protein from *Electrophorus* by affinity chromatography, *FEBS Lett.* **28**:96–100.

Patrick, J., Lindstom, J., Culp, B., and McMillan, J., 1973, Studies on purified eel acetylcholine receptor and anti-acetylcholine receptor antibody, *Proc. Natl. Acad. Sci. USA* **70**:3334–3338.

Peterson, R. G., and Gruener, R. W., 1978, Morphological localization of PNS myelin proteins, *Brain Res.* **152**:17–29.

Peterson, R. G., and Pease, D. C., 1972, Myelin imbedded in polymerized glutaraldehyde urea, *J. Ultrastruct. Res* **41**:115–132.

Peterson, R. G., and Sea, C. P., 1975, Localization of the main myelin protein in PNS, *Trans. Am. Soc. Neurochem.* **6:**211.

Pfenninger, K. H., 1973, Synaptic morphology and cytochemistry, *Progr. Histochem. Cytochem.* **5:**1–86.

Pfenninger, K. H., 1978, Organization of neuronal membranes, *Annu. Rev. Neurosci.* **1:**445–471.

Pfenninger, K. H., and Maylie-Pfenninger, M.-F., 1981a, Lectin labeling of sprouting neurons. I. Regional distribution of surface glycoconjugates, *J. Cell Biol.* **89:**536–546.

Pfenninger, K. H., and Maylie-Pfenninger, M.-F., 1981b, Lectin labeling of sprouting neurons. II. Relative movement and appearance of glycoconjugates during plasmalemmal expansion, *J. Cell Biol.* **89:**547–559.

Pfenninger, K. H., Ellis, L., Friedman, L. B., Johnson, M. P., and Somlo, S., 1982, Nerve growth cones isolated by subcellular fractionation from fetal rat brain, *J. Cell Biol.* **95:**95a.

Prives, J., Fulton, A. B., Penman, S., Daniels, M. P., and Christian, C. N., 1982, Interaction of the cytoskeletal framework with acetylcholine receptor on the surface of embryonic muscle cells in culture, *J. Cell Biol.* **92:**231–236.

Quarles, R. H., 1976, Effects of Pronase and neuraminidase treatment on a myelin-associated glycoprotein in developing brain, *Biochem. J.* **156:**143–150.

Quarles, R. H., 1979, Glycoproteins in myelin and myelin-related membranes, in: *Complex Carbohydrates of Nervous Tissue* (R. U. Margolis and R. K. Margolis, eds.), pp. 209–233, Plenum Press, New York.

Quarles, R. H., and Everly, J. L., 1977, Glycopeptide fractions prepared from purified central and peripheral rat myelin, *Biochim. Biophys. Acta* **466:**176–186.

Quarles, R. H., and Pasnak, C. F., 1977, A rapid procedure for selectively isolating the major glycoprotein from purified rat brain myelin, *Biochem. J.* **163:**635–637.

Quarles, R. H., Everly, J. L., and Brady, R. O., 1973, Evidence for the close association of a glycoprotein with myelin in rat brain, *J. Neurochem.* **21:**1177–1191.

Quarles, R. H., Foreman, C. F., Poduslo, J. F., and McIntyre, L. J., 1977, Interactions of myelin-associated glycoproteins with immoblized lectins, *Trans. Am. Soc. Neurochem.* **8:**201.

Quarles, R. H., Johnson, D., Brady, R. O., and Sternberger, N. H., 1981, Preparation and characterization of antisera to the myelin-associated glycoprotein, *Neurochem. Res.* **6:**1115–1127.

Raftery, M. A., 1973, Isolation of acetylcholine receptor–α-bungarotoxin complexes from *Torpedo californica* electroplax, *Arch. Biochem. Biophys.* **154:**270–276.

Raftery, M. A., Hunkapiller, M. W., Strader, C. D., and Hood, L. E., 1980, Acetylcholine receptor: Complex of homologous subunits, *Science* **208:**1454–1457.

Rambourg, A., and Leblond, C. P., 1967, Electron microscopic observations on the carbohydrate-rich coat present at the surface of cells in the rat, *J. Cell Biol* **32:**27–53.

Ramón y Cajal, S., 1905, *Trab. Lab. Invest. Biol. Univ. Madrid* **4:**219, in: *Studies on Vertebrate Neurogenesis* (L. Guth, transl.), pp. 5–70, Thomas, Springfield, Ill.

Ramon y Cajal, S., 1908, *Anat. Anz.* **32:**1–65, in: *Studies on Vertebrate Neurogenesis* (L. Guth, transl.), pp. 71–116, Thomas, Springfield, Ill.

Roomi, M. W., and Eylar, E. H., 1978, Isolation of a product from the trypsin-digested glycoprotein of sciatic nerve myelin, *Biochim. Biophys. Acta* **536:**122–133.

Roomi, M. W., Ishaque, A., Kahn, N. R., and Eylar, E. H., 1978a, Glycoproteins and albumin in peripheral nerve myelin, *J. Neurochem.* **31:**375–380.

Roomi, M. W., Ishaque, A., Khan, N. R., and Eylar, E. H., 1978b, The P_o protein: The major glycoprotein of peripheral nerve myelin, *Biochim. Biophys. Acta* **536:**112–121.

Rostas, J. A. P., Kelly, P. T., Pesin, R. H., and Cotman, C. W., 1979, Protein and glycoprotein composition of synaptic junctions prepared from discrete synaptic regions and different species, *Brain Res.* **168:**151–167.

Rutishauser, U., Thiery, J.-P., Brackenbury, R., Sela, B.-A., and Edelman, G. M., 1976, Mechanisms of adhesion among cells from neural tissues of the chick embryo, *Proc. Natl. Acad. Sci. USA* **73:**577–581.

Rutishauser, U., Gall, W., and Edelman, G. M., 1978, Adhesion among neural cells of the chick embryo. IV. Role of the cell surface molecule CAM in the formation of neurite bundles in cultures of spinal ganglia, *J. Cell Biol.* **79:**382–393.

Rutishauser, U., Hoffman, S., and Edelman, G. M., 1982, Binding properties of a cell adhesion molecule from neural tissue, *Proc. Natl. Acad. Sci. USA* **79:**685–689.

Saitoh, T., Oswald, R., Wennogle, L. P., and Changeux, J.-P., 1980, Conditions for the selective labeling of the acetylcholine receptor by the covalent non-competitive blocker 5-azido-[^{3}H]trimethisoquin, *FEBS Lett.* **116:**30–36.

Salvaterra, P. M., Gurd, J. M., and Mahler, H. R., 1977, Interactions of the nicotinic acetylcholine receptor from rat brain with lectins, *J. Neurochem.* **29:**345–348.

Sato, S., Quarles, R. H., and Brady, R. O., 1982, Susceptibility of the myelin-associated glycoprotein and basic protein to a neutral protease in highly purified myelin from human and rat brain, *J. Neurochem.* **39:**97–105.

Schiebler, W., and Hucho, F., 1978, Membranes rich in acetylcholine receptor: Characterization and reconstitution to excitable membranes from exogenous lipids, *Eur. J. Biochem.* **85:**55–63.

Schlessinger, J., Elson, E. L., Webb, W. W., Yahara, I., Rutishauser, U., and Edelman, G. M., 1977, Receptor diffusion on cell surfaces modulated by locally bound concanavalin A, *Proc. Natl. Acad. Sci. USA* **74:**1110–1114.

Schmidt, J., and Raftery, M. A., 1972, Use of affinity chromatography of acetylcholine receptor purification, *Biochem. Biophys. Res. Commun* **49:**572–578.

Sea, C. P., and Peterson, R. G., 1975, Ultrastructure and biochemistry of myelin after isoniazid-induced nerve degeneration in rats, *Exp. Neurol.* **48:**252–260.

Simpson, D. L., Thorne, D. R., and Loh, H. H., 1976, Sulfated glycoproteins, glycolipids and glycosaminoglycans from synaptic plasma and myelin membranes: Isolation and characterization of sulfated glycopeptides, *Biochemistry* **15:**5449–5457.

Simpson, D. L., Thorne, D. R., and Loh, H. H., 1977, Developmentally regulated lectin in neonatal rat brain, *Nature* (*London*) **266:**367–369.

Simpson, D. L., Thorne, D. R., and Loh, H. H., 1978, Lectins: Endogenous carbohydrate-binding proteins from vertebrate tissues: Functional roles in recognition processes, *Life Sci.* **22:**727–748.

Singh, H., Silberlicht, I., and Sinch, J., 1978, A comparative study of the polypeptides of mammalian peripheral nerve myelin, *Brain Res.* **144:**303–311.

Slater, E. P., Zaremba, S., and Hogue-Angeletti, R. A., 1981, Purification of membrane-bound dopamine β-monooxygenase from chromaffin granules: Relation to soluble dopamine β-monooxygenase, *Arch. Biochem. Biophys.* **211:**288–296.

Spencer, P. S., 1979, Neuronal regulation of myelinating cell function, (Soc. Neurosci. Symp. **4:**275–321.

Sperry, R. W., 1963, Chemoaffinity in the orderly growth of nerve fiber patterns and connections, *Proc. Natl. Acad. Sci. USA* **50:**703–707.

Stefanovic, V., Mandel, P., and Rosenberg, A., 1975, Activation of acetyl- and butyrylcholinesterase by enzymatic removal of sialic acid from intact neuroblastoma and astroblastoma cells in culture, *Biochemistry* **14:**5257–5260.

Sternberger, N. H., Quarles, R. H., Itoyama, Y., and Webster, H. D., 1979, Myelin-associated glycoprotein demonstrated immunocytochemically in myelin and myelin-forming cells of developing rat, *Proc. Natl. Acad. Sci. USA* **76:**1510–1514.

Stromer, M. H., and Goll, D. E., 1972, Studies on purified alpha-actinin. II. Electron microscopic studies on the competitive binding of alpha-actinin and tropomyosin to Z-line extracted myofibrils, *J. Mol. Biol.* **67**:489–494.

Thiery, J.-P., Brackenbury, R., Rutishauser, U., and Edelman, G. M., 1977, Adhesion among neural cells of the chick embryo. II. Purification and characterization of a cell adhesion molecule from neural retina, *J. Biol. Chem.* **252**:6841–6845.

Trapp, B. D., Itoyama, Y., Sternberger, N. H., Quarles, R. H., and Webster, H. F., 1981, Immunocytochemical localization of P_0 protein in Golgi complex membranes and myelin of developing rat Schwann cells, *J. Cell Biol.* **90**:1–6.

Tzartos, S. J., and Lindstrom, J. M., 1980, Monoclonal antibodies used to probe acetylcholine receptor structure: Localization of the main immunogenic region and detection of similarities between subunits, *Proc. Natl. Acad. Sci. USA* **77**:755–759.

Tzartos, S. J., Rand, D. E., Einarson, B. L., and Lindstrom, J. M., 1981, Mapping of surface structures of *Electrophorus* acetylcholine receptor using monoclonal antibodies, *J. Biol. Chem.* **256**:8635–8645.

Van der Loos, H., 1963, The fine structure of synapses in the cerebral cortex, *Z. Zellforsch. Mikrosk. Anat.* **60**:815–824.

Vandlen, R., Wu, W. C.-S., Eisenach, J. C., and Raftery, M. A., 1979, Studies of the composition of purified *Torpedo californica* acetylcholine receptor and of its subunits, *Biochemistry* **18**:1845–1854.

Van Heyningen, S., 1974, Cholera toxin: Interaction of its subunits with ganglioside G_{M1}, *Science* **183**:656–657.

Vincendon, G., Gombos, G., and Morgan, I. G., 1973, The interest of studying glycoproteins in the central nervous system, in: *Methodologie de la Structure et du Metabolisme des Glycoconjugates (Glycoproteins et Glycolipids)*, Vol. 2, CNRS, Paris.

von Wedel, R. J., Carlson, S. S., and Kelly, R. B., 1981, Transfer of synaptic vesicle antigens to the presynaptic plasma membrane during exocytosis, *Proc. Natl. Acad. Sci. USA* **78**:1014–1018.

Waehneldt, T. V., Matthieu, J.-M., and Neuhoff, V., 1977, Characterization of a myelin related fraction (SN 4) isolated from rat forebrain at two developmental stages, *Brain Res.* **138**:29–43.

Weinberg, C. B., Sanes, J. R., and Hall, Z. W., 1981, Formation of neuromuscular junctions in adult rats: Accumulation of acetylcholine receptors, acetylcholinesterase, and components of synaptic basal lamina, *Dev. Biol.* **84**:255–266.

Weiss, P., 1947, The problem of specificity in growth and development, *Yale J. Biol. Med.*, **19**:235–259.

Wenthold, R. J., Mahler, H. R., and Moore, W. J., 1974, Properties of rat brain acetylcholinesterase, *J. Neurochem.* **22**:945–949.

Wiggins, R. C., Benjamin, J. A., and Morell, P., 1975, Appearance of myelin proteins in rat sciatic nerve during development, *Brain Res.* **89**:99–106.

Winkler, H., and Westhead, E., 1980, The molecular organization of adrenal chromaffin granules, *Neuroscience* **5**:1803–1823.

Wood, J. G., and Dawson, R. M. C., 1973, A major myelin glycoprotein of sciatic nerve, *J. Neurochem.* **21**:717–719.

Wood, J. G., and Dawson, R. M. C., 1974a, Some properties of a major structural glycoprotein of sciatic nerve, *J. Neurochem.* **22**:627–630.

Wood, J. G., and Dawson, R. M. C., 1974b, Lipid and protein changes in sciatic nerve during Wallerian degeneration, *J. Neurochem.* **22**:631–635.

Wood, J. G., and Engel, E. L., 1976, Peripheral myelin glycoproteins and myelin fine structure during development of rat sciatic nerve, *J. Neurocytol.* **5**:605–615.

Wood, J. G., and McLaughlin, B. J., 1975, The visualization of concanavalin-A binding sites in the intraperiod line of rat sciatic nerve myelin, *J. Neurochem.* **24**:233–235.

Zanetta, J. P., Sarlieve, L. L., Mandel, P., Vincendon, G., and Gombos, G., 1977, Fractionation of glycoproteins associate to adult rat brain myelin fractions, *J. Neurochem.* **29:**827.

Zanetta, J. P., Reeber, A., and Vincendon, G., 1981, Glycoproteins from adult rat brain synaptic vesicles: Fractionation on four immobilized lectins, *Biochim. Biophys. Acta* **670:**393–400.

Zanini, A., Giannattasio, G., Nussdorfer, G., Margolis, R. K., Margolis, R. U., and Meldolesi, J., 1980, Molecular organization of prolactin granules. II. Characterization of glycosaminoglycans and glycoproteins of the bovine prolactin matrix, *J. Cell Biol.* **86:**260–272.

7

The Role of Glycoproteins in the Life Cycle of the Cellular Slime Mold Dictyostelium discoideum

Ellen J. Henderson

1. INTRODUCTION

The cellular slime molds have been studied for a number of years as a model for analysis of developmental processes. The most commonly used species is *Dictyostelium discoideum*, which was first discovered in 1935 by Kenneth Raper, whose elegant pioneering studies soon attracted others to work with this organism. Raper's 1940 report in the *Journal of the Elisha Mitchell Scientific Society* remains a classic. In the 1950s and 1960s, there were a handful of laboratories around the world working on *D. discoideum*, and they proved the usefulness of this model organism as a source of biochemical and molecular information about developmental processes, and generated an essentially exponential growth in the number of laboratories working on slime molds. The organism's popularity was spread considerably with Bonner's excellent monograph in 1967 and with the more recent volumes from Loomis (1975, 1982), which are the scientific equivalents of the Bible in many a laboratory. Currently, there is a vast literature on this organism; Loomis (1982) has assembled a bibliography of 1800 publications. The following description of the life cycle

Ellen J. Henderson • Department of Biology, Georgetown University, Washington, D.C. 20057.

371

of *D. discoideum* is necessarily brief and is intended as a summary for those not acquainted with this organism. My apologies go to colleagues for the absence of individual citations of their contributions in this section. The non-Dictyosteliac can find the original citations in the volumes noted above.

1.1. The Life Cycle of Dictyostelium discoideum

Scientists who study an organism with a name like "slime mold" have an instant public relations problem which is most effectively overcome by a 5-min time-lapse film of the developmental cycle in which the organism itself charms the viewer. The life cycle is schematically illustrated in Fig. 1 and selected stages shown as the photographs of Fig. 2.

1.1.1. The Vegetative Phase

D. discoideum is a soil organism. The vegetative phase amoebae are haploid, contain seven linkage groups, are about 10 μm in diameter, and feed on bacteria, growing optimally at 22°C. Vegetative amoebae display positive chemotaxis to sources of folic acid. Therefore, it is currently believed that folate release by bacteria provides a sensing device for the amoebae. After phagocytosis of about 1000 bacteria, an amoeba divides by binary fission. By mutation of the wild type NC4, strains AX2 and AX3 have been derived, which are capable of growth in an axenic medium of peptone, yeast extract, and salts as well as the completely defined medium of Franke and Kessin. In the laboratory, amoebae can be grown in association with bacteria on agar medium containing nutrients for the bacteria or as shaken suspensions with either live or autoclaved bacteria or in axenic medium. When nutrients are depleted or removed, the population enters a relatively simple developmental cycle in which the initially independent cells aggregate into a multicellular tissuelike structure that reorganizes into a migratory, sluglike pseudoplasmodium, i.e., the cells of the slug do not fuse into a true plasmodium as is the case with the acellular slime molds. Eventually, slug migration stops, and the culmination phase ensues in which a sorocarp or fruiting body is generated, consisting of a spore sac supported by a slender cellular stalk. Each of these developmental stages has been studied in considerable detail as summarized in the following sections.

1.1.2. Aggregation

The developmental cycle is studied with cells plated on a nonnutrient surface, which can be agar or a filter resting on a moist support pad. For

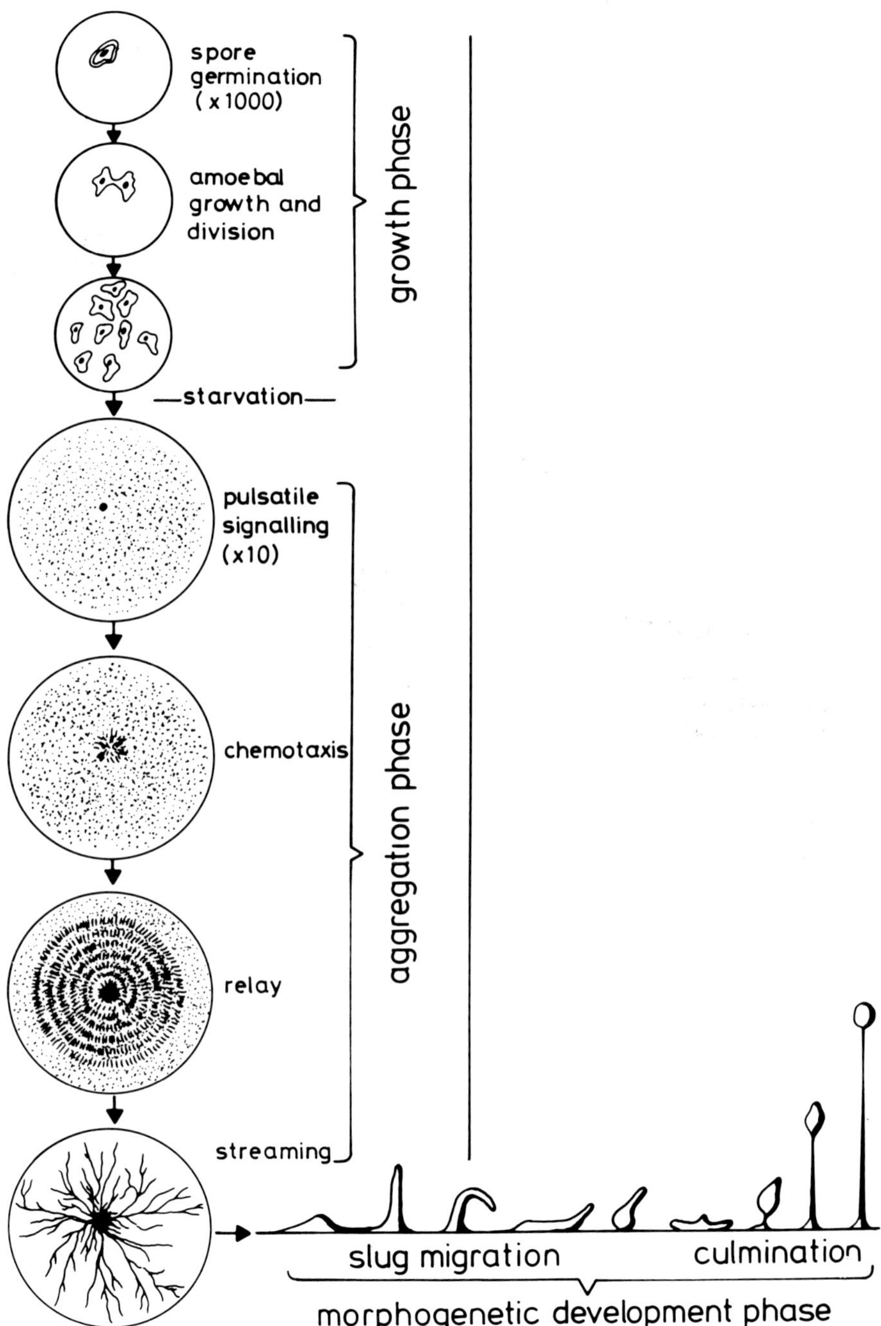

Figure 1. The life cycle of *Dictyostelium discoideum* showing the growth phase, aggregation phase, and morphogenetic developmental phase. (From Newell, 1977, with permission.)

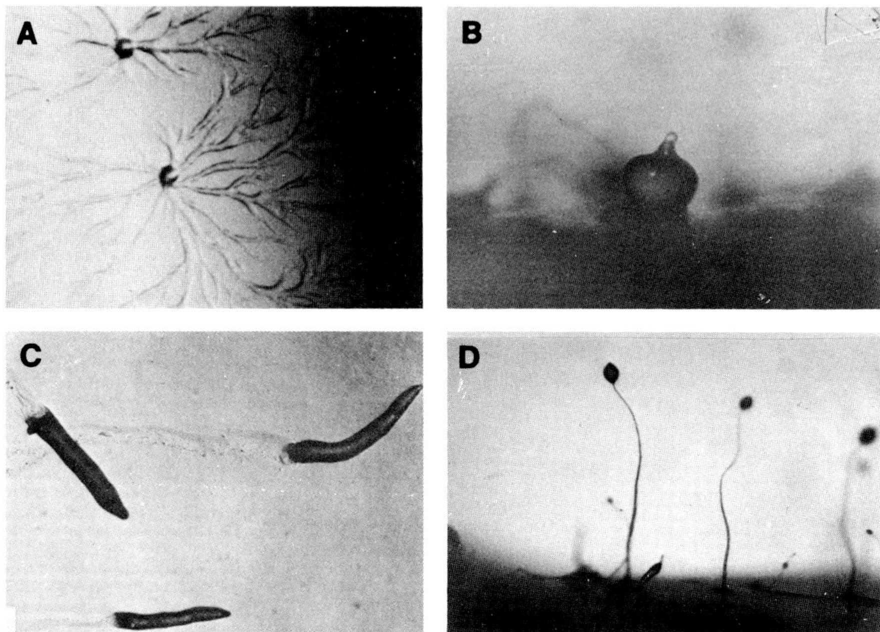

Figure 2. Photographic illustrations of major stages in the developmental program of *D. discoideum*. (A) Streaming aggregates; (B) tip emergence; (C) slug migration; (D) fruiting bodies. [(C) from Raper, 1951, with permission.]

the first few hours, during what is called interphase, the amoebae move in an apparently random manner, but, by about 6 hr, they move in oriented fashion toward autonomously generated aggregation centers, which are a source of the chemotactic "acrasin" now known to be cAMP. By this stage the amebae have expressed new cell surface receptors for cAMP. The aggregation centers release cAMP in a pulsatile fashion. Amoebae move toward the cAMP source and also relay the signal, that is, in response to cAMP, they release cAMP, which then serves to attract other cells, more distant from the signaling center. This relay response, therefore, propagates the initial signal from cell to cell across an aggregation territory. By this stage of development, the amoebae have also produced substantial amounts of cAMP phosphodiesterases, which destroy the signal between pulses and, thus, maintain the signal-to-noise ratio.

During early aggregation, the interplay between these factors generates a pattern in which concentric rings of moving and stationary cells surround aggregation centers (Fig. 1). Quantitative measures of pulsatile signaling, signal relay and timing were first made by Gerisch and co-workers who showed that cells at this stage, when placed in aerated suspensions, gave a biphasic light-scattering response to stimulation with exogenous cAMP and that such suspensions could generate autonomous oscillations in this light-scattering response with a period of a few minutes. This established an ability to perform solution biochemistry on cells during the course of cellular excitation–response–relaxation cycles. Measurement of the concentrations and timing of intracellular and extracellular cAMP provided the first proof of the pulsations and relays inferred from cell behavior during aggregation. These and other studies initially directed toward developmental issues have broadened interest in *D. discoideum* as a model for general cellular physiology and biology. Finally, on the issue of cAMP, there is a substantial body of evidence implicating cAMP signaling as a regulator of gene expression. This evidence will be considered at appropriate points in the text.

During this period of development, the cells differentiate more extensively than just elaboration of the cAMP chemosensory system. There are many changes in the profile of plasma membrane proteins and glycoproteins (Section 2.2). A significant alteration is that the cells develop very strong mutual cohesiveness (Section 6). The existence of cohesion mechanisms contributes to altering the pattern of aggregation from that of concentric circles of cells to one of streams of cells flowing toward aggregation centers (Figs. 1, 2A). Most cells are not responding directly to an aggregation center but to a relayed signal provided by a neighbor. Thus, the ring pattern is transient and collapses to streams as cohesive cells reach a signaling neighbor and form a stable association. By the end of aggregation, about 12 hr after the onset of starvation, all streams have been drawn into centers to produce a mound of cells, each mound containing about 10^5 cells. Signaling centers compete for amoebae, and if 10^8 cells are plated on a standard petri dish, about 10^3 cell mounds are produced. Finally, for excellent descriptions of the behavior and biochemistry of aggregation, see Gerisch (1968), Loomis (1979), and Devreotes (1982).

1.1.3. Tip Emergence on Cell Mounds

After aggregation and generation of cell mounds, the cell mass appears to tighten or "contract" and the upper surface of the mound generates (by 13–14 hr) a morphologically distinct tip region (Fig. 2B), which acts much as a classic "embryonic organizer." Amputation of this tip blocks subsequent development until a new tip forms. Amoebae can aggregate under a thin film of liquid, but no tips form without an air–water interface, in which circumstance development proceeds no further.

This is the first clear stage of production of the slime sheath (Section 7) from which these organisms take their name. It is also the stage at which the vast majority of developmental changes in gene expression occur, as shown by Lodish and co-workers. It is currently believed that most of the genes essential to final spore and stalk cell maturation are turned on at this stage as will be considered in more detail in Section 3.1. An illustrative example is provided by the work of Takeuchi who generated an antibody that is specific for spores and a "prespore" vacuole unique to cells of the posterior portion of slugs (see Section 8). This spore-specific antigen is first expressed at the tip stage.

1.1.4. The Slug Stage

After tip emergence, the cell mass contracts radially and elongates to generate a slender cell column surmounted by the tip. Under conditions of relatively high pH and ionic strength, this structure can proceed directly with the culmination process. However, under many common laboratory conditions, the column bends over to form a migratory pseudoplasmodium, or slug (Figs. 1, 2C). During migration the slug secretes slime from the tip, which encases the migrating structure, hardens toward the posterior end, and is deposited as a trail behind the slug. Slugs display both phototaxis and thermotaxis. From a teleological viewpoint, this explains many factors of *D. discoideum* development. Formation of spores directly in food-depleted soil would preclude an efficient dispersal mechanism. Therefore, individual cells aggregate into these multicellular structures, and the tactic behaviors direct them to the soil surface where, following culmination, environmental carriers can disperse the spores to new areas.

This stage is a particularly interesting one because in slugs the two cell types that comprise the bulk of the fruiting body are distinguishable by a number of criteria. They are also largely physically separated. Experiments in which transverse sections of slugs are induced to complete differentiation have shown that the anterior 20–25% of the slug cells are "prestalk" in that this section will generate a sporeless stalk. Similarly, the posterior 75–80% of the slug cells are "prespore." However, if transverse slug sections are incubated under conditions that induce slug migration rather than culmination, when later induced to culminate they form small fruiting bodies with the normal stalk-to-spore cell ratio.

Thus, the prestalk and prespore distribution character of a slug is not irreversible, and this fact has contributed greatly to the usefulness of this organism as a model for development. In most developmental systems, many different cell types differentiate simultaneously. In *D. discoideum* there are a small number of base plate cells supporting the fruiting body, but the vast majority of cells have a single decision to make: to become a stalk cell or a spore cell. Under normal conditions, the ratio of these two cell types is fixed within narrow limits, independent of size of the fruiting body, even for a fruiting body containing only 12 cells rather than the usual 10^5. This size invariance of tissue proportioning is, of course, a general property of embryonic systems. The distribution of prestalk and prespore cells in the slugs and the redifferentiation in bisected sections open many experimental avenues for investigation of factors controlling the differentiation pathway selected by a cell. Also, the ease with which developmental mutants can be obtained in this haploid organism allows genetic analyses of factors involved in differentiation.

1.1.5. The Culmination Stage

When the slug ceases migration, the tip region essentially sits back on the prespore region to restore the earlier vertical axis. The prestalk cells actively stream downward through the prespore cell mass becoming encased in cellulose and progressively vacuolated as the stalk forms. Ultimately, the stalk cells lose viability.

As the stalk forms, the prespore cell mass is raised along its length, and the cells lose their ameboid appearance as the spore coat forms. Fully differentiated spores are eliptical in shape and about 6 μm in length. The final fruiting body is about 3–4 mm high (Fig. 2D).

1.1.6. Spore Germination

Spores can be stored in the laboratory in refrigerated buffer or, more commonly, desiccated on silica gels for at least 1–2 years. In nature, the spore mass of the fruiting body contains germination inhibitors, which are water soluble. Therefore, a moist environment is required for germination, which can be triggered by nutrients or a brief heat shock. The spore swells, the coat cracks, and an amoeba emerges to resume the vegetative stage of the life cycle.

2. DEVELOPMENTAL REGULATION OF GLYCOPROTEINS

As in other systems, it is generally believed that glycoproteins and their receptors will provide the bases for many recognition processes in *D. discoideum*. Two complementary, sometimes overlapping, approaches have been used to address this issue. One is to focus on a function or process and then attempt identification of the participating molecules. Many such studies have been performed and will be considered later (Sections 5–8). The other approach is to identify glycoproteins present in different subcellular compartments or at different developmental stages and subsequently attempt to identify functions and regulatory mechanisms. In such studies, there has been an emphasis on plasma membrane/ cell surface glycoproteins due to the obvious interest in mechanisms of cell–cell cohesion.

2.1. Cell Surface Lectin Receptors

2.1.1. Con A Receptors

Weeks (1973) first reported the ability of Con A to agglutinate growing and aggregated amoebae of strain AX2, and subsequently, Weeks and Weeks (1975) compared the effects of Con A on NC4 and AX2.

It is widely observed with axenic strains that, after entry into late stationary phase, subsequently harvested cells develop poorly, if at all. For stationary phase AX2, substantially greater Con A concentrations were required for agglutination, and background, Con A-independent agglutination dropped. Thus, the failure of such late cultures to develop may be due in part to irreversible cell surface changes, as also indicated by Jaffe *et al.* (1979) (see Section 5.1).

At vegetative stages, agglutination of NC4 required 10-fold greater Con A concentrations than did AX2. At the same time, Con A-independent agglutination of NC4 was less than that of AX2 under identical conditions. Therefore, at "vegetative" stages these strains have significant cell surface distinctions. Since this report, it has been routinely observed in many laboratories that most axenic strains, when grown without bacteria, have increased endogenous agglutinability, suggesting partial starvation during growth in axenic broth. Indeed, the generation time is approximately doubled in broth, and a few proteins that do not normally appear until aggregation stage appear precociously in broth cultures. A particularly relevant example is the lectin discoidin (Simpson *et al.*, 1974). We have obtained preliminary data indicating that the oligosaccharide profiles of glycoproteins are not identical between axenically and bacterially grown AX3 (Section 4). This may explain the different responses to Con A.

Detection of Con A-binding glycoproteins led to attempts to identify possible functions for them by Con A-induced perturbation of early development. Gillette and Filosa (1973) observed a substantial delay in aggregation of strain NC4 as well as a dramatic elevation of cell-associated cAMP phosphodiesterase after 1 hr of starvation in suspension, levels of this enzyme normally not obtained prior to 8–10 hr of development. They suggested that the delay in development could be due to a Con A-induced distortion of a critical balance between cAMP release and destruction. The results of Weeks and Weeks (1975) indicate, however, that perturbation of development requires multivalent Con A, suggesting that the effects on development are indirect and may be due to patching and capping of the Con A receptors.

2.1.2. Wheat Germ Agglutinin (WGA) Receptors

Scanning (Molday *et al.*, 1976) and transmission (Ryter and Hellio, 1980) electron microscopy have shown the presence of cell surface WGA receptors at all stages of the life cycle. However, agglutination of vegetative cells required much less WGA than aggregation stage cells (Wilhelms *et al.*, 1974; Reitherman *et al.*, 1975), indicating the qualitative and quantitative changes in WGA receptors that were later directly demonstrated (West *et al.*, 1978).

2.1.3. Other Lectins

A number of reports have demonstrated the absence of detectable cell surface receptors for lectins specific for L-fucose, D-galactose, or *N*-acetyl-D-galactosamine using lectins from *Lotus tetragonolobus, Ulex europaeus, Ricinus communis* (RCA-120 and RCA-60), and soybean. For example, West *et al.* (1978) using either fluorescent or microsphere-conjugated lectins failed to detect any binding of ricin to AX3 amoebae whereas red blood cells as internal controls were abundantly labeled (Fig. 3).

2.1.4. Cell Surface Distribution of Lectin Receptors

Gillette *et al.* (1974) reported that Con A induced capping of its receptors. Molday *et al.* (1976) expanded on this observation by use of lectin-conjugated microspheres and scanning EM. AX3 cells were glutaraldehyde-fixed and subsequently labeled with Con A- or WGA-microspheres. Con A receptors were uniformly distributed at early stages. In contrast, cells from slugs had patchy distribution with some areas densely labeled and some unlabeled. WGA receptors at all stages showed densely packed distribution over the whole cell surface. Additionally, the WGA and Con A receptors were largely distinct. After capping of Con A-microspheres had occurred on living cells, cells were fixed and challenged with WGA-microspheres of different size. These generated the same uniform labeling of cell surfaces. Therefore, the two classes of receptors do not "cocap." Similar results were obtained by Ryter and Hellio (1980; Hellio and Ryter, 1980).

Agglutination of cells by lectins, developmental variations in agglutinability, and effects of lectins on cell behavior have also been complemented by analysis of individual glycoprotein constituents and their changes during the life cycle, as described below.

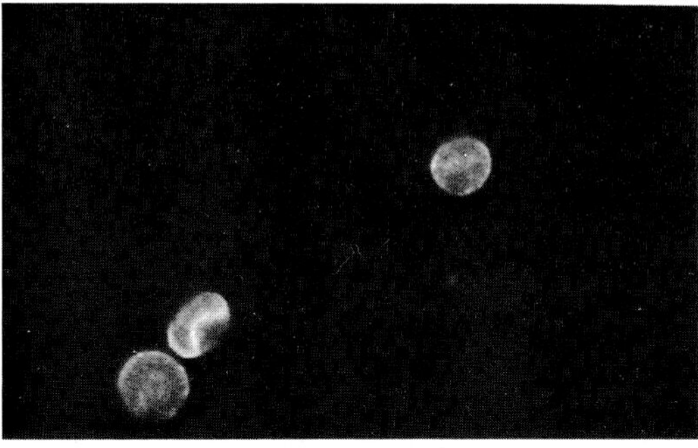

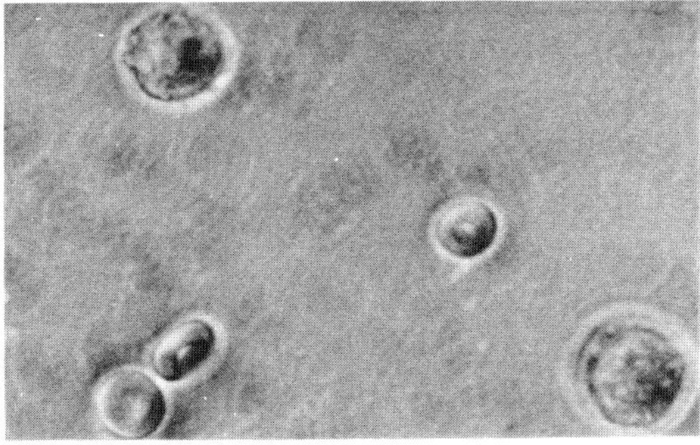

Figure 3. A representative mixture of aggregation-stage *D. discoideum* cells and human red blood cells treated with fluorescein-tagged microspheres conjugated to RCA. (Bottom) Phase-contrast micrograph showing both cell types (red blood cells are smaller and discoid). (Top) Fluorescent micrograph of the same field. Only the red blood cells are labeled and visible. (From West *et al.*, 1978, with permission.)

2.2. Developmental Changes in Glycoprotein Composition

At any developmental stage, the glycoprotein constituents will be composed of species synthesized during that stage and species conserved from preceding stages. If the stage-specific species are produced in sufficient quantity, their *de novo* appearance will be detected by any general glycoprotein detection method. However, if they are produced in low abundance, they might only be observed by pulse metabolic labeling with a radioactive precursor. Results obtained with both of these approaches are considered below.

2.2.1. "Total" Glycoprotein Profiles

Three detection methods have been used here for glycoproteins fractionated by polyacrylamide gel electrophoresis (PAGE): (1) periodic acid–

Schiff (PAS) staining, (2) staining with derivatives of lectins, and (3) covalent chemical modification of intact cells with membrane-impermeant, radioactive labels prior to PAGE. While the slime mold has not changed, the extent of developmental change in its glycoprotein constituents gets greater every year, as new techniques are used.

The first reports on this subject (Smart and Hynes, 1974; Geltosky *et al.*, 1976; Siu *et al.*, 1976, 1977) utilized lactoperoxidase-catalyzed radioiodination of cell surface and, therefore, presumably glycosylated proteins at stages through tip formation, with analysis by SDS–PAGE. Plasma membranes were purified about 10-fold prior to electrophoresis except by Geltosky and co-workers who solubilized cells and analyzed proteins that bound Con A–Sepharose columns and were eluted by 0.1 M α-methylmannoside. Several conclusions can be reached based on the results of these four papers. The most significant is that conservation seemed more common than change. Two or three cell surface species increased during aggregation, one or two at the tip stage, and one or two decreased over these times. Most of the 15–20 bands detected were apparently unchanged in amount. While it is difficult to compare the species between these four reports because of the use of different gel systems, they appear generally compatible. In all of these reports, the radioiodinated surface proteins did not comigrate with any of the major Coomassie blue-stained bands, suggesting that surface species are of minor abundance. As glycoproteins, however, do not stain well or in some cases at all with Coomassie blue, this need not be the case. However, when iodinatable species are compared with proteins metabolically labeled with precursors such as [^{35}S]methionine (Siu *et al.*, 1977; Das and Henderson, 1983a) or [^{14}C]acetate (Parish, 1979), the minor abundance of surface species is confirmed.

In 1977 a report appeared from Hoffman and McMahon who used strain AX3 and analyzed highly purified plasma membranes by SDS–PAGE. PAS stain was used to detect glycoproteins, and once again the glycoprotein profiles were essentially completely distinct from Coomassie blue stain patterns. However, this paper was the first to report major developmental regulation of glycoproteins. Of 25 glycoproteins detected between vegetative and slug stages, 24 suffered some degree of developmental regulation, ranging from on/off stage specificity to quantitative modulation. The major changes occurred during the transition from aggregation to slug stage. Similar results through the aggregation stage (later stages were not tested) were reported in 1979 by Gilkes *et al.* who also used PAS staining of gels and highly purified plasma membranes. The dramatic difference in the extent of developmental change in glycoproteins reported in these papers is initially somewhat difficult to reconcile with the more limited changes of earlier reports based on cell surface iodination. There are, however, several possibilities. One is that many surface glycoproteins may not be accessible to lactoperoxidase and escape detection by that test. Another is that the more highly purified membranes may have enriched more for glycoproteins that were below detection limits in previous reports.

In 1977, West and McMahon reported a new technique for detection of glycoproteins in gels, namely staining with fluorescein isothiocyanate (FITC)–Con A. Purified plasma membranes were shown to contain 41 Con A receptors of which 31 were either stage-specific (not present at all stages) or changed in relative content over the course of vegetative to preculmination (slug stage) development. This report also demonstrated the presence of the *D. discoideum* lectins, discoidins I and II (see Section 6.4), in purified plasma membranes. Use of this gel staining method was soon extended to include analysis of receptors for additional lectins (West *et al.*, 1978). Purified plasma membranes did not contain detectable receptors for *R. communis* lectins 60 and 120 or soybean agglutinin (specificity for galactose/galactosamine), nor for *U. europaeus* agglutinin I (fucose specificity). However, FITC–WGA staining of SDS gels revealed 25 WGA receptors and many were subject to developmental regulation; further, for the most part the WGA and Con A receptors were distinct, as previously reported. Burridge and Jordan (1979) reported similar studies on total cellular proteins using radioiodinated lectins to stain gels. In contrast to the results of West and co-workers, they observed fairly limited regulation of Con A-binding proteins. The implication is that most plasma membrane receptors were below detection limits and that receptors confined to cell-interior domains suffer less drastic developmental regulation than those destined for the cell surface. This phenomenon has been shown directly for proteins of whole cells and plasma membranes after [^{35}S]methionine pulse labeling of the cells (Das and Henderson, 1983a).

Burridge and Jordan were also the first to detect receptors for fucose-binding proteins (FBPs). An FBP receptor in vegetative cells was lost in early development, and four minor staining bands appeared during culmination, a stage later than those tested by West *et al.* (1978) for plasma membranes. These bands may be related to fucosylated spore coat proteins (Lam and Siu, 1981; Devine *et al.*, 1982). Finally, Burridge and Jordan noted that phytohemagglutinin (specific for *N*-acetylgalactosamine) and the *D. discoideum* lectins, discoidins I and II (highest affinity for galactose and *N*-acetylgalactosamine), failed to detect receptors.

The most recent study of glycoprotein changes was by Toda *et al.* (1980) in strain NC4. Cell surface sugars were radiolabeled from vegetative to aggregation stages by periodate oxidation and borotritide reduction. Plasma membranes were purified about eightfold, fractionated by PAGE, and analyzed by fluorography. There were several interesting results in this work. The first was that the mild conditions for periodate oxidation of sialic acid failed to result in subsequent radiolabel incorporation on borotritide reduction, supporting the numerous reports of failure to detect sialic acid in this organism by direct chemical means. They also noted that use of galactose oxidase as the oxidizing agent was unsuccessful, supporting other reports that galactose and galactosamine are absent or are of very limited abundance.

These authors were the first to use two-dimensional PAGE to fractionate glycoproteins, detecting 63 radiolabeled species of which two-thirds were developmentally regulated. The most abundant species were in the high-molecular-weight, acidic quadrant of the gels and streaked/stuttered in the isoelectric focusing (IEF) dimension. The basis for such isoelectric heterogeneity in an organism lacking sialic acid is not clear. There are sulfated and phosphorylated oligosaccharide substituents in this organism, but these are found in only low amounts in highly purified plasma membranes and at least some may be confined to lysosomal enzymes (Sections 4.2 and 5.3).

In summary, methods that test directly for glycoproteins, whether they be PAS or lectin staining of gels or radiolabeling of cell surface sugars, all establish that dramatic changes occur in plasma membrane glycoproteins. At the same time, these methods all test the total profile of such glycoproteins and, therefore, will preferentially detect the most abundant species. To test the possibility that there were minor, possibly high stage-specific, species, workers resorted to pulse labeling with metabolic precursors.

2.2.2. Metabolic Labeling Studies

A number of studies using labeling with radioactive amino acids have reported substantial changes in the profiles of newly synthesized proteins of whole cells and purified plasma membranes during the course of development (Siu *et al.*, 1977; Parish and Schmidlin, 1979a; Ono *et al.*, 1981; Das and Henderson, 1983a). For the plasma membranes it is likely that many of these species are glycoproteins. This issue has been addressed directly by metabolic labeling with precursor sugars. Parish and Schmidlin (1979a) using glucosamine and fucose, identified candidates for glycoproteins thought to participate in cell–cell cohesion. Lam and Siu (1981) used solely metabolic fucose incorporation to study developmental changes in glycoprotein synthesis. Net fucose incorporation declined over the first 10 hr of development but increased substantially during morphogenesis of the multicellular aggregates, to a maximum level during culmination. The glycoprotein products of fucosylation displayed a dra-

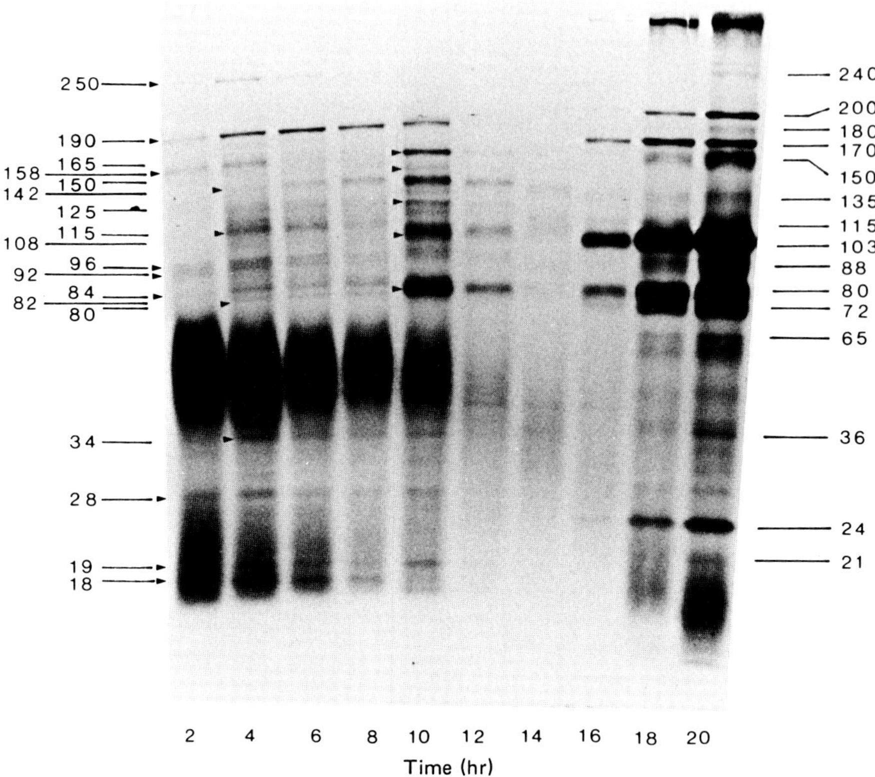

Figure 4. Fluorograms of ^{3}H-fucose-labeled total cellular glycoproteins at different developmental stages. NC4 cells were pulse-labeled at 2-hr intervals during development with [^{3}H]fucose at 400 μCi/10^8 cells. Cells were then washed and solubilized with detergent, and approximately 50 μg of proteins from each sample was loaded on a 5–15% gradient polyacrylamide slab gel for electrophoresis. The gel was processed for fluorography. Apparent molecular weights of the labeled bands were estimated from the molecular weight markers. Molecular weight ($\times 10^{-3}$) of glycoproteins synthesized during the first 12 hr of development are marked on the left with arrowheads to also show the timing of changes. Molecular weights of those that appeared between 16 and 20 hr are indicated on the right. (From Lam *et al.*, 1981, with permission.)

matic stage specificity as shown by Fig. 4. Two general categories of changes are apparent. First, the species synthesized in early development are no longer made by the tight cell mound stage. This may be due in part at least to disappearance of the responsible fucosyltransferase(s) (Section 4.2.1). In late development a largely, possibly completely, different profile was observed. While some species migrated with the same apparent molecular weights at early and late stages, this need not mean

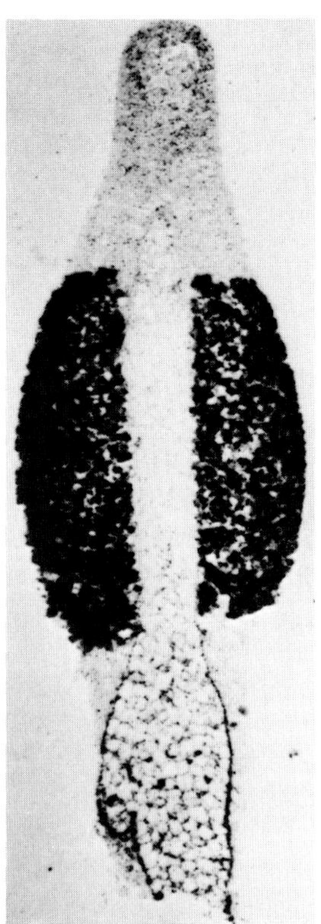

Figure 5. Autoradiogram of a culminating fruiting body after metabolic pulse-labeling with [^{3}H]fucose. Note the intense grains in the prespore/spore cell mass. (From Gregg and Karp, 1978, with permission.)

they are the same glycoproteins. For example, the gp80 in early stages was shown to be on the cell surface while the species present at late stages was not. The second category of change was the regulation of individual species within early and late glycoprotein pools. New species appeared between 2 and 4 hr, and during streaming aggregation (8–10 hr) a number of high-molecular-weight glycoproteins appeared in considerable abundance. Not all of the species expressed after 14 hr appeared at the same stage of morphogenesis, and total incorporation progressively increased. Three species (gp72, 80, and 103) were the most abundant, accounting for 70% of incorporation.

Cytological studies by Gregg and Karp (1978) had demonstrated preferential incorporation of tritiated fucose into prespore cells (Fig. 5). Based on this report, Lam and Siu separated prestalk and prespore cells after fucose labeling of pseudoplasmodia and observed that most, possibly all, of the fucosylated glycoproteins of pseudoplasmodia were in the prespore cells. Several of the high-abundance glycoproteins appear to be precursors of spore coat proteins (Section 8). After the fucose pulse, these species could be chased and recovered in mature spores, suggesting that they are also unusually stable glycoproteins. In this context, Hoffman and McMahon (1978) reported that cycloheximide induced selective loss of some plasma membrane glycoproteins, indicating that these species have rapid turnover rates. Indeed, turnover is an alternate regulatory device for controlling protein and glycoprotein composition and one that has not been investigated very thoroughly.

2.3. Summary

The take-home messages from all the studies summarized in this section are several. First, it is very difficult to obtain really quantitative data on the total developmental changes in glycoprotein constituents and many approaches are needed, as each has inherent limitations. If a general glycoprotein detection method is used, e.g., PAS staining of gels, only the most abundant species will be detected. Some of these may have been conserved from vegetative cells and have no specific developmental role, whereas minor undetected species may change more dramatically and have important developmental functions. Further, detection is dependent not only on abundance of the glycoprotein but on the nature and extent of glycosylation. The same problems of detection limitations exist for any technique. Each lectin probe will identify only a subset of all species and, again, the most abundant. Pulse metabolic labeling will depend on the specific precursor used and is also restricted to detection of the most abundant of even the more minor species. Fractionation techniques are also important. As apparent molecular weight of glycoproteins is a critical function of the amount of glycosylation and pore size of polyacrylamide gels, it is extremely difficult to relate results from different laboratories. In all cases the molecular weight standards are lowly or nonglycosylated proteins. Further, it is now clear that multiple species often comigrate on SDS–PAGE and can suppress the extent of developmental change detected by two-dimensional analysis. Additionally, different authors define changes in different ways. Modulation of the level of a glycoprotein can be considered developmental change and in many instances this will be true. At the same time, dramatic changes in amounts of a few species can generate apparent modulation of other species when similar amounts of protein or radioactivity are used per lane of a gel, even if the absolute amount of the glycoprotein per cell has not changed. The situation is alleviated only slightly when scoring is for on/off changes. A species diagnosed as absent may simply be below detection limits and such data cannot constitute proof for presence or absence of a messenger RNA, i.e., on/off changes in gene expression.

While a summary of the work that has been done illustrates the problems, it also emphasizes that there is dramatic regulation of the glycoprotein constituents of cells and especially of their plasma membranes during development of *D. discoideum*. Despite the essentially descriptive nature of such studies, this implies that these regulated glycoproteins have developmentally important functions. This is supported by Lam and Siu's (1981) demonstration that at least some of the late fucosylated species are cell-type specific, i.e., in prespore cells as precursors to spore coat proteins. Finally, as will be considered later, functional roles for a number of individual glycoproteins have now been established or strongly implicated and mechanistic rather than descriptive analyses are now possible.

2.4. Effects of Inhibitors on Development

If the catalog of changes in glycoproteins observed during development reflects glycoprotein roles important to this process, blockade of glycoprotein production should disrupt the normal course of development, and this has been observed. Several laboratories have tested effects of tunicamycin on development and macromolecular synthesis, as this drug blocks assembly of the lipid-linked precursor of glycoprotein *N*-linked oligosaccharides. The general features of these studies are in agreement: tunicamycin blocks both growth and development. The drug also completely inhibited the appearance of specific glycoproteins of known function, i.e., gp80 (contact sites A, Ochiai *et al.*, 1982a; Lam and Siu, 1982) and alkaline phosphatase (Brian W. Poor, personal communication). Correlated with the lack of expression of gp80 was a lack of expression of the EDTA-resistant cell–cell cohesion associated with these sites as established by Gerisch and co-workers (see Section 6.1.1). This suggests at least one molecular basis for the blockade of normal development. Interestingly, tunicamycin appeared to have essentially no effect on EDTA-sensitive contact sites B, which are present in vegetative cells and throughout at least aggregation. However, the effects of tunicamycin were only tested after vegetative growth so that contact sites B (Section 5.1) could be a stable glycoprotein, not requiring continual synthesis. It remains to be established that all developmentally regulated glycoproteins have important functions and, further, whether their oligosaccharide substituents are essential to these functions. For example, inhibition of development by tunicamycin could occur if an otherwise functional polypeptide was unable to be shipped to its proper cellular locale when oligosaccharides are absent.

3. REGULATION OF GLYCOPROTEIN BIOSYNTHESIS

There can, of course, be a wide array of independent regulatory mechanisms that, in a developmental system, control gene expression.

For glycoproteins, this involves two subtypes of regulation. First, there is the type of RNA transcripts and primary translation products (polypeptides) available for glycosylations. Second, there could be regulation of expression of the genes encoding enzymes that catalyze these post-translational modifications. This section will consider general mechanisms regulating gene expression and their relevance to synthesis of the polypeptides of glycoproteins. Section 4 will consider developmental regulation of glycosylations.

3.1. Regulation of Gene Expression in D. discoideum

From the work of Loomis and co-workers (Loomis *et al.*, 1976), it is known that progression through the developmental cycle is accompanied by an orderly, sequential expression of stage-specific enzymes and that, for the most part, mutations that block development at any stage also prevent expression of enzymes of the later stages.

Based on well-studied hallmarks of the various stages in the development of *D. discoideum*, a number of obvious candidates for stage-specific regulators of gene expression exist. These include starvation in very early development, cAMP intercellular signaling during aggregation, formation of stable cell–cell contacts during establishment of multicellularity, subsequent tightening of these contacts and/or invigoration of endogenous signals by this tightening during tip emergence, and, finally, cell sorting or positional signals that may control cell-type-specific gene expression. A detailed review of these areas is not possible here. Fortunately, an excellent review has been published recently (Lodish *et al.*, 1982).

Starvation is an obvious candidate for an inducer of very early changes in gene expression. Marin (1976) reported that addition of certain amino acids to the starvation medium blocked development and (Marin, 1977) induction of cAMP chemotaxis and EDTA-resistant cohesion. Darmon and Klein (1978) showed that amino acid addition inhibited induction of adenylate cyclase and phosphodiesterase activities. Margolskee *et al.* (1980) showed that removal of trypticase and yeast extract from axenic medium was sufficient to induce the normal ribosome runoff from polysomes and expression of cAMP receptors. However, the synthesis of three early proteins, including actin, occurred independent of medium composition if the cells were plated at sufficiently high density. Grabel and Loomis (1978) reported a similar density requirement for expression of N-acetylglucosaminidase. It is not certain currently whether this reflects a need for cell–cell contacts or a need for a sufficient concentration of secreted factors.

Prior to and during aggregation, a number of components required for that process are elevated or appear apparently *de novo*. These include, at least, expression of the cAMP chemosensory system and EDTA-resistant cell–cell cohesion sites. It is currently thought that early cAMP signals induce these events. When cells are harvested from stationary rather than exponential growth, development is blocked. This block is overcome by pulsatile but not continuous application of cAMP (Gerisch *et al.*, 1975). Some mutants that fail to aggregate are incapable of producing autonomous cAMP signals but are induced to aggregate when these are experimentally supplied (Darmon *et al.*, 1975). Recently, a mutant has been described (Barclay and Henderson, 1982) that is temperature sensitive for chemotaxis toward cAMP and also temperature sensitive for normal elevation of cell surface cAMP receptors, which are thermostable once expressed. The simplest interpretation of these observations is that chemosensory function plays a positive feedback role in controlling expression of chemosensory and cohesive elements. cAMP may also be a negative regulator in some cases. Amounts of cAMP produced by amoebae increase substantially during aggregation. In early development of bacterially grown cells, there is a large increase in levels of mRNA for the lectin, discoidin I, which is suppressed by late aggregation when cAMP signaling is most intense. Williams *et al.* (1980) have shown that when developing cells were shaken in suspension, subsequently isolated nuclei continued to synthesize this mRNA at high levels. However, addition of cAMP to the cell suspensions caused a 10-fold depression of discoidin I mRNA in the nuclei.

As the biological job of early development is establishment of multicellularity, it might be expected that this would require relatively limited changes in gene expression. From mutational target size, Loomis (1978) has estimated that about 150 genes are required for aggregation. Blumberg and Lodish (1980) could not detect mRNAs for these above the background of the approximately 4800 genes expressed in mRNA of vegetative cells. However, one might also expect that genes required for aggregation would preferentially encode plasma membrane proteins, such as those of the cAMP chemosensory and cell–cell cohesion systems, which are minor abundance species in the cell. This expectation was confirmed when cells at different stages of early development are metabolically pulse labeled with [^{35}S]methionine and 30-fold purified plasma membranes analyzed by two-dimensional PAGE (Das and Henderson, 1983a,b). The vast majority of plasma membrane proteins were below detection limits in whole cell samples. The newly synthesized plasma membrane proteins were dramatically different at the four early stages tested and could account for a substantial proportion of the estimated 150 genes.

Next, it might be expected that the largest number of changes in gene expression would be required for differentiation of the two terminal cell types. As precursors of these cells, i.e., prestalk and prespore cells, exist at the slug stage, it was possible that many of these genes would be expressed at the end of aggregation. Blumberg and Lodish (1980) showed that 2000–3000 new mRNA species were expressed in late aggregation. As under some conditions the slug stage can be bypassed, it is not surprising that a large proportion of the "late" genes are expressed at late aggregation. Finally, during culmination, when morphological differentiation of cell types occurs, a more limited number of new species are synthesized that are cell-type specific (Coloma and Lodish, 1981) such as the spore coat proteins and glycoproteins (Orlowski and Loomis, 1979).

The large number of new mRNA species produced at the end of aggregation is compatible with the expression of many "late" enzymes at this stage and, accordingly, this stage has been the most studied in the search for the regulatory mechanisms. In particular, roles of cAMP versus formation of cell–cell contacts have received much attention. From work by Sussman and co-workers (see, e.g., Newell *et al.*, 1972), it has been known that expression of late enzymes required cellular interactions. Dissociation of slug stage cells caused cessation of enzyme synthesis, but reaggregation of these cells was accompanied by renewed synthesis. These studies suggest that formation of stable cell–cell contacts may be the regulatory signal. However, it is also possible that when the cells are closer together, the local concentration of diffusible factors could be raised above some critical threshold level. Alternatively, some signal produced *de novo* by the tip could be critical. Blumberg *et al.* (1982) reported that 40% of the normal level of new mRNAs was produced by a mutant that completed aggregation but did not form a tip. Another mutant that was blocked in development after tip emergence produced the full quota of mRNAs. This suggests that a unique tip-derived signal is probably not required, but it is compatible with possible invigoration of an endogenous signal. The most obvious candidate for such a signal is cAMP. Lodish *et al.* (1982) have reviewed a number of studies on the effects of cAMP on cells shaken in suspension at slow speeds (which allow formation of agglutinates) and fast speeds (which do not) and on cells dissociated after normal aggregation. The general conclusion from these studies is that expression of the late aggregation genes requires at least a period of actual cell–cell contact and that most also require cAMP. For example, disaggregation of slug cells resulted in selective and rapid decay of those mRNAs whereas constitutive mRNAs were conserved. While cAMP alone was unable to induce late mRNA expression in cells rapidly shaken in suspension from the initiation of starvation, it could restore their expression in the disaggregated slug cells (Chung *et al.*, 1981).

There remain certain observations not totally explained by this analysis. Bonner (1970) showed that when plated on high levels of cAMP, a small proportion of NC4 cells were induced to differentiate, without morphogenesis, into stalk cells. Gross and co-workers later observed that, in strain V12/M2, cAMP could induce 100% stalk cell differentiation (Town *et al.*, 1976) at sufficiently high cell densities. It was later reported that a dialyzable factor (differentiation-inducing factor, DIF) together with cAMP could produce the same result at low cell densities (Town and Stanford, 1979). Further, as stalk cells are not viable, this provided a convenient selection for mutants that differentiated into spores under these conditions, with no cell contact formation (Kay and Trevan, 1981). This harkens back to models in which diffusible factors are maintained at high local concentrations when cells are in tight contacts. Such factors may include cAMP, DIF, and ammonia (the latter reviewed in Sussman, 1982).

3.2. Mechanisms Regulating Glycoprotein Synthesis

There appear to be no surprises here, and the data are all compatible with mechanisms considered in the previous section. Parish *et al.* (1978a) studied two mutant strains, isolated in Brachet's laboratory (Darmon *et al.*, 1975), which fail to generate autonomous cAMP pulses and fail to aggregate. Exogenous pulses of cAMP induce biochemical differentiation and aggregation of these amoebae (Darmon *et al.*, 1977; Klein and Darmon, 1977; Juliani and Klein, 1978). When Parish and co-workers pulsed these mutants with cAMP, this induced incorporation of [^{14}C]acetate and glucosamine into a glycoprotein with the apparent molecular weight of the contact sites A candidate, gp80. These sites had been previously known to be at least functionally activated by cAMP pulses (Gerisch *et al.*, 1975). Toda *et al.* (1981) observed a similar result in that exogenous cAMP pulses of suspensions of wild-type cells accelerated the appearance of a gp89 which, on their gel system, may correspond to contact sites A. This paper also reported different changes of cell surface glycoproteins between cells starved in shaken suspension versus cells developing normally on a filter. Some glycoproteins of vegetative cells were lost during normal development but not lost by suspended cells. Cell suspensions also produced lesser amounts of some glycoproteins than cells on filters. Both of these ''defects'' in suspended cells were corrected by subsequent plating of the cells, suggesting that some form of cellular interaction is involved.

Parish and Schmidlin (1979b) have reported data compatible with resynthesis of some plasma membrane glycoproteins during recapitulation of morphogenesis by dissociated slug cells, i.e., [^{14}C]acetate was incorporated into polypeptides with a mobility on SDS gels that was that of glycoproteins detected by con A staining of the gels. In a subsequent report, Parish (1979) plated dissociated slug cells as monolayers under cellophane, which blocked reaggregation. Under these conditions there was resynthesis of one 110K plasma membrane glycoprotein but not of 90 and 95K glycoproteins resynthesized during reaggregation. However, inclusion of 1 mM cAMP (but not 5'-AMP) induced synthesis of these glycoproteins by the monolayer cells.

4. SYNTHESIS AND PROCESSING OF GLYCOPROTEIN-LINKED OLIGOSACCHARIDES

As polypeptide determinants can influence protein glycosylations, changes in the profile of glycoproteins during development could be due solely to changes in expression of primary polypeptide sequences available to the enzymes responsible for these posttranslational modifications. However, developmental regulation of glycosyltransferases and glycohydrolases could also contribute significantly to variations of the glycoprotein species observed. Recent studies provide clear evidence for such a contribution, as well as establishing that the vast majority of oligosaccharides are *N*-linked to proteins and initially assembled via a lipid-linked precursor as in other eukaryotes. However, the processing routes for this oligosaccharide after transfer to polypeptides are dramatically dependent on the developmental stage.

4.1. Lipid-Linked Precursors

Polyisoprenyl-like units with bound mannose were first reported in vegetative amebae by Crean and Rossomando (1977) and in developing cells by Rössler *et al.* (1978). More recently, Rössler *et al.* (1981) tested a series of polyisoprenylphosphates for their ability to stimulate incorporation of mannose from GDP-mannose into glycolipids and proteins in a particulate membrane fraction. The only structure that was effective was a C_{55}-dihydropolyprenylphosphate. Larger, smaller, and fully unsaturated polyprenylphosphates were inactive. Further, they cited unpublished observations from mass spectral analyses that the predominant endogenous mannolipid contained 11 isoprenoid units, the terminal unit of which was saturated, i.e., the same structure as the only exogenous material that was effective. Thus, the authors have identified a polyisoprenol that appears to be identical to an endogenous acceptor and has properties that span prokaryotic and higher eukaryotic systems; namely, the polyisoprenol has a short length characteristic of prokaryotes but a terminal saturated isoprene unit characteristic of higher eukaryotes. Rössler *et al.* (1982) have extended these studies to the identification of an activity that phosphorylates exogenous dolichols of C_{80-105} length in membrane fractions. The activity was maximal after 4–6 hr of development and depressed at 7–8 hr. As these authors have shown that the C_{55}-dolichol form appears to be the endogenous species, reports on that form are now needed. Nonetheless, the authors appropriately speculate that a balance between dolichol kinase and phosphodolichol phosphatase activities could be an important regulatory mechanism. These results are compatible with a role for these activities in *de novo* glycoprotein biosynthesis by assembly of a lipid-linked precursor essentially identical to that of other eukaryotes (for a review, see Hubbard and Ivatt, 1981).

A simplified structure of this precursor is:

$$\text{dolichol-P-P-GN-GN-M}_9\text{-G}_3$$

The oligosaccharide contains three glucose (G) residues, nine mannose (M) residues, two GlcNAc (GN) units and is linked to dolichol via the pyrophosphate (P-P). After metabolic labeling of unbroken cells with precursor sugars, we (Ivatt *et al.*, 1984b) have isolated a lipid-linked oligosaccharide that contains GlcNAc, mannose, and glucose, and mild acid treatment releases a free oligosaccharide that coelutes on calibrated gel filtration columns with the mammalian precursor oligosaccharide. The *Dictyostelium* and mammalian precursor oligosaccharides are both substrates for the enzyme endo-β-*N*-acetylglucosaminidase (endo H). Endo H cleaves high-mannose oligosaccharides between the two core GlcNAc residues unless they are blocked by other substituents such as fucose (Tarentino and Maley, 1975). Therefore, the *Dictyostelium* precursor appears to be identical to the precursor assembled by all other eukaryotes tested. The precursor does not contain fucose, galactose, galactosamine, or sulfate. Radioactivity is incorporated from galactose, but this appears to occur only after epimerization of the nucleotide-sugar to glucose. After strong acid hydrolysis, all radioactivity migrated on paper chromatography as glucose.

This precursor is transferred *en bloc* to polypeptides after which processed species accumulate. The precursor is synthesized at all stages of development.

4.2. Oligosaccharide-Processing Pathways

In mammals, processing is by two general routes that can coexist in the same cells (Hubbard and Ivatt, 1981). In both, the three glucoses are rapidly removed. In one pathway, a few mannose residues are also trimmed to yield mature high-mannose oligosaccharides. In the other pathway, six mannoses are removed, and the mature species contain additional GlcNAc, Gal, Fuc, and sialic acid in complex structures.

As noted previously, *D. discoideum* does not contain sialic acid. Also, we have been unable to detect incorporation of galactose (except after epimerization to glucose) or galactosamine into glycoproteins at any stage of the life cycle, which explains the failure of lectins that bind these sugars to detect glycoprotein receptors in this organism. However, the pathway by which the precursor oligosaccharide is processed depends critically on the developmental stage (Ivatt *et al.*, 1981, 1984a). The pathways have been defined largely by a single type of experimental regime as follows. Cells at a series of developmental stages were pulse labeled with appropriate precursors and were Folch extracted to obtain delipidated proteins, which were exhaustively digested with Pronase and, subsequently, with endo H. The samples were then fractionated on gel filtration columns (Bio-Gel P4-400) previously calibrated with fibroblast oligosaccharides of known structure.

4.2.1. The Early Pathway

This processing route is present at the earliest stage of starvation tested for axenically grown AX3, i.e., metabolic labeling is initiated 30 min after removal of mid-log-phase cells from the growth medium. The activity of this pathway elevates (in terms of label uptake per cell) at 4 hr, declines during aggregation, and is largely abolished at the tip formation stage. This pathway is summarized schematically in Fig. 6A. Data for the earliest time point are shown in Fig. 7. There is polysaccharide material in the void volume (fraction 60), which incorporates label from all precursors as well as (not shown) phosphate.

Endo H-sensitive oligosaccharides (those that would have coeluted with resistant material without this endo H digestion but that elute at lower molecular weight after digestion) contain glucosamine, mannose, and glucose (fractions 90–110) and elute as though they have lost either one or two hexose units of the precursor (not shown). Endo H-resistant glycopeptides, which elute as a broad peak with substructure (fractions 70–90), contain glucosamine, mannose, fucose, sulfate, and a small amount of what may be glucose. The endo H resistance could be correlated with fucosylation of probably the GN_2 core region, as mild acid conditions that remove fucose (but not sulfate) rendered them sensitive to endo H (Ivatt *et al.*, 1984a). This indicates that fucosylation is of core sugars, and sulfation occurs on peripheral regions which do not block access by endo H. Also, all endo H-resistant species contained fucose and only some contained sulfate, as determined by fractionation on Con

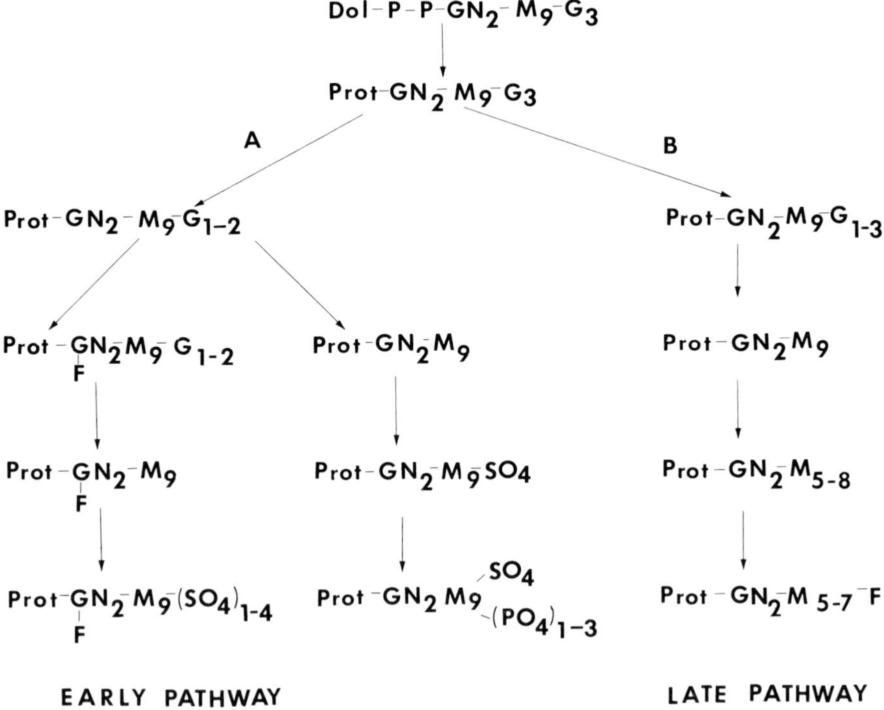

EARLY PATHWAY   LATE PATHWAY

Figure 6. Schematic representation of the early (A) and late (B) developmental oligosac-charide-processing pathways. All structures are found on mature glycoproteins (except GN_2-M_9-G_3 in the early pathway) and, therefore, are not merely intermediates.

A–Sepharose columns (Ivatt *et al.*, 1984b). The number of sulfate residues per oligosaccharide is not certain, but column profiles suggest at least three sulfated species besides the polysaccharide (Fig. 7).

We (Das and Henderson, 1983c) have recently detected an additional class of oligosaccharides, which are of low abundance in whole cell samples but are enriched in secreted and some intracellular glycoprotein pools. These are not fucosylated but are sulfated and contain up to at least three phosphates (Fig. 6A). They are absent or are at only very low levels in purified plasma membranes (Das and Henderson, 1983c; Matarazzo and Henderson, 1983a). However, it is known that lysosomal enzymes contain sulfated and phosphorylated oligosaccharides (Freeze and Miller, 1980; Knecht *et al.*, 1982) and that the phosphomannosyl groups can serve as recognition markers (Freeze *et al.*, 1980). Thus, func-

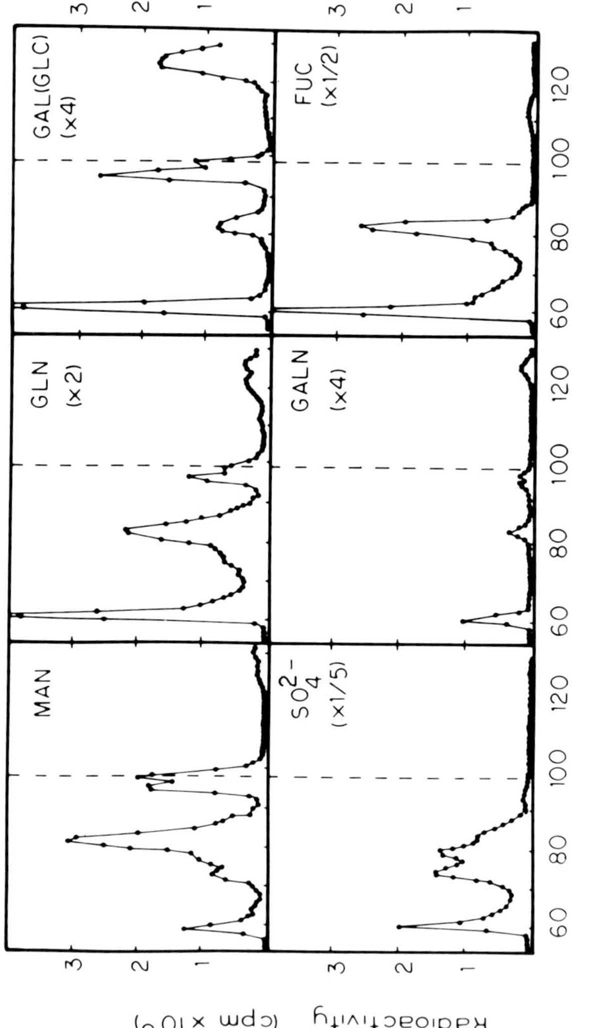

Figure 7. Glycopeptides from *D. discoideum* labeled during the first hour of starvation. Amoebae were labeled with mannose or glucosamine for 15 min and chased for 45 min with unlabeled sugars, or labeled with galactose, sulfate, galactosamine, or fucose for 60 min. Radiolabeled glycoproteins were prepared from the amoebae by exhaustive solvent extraction (chloroform:methanol:water, 10:10:3 v/v) of the protein residue prepared by the Folch procedure. The radiolabeled glycopeptides and oligosaccharides were subjected to gel-filtration chromatography on Bio-Gel P4. MAN, [³H]mannose; GLN, [³H]glucosamine; GAL, [³H]galactose; SO₄²⁻, [³⁵S]sulfate; GALN, [³H]galactosamine; FUC, [³H]fucose. The scales were adjusted by the factors indicated on the individual panels. (From Ivatt *et al.*, 1981, with permission.)

tion of these oligosaccharides as recognition for organelle-specific intracellular sorting of glycoproteins may be implicated (see Section 5.3). We have also observed (Das and Henderson, 1983c) that, of the two endo H–sensitive oligosaccharides (Fig. 7), the larger form is found only on glycoproteins shipped to external destinations (plasma membrane and secreted glycoproteins) while the smaller seems confined to intracellular membranes.

The observation that the activity of the early developmental pathway increased over the first 4 hr of starvation raised the question as to whether this was the usual processing pathway in vegetative amoebae or was precociously induced by axenic growth. We (Tschursin and Henderson, 1983) have recently shown that bacterially grown AX3 synthesize very few oligosaccharides more complex than $GN_2M_9G_{1-2}$ during early exponential growth. The more complex structures appear at mid-log cell densities, and only by stationary phase are these species synthesized in amounts comparable to axenically grown amoebae at all stages of the growth curve. We are currently pursuing these studies since axenic growth appears to precociously induce appearance of not only some fucosylated oligosaccharide species but of the carbohydrate-binding lectin, discoidin, for which L-fucose is a good ligand.

The "early" developmental pathway progressively declines during aggregation. At the tip construction stage after aggregation, the stage at which the major developmental changes in gene expression occur, the early pathway is essentially gone and a new oligosaccharide-processing pathway emerges.

4.2.2. The Late Pathway

This pathway has been defined by experiments analogous to those described for the early route and is summarized in Fig. 6B. Polysaccharide material is present, but it no longer incorporates sulfate and other precursors are incorporated in ratios that differ from early stages (Ivatt *et al.*, 1981). The vast majority of oligosaccharides are endo H sensitive. Three of the new species are fucosylated, and their endo H sensitivity indicates that the fucosylations are now on peripheral rather than core sugars. A wide size spectrum appears, the largest oligosaccharide eluting at the same position as the endo H-digested precursor. The size indicates 12 hexose units attached to GN. This could suggest that the enzyme that removes the first glucose has been depressed. However, this species also does not contain label from galactose. This could indicate that either the galactose-glucose epimerase has disappeared or that a mannosyltransferase now adds mannose residues to the glucose-trimmed form. The former is known to occur (Telser and Sussman, 1971). The latter is an interesting possibility in that yeasts also assemble a lipid-linked precursor but process it in part to a protein-GN_2-M_{12} structure after glucose trimming (Nakajima and Ballou, 1974).

4.2.3. Developmental Regulation of Synthetic and Processing Enzymes

We have identified synthesis of the same lipid-linked precursor and its transfer to proteins at all stages of development. Therefore, the participating enzymes must always be present. The dramatic developmental change in the processing of protein-linked oligosaccharides between two almost mutually exclusive pathways implies coordinate developmental regulation of the processing enzymes (Ivatt *et al.*, 1981, 1984a). Figure 8 shows simultaneous suppression of the early transferases responsible for core fucosylation and sulfations. Similarly, the late mannohydrolase(s) and peripheral fucosylations appear and increase simultaneously. The hydrolases that remove the glucose residues clearly span both processing pathways, as some species have lost all glucoses even at the earliest stages tested. There is, however, a change in at least the second glucohydrolase, possibly by the same mechanisms that control the early pathway, as this activity increases in parallel with the decreases in core fucosylation and sulfation. Depression of the early pathway precedes expression of the late pathway, which appears *de novo* at tip formation. Therefore, all these temporal changes could be accommodated by a model in which there are only two regulatory switches, one that switches off expression of the early processing enzymes and elevates the second glycohydrolase, and another mechanism that initiates expression of the late processing enzymes.

However, it appears that there may be multiple regulatory signals controlling at least the early pathway, as decreases in sulfation and core fucosylation can be uncoupled experimentally. As Fig. 8 shows, suppression of core fucosylation does not occur when cells are starved in shaken suspension and cell–cell contact thus prevented, but sulfation reactions are suppressed with approximately normal developmental kinetics. Addition of cAMP to these shaken suspensions has very little effect on core fucosylations but prematurely depresses sulfations despite the presence of core fucosylated substrates for these reactions. The implications are, therefore, that cell contacts and cAMP are the regulators of the core fucosyltransferase(s) and sulfate transferases, respectively.

It is possible that not all of the processing enzymes suffer developmental regulation of level, as their activity could be controlled by substrate availability. For example, the transferase that adds fucose to peripheral sugars in the late pathway could be constitutively expressed but unable to function catalytically until mannose-trimmed substrates become

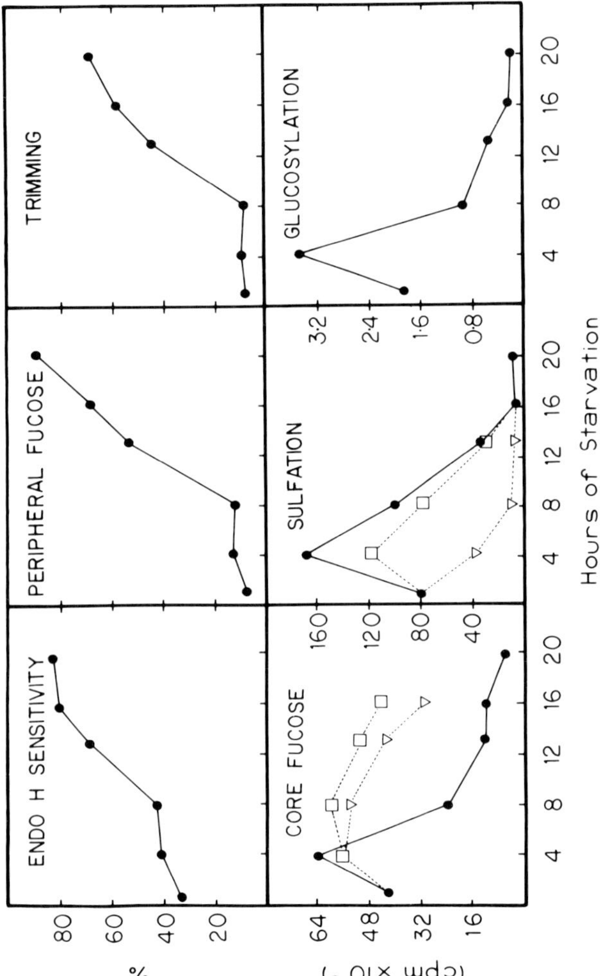

Figure 8. Time courses for developmentally regulated oligosaccharide-processing pathways. The solid lines indicate the normal developmental increases in endo H sensitivity, peripheral fucosylation, and mannose trimming that appear at the tip stage (top three panels) and the suppression of core fucosylations, sulfations, and the increase in glucohydrolase activity that occur during aggregation. The dashed lines indicate the behavior of cells starved in fast-shaken suspension to prevent cell contact formation and in the absence (□) or presence (▽) of 1 mM cAMP. These suspension-starved cells were harvested and plated prior to an immediate 60-min pulse with [³H]fucose or [³⁵S]sulfate.

available. Indeed, a nonlysosomal α-mannosidase activity has been shown to be expressed at the tip formation stage (Free and Loomis, 1974).

Finally, it is provocative to consider that in early development, when amoebae behave as autonomous cells, processing is yeastlike in that trimming of mannose units does not occur. With the establishment of tissuelike multicellularity, mannose trimming occurs as is true in mammalian cells.

4.3. Other Glycosylation Mechanisms

Glycopeptides obtained by Pronase digestion of Folch-extracted glycoproteins were not cleaved by mild reductive alkali, suggesting either the absence of O-linked oligosaccharides or that they are of substantially lower abundance than *N*-linked species when total glycoproteins are studied.

Wilhelms *et al.* (1974) identified an antigen in phenol–water extracts of cells that appeared to contain GlcNAc in *O*-linkages to serine and had a GlcNAc: fucose:mannose ratio of 12:5:6. Possibly this material is a proteoglycan with polysaccharide bound via serine residues, for the polysaccharide material reported by Ivatt *et al.* (1981) is rich in GlcNAc and fucose relative to mannose.

While proteoglycans are not the focus of this review, a glycosaminoglycan reported by White and Sussman (1963a,b) was rich in GalNAc, Gal, and galacturonic acid. One of the synthetic enzymes, UDP-Gal polysaccharide transferase, was the first enzyme in *D. discoideum* shown to be stage specific (Sussman and Osborn, 1964) and was also the first enzyme shown to be cell-type (prespore) specific (Newell *et al.*, 1969). The polysaccharide product may be one of the prespore vesicle contents, as it is eventually externalized and may provide "glue" for the spore coat.

Lysosomal enzymes have a number of interesting glycosylations that are compatible with the early processing pathway considered above (Section 5.3). However, Gustafson and Milner (1980a) reported an independent type of glycosylation of one lysosomal enzyme, proteinase I, which contained *N*-acetyl-α-D-glucosamine-1-phosphorylserine.

4.4. Summary and Implications

It is clear that the major form of glycoprotein glycosylation in *D. discoideum* is due to transfer of a large, lipid-linked precursor to an *N*-linked form on protein acceptors. As transfer to polypeptides is generally cotranslational, this suggests that metabolic labeling with mannose or glucosamine may be a valid indicator of polypeptide synthesis. The nature of posttransfer processing is critically dependent on the developmental stage. Processing in early development leads to incorporation of charged moieties (sulfate, phosphate) into some oligosaccharides, and this pathway in essence disappears by tip formation. At this stage, a new processing system appears and generates endo H-sensitive, trimmed, but neutral, oligosaccharides.

As diagnosed by electrophoresis on one- and two-dimensional gels, the tip stage shows the most dominant changes in protein profiles of whole cells and purified plasma membranes. It is, therefore, appropriate to consider the extent to which these changes might be due to changes in expression of polypeptide species or in the nature of glycosylations.

Lysosomal enzymes have been shown to have less acidic isoelectric points after the early-to-late glycosylation change (Cardelli *et al.*, 1982). In most of these cases, it does not appear that a new enzyme structural gene is expressed, and glycosylations are a candidate for the changes. At the same time, Alton and Lodish (1977) showed that most of the proteins that appeared and disappeared during development were reflected by changes in the mRNA population when translated *in vitro*. We (Das and Henderson, 1983a) have shown that there are dramatic changes at this stage in the two-dimensional gel profiles of newly synthesized proteins in highly purified plasma membranes. However, we do not detect significant amounts of sulfated or phosphorylated oligosaccharides in plasma membranes so that most of the changes are not due to altered oligosaccharide charge. Thus, it would appear that most of the apparent changes in protein composition at this critical developmental stage reflect *de novo* changes in polypeptide expression as expected from the large number of new mRNA species. There is, however, also a dramatic change in the nature of oligosaccharide processing at this stage. The biological consequence of this remains to be determined. The systems responsible for intracellular cohesion also change at this stage, and possibly these systems involve oligosaccharide recognition (Section 6).

Finally, having considered, thus far, the extent to which glycoproteins are developmentally regulated, possible mechanisms for such regulation, and the nature of the carbohydrate moieties, it is time to consider what biological jobs glycoproteins have.

5. GLYCOPROTEIN ROLES IN VEGETATIVE CELLS

5.1. Contact Sites B

While vegetative amoebae appear to function autonomously, they are mutually cohesive and will agglutinate in buffered suspensions. Gerisch (1961) showed that this agglutination was blocked by EDTA and, subsequently (Beug *et al.*, 1970, 1973a,b; Gerisch, 1980), that the molecular entity (or entities), called contact sites B, was antigenically distinct from the EDTA-resistant cohesion system (contact sites A) produced during aggregation. Contact sites B are not lost during aggregation but contribute lateral interactions between aggregating cells, as opposed to the end-to-end contacts mediated by contact sites A. Jaffe *et al.* (1979) have proposed a model in which cell–cell cohesion mediated by contact sites B is not a simple homodimer formation of these sites between two cells but functions as part of a ligand–receptor system. They showed that log-phase amoebae were mutually cohesive whereas cells in late stationary phase were not. However, these stationary-phase cells formed mixed agglutinates with both vegetative and aggregation-competent cells, suggesting that they have lost either the ligand or the receptor, but not both. Further, a dialyzable material, which appears to be a small carbohydrate and accumulates in stationary culture medium, blocked the mutual cohesion of vegetative cells. This inhibitor presumably corresponds to the release of part of the ligand or receptor into the medium during entry into the stationary phase. A molecular candidate for contact sites B has recently been identified (Chadwick and Garrod, 1983) by serological criteria. It was a 126K cell surface protein that adsorbed the adhesion-blocking activity of a polyclonal antiserum prepared against vegetative cells, and a polyclonal antiserum prepared against purified p126 blocked contact sites B-mediated cohesion.

Contact sites B may play an essential developmental role in addition to providing lateral contacts. If EDTA is present from the onset of starvation, contact sites A are not expressed (Beug *et al.*, 1973a), and aggregation is inhibited (Gingell and Garrod, 1969; Mason *et al.*, 1971). At least a period of high cell density is required for early expression of *N*-acetylglucosaminidase (Grabel and Loomis, 1978) and contact sites A (Marin, 1977; Marin *et al.*, 1980), and plating at high density is sufficient to induce a restricted set of three early developmental proteins (Margolskee *et al.*, 1980). Glucose, at fairly high concentrations (ca. 50 mM), inhibits development of aggregation competence (Rickenberg *et al.*, 1975; Rahmsdorf *et al.*, 1976; Marin, 1976; Darmon and Klein, 1978), blocks EDTA-sensitive agglutination and expression of contact sites A (Marin *et al.*, 1980). GlcNAc blocked EDTA-sensitive agglutination as effectively as glucose (Marin *et al.*, 1980); effects on expression of contact sites A were not tested. These observations, while indirect, are compatible with the ligand–receptor model above, a need for contact sites B-mediated cellular interactions in the initiation of early development, and suggest that glucose and glucosamine contribute to the recognition mechanism. Possibly the inhibitor of Jaffe *et al.* (1979) has such an exposed residue, though effects of their inhibitor on development have not been reported.

5.2. Phagocytosis and Pinocytosis

It is interesting that these activities have been studied nearly exclusively in axenic strains rather than wild type, and the nature of the two mutations (Williams *et al.*, 1974) required for good axenic growth is not known. The axenic strains retain their ability to grow on a bacterial food associate but have the added ability to internalize a nutrient-rich fluid phase in the absence of particles. Two general approaches to roles of receptors and glycoproteins in phagocytosis and pinocytosis have been used. One has been to determine properties and fates of internalized membranes. The other has been to probe directly for receptors for particles to be internalized.

Based on morphometric analysis, Ryter and de Chastellier (1977) showed that during yeast phagocytosis, intracellular phagosomal membranes had a surface area 40% that of the plasma membrane area, which remained constant. They concluded that the loss of cell surface membrane into phagosomes was compensated by outflow of internal membrane. This was directly and elegantly shown for pinocytosis in vegetative amoebae by Thilo and Vogel (1980). They covalently attached [^{3}H]galactose to cell surfaces with a galactosyltransferase and showed that this galactose could be released by a β-galactosidase. SDS gels showed a heterogeneous profile of glycoprotein acceptors. During pinocytosis, as monitored by fluid-phase uptake, a portion of the radiolabeled glycoproteins became resistant to β-galactosidase. Subsequently, these species became sensitive to hydrolysis, indicating their reappearance from a masked pool, presumably the pinosome. The glycoprotein profile of the pinosomes was approximately the same as that of the cell surface prior to pinocytosis, suggesting that there is little selectivity during internalization.

As amoebae can internalize a variety of different types of particles by phagocytosis (yeast, bacteria, red blood cells, latex beads), the question arises as to whether a single receptor recognizes all these diverse particle types or whether each has a unique receptor. Several reports indicate that selective receptors exist. Ryter and co-workers (Ryter and Hellio, 1980; Hellio and Ryter, 1980; Favard-Séréno *et al.*, 1981) have studied cell surface distribution and possible recognition roles, during phagocytosis, of Con A and WGA receptors. The general conclusions from their studies were that Con A receptors are probably not involved in phagocytosis, but WGA receptors appear to be required, specifically for phagocytosis of yeast.

Vogel and co-workers used a genetic approach for identification of recognition systems for phagocytosis. Mutants were selected for their inability to internalize tungsten beads (Vogel *et al.*, 1980). The one mutant category analyzed had properties consistent with the following hypothesis

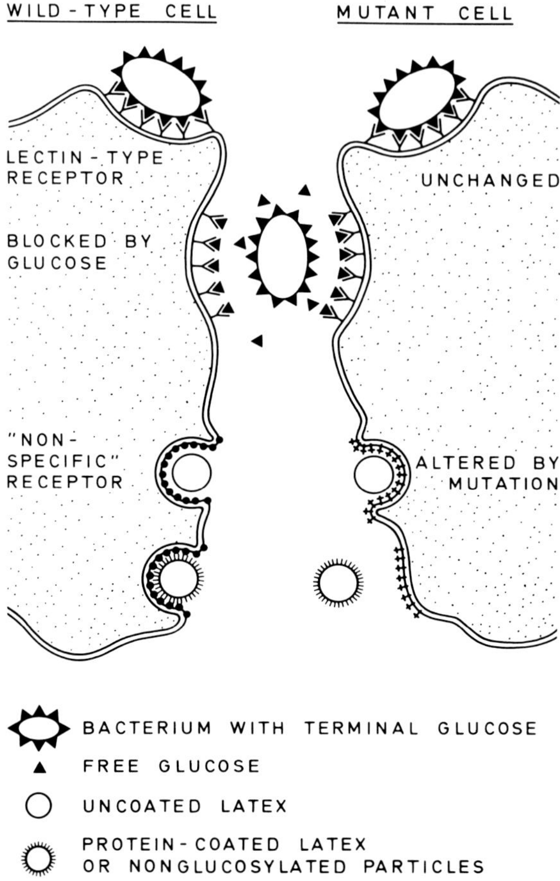

Figure 9. Model depicting phagocytic receptors on wild-type and defective mutant amoebae. (From Vogel *et al.*, 1980, with permission.)

by these authors. There are at least two classes of recognition receptors for phagocytosis. One class is nonspecific and recognizes both hydrophobic and hydrophilic derivatives of latex beads, bacteria, erythrocytes, etc. The mutation has altered this receptor class so that only strongly hydrophobic materials (latex beads) are internalized. Hydrophilic particles are no longer recognized by this receptor, allowing identification of a glucose-specific receptor as the second class. The mutants could phagocytose *E. coli* B/r (which have exposed glucose residues), and glucose and its analogs could block this phagocytosis. An *E. coli* mutant lacking terminal glucose residues was not phagocytosed. The general model for these receptors is shown in Fig. 9.

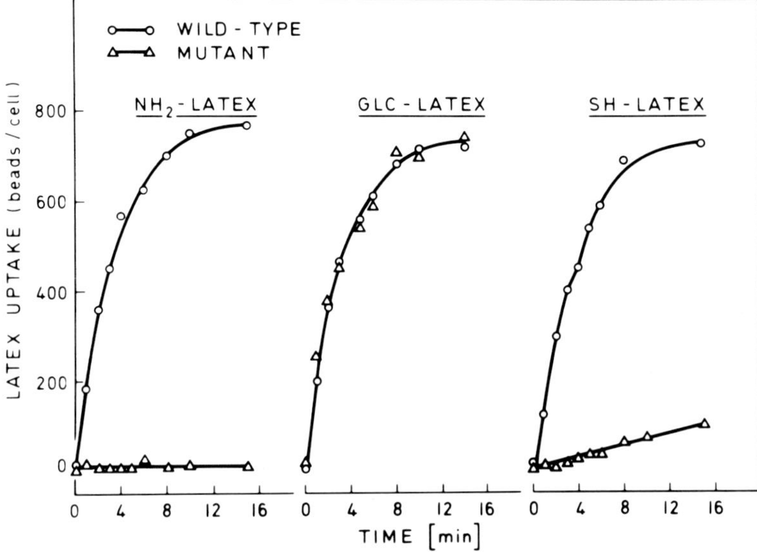

Figure 10. Comparison of uptake rates of amino-, glucose-, and thiol-containing latex spheres (diameter 0.45 μm) at 20°C by wild-type (○) and mutant (△) amoebae in shaken cultures. (From Vogel, 1981, with permission.)

Further support for this model was provided (Vogel, 1981) by comparison of uptake of latex beads derivatized with amino or sulfhydryl groups versus glucose residues. The mutants could internalize the latter at the same rate as wild type, and free glucose was an efficient competitor. The two nonglucosylated, hydrophilic beads were actively internalized by wild type but not by the mutant (Fig. 10). The mutants, therefore, appear to retain a glucose recognition mechanism for phagocytosis that is largely masked in wild type by the nonspecific hydrophilic system. The mutants also failed to display contact sites B-mediated cohesion in shaken suspension, though they developed normally on surfaces, and could also phagocytose all particle types on unperturbed surfaces. This indicates that the mutation may have only weakened the nonspecific hydrophilic site and not totally abolished its activity.

A number of interesting questions remain to be answered about these mutants. *E. coli* B/r are phagocytosed by the mutant and wild type at the same rates. If wild type uses both the glucose-specific and the hydrophilic nonspecific receptors, then added glucose should inhibit that proportion of uptake due to the glucose receptor, whereas in fact glucose had a slightly stimulatory effect. If the absence of obvious inhibition by glucose is due to a very low glucose-to-hydrophilic receptor ratio, then mutants without the latter receptor should internalize bacteria at a much lower rate than wild type, unless they overproduce the glucose receptor.

The parallel effects of the mutation on weakening of contact sites B agglutination and phagocytosis suggest that a common molecular entity may be affected, though other alternatives are possible. The competition experiments indicate that glucose and GlcNAc participate in the ligand–receptor system of contact sites B. The GlcNAc competition could be a reflection of the WGA receptor implicated in phagocytosis of yeast, presumably as part of the nonspecific hydrophilic system. If the mutation alters this molecule, it could simultaneously have weakened the ability to interact with contact sites B and hydrophilic particles. These mutants may provide an important first step in molecular analysis of both systems.

5.3. Protein Sorting and the Lysosomal Common Antigen

The issue of intracellular protein sorting has received considerable interest lately. The question is how proteins synthesized at a common site, the rough endoplasmic reticulum, can be correctly sorted and shipped to diverse locations/organelles. One hypothesis is that specific glycosylations create recognition signals for the mechanisms that monitor correct sorting. However, even this information would have to be contained in the genome, i.e., in the specificity of oligosaccharide-processing enzymes and the primary amino acid sequence of the substrates to generate "organelle-specific" oligosaccharides. It has already been noted that at least some glycoprotein-linked oligosaccharides are nonrandomly distributed among cellular compartments (Section 4.2.1).

The best current case for a common recognition system involving oligosaccharides is for lysosomal enzymes in *D. discoideum*. These enzymes have a common posttranslational modification mechanism initially identified in a strain carrying a mutation (*mod* A) that caused changes in physical properties of the several lysosomal enzymes studied; in particular, the enzymes had partial deficiencies in catalytic activity and were electrophoretically less negative (Free *et al.*, 1978; Free and Schimke, 1978). Lysosomal enzymes are known to be glycoproteins (Free *et al.*, 1978; Every and Ashworth, 1973). Freeze and Miller (1980) showed that *mod* A did not reduce incorporation of mannose, galactose (presumably as glucose; see Section 4), or glucosamine but caused a large depression of incorporation of sulfate and phosphate into glycopeptides (obtained after Pronase digestion), indicating that *mod* A blocks a relatively late event in oligosaccharide processing. Finally, Freeze and Miller demonstrated the presence of mannose-6-phosphate in acidic glycopeptides of the parent AX3 but could not detect it in the *mod* A mutant. Indeed, *D*.

discoideum lysosomal enzymes are actively internalized by human fibroblasts via their mannose-6-phosphate receptors (Freeze *et al.*, 1980).

Additional evidence for common determinants among *D. discoideum* lysosomal enzymes has come from serological analyses. Dimond and Loomis (1976) showed that rabbit antisera to a purified lysosomal enzyme could recognize several other such enzymes and that the antigen was periodate sensitive. Knecht and Dimond (1981) showed that antibodies to the common antigen precipitated a large number of cellular proteins but noted that these antibodies did not recognize any proteins when whole cell mRNA was translated *in vitro* in a rabbit reticulocyte system. More recently, Dimond and co-workers have obtained monoclonal antibodies to lysosomal enzymes. As expected from the immunological potency of the common antigen, most of these antibodies recognized multiple proteins, though these were a subset of the total common antigen proteins (Cardelli *et al.*, 1982). The proteins recognized by the monoclonal antibodies had very acidic isoelectric points and were sulfated and phosphorylated. As opposed to the rabbit antisera, binding of monoclonal antibodies to their antigens was inhibited by sulfated polysaccharides such as dextran and chondroitin sulfates and heparin (Knecht *et al.*, 1982). Further evidence for a role of glycosylations in the generation of the common antigen is provided by the recent result that during postaggregation development, newly synthesized lysosomal enzymes are less acidic and seem to lack the common antigen (Cardelli *et al.*, 1982). This correlates well with the previously described transition in oligosaccharide-processing routes.

The exact role of the common antigen in intracellular protein sorting is not clear to date. Under some circumstances, most, but not all, of the protein species bearing the common antigen are secreted; a few species remain totally intracellular (Knecht and Dimond, 1981). Mechanisms that regulate secretion versus intracellular packaging appear to be complex (Burns *et al.*, 1981), and kinetic studies showed that not all lysosomal enzymes were simultaneously secreted and that none correlated temporally with egestion of residual bodies as monitored by release of internalized latex beads (Dimond *et al.*, 1981). These observations suggest both structural and functional complexity of recognition processes in the digestive apparatus, as also indicated by EM observations (Ryter and de Chastellier, 1977; de Chastellier and Ryter, 1977). If more than one recognition system is operative, this could explain otherwise apparently contradictory observations. Gustafson and Milner (1980a) have reported data consistent with occurrence of *N*-acetylglucosamine-1-phosphate in proteinase I (Gustafson and Milner, 1980b) and immunological cross-reactivity between proteinase I and *N*-acetylglucosaminidase, the latter being a common antigen protein. Therefore, it is possible that each lysosomal enzyme contains multiple recognition determinants, some generating sorting to distinct "prelysosomal" compartments. This could explain different levels and kinetics of secretion of these enzymes and the "imbrication" of digestive vacuoles, one into the other, reported by de Chastellier and Ryter (1977) in at least early starvation.

In summary, lysosomal enzymes of *D. discoideum* are clearly close relatives of the mammalian enzymes. They possess a common recognition device that is generated by posttranslational glycosylations. It would also appear that the role of such modifications is more complex than previously believed, as not all common antigen proteins share the same fate in terms of ultimate locale. Given the existence of *mod* A mutants, mutants defective in structural genes for several lysosomal enzymes, the relative ease with which additional mutants can be obtained, and availability of the antisera, studies with *D. discoideum* should provide a most efficient system for dissection of recognition and regulatory mechanisms.

6. DEVELOPMENTAL REGULATION OF CELLULAR COHESION

The strategies and pitfalls encountered in pursuit of molecular elements of cell–cell cohesion systems are reviewed elsewhere in this volume (Damsky *et al.*). All of these strategies have been exploited in the slime mold and have been reviewed recently by Barondes *et al.* (1982) who also considers the possibilities that all of the pitfalls have been or are being encountered. Nonetheless, the relative simplicity of the slime molds and the ease of mutant isolation generate a good probability that studies with them will contribute important mechanistic information on this nearly universal biological phenomenon of cellular recognition.

Simple observation of the developmental cycle suggests that there may be several sequentially expressed cohesion systems. At least one system would be needed to accompany acquisition of a multicellular state. In the slug stage the prestalk and prespore cells occupy anterior and posterior positions, respectively. If slugs are dissociated, the cells will reaggregate and these cell types sort out to assume their previous positions. During culmination, prestalk cells migrate as a "tissue" through the prespore cell mass and do not get stuck to it. These observations suggest that after aggregation there may be cell-type-specific cohesion systems superimposed on a "nonspecific" system. Evidence has accumulated in support of all of these expectations.

6.1. Aggregation-Stage Cohesion

6.1.1. The Saga of Contact Sites A (gp80)

As noted in the preceding section, vegetative cells have cohesion sites that are blocked by EDTA and that persist through the aggregation stage. This was demonstrated by Gerisch (1961) who established the first quantitative assay for cellular cohesiveness in the slime mold and showed that acquisition of aggregation competence was accompanied by appearance of an EDTA-resistant cohesion system. The existence of these two distinct systems was further documented by serological criteria (for reviews see Gerisch, 1980; Gerisch *et al.*, 1980). Beug *et al.* (1970) showed that Fab fragments from rabbit antisera against homogenates of aggregation-competent cells could block formation of contacts between aggregating amoebae but did not interfere with motility or chemotaxis. Further, this was not due to a nonspecific coating of cells by Fab, for Fab of an anticarbohydrate antiserum did coat the cell surface but did not block formation of cell contacts. Subsequently, Beug *et al.* (1973a,b) demonstrated that the EDTA-sensitive (contact sites B) and -resistant (contact sites A) systems played separate roles in the architecture of aggregation. During aggregation, cells in streams form loose side-to-side and strong end-to-end contacts. Fab against aggregation-competent cells blocked both types of contacts. If the Fab was preadsorbed with vegetative cells, it could no longer block side-to-side B contacts but end-to-end contacts (sites A) were still inhibited. Further, the side-to-side contacts in streams would be blocked by either EDTA or Fab fragments against growth-phase cells, indicating that these are due to contact sites B conserved from vegetative cells. The antigen mediating end-to-end contacts was absent in growth-phase cells and appeared coincident with EDTA-resistant cohesiveness.

The availability of an antiserum that blocked contact sites A after adsorption against vegetative cells allowed purification of the antigen, for it could be assayed by its ability to remove adhesion-blocking activity from the Fab preparation. The antigen eventually purified is commonly called gp80, as it is a glycoprotein that binds Con A (Eitle and Gerisch, 1977) and has an electrophoretic mobility in 7.5% acrylamide SDS gels that indicates an apparent M_r of 80K but which varies with acrylamide concentration as expected for a glycoprotein (Müller *et al.*, 1979). This name also eliminates any functional connotation, which, as considered below, might be fortunate.

Gp80 is an integral membrane protein (Stadler *et al.*, 1982) that can be extracted from membranes with *n*-butanol (Huesgen and Gerisch, 1975) and purified to very near homogeneity by conventional fractionations (Müller *et al.*, 1979; Stadler *et al.*, 1982). Its molecular weight varies with the method of determination. Overall, the data are compatible with a multimeric native state (Huesgen and Gerisch, 1975; Müller *et al.*, 1979), which is also compatible with staining patterns on cells treated with a ferritin–monoclonal antibody conjugate (Ochiai *et al.*, 1982b).

Purified gp80 was reported to be composed of 20–33% sugars including approximately equimolar amounts of mannose, glucosamine, and glucose and a small amount of fucose (Müller *et al.*, 1979). The authors suggest that the glucose may represent contamination. Currently, no oligosaccharides have been detected (Section 4) that could account for equimolar amounts of mannose and glucosamine. However, presence of both standard high-mannose oligosaccharides and components similar to the GlcNAc-phosphoserine of proteinase I could account for these data. The amino acid composition is unusual. The data show high content of proline, serine, threonine, aspartate/asparagine, and glutamate/glutamine (Müller *et al.*, 1979; Stadler *et al.*, 1982).

Purified gp80 is acidic and displays substantial isoelectric heterogeneity, producing a series of stuttering spots in the IEF dimension of two-dimensional gels (Murray *et al.*, 1981; Coffman *et al.*, 1982; Ochiai *et al.*, 1982b). Glycosylations probably do not contribute to this heterogeneity (Ochiai *et al.*, 1982a). However, gp80 is also a phosphoprotein (Coffman *et al.*, 1981, 1982). Schmidt and Loomis (1982) showed that the phosphate was on serine residues and that all multiple isoelectric species were phosphorylated. Further, limited digestion with three different proteases generated, in each case, a single phosphorylated peptide. Thus, it is not certain that heterogeneity is due to successive phosphorylations of serine residues. If this is the case, the sites must be clustered.

Gp80 is a minor constituent of the plasma membrane, approximately 1% of the total membrane proteins. Estimates of the number of copies per cell are in good agreement with the number of molecules required to adsorb the adhesion-blocking activity of the Fab preparation (Müller and Gerisch, 1978).

Given the ability to purify gp80 by straightforward methods, it became possible to obtain antibodies to the purified protein rather than crude preparations. Both rabbit polyclonal and mouse monoclonal antibodies have been obtained and have been used to test the expectation that gp80 should appear coordinately with the acquisition of EDTA-resistant cohesion. This has been shown to be true both by tests for the presence of gp80 in membranes (Murray *et al.*, 1981; Ochiai *et al.*, 1982b) and *de novo* synthesis by metabolic incorporation of [^{35}S]methionine (Murray *et al.*, 1981).

At this point, all of the evidence indicated that gp80 is contact site A, a bona fide, developmentally regulated cohesion molecule participating directly in binding of cells to each other. However, firm proof has remained somewhat elusive, and what has emerged is a potent reminder of how complex and difficult this area of study can be even in a simple system with all its attendant experimental advantages. Whatever the ultimate solution, the pioneering studies of Gerisch and his co-workers will have laid the foundation by establishment of quantitative assays and serological approaches. Indeed, the purification of gp80 in Gerisch's laboratory made possible the further serological approaches that have raised some doubts, as well as questions pertinent to all systems as to the definition of acceptable criteria for "proof."

In the ideal case, availability of pure gp80 would allow preparation of an antiserum that binds only gp80, blocks contact sites A, stains only at the poles of cells and could be used to identify or select mutants devoid of gp80 and contact sites A function. To date, none of the reported polyclonal or monoclonal antibodies meet these criteria. None of the antisera are monospecific. A polyclonal antibody preparation (Murray *et al.*, 1981) bound a number of different proteins and at all developmental stages, not just aggregation. Preadsorption with vegetative cells greatly enhanced specificity for gp80 without removal of adhesion-blocking activity of the antiserum. The adsorbed serum still bound several other proteins although these proteins either were not present in isolated plasma membranes or appeared too late to be contact sites A. Preadsorption with gp80 removed all adhesion blocking activity of the IgG as well as all specificities at all developmental stages. However, Murray and Loomis (1982) reported that a more exhaustive preadsorption with vegetative cells removed adhesion-blocking ability and yet the IgG retained the ability to bind gp80. In summary, antigenic determinants of gp80 were not unique to aggregating cells and the determinant which survived exhaustive preadsorption was not critical to cohesion. However, gp80 was the only detectable protein that appeared at the right place (cell surface) and the right time, and it carried all the antigenic determinants recognized by the IgG.

Reported monoclonal antibodies (Ochiai *et al.*, 1982b; Murray *et al.*, 1983) are also not monospecific although they are much improved over the unadsorbed polyclonal preparation and bind very few species (Fig. 11) whereas Coomassie blue or Con A staining detect vast numbers of bands. However, while the monoclonals bind gp80, they do not block EDTA-resistant contact sites A. Murray *et al.* (1983) considered a series of possible explanations for this. First, the monoclonal antibodies reported may be against a determinant not essential to cohesion. Second, antibodies may need to mask more than one determinant. Third, binding of the antibody may not be strong enough to compete with endogenous cohesion ligands. Finally, gp80 may not be involved, though it shares determinants with those molecules that are involved.

This latter possibility is compatible with the observation that monoclonal antibodies (Ochiai *et al.*, 1982b), as well as all antibodies tested, bind over the entire cell surface. As the A contacts between cells are end-to-end, this suggests that either the predominant antigen is not the cohesion molecule or it is selectively activated for cohesion at the poles of

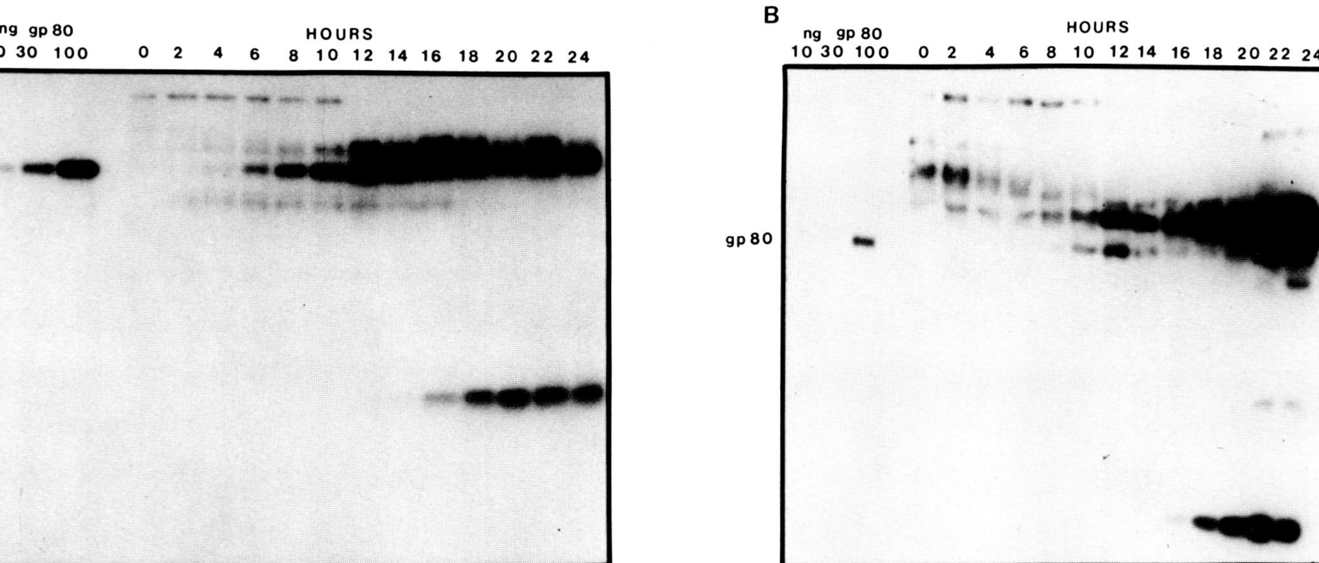

Figure 11. Immune staining of *D. discoideum* proteins. Extracts equivalent to 5×10^5 cells from the indicated hours of development were electrophoresed in 10% (panel A) or 8% (panel B) polyacrylamide gels in the presence of SDS. After transfer to nitrocellulose, the separated proteins were stained with polyclonal rabbit IgG raised against purified gp80 and absorbed five times with vegetative AX3 cells (panel A) or with monoclonal antibody followed by rabbit anti-mouse κ-chain serum (panel B). Bound antibody was visualized by autoradiography after incubation with ^{125}I-labeled staphylococcal protein A. Positions of purified gp80 are indicated. (From Murray *et al.*, 1983, with permission.)

the cell. Further (Murray *et al.*, 1983), after mutagenesis and screening, two independent mutants were obtained that did not bind the monoclonal antibodies and did not make a gp80 form detectable on gels. Both mutations were reported to map in a *mod* B locus, i.e., they do not appear to be in a structural gene for gp80. Both mutants expressed EDTA-resistant cohesion that could be blocked by polyclonal antibodies. These mutants may aid in identification of the determinants that are critical for cohesion. Indeed, the existence of these mutants need not be interpreted as evidence against a role for gp80. The mutations may serve only to eliminate a posttranslational modification that generates a determinant of gp80 that is not essential to cohesion and may help reveal those determinants that are essential. However, an independent monoclonal antibody against gp80 also binds a 66 kd protein produced in the presence of tunicamycin, and this antibody does not inhibit EDTA-resistant cohesion (Ochiai *et al.*, 1982a).

Recently we (Matarazzo and Henderson, 1983b) have shown that endo H-resistant, but not endo H-sensitive, oligosaccharides obtained from plasma membrane glycoproteins of aggregating cells block agglutination of aggregation-stage cells but not of cells from vegetative or tip formation stages. The specific glycoprotein(s) from which these oligosaccharides are derived is currently unknown.

To summarize, Fab fragments of a polyclonal antiserum against aggregation stage cells, after adsorption with vegetative cells, can block the EDTA-resistant, end-to-end contacts of aggregating cells, contact sites A. Using adsorption of this blocking ability as an assay, the most abundant antigenic species, gp80, was purified and characterized. Polyclonal antibodies against purified gp80 also can block EDTA-resistant cohesion but also detect cross-reacting antigens at all stages of development, although gp80 is the only cell surface species detected which appears with the developmental kinetics of contact sites A. Monoclonal antibodies to gp80 also detect other cross-reacting antigens, although the specificity for gp80 is considerably enhanced over the polyclonals. However, no monoclonal antibody tested has been reported to block EDTA-resistant cohesion, and two mutants with altered gp80 still express this cohesion. Clearly, wild type gp80 carries determinants which participate in cohesion, but it remains to be established that gp80 is the actual molecule which binds cells together. Overall, the gp80/contact site A story well illustrates the dangers in analysis of cohesion even in a relatively simple system where cellular, biochemical and genetic analyses can be applied. Those cautionary notes certainly must be borne in mind in considering other antigens implicated in cohesion.

6.1.2. A Polysaccharide Candidate

Springer and Barondes (1982) prepared a rabbit antiserum against aggregating NC4 cells and showed that both Fab fragments and IgG could completely block EDTA-resistant cohesion. The latter assay (Springer and Barondes, 1980) uses the bivalent primary antibody with Fab fragments of goat anti-rabbit IgG to block antibody-induced cell agglutination. The antigen could be detected on cell surfaces and in the extracellular medium of all developmental stages but increased during aggregation and declined during postaggregation stages. However, solubilized preparations of vegetative cells contained as much antigenic material as did aggregating cells. The antigen had an M_r of 2×10^6 under nondenaturing conditions. The ability of the antigen to adsorb adhesion-blocking activity of the antiserum was completely resistant to Pronase digestion or heat but was totally abolished by periodate oxidation. The Pronase-treated antigen was purified 15-fold. By comparing binding of [^{125}I]-IgG to cells before and after adsorption with the purified antigen, it was estimated that there were about 5×10^5 molecules of that specific antigen per cell. However, this was only a small fraction of the IgG molecules in the adsorbed IgG capable of binding the cells without blocking cell cohesion. Therefore, blockade of cohesion was not simply due to coating of the cell surface.

The authors also noted that the partially purified antigen inhibited the binding of the lectin discoidin I to a cognate glycoprotein. While the antigen was not completely pure, it is of note that a similar carbohydrate-rich macromolecule has been found in the medium of another cellular slime mold, *Polysphondylium pallidum*, and been shown to interact with the endogenous lectin, pallidin, of that species (Drake and Rosen, 1982). It is interesting in this context that Ivatt *et al.* (1981) reported synthesis by vegetative and aggregation-stage AX3 cells of a polysaccharide that incorporates radiolabel from mannose, glucosamine, sulfate, fucose, and galactose. This polysaccharide might be a ligand for discoidin as the latter two sugars are the best monosaccharide ligands for discoidin (Rosen *et al.*, 1973). Finally, Springer and Barondes point out that detection of their antigen was possible because they did not preadsorb the serum with vegetative cells, as was done for identification of gp80.

6.2. Postaggregation Cohesion

6.2.1. gp95

In a manner reminiscent of gp80, slug plasma membranes contain a potently antigenic glycoprotein, gp95, which is a very minor membrane species. Parish *et al.* (1978b) showed that an antiserum against slug plasma membranes bound predominantly to a 95K glycoprotein on gels whereas a Con A–peroxidase stain for glycoproteins showed a very minor band at this position. Synthesis of this antigen occurred throughout postaggregation development (Parish and Schmidlin, 1979a). Fab against slug plasma membranes blocked reagglutination of dissociated slug cells but failed to block EDTA-resistant cohesion of aggregating cells or EDTA-sensitive cohesion of vegetative cells (Steinemann and Parish, 1980). The blockade of slug cell agglutination by this Fab was not removed by adsorption with aggregation-competent cells but could be removed by adsorption with a gel slice containing gp95 but not by another region of the gel. These results indicate that two different cohesion systems operate during and after aggregation and suggest that gp95 may be a component of the latter.

Evidence in support of these observations was subsequently generated by Wilcox and Sussman in studies with an unusual temperature-sensitive mutant, JC-5. This mutant develops normally at both restrictive (27°C) and permissive (22°C) temperatures through the stage of tip formation on cell mounds. After this, at the restrictive temperature, it literally "slumps" in that the morphology collapses into an amorphous cell mound, and the cells eventually disperse into a smooth lawn (Wilcox and Sussman, 1981a). This behavior suggested a cohesion defect, and, as the defect was only observed after aggregation, it further suggested that a new cohesion system, thermolabile in JC-5, had supplanted the contact sites A system. This was tested serologically (Fig. 12) with results in essence identical to those of Steinemann and Parish. However, they also showed that postaggregation adhesion-blocking activity of their Fab could be removed by adsorption against late wild-type or JC-5 cells developed at the permissive but not the restrictive temperature. Therefore, noncohesive cells also lacked expression of the critical antigen (Wilcox and Sussman, 1981b). Similar results were obtained using preadsorption with isolated plasma membranes.

The more detailed characterization of JC-5 (Wilcox and Sussman, 1981a) showed that its expression of EDTA-resistant, aggregation-stage cohesion was normal. Further, during only a short period of time after morphology began to collapse at 27°C could subsequent developmental morphologies be rescued by a shift to 22°C. Cells harvested from 27°C during this period were noncohesive when assayed for agglutinability at 27°C but recovered within 20 min when assayed at 22°C. Cells harvested at the same time period after development at 22°C agglutinated promptly at 22°C but at 27°C aborted a prompt partial agglutination reaction and became dispersed by 40 min. These observations suggested that either the cohesion element was reversibly thermolabile or that a metabolic interconversion or turnover was occurring. These alternatives were tested

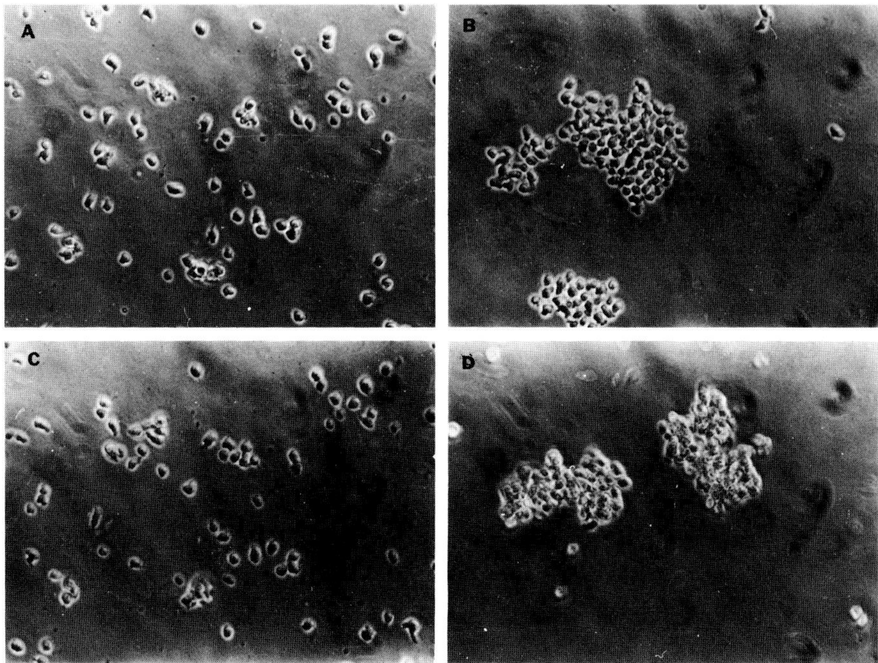

Figure 12. Photomicrographs of cohesion assays performed in the presence of Fab. (Upper) Mutant test cells incubated 9 hr and exposed to early Fab (9-hr, 22°C wild type) (A) and late Fab (17-hr, 22°C wild type) (B). (Lower) Mutant test cells incubated 13 hr and exposed to late Fab (C) and early Fab (D). × 400. (From Wilcox and Sussman, 1981b, with permission.)

by examining agglutinability of cell ghosts or of cells in the presence of cycloheximide. Ghosts neither recovered at 22°C nor lost agglutinability at 27°C, indicating that the cohesion component per se was not thermolabile. Further, in the presence of cycloheximide, cohesive cells lost their agglutinability and noncohesive cells could not recover it. This suggests that there is turnover of the cohesion element and that protein synthesis and/or posttranslation modifications of the newly synthesized protein were required for its continual replacement.

In efforts to identify the cohesion molecule, plasma membrane proteins were analyzed on SDS gels. No significant differences could be distinguished between 22°C and 27°C cell membranes by Coomassie blue staining or autoradiography after metabolic labeling of cells with [^{14}C]acetate. As the cohesion element was probably a minor membrane constituent, subsets of total membrane proteins were probed by staining gels with lectins. During loss and recovery of cohesiveness, this assay detected a loss and partial recovery of a WGA-binding glycoprotein at approximately 95K. This gp95 is probably the same as that of Steinemann and Parish (1980), but the gel systems were not the same, and they used metabolic labeling with glucosamine and Con A staining to identify gp95.

Finally, the ultimate developmental fate of JC-5 cells at 27°C may be highly pertinent to the role of cell contacts in regulation of gene expression. Even when developed at 27°C, JC-5 synthesizes postaggregation, developmentally regulated enzymes with the same kinetics as its parent. Further, despite the fact that the cells disperse into a lawn, they differentiate into stalk and spore cells in the same proportions as their 22°C relatives. If cell–cell contact is required for expression of late developmental genes, this suggests that the "late" cohesion system is not critical. Aggregation-stage contacts are an alternative, or possibly some unknown event of tip emergence per se, the latter of which appears to require a response to cAMP signals (Barclay and Henderson, 1982), which have also been implicated in control of late gene expression (Lodish *et al.*, 1982). Also, the serological evidence indicates that the contact sites A cohesion system is no longer functional in postaggregation development. As gp80 persists, though at a reduced level, during this period (see Section 6.1.1), if it does correspond to contact sites A, this suggests that a receptor could be lost or that there is a change in the molecule that is not detectable by antibodies against the purified glycoprotein.

To return to gp95, further work in Sussman's laboratory has provided considerable support for its possible direct function in cellular cohesion. Saxe and Sussman (1982) solubilized plasma membrane proteins and fractionated them on a WGA-affinity column. Material eluted by GlcNAc stimulated cohesion of JC-5 at 27°C and enhanced the size of wild-type agglutinates. This activity could only be isolated from cells expressing the correct, stage-specific cohesion and was only effective with cells of the corresponding stage. For example, this fraction did not enhance the size of agglutinates of aggregation-competent, pretip cells. The implication is that there is a receptor for this material that is also stage specific (Williams and Sussman, 1982). After WGA-affinity column enrichment, the active material was fractionated on SDS gels. Slices of the gels were eluted and dialyzed, and agglutination-stimulating activity could be recovered from regions corresponding to M_r of 95 and 40–50K.

Clearly, pursuit of roles for gp95, as well as gp80, are hot topics. If Steinemann and Parish's gp95 is the same as that of Sussman and co-workers, an interesting pattern emerges. Both glycoproteins appear to possess all antigens required for cellular cohesion at the appropriate stages. Both are very minor membrane species, and both are potently antigenic. Possibly neither are direct cohesion components but the current

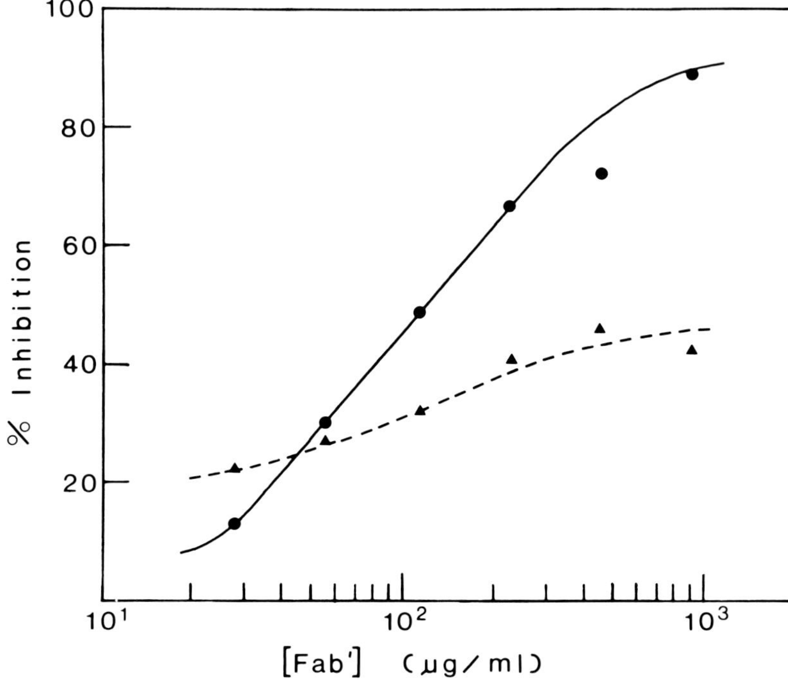

Figure 13. Effect of Fab' directed against gp150 on cell cohesiveness of 18-hr prespore cells and prestalk cells. NC4 cells were developed for 18 hr and then separated into prespore cells and prestalk cells. Cohesion assay was performed on these two cell types in 10 mM EDTA and varying concentrations of Fab' against gp150. ●, Prespore cells; ▲, prestalk cells. (From Lam *et al.*, 1981, with permission.)

Finally for the prespore case, is the critical antigen really gp150? Probably. However, experiments have not been reported that test the ability of homogeneously pure gp150 to adsorb the agglutination-inhibiting activity of the Fab, similar to those with gp80 and gp95. While those experiments are needed, their absence does not negate the value of the antiserum in identifying a cohesion device.

6.3. Summary of gp80, the Polysaccharide, gp95, and gp150

As stated in the introduction to this section, the biological behavior of this organism predicts a general cohesion system to establish multi-cellularity, a nonspecific "glue" system in slugs and cell-type-specific systems to mediate sorting and tissue migration. Contact sites A, possibly gp80, define the first expected system. The relationship between contact sites A and the polysaccharide antigen, if any, remains to be determined.

gp95 may be the general glue in slugs. Gp150 may be part of a prespore-specific system, but molecular candidates for a prestalk-specific mechanism remain to be identified. The molecular mechanisms of gp80 and gp95 are unknown. They could simply dimerize between adjacent cell surfaces or they could, and probably do, have receptors so that at least a heterodimer would be required. This latter probably is the case for gp150 as it is found on all cell types.

Another mechanistic element may also be important, particularly for contact sites A, namely, proper display or presentation of the cohesion molecules. During aggregation on surfaces, cells retain their typical amoeboid morphology only until they form a head-to-tail dock with a stream at which time they rapidly elongate to a polar appearance. In time-lapse films, cells are observed which attempt to dock and fail, but succeed at a later attempt. Possibly such cells have not yet produced some threshold level of contact sites A. An alternative is that there is an equilibrium between biologically active "displayed" and inactive conformations of cohesion elements, and cells that fail on first attempt did not have enough "active" molecules. Regulation of such display could provide a basis for polar cell contacts, and, as suggested by Gerisch, could explain the lack of polar staining by antibodies. If display is important in cohesion systems, it may be important to note that, in all cases except contact sites A, cohesion blockade by antibodies has been assayed on presumably symmetric cells in shaken suspensions. Finally, it is not certain that these cohesion systems are the only ones operating at a given stage. The preceding discussion has, for example, largely ignored the possible contribution of the *D. discoideum* lectins. The lectins certainly exist and presumably have important fuctions. The questions are in regard to what those functions might be.

6.4. Discoidins I and II, the D. discoideum Lectins

The discoidins were first discovered and studied by Barondes and co-workers. The first report (Rosen *et al.*, 1973) demonstrated that extracts of aggregating, but not vegetative, amoebae could agglutinate formalinized sheep erythrocytes and that agglutination could be blocked by specific sugars. This discovery of developmentally regulated lectins generated great excitement that they might contribute to or be the basis for aggregation-stage cohesion, i.e., cell surface lectins could bind specific oligosaccharides of neighbor cells. Since this report, the discoidins have become the most thoroughly studied proteins from this organism, and there is a vast literature about them. Amino acid sequences are known, mutants have been isolated, genes have been cloned, developmental time courses quantified, sugar-binding specificities carefully analyzed, etc. A review of this literature is well beyond the current scope. Fortunately, an excellent and detailed review was recently published (Bartles *et al.*, 1982a) as well as a more brief summary of possible cohesion roles (Barondes *et al.*, 1982). Therefore, the following will be a curtailed discussion summarizing some highlights with references only to more recent reports. Also, all slime mold species produce lectins, and those from *D. purpureum* (purpurins) and *P. pallidum* (pallidins) have also been studied.

As basic background, discoidin I is a tetramer of 26K subunits; discoidin II has 24K subunits. In late aggregation, the I:II ratio is 9:1 but earlier it is 3:1. Together they comprise about 1% of cellular proteins at their maximum. Discoidin I is isoelectrically heterogeneous, presumably reflecting the fact that it is encoded by a family of multiple genes.

Speculation on a possible role for these lectins in cohesion generates a number of predictions about expected properties and behaviors:

1. They should be expressed coincident with cohesivity.
2. They should be found on the cell surface.
3. Appropriate ligands should alter cell cohesion.
4. Mutants lacking them should be noncohesive.

In each case, supportive evidence is available but is also open to alternative interpretations, as considered below.

Developmental Regulation. In wild-type NC4, discoidins are synthesized coincident with appearance of EDTA-resistant cohesiveness. However, in axenic strains, grown axenically, vegetative amoebae have substantial amounts of the discoidins. In some axenic strains this appears to correlate with precocious acquisition of cohesiveness. However, Gerisch and co-workers (Ochiai *et al.*, 1982b), working with another axenic strain (AX2-214), reported tight coordination between expression of gp80 and EDTA-resistant cohesion with no increase in the hemagglutination titer of cell extracts. It could be argued that gp80 is a lectin receptor and, therefore, the lectin need not be developmentally regulated. However, partially purified gp80 did not inhibit hemagglutination activity. Other receptor candidates will be considered later.

Cell Surface Locale. Over 90% of lectin activity is in the cytoplasm of the cells and is soluble in lysates. Inclusion of the cognate sugar during lysis somewhat enhances the yield of soluble lectin, and, for the pallidins, there is evidence that there is binding to the cytoplasmic face of the endoplasmic reticulum in permeabilized, fixed cells. Erythrocytes, however, can bind intact amoebae and only in the absence of the competing hapten sugar. Immunological studies, cell surface iodinations, and elution of lectins with cognate sugars all show the presence of lectin on the cell surface. Recently, it has been shown that this surface detectability includes discoidin II as well as the more abundant discoidin I (Armant and Berger, 1982).

However, only about 2% of total cellular lectin is on the cell surface, and this has been a major conceptual barrier to general acceptance of lectins as cohesion elements. Several possibilities present themselves. First, exogenously added lectins bind well to cell surfaces. Lysis of only 2% of cells during culture or experimental manipulations could account for all the cell surface molecules. Support for the cell lysis model is provided by EM (Erdos and Whitaker, 1983). Transmission microscopy of cells labeled immunocytochemically detected discoidins on intracellular membrane populations but not on cell surfaces unless a nearby cell was clearly broken.

The second possibility is that binding of ligands to the small amount of cell surface lectin provides a trigger for externalization of additional lectin from the intracellular reservoir. This "elicitation" response has been demonstrated by Barondes and co-workers for the purpurins for which binding of bivalent (but not univalent) antibodies to the lectin or multivalent ligands for the lectin enhanced the amount of lectin subsequently detected on the cell surface by 10-fold, i.e., 20% of the total pool. This response did not require direct interaction of lectins and ligands but was also induced to a lesser extent by multivalent (but not univalent, succinyl) Con A and by exogenous purpurin itself. Thus, elicitation appears to require only cross-linking of some set cell surface receptors by multivalent ligands. If lectins participate in cell cohesion, elicitation could suggest that cohesion is an autocatalytic process, i.e., the more lectin bound, the more becomes available for binding. However, as some elicitation can be induced by molecules that do not interact directly in an obvious fashion with the lectin, the relevance to cohesion is not certain. For a review of the elicitation phenomenon in both chicken tissues and slime molds, see Barondes *et al.* (1981).

Effects of Lectin–Ligand Interactions on Cohesion. Under normal physiological conditions, antibodies to lectins have not been able to block EDTA-resistant cohesion. Since this might be due to a higher affinity of the lectin for endogenous acceptors than for the Fab, Springer and Barondes (1980) used a combination of bivalent rabbit antidiscoidin plus saturating goat anti-rabbit Fab (to prevent agglutination by the primary antibody). Very little inhibition of cohesion was observed. It is possible that the bivalent antibody may have elicited more discoidin, which escaped blockade.

The only case in which ligands and antilectin Fab have been shown to block cohesion is for the pallidins, but these effects could only be observed under nonphysiological conditions, i.e., in hypertonic solutions or in the presence of metabolic inhibitors such as dinitrophenol or azide.

One possibility is that these conditions prevent elicitation in this species. Another is that endogenous acceptors/receptors have a higher affinity for the lectin than competing ligands and Fab under physiological conditions. If so, the isolation of these receptors might show that their exogenous addition could alter cell cohesion.

Staining of SDS gels of axenic cell proteins with derivatives of discoidin failed to detect any receptors whereas Con A and WGA bound many bands (Burridge and Jordan, 1979). Therefore, an alternative approach and wild-type strains have been used subsequently. Breuer and Siu (1981) found 11 cell surface proteins that bound a discoidin–Bio-Gel affinity matrix and could be eluted with galactose. This mixture of proteins inhibited the hemagglutination activity of discoidin, and their addition enhanced the EDTA-resistant agglutinability of aggregation-stage cells. Three of these proteins showed developmental regulation in NC4, one decreasing and two increasing during aggregation. One of these (p28) could be a discoidin subunit itself and another (p31) could not be detected in cohesive AX3. Using discoidin–Sepharose chromatography, and low pH elution, Ray and Lerner (1982) obtained material highly enriched in a gp80 (which was distinct from the gp80 associated with contact sites A). Breuer and Siu (1981) detected a band of 78K in their species that could correspond. However, the material obtained by Ray and Lerner inhibited rather than stimulated EDTA-resistant cohesion. Therefore, the role of these putative discoidin receptors remains uncertain.

Bartles and Frazier (1982) presented evidence for two distinct types of binding of discoidin I to glutaraldehyde-fixed cells, one that could be competed by sugars. The number of these sites increased fourfold during early development. The other, nonsaturable type could be competed by high ionic strength or negatively charged polyelectrolytes and was reduced by chloroform–methanol extraction. The implication that the latter class were negatively charged lipids was supported by the demonstration that discoidin I bound and agglutinated negatively charged lipid vesicles (Bartles *et al.*, 1982b). Similar binding sites were also detected in living, unfixed cells (Bartles *et al.*, 1982c). It was suggested, thus, that binding to the cell could be via lipid sites leaving the sugar-binding sites exposed.

One possibility already noted (Section 6.1.2) is that discoidin's carbohydrate-binding sites have specificity for polysaccharides. Drake and Rosen (1982) have purified a water-soluble, high-molecular-weight proteoglycan (80% carbohydrate) from *P. pallidum* that can be precipitated by pallidin. As it also is a potent inhibitor of hemagglutination by pallidin, it has been named hemagglutination inhibitor (HAI). HAI could be metabolically labeled, indicating it is not a growth medium contaminant. Some HAI was released by the cells, and this could be enhanced by galactose but not mannose, compatible with the effects of these sugars on hemagglutination by pallidin. In the presence of dinitrophenol, to block endogenous cohesiveness, added HAI enhanced EDTA-resistant agglutination.

Therefore, a number of different types of potential receptors for these lectins have been identified. The biological roles of these receptors, and what they may tell us about lectin functions, remain to be determined.

Discoidin Mutants. A single mutant with altered discoidin I (but not II) has been isolated that, along with its full and partial revertants, has provided the most direct evidence for the importance of this lectin in cohesion. Unfortunately, even here, alternate interpretations are possible. The mutant, HJR-1, was isolated from an unmutagenized NC4 population of starved cells based on its inability to adhere to an asialofetuin–Sepharose affinity column and was chosen for subsequent study because of its failure to develop normally (Ray *et al.*, 1979). The developmental defect was relatively early. Aggregation of amoebae, presumably by normal chemotaxis, into loose mounds occurred, but no tight cell–cell contacts were observed. HJR-1 also failed to express both EDTA-resistant cohesiveness and, in hemagglutination assays, discoidin I activity. It did produce normal amounts of discoidin II and material that could be immunoprecipitated with antidiscoidin antibodies and had a SDS gel mobility like that of discoidin I. Nitrosoguanidine-induced revertants fell into three classes for which the extent of recovery of development was correlated with that of hemagglutination activity.

Unquestionably, the most direct interpretation of these results is that discoidin I is a cohesion molecule, a structural component that links cells together. Further, one revertant was found (Shinnick and Lerner, 1980) that recovered in cohesivity and developmental behavior but did not recover hemagglutination activity and could not synergize with wild type. This particular revertant was presumed to have retained its nonfunctional lectin but to have a mutated receptor, complementary to the altered lectin. Its lack of synergy with wild type could, thus, be explained by having both lectin and receptor incapable of recognition of the wild-type constituents, but uniquely capable of self-recognition.

This evidence in favor of a cohesion role for discoidin I has seemed powerful but did not completely rule out other alternatives. First, there are at least three, maybe four, genes that are transcribed and translated into discoidin I polypeptides and probably account for the isoelectric heterogeneity of the tetrameric protein. (For a recent review of the *D. discoideum* genome and the discoidin genes, see Kimmel and Firtel, 1982). Reversion frequencies of HJR-1 were compatible with a single point mutation. Possibly, a point mutation in one gene could generate an altered polypeptide that, on coassembly with normal products of the unaltered genes, generates functionally inactive but serologically recognizable discoidin I tetramers. Even this possibility, if true, need not mean that the lectin is a structural component in cohesion. As suggested by Marin *et al.* (1980), it could be a regulatory element required for expression or activation of the actual structural components of cohesion. A similar argument can, of course, be made for the effects of antibodies against other cohesion candidates.

Finally, Blumberg *et al.* (1982) have cited a personal communication to the effect that current subcultures of HJR-1 do make a discoidin I capable of binding specifically to immobilized galactose. Other phenotypic properties of these current cultures were not noted but the implication is that the developmental defects are still displayed, i.e., did not corevert with the sugar recognition of discoidin I.

If the discoidins are not involved directly in cohesion, what alternative function(s) might they have? Bartles *et al.* (1982a) have discussed interesting analogies with mammalian carbohydrate-binding proteins such as the asialoglycoprotein and the phosphomannosyl-lysosomal enzyme receptors (for a review of these receptors, see Neufeld and Ashwell, 1980) for which only a fraction is on the cell surface and which appear to function in glycoprotein endocytosis and intracellular transport. Further studies are necessary to clarify the biological job(s) of these lectins.

6.5. Summary

Three glycoproteins and a polysaccharide have been identified by immunological methods as excellent candidates for direct structural elements of cohesion. During aggregation, contact sites A, which may be gp80, appears to contribute to cohesion, possibly in some concert with the polysaccharide. After aggregation, a new system takes over (gp95), which may not be cell-type specific. As the two cell types emerge, gp150 appears to contribute to specific cohesion of prespore cells. Additionally, there is evidence supporting a role during aggregation-stage cohesion for surface lectin molecules. This role could be direct but is more likely to be regulatory, as, of course, could also be the case for any candidate molecule at this time. It is also possible that the lectins have other functions, such as intracellular glycoprotein transport, unrelated to cohesion.

7. THE SLIME SHEATH

This is the structural component from which these organisms take their (informal) name. The sheath appears at the end of aggregation and encases subsequent stages such as the slug (Raper, 1935, 1940). During slug migration, sheath material is continually produced, apparently over the entire slug surface (Loomis, 1972). The sheath does not move over the substratum, but the cells move with respect to the sheath and leave behind the slug a slime trail (Shaffer, 1965), a bit like an empty sausage case (Fig. 2C). As production is over the surface of the slug and migration is polar, i.e., only in the direction of the tip, the sheath thickness increases linearly from tip to tail of the slug (Loomis, 1972; Farnsworth and Loomis, 1975).

7.1. Structure and Composition of the Sheath

The sheath appears in the EM to have two components, a smooth amorphous layer interspersed with fibrillar material (Farnsworth and Loomis, 1975; Hohl and Jehli, 1973). The fibrillar component was isolated by exhaustive extraction of slugs and slime trails with 9 M urea in 2% SDS (to remove contaminating cellular material), and the insoluble residue was characterized (Freeze and Loomis, 1977a). On a dry weight basis the material was: (1) 60% cellulose (which contributed 95% of the carbohydrate), (2) 15% polypeptides ranging from 1 to 15K in size, (3) 3% heteropolymeric carbohydrate containing mannose, GlcNAc, galactose, fucose, and xylose in progressively decreasing relative amounts, (4) 3–5% fatty acids, largely palmitic and oleic, and (5) 1% sulfate. No phosphate or uronic acid was detected. The carbohydrate polymers were very tightly complexed with each other. Release of the polypeptides required a mild base treatment, suggestive of hydrolysis of *O*-glycosidic bonds to serine or threonine, but expected products of such hydrolysis were not detectable. Therefore, the nature of the protein–matrix association is not certain. The amino acid composition showed very high content of aspartate/asparagine, serine, threonine, and cysteine and a low content of alanine, glutamate/glutamine, tyrosine, phenylalanine, lysine, and arginine, compared to the slug cells.

Smith and Williams (1979) worked with a *D. discoideum* strain in which very few cells are lost and entrapped in the slime trail during slug migration, compared to the reputed significant entrapment of such cells in the axenic strain used by Freeze and Loomis. They harvested entire slime trail material, for urea/SDS extraction was not needed to remove cellular material. On SDS gels this material generated many protein bands with a wide range of molecular weights, i.e., several were larger than bovine serum albumin. From elemental analysis and calculations of the contribution of nitrogen in proteins, they estimated that the native sheath could be up to 60% polypeptide compared to the 15% after urea/SDS extraction. As they were examining many more proteins than Freeze and Loomis, it is interesting that all of the trends in amino acid composition, i.e., enrichment or depletion in the sheath, reported by Freeze and Loomis were present in their preparation except for alanine and tyrosine. Cysteine was not tested.

After culmination, the sheath appears to surround the stalk but not spores (Raper and Fennell, 1952; Murata and Ohnishi, 1980). Accordingly, Freeze and Loomis (1978) analyzed the urea/SDS-insoluble material from stalks and found its composition to be essentially identical to the slug sheath.

7.2. Functions of the Sheath

One obvious function is the control of polarity of slug migration. The only substantial extensibility/flexibility of the sheath is at the tip region (Francis, 1962), and that is the only direction of movement available to the encased cells, for the sheath collapses at the rear.

Another function is in maintenance of integrity of the postaggregation morphologies, which appears to be related to the crystallinity of the sheath cellulose as assayed by glucose release during methanolysis. More crystalline material is more resistant to methanolysis. Using this assay, Freeze and Loomis (1977b) showed that mutants lacking developmentally regulated N-acetylglucosaminidase produced a less crystalline sheath. These mutants also cannot migrate as normal-sized slugs but fragment into very small slugs (Dimond et $al.$, 1973). Mutants defective in UDP-glucose pyrophosphorylase could not produce cellulose, the fibrillar component of the sheath by EM, or any detectable urea/SDS-insoluble sheath. These mutants developed normally up to the slug stage, but slugs were fragile and unable to migrate. These strains also could not culminate and could not produce stalk or spore cells.

A consequence of the gradation in thickness is that the sheath is most permeable at the thinnest region, surrounding the slug tip (Farnsworth and Loomis, 1974). As tip and posterior cells differ in their ultimate developmental fates, differential ability of molecules to enter or escape from the slug along its length could regulate local concentrations of morphogens which determine the path of cytodifferentiation, as originally proposed by Loomis (1972). Based on effects of cAMP and ammonia, Sussman has proposed a specific model incorporating this postulate (reviewed in Sussman, 1982).

Finally, as the sheath is clearly a close, possibly identical-twin, relative of structural material of the stalk after culmination, a single type of structural material seems competent for several biological functions (Freeze and Loomis, 1978). It is very interesting in this context to note that, while $D.$ $discoideum$ produces mature stalk cells only during culmination, some slime mold species, e.g., $D.$ $mucoroides$, continually produce a mature stalk during slug migration (Bonner et $al.$, 1955). During migration, the prestalk region of the slug is replenished by conversion of prespore cells that, in turn, depletes the population of the cell type that will retain ultimate viability (Gregg and Davis, 1982). Possibly other species, such as *D. discoideum*, have "found" a way to block stalk cell differentiation during slug migration, to retain the greatest number of spore-forming cells in the "hope" of enhancing survival probabilities.

8. GLYCOPROTEINS SPECIFIC FOR PRESTALK OR PRESPORE CELLS

The major developmentally regulated changes in gene expression occur at about the stage of tip emergence, and there is limited expression of new genes thereafter (Blumberg and Lodish, 1981). Therefore, it would be expected that most of the gene products needed for terminal differentiation of stalk and spore cells would be present in slugs and would be differentially present in prestalk or prespore cells unless a gene product was required for both terminal cell types.

Alton and Brenner (1979) used [^{35}S]methionine to metabolically label slugs and analyzed radioactive proteins on two-dimensional gels after manual dissection of anterior and posterior portions. Of about 400 total proteins detected, 57 displayed at least some degree of cell-type specificity. Barklis and Lodish (1983) fractionated prestalk and prespore cells on Percoll density gradients and isolated mRNAs for genes only expressed after aggregation. Of 13 late-stage mRNA species analyzed in detail, 12 showed some degree of cell-type specificity. In several cases the specificity was almost absolute. In contrast, mRNAs that were expressed early in development, and in most cases suppressed later, did not show cell-type specificity. However, discoidin I mRNA, classified as nonspecific, did appear to be somewhat more abundant in the densest prespore fraction, and Lam *et al.* (1981) have shown that prespore cells contain twice the amount of discoidin I in prestalk cells.

The earliest cell-specific antigen was detected by Takeuchi (1963, 1972) who showed that antisera produced against *D. mucoroides* spores, and adsorbed with vegetative cells, specifically stained organelles of posterior cells of *D. discoideum* slugs. The antigenic determinant is unknown. However, the enzyme UDP-galactose polysaccharide transferase is uniquely localized to prespore cells (Newell *et al.*, 1969) and transfers galactose to a polysaccharide found in spore but not stalk cells (Sussman and Osborn, 1964). This polysaccharide is found in slug prespore cells in association with the organelle, called the prespore vacuole, that reacts with the antiserum (Maeda and Takeuchi, 1969; Takeuchi, 1972). Hohl and Hamamoto (1969) have reported exocytosis of the prespore vacuole during culmination, and it has been suggested (Farnsworth and Loomis, 1976) that the released contents may help the spores cohere, though other roles are also possible (Morrissey, 1982).

Other elements of the spore are also obvious candidates for slug cell specificity. Orlowski and Loomis (1979) isolated the spore coats left behind after spore germination and found that they contained five major proteins, designated SP60, SP68, SP70, SP96, and SP200. Two, SP96 and SP70, were glycoproteins. Devine *et al.* (1982) showed that they incorporated substantial amounts of fucose. They appear to be the same fucosylated species observed by Lam and Siu (1981). All the spore coat proteins were synthesized exclusively in prespore cells of slugs (Lam and Siu, 1981; Devine *et al.*, 1982). The spore coat proteins are not the antigenic determinants recognized by Takeuchi's antiserum, as the determinants but not these proteins are present in a sporeless mutant (Loomis, 1982).

The use of fucose for metabolic labeling of these glycoproteins was prompted by the earlier report of Gregg and Karp (1978) who showed a highly preferential incorporation of [^{3}H]fucose into prespore cells and spores (Fig. 7). Synthesis of the spore coat proteins appears to account for most of this fucose incorporation (Lam and Siu, 1981; Devine *et al.*, 1982), though some less abundant fucosylated proteins are not found in the spore coat, i.e., gp150 (Lam and Siu, 1981).

There are fewer specific markers of prestalk cells. Oohata and Takeuchi (1977) reported electrophoretically distinct isozymes of β-galactosidase and acid phosphatase in prestalk cells. Other markers have been identified in studies of the axial distribution of cell surface glycoproteins. There are a number of reports consistent with differences in cohesivity of prestalk and prespore cells (Feinberg *et al.*, 1979; Tasaka and Takeuchi, 1981; Lam *et al.*, 1981). This predicts molecular differences at the cell surfaces, a prediction confirmed by West and McMahon (1979). Slugs were manually dissected into prestalk and prespore regions, which were then subjected to SDS electrophoresis. The gels were stained with WGA, Con A, and an antiserum raised against slug plasma membranes (and preadsorbed against vegetative cells). Con A staining did not reveal reproducible differences between the cell types. However, WGA detected glycoproteins of apparent M_r of 100 and 110K and a glycoconjugate smeared over the 35–50K region that were only in the prestalk region. WGA also detected glycoproteins of 21 and 80K that were specific to prespore cells. It was not possible to obtain sufficient material from dissected slugs to isolate plasma membranes. However, all of these glycoproteins were present in whole slug plasma membranes but not those of vegetative cells. These five species accounted for over 80% of the WGA binding to the gels. The antiserum also detected the 110, 100, and 80K species and appeared to also reveal their cell-type specificity.

West and McMahon (1981) followed up these observations by examining the effects of WGA and the antiserum on reaggregation of dissociated slug cells. In the presence of 2.5 (but not 1.0) mg/ml WGA, the cells formed shallow aggregates but did not regenerate the standing slugs that formed within 3–4 hr in the absence of the lectin (Fig. 14). When the blocked cells were harvested and the WGA eluted, standing slugs were able to form in the normal time, suggesting that the action of the lectin was at the cell surface. The inhibitory effect of WGA was not simply due to agglutination of the cells, as the antibodies also initially agglutinated cells but only slightly retarded reestablishment of morphology. Finally, as WGA does not induce redistribution of its receptors, the simplest interpretation is that by steric blockade the lectin prevents recognition processes required for recapitulation of proper morphology. As the antiserum also binds to two prestalk- and one prespore-specific WGA receptor, this suggests that the critical species may be the heterogeneous glycoconjugate of prestalk cells and/or the 21K prespore marker. The relationships of all the species with gp95 and gp150 remain to be determined.

Additional effects of lectins on late development have been reported by Laroy and Weeks (1982) in studies with the sporogenous mutants of strain V12/M2 described by Kay *et al.* (1978). These mutants were obtained by virtue of their survival under conditions that induce quantitative differentiation of monolayers of this strain into nonviable stalk cells (Town *et al.*, 1976). The sporogenous mutants under these conditions yield about 25% spores. Laroy and Weeks (1982) showed that low concentrations (30–40 μg/ml) of WGA or Con A completely inhibited spore cell differentiation in the monolayers and reduced the yield of stalk cells, but less dramatically. The lectins had no significant effects on the time courses of four developmentally regulated enzymes, including prespore-specific UDP-galactose polysaccharide transferase. Discoidin, at 200 μg/ml, reduced spore differentiation threefold, which is interesting in view of the report (Lam *et al.*, 1981) that prespore cells have twofold more discoidin than prestalk cells. The relationship of these results to the axial distribution of WGA receptors in prestalk and prespore cells (West and McMahon, 1979) is not clear, as Con A receptors did not appear to be cell-type specific, and WGA and Con A were equally effective in blocking spore differentiation. Finally, the mechanism of inhibition is unknown, but protease treatment (Peacey and Gross, 1981) also blocked spore cell differentiation. Both treatments may interfere with cell–cell or cell–substrate interactions or with some aspect of spore coat formation.

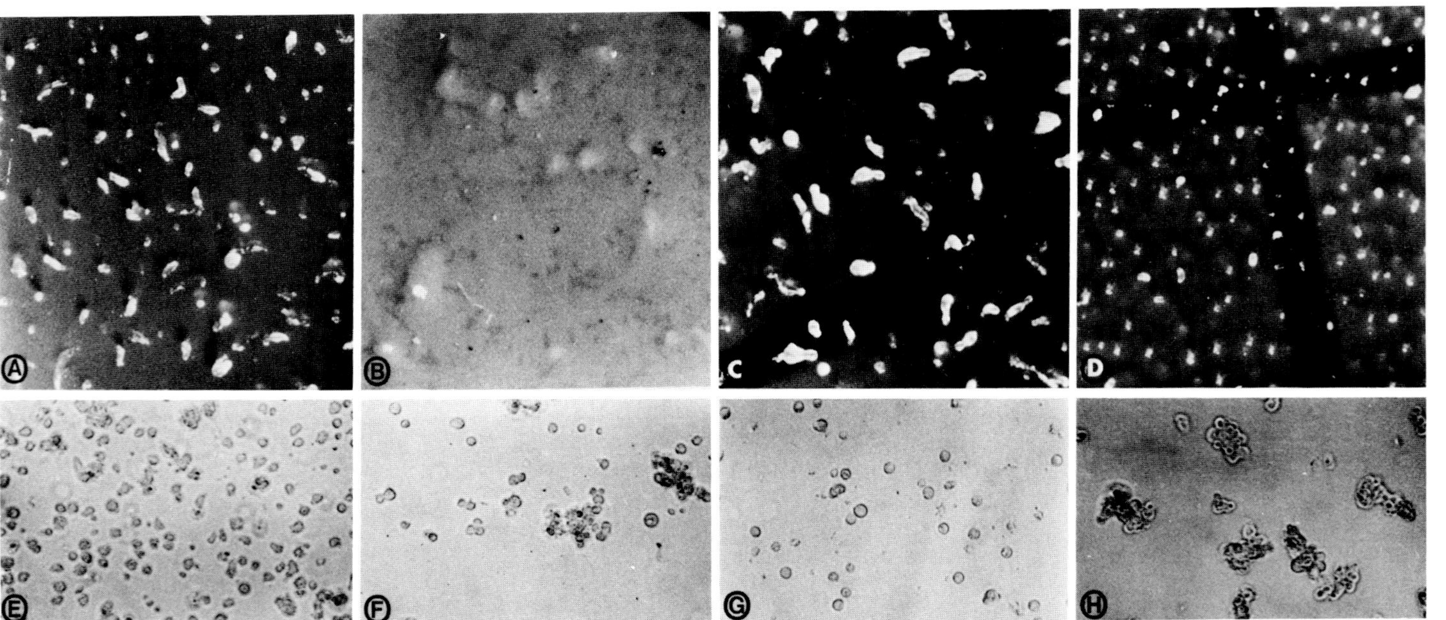

Figure 14. Inhibition of development with wheat germ agglutinin. Pseudoplasmodia (18 hr) were dissociated into single cells and incubated for 15 min in 2.5 mg/ml of bovine serum albumin (A, E), WGA (B, F), WGA plus a hapten inhibitor, 10 mM NAG_5 (C, G), or IgG produced against pseudoplasmodia plasma membranes (D, H). Following incubation, cells were replated on Millipore filters. After $5\frac{1}{2}$ hr, cells were photographed to evaluate their ability to re-form pseudoplasmodia (A–D). Immediately following incubation, a sample of cells was also diluted and observed in a phase-contrast microscope for agglutination (E–H). Scale for (A–D) is given in (C) and (D), where the grid bar is 0.22 mm wide. Magnification for (E–H) is $\times$ 80. (From West and McMahon, 1981, with permission.)

9. CONCLUDING REMARKS

A major goal of this chapter, in addition to a review of the glyco-protein literature, has been to introduce this organism and its status as an experimental system to readers previously unacquainted with it. The behaviors of the organism in the laboratory appear to be a faithful reflection of those in nature. It is amenable to studies at cellular, biochemical, and genetic levels, and important areas of cellular physiological responses and developmental processes are accessible to a degree seldom observed in more complex organisms.

For a specific subject of this chapter, roles of glycoproteins and their oligosaccharide substituents are under study in many laboratories. Both polypeptide and oligosaccharide elements of glycoproteins suffer dramatic developmental regulation, and the mechanisms regulating these changes are being analyzed by studies of extracellular signals and mechanisms by which these are transduced into changes in gene expression. Recognition functions for specific glycoproteins are being probed for individual cellular activities such as intracellular protein sorting and phagocytosis. A number of specific molecules implicated in developmentally regulated cellular cohesion are being studied at levels not available in many other systems, as behavioral mutants can be obtained and analyzed not only by immunology and biochemistry but by genetics and molecular biology. Indeed, current results with *D. discoideum* should provide a strong cautionary note for systems where such approaches are not easily available. Finally, the relatively frequent international meetings on this organism attract a blend of developmentalists, cell biologists, biochemists, geneticists, and molecular biologists with nothing in common but the organism and send one home believing *D. discoideum* will be the first eukaryotic organism to be solved at the level, say, of current understanding of *E. coli*.

ACKNOWLEDGMENTS. The author expresses thanks to Ms. Mary Creel for assistance in assembly of the references and, very specially, to Ms. Carmen Felix for valiance and dedication in typing of the manuscript, particularly in the face of a word processor defect that resulted in loss of the entire first typed copy!

REFERENCES

Alton, T. H., and Brenner, M., 1979, Comparison of proteins synthesized by anterior and posterior regions of *Dictyostelium discoideum* pseudoplasmodia, *Dev. Biol.* **71:**1–9.

Dimond, R. L., Burns, R. A., and Jordan, K. B., 1981, Secretion of lysosomal enzymes in the cellular slime mold, *Dictyostelium discoideum*, *J. Biol. Chem.* **256**:6565–6572.

Drake, D. K., and Rosen, S. D., 1982, Identification and purification of an endogenous receptor for the lectin pallidin from *Polysphondylium pallidum*, *J. Cell Biol.* **93**:383–389.

Eitle, E., and Gerisch, G., 1977, Implication of developmentally regulated concanavalin A binding proteins of *Dictyostelium* in cell adhesion and cyclic AMP regulation, *Cell Differ.* **6**:339–346.

Erdos, G. W., and Whitaker, D., 1983, Failure to detect immuno-cytochemically reactive endogenous lectin on the cell surface of *Dictyostelium discoideum*, *J. Cell Biol.* **97**:993–1000.

Every, D., and Ashworth, J. M., 1973, The purification and properties of extracellular glycosidases of the cellular slime mould *Dictyostelium discoideum*, *Biochem. J.* **133**:37–47.

Farnsworth, P., and Loomis, W. F., 1974, A barrier to diffusion in pseudoplasmodia of *Dictyostelium discoideum*, *Dev. Biol.* **41**:77–83.

Farnsworth, P., and Loomis, W. F., 1975, A gradient in the thickness of the surface sheath in pseudoplasmodia of *Dictyostelium discoideum*, *Dev. Biol.* **46**:349–357.

Farnsworth, P. A., and Loomis, W. F., 1976, Quantitation of the spatial distribution of prespore vacuoles in pseudoplasmodia of *Dictyostelium discoideum*, *J. Embryol. Exp. Morphol.* **35**:499–505.

Favard-Séréno, C., Ludosky, M. A., and Ryter, A., 1981, Freeze-fracture study of phagocytosis in *Dictyostelium discoideum*, *J. Cell Sci.* **51**:63–84.

Feinberg, A. P., Springer, W. R., and Barondes, S. H., 1979, Segregation of prestalk and prespore cells of *Dictyostelium discoideum*: Observations consistent with selective cell cohesion, *Proc. Natl. Acad. Sci. USA* **76**:3977–3981.

Francis, D. W., 1962, The movement of pseudoplasmodia of *Dictyostelium discoideum*, Ph.D. thesis, University of Wisconsin, Madison.

Free, S. J., and Loomis, W. F., 1974, Isolation of mutations in *Dictyostelium discoideum* affecting alpha-mannosidase, *Biochimie* **56**:1525–1528.

Free, S. J., and Schimke, R. T., 1978, Effects of a post-translational modification mutation on different developmentally regulated glycosidases in *Dictyostelium discoideum*, *J. Biol. Chem.* **253**:4107–4111.

Free, S. J., Schimke, R. T., Freeze, H., and Loomis, W. F., 1978, Characterization and genetic mapping of *modA*, a mutation in the post-translational modification of the glycosidases of *Dictyostelium discoideum*, *J. Biol. Chem* **253**:4102–4106.

Freeze, H., and Loomis, W. F., 1977a, Isolation and characterization of a component of the surface sheath of *Dictyostelium discoideum*, *J. Biol. Chem.* **252**:820–824.

Freeze, H., and Loomis, W. F., 1977b, The role of the fibrillar component of the surface sheath in the morphogenesis of *Dictyostelium discoideum*, *Dev. Biol.* **56**:184–194.

Freeze, H., and Loomis, W. F., 1978, Chemical analysis of stalk components of *Dictyostelium discoideum*, *Biochim. Biophys. Acta* **539**:529–537.

Freeze, H. H., and Miller, A. L., 1980, *Mod*A, a posttranslational mutation affecting phosphorylated and sulfated glycopeptides in *Dictyostelium discoideum*, *Mol. Cell. Biochem.* **35**:17–27.

Freeze, H. H., Miller, A. L., and Kaplan, A., 1980, Acid hydrolases from *Dictyostelium discoideum* contain phosphomannosyl recognition markers, *J. Biol. Chem.* **255**:11081–11084.

Geltosky, J., Siu, C.-H., and Lerner, R., 1976, Glycoproteins of the plasma membrane of *Dictyostelium discoideum* during development, *Cell* **8**:391–396.

Geltosky, J. E., Weseman, J., Bakke, A., and Lerner, R. A., 1979, Identification of a cell surface glycoprotein involved in cell aggregation in *Dictyostelium discoideum, Cell* **18**:391–398.

Geltosky, J. E., Bidwell, C. R., Weseman, J., and Lerner, R. A., 1980, A glycoprotein involved in aggregation of *Dictyostelium discoideum* is distributed on the cell surface in a nonrandom fashion favoring cell junctions, *Cell* **21**:339–345.

Gerisch, G., 1961, Zellfunktionen und Zellfunktionswechsel in der Entwicklung von *Dictyostelium discoideum.* V. Stadienspezifische Zelkontalkbidung und ihre quantitative Erfassung, *Exp. Cell Res* **25**:535–554.

Gerisch, G., 1968, Cell aggregation and differentiation in *Dictyostelium discoideum, Curr. Top. Dev. Biol.* **3**:157–197.

Gerisch, G., 1980, Univalent antibody fragments as tools for the analysis of cell interactions in *Dictyostelium, Curr. Top. Dev. Biol.* **14**:243–269.

Gerisch, G., Fromm, H., Huesgen, A., and Wick, U., 1975, Control of cell contact sites by cAMP pulses in differentiating *Dictyostelium discoideum* cells, *Nature* (*London*) **255**:547–549.

Gerisch, G., Krelle, H., Bozzaro, S., Eitle, E., and Guggenheim, R., 1980, Analysis of cell adhesion in *Dictyostelium* and *Polysphondylium* by the use of Fab, in: *Cell Adhesion and Motility* (A. S. G. Curtis and J. D. Pitts, eds.), Cambridge University Press, London.

Gilkes, N. R., and Weeks, G., 1977, The purification and characterization of *Dictyostelium discoideum* plasma membrane, *Biochim. Biophys. Acta* **464**:142–156.

Gilkes, N. R., Laroy, K., and Weeks, G., 1979, An analysis of the protein, glycoprotein and monosaccharide composition of *Dictyostelium discoideum* plasma membranes during development, *Biochim. Biophys. Acta* **551**:349–362.

Gillette, M. U., and Filosa, M. F., 1973, Effect of concanavalin A on cellular slime mold development: Premature appearance of membrane-bound cyclic AMP phosphodiesterase, *Biochem. Biophys. Res. Commun.* **53**:1159–1166.

Gillette, M. U., Dengler, R. E., and Filosa, M. F., 1974, The localization and fate of concanavalin A in amoebae of the cellular slime mold *Dictyostelium discoideum, J. Exp. Zool.* **190**:243–248.

Gingell, D., and Garrod, D. R., 1969, Effect of EDTA on electrophoretic mobility of slime mould cells and its relationship to current theories of cell adhesion, *Nature* (*London*) **221**:192–193.

Grabel, L., and Loomis, W. F., 1978, Effector controlling accumulation of N-acetylglucosaminidase during development of *Dictyostelium discoideum, Dev. Biol.* **64**:203–209.

Gregg, J. H., and Davis, R. W., 1982, Dynamics of cell redifferentiation in *Dictyostelium mucoroides, Differentiation* **21**:200–205.

Gregg, J. H., and Karp, G. C., 1978, Patterns of cell differentiation revealed by tritiated fucose incorporation in *Dictyostelium discoideum, Exp. Cell Res.* **112**:31–46.

Gustafson, G. L., and Milner, L. A., 1980a, Occurrence of N-acetylglucosamine-1-phosphate in proteinase I from *Dictyostelium discoideum, J. Biol. Chem.* **255**:7208–7210.

Gustafson, G. L., and Milner, L. A., 1980b, Immunological relationship between beta-N acetylglucosaminidase and proteinase I from *Dictyostelium discoideum, Biochem. Biophys. Res. Commun.* **94**:1439–1444.

Hellio, R., and Ryter, A., 1980, Relationship between anionic sites and lectin receptors in the plasma membrane of *Dictyostelium discoideum* and their role in phagocytosis, *J. Cell Sci.* **41**:89–104.

Hoffman, S., and McMahon, D., 1977, The role of the plasma membrane in the development of *Dictyostelium discoideum.* II. Developmental and topographic analysis of polypeptide and glycoprotein composition, *Biochim. Biophys. Acta* **465**:242–259.

Hoffman, S., and McMahon, D., 1978, Defective glycoproteins in the plasma membrane of an aggregation minus mutant of *Dictyostelium discoideum* with abnormal cellular interactions, *J. Biol. Chem.* **253:**278–287.

Hohl, H. R., and Hamamoto, S. T., 1969, Ultrastructure of spore differentiation in *Dictyostelium*: The prespore vacuole, *J. Ultrastruct. Res.* **26:**442–453.

Hohl, H. R., and Jehli, J., 1973, The presence of cellulose microfibrils in the proteinaceous slime track of *Dictyostelium discoideum, Arch. Mikrobiol.* **92:**179–187.

Hubbard, S. C., and Ivatt, R. J., 1981, Synthesis and processing of asparagine-linked oligosaccharides, *Annu. Rev. Biochem.* **50:**555–583.

Huesgen, A., and Gerisch, G., 1975, Solubilized contact sites A from cell membranes of *Dictyostelium discoideum, FEBS Lett.* **56:**46–49.

Ivatt, R. J., Das, O. P., Henderson, E. J., and Robbins, P. W., 1981, Developmental regulation of glycoprotein biosynthesis in *Dictyostelium, J. Supramol. Struct. Cell. Biochem.* **17:**359–368.

Ivatt, R. J., Das, O. P., Henderson, E. J., and Robbins, P. W., 1984a, Glycoprotein biosynthesis in *Dictyostelium discoideum*: Developmental regulation of the protein–linked glycans, *Cell,* submitted.

Ivatt, R. J., Das, O. P., Henderson, E. J., and Robbins, P. W., 1984b, Glycoprotein biosynthesis in *Dictyostelium discoideum*: Assembly and processing of the oligosaccharides, in preparation.

Jaffe, A. R., Swan, A. P., and Garrod, D. R., 1979, A ligand receptor model for the cohesive behavior of *Dictyostelium discoideum* axenic cells, *J. Cell Sci.* **37:**157–167.

Juliani, M. H., and Klein, C., 1978, A biochemical study of the effects of cyclic AMP pulses on aggregateless mutants of *Dictyostelium discoideum, Dev. Biol.* **62:**162–172.

Kay, R. R., and Trevan, D. J., 1981, *Dictyostelium* amoebae can differentiate into spores without cell-to-cell contact, *J. Embryol. Exp. Morphol.* **62:**369–378.

Kay, R. R., Garrod, D., and Tilly, R., 1978, Requirement for cell differentiation in *Dictyostelium discoideum, Nature (London)* **271:**58–60.

Kimmel, A. R., and Firtel, R. A., 1982, The organization and expression of the *Dictyostelium* genome, in: *The Development of Dictyostelium discoideum* (W. F. Loomis, ed.), Academic Press, New York.

Klein, C., and Darmon, M., 1977, Effects of cyclic AMP pulses on adenylate cyclase and the phosphodiesterase inhibitor of *Dictyostelium discoideum, Nature (London)* **268:**76–78.

Knecht, D. A., and Dimond, R. L., 1981, Lysosomal enzymes possess a common antigenic determinant in the cellular slime mold, *Dictyostelium discoideum, J. Biol. Chem.* **256:**3564–3575.

Knecht, D. A., Mierendorf, R. C., and Dimond, R. L., 1982, Monoclonal antibodies recognizing common antigenic determinants on lysosomal enzymes, International Cellular Slime Mold Conference, Hartford, Conn.

Lam, T. Y., and Siu, C.-H., 1981, Synthesis of stage-specific glycoproteins in *Dictyostelium discoideum* during development, *Dev. Biol.* **83:**127–137.

Lam, T. Y., and Siu, C.-H., 1982, Inhibition of cell differentiation and cell cohesion by tunicamycin in *Dictyostelium discoideum, Dev. Biol.* **92:**398–407.

Lam, T. Y., Pickering, G., Geltosky, J., and Siu, C.-H., 1981, Differential cell cohesiveness expressed by prespore and prestalk cells of *Dictyostelium discoideum, Differentiation* **20:**22–28.

Landfear, S. M., and Lodish, H. F., 1980, A role for cyclic AMP in expression of developmentally regulated genes in *Dictyostelium discoideum, Proc. Natl. Acad. Sci. USA* **77:**1044–1048.

Laroy, K., and Weeks, G., 1982, Inhibition of *Dictyostelium discoideum* differentiation in monolayers *in vitro* by endogenous and exogenous lectins, *J. Cell Sci.,* **59:**203–212.

Lodish, H. F., Blumberg, D. D., Chisholm, R., Chung, S., Coloma, A., Landfear, S., Barklis, E., Lefebvre, P., Zuker, C., and Mangiarotti, G., 1982, Control of gene expression, in: *The Development of Dictyostelium discoideum* (W. F. Loomis, ed.), Academic Press, New York.

Loomis, W. F., 1972, Role of the surface sheath in the control of morphogenesis in *Dictyostelium discoideum*, *Nature New Biol.* **240**:6–9.

Loomis, W. F., 1975, *Dictyostelium discoideum: A Developmental System*, Academic Press, New York.

Loomis, W. F., 1978, The number of developmental genes in *Dictyostelium*, *Birth Defects Orig. Artic. Ser.* **XIV**:497–505.

Loomis, W. F., 1979, Biochemistry of aggregation in *Dictyostelium*: A review, *Dev. Biol.* **70**:1–12.

Loomis, W. F. (ed.), 1982, *The Development of Dictyostelium discoideum*, Academic Press, New York.

Loomis, W. F., White, S., and Dimond, R., 1976, A sequence of dependent stages in development of *Dictyostelium discoideum*, *Dev. Biol.* **53**:171–177.

Maeda, Y., and Takeuchi, I., 1969, Cell differentiation and fine structures in the development of the cellular slime molds, *Dev. Growth Differ.* **11**:232–245.

Margolskee, J. P., Froshauer, S., Skrinska, R., and Lodish, H. F., 1980, The effects of cell density and starvation on early developmental events in *Dictyostelium discoideum*, *Dev. Biol.* **74**:409–421.

Marin, F. T., 1976, Regulation of development in *Dictyostelium discoideum*. I. Initiation of the growth to development transition by amino-acid starvation, *Dev. Biol.* **48**:110–117.

Marin, F. T., 1977, Regulation of development in *Dictyostelium discoideum*. II. Regulation of early cell differentiation by amino-acid starvation and intercellular interaction, *Dev. Biol.* **60**:389–395.

Marin, F. T., Goyette-Boulay, M., and Rothman, F. G., 1980, Regulation of development of *Dictyostelium discoideum*. III. Carbohydrate specific intercellular interactions in early development, *Dev. Biol.* **80**:301–312.

Mason, J. W., Rasmussen, H., and Dibella, F., 1971, 3′,5′ AMP and Ca^{2+} in slime mold aggregation, *Exp. Cell Res,* **67**:156–160.

Matarazzo, S. Z., and Henderson, E. J., 1983a, Phosphorylated oligosaccharides of *Dictyostelium discoideum* glycoproteins are absent in plasma membranes, (in preparation).

Matarazzo, S. Z., and Henderson, E. J., 1983b, A role for glycoprotein-linked oligosaccharides in cohesion of aggregating *Dictyostelium discoideum* amoebae, (in preparation).

Molday, R., Jaffe, R., and McMahon, D., 1976, Concanavalin A and wheat germ agglutinin receptors on *Dictyostelium discoideum*, *J. Cell Biol.* **71**:314–322.

Morrissey, J. H., 1982, Cell proportioning and pattern formation, in: *The Development of Dictyostelium discoideum* (W. F. Loomis, ed.), Academic Press, New York.

Müller, K., and Gerisch, G., 1978, A specific glycoprotein as the target site of adhesion blocking Fab in aggregating *Dictyostelium* cells, *Nature (London)* **274**:445–449.

Müller, K., Gerisch, G., Fromme, I., Mayer, H., and Tsugita, A., 1979, A membrane glycoprotein of aggregating *Dictyostelium* cells with the properties of contact sites A, *Eur. J. Biochem.* **99**:419–426.

Murata, Y., and Ohnishi, T., 1980, *Dictyostelium discoideum* fruiting bodies observed by scanning electron microscopy, *J. Bacteriol.* **141**:956–958.

Murray, B. A. and Loomis, W. F., 1982, Immunological analysis of cell–cell adhesion, International Cellular Slime Mold Conference, Hartford, Conn.

Murray, B. A., Yee, L. D., and Loomis, W. F., 1981, Immunological analysis of a glycoprotein (contact sites A) involved in intercellular adhesion of *Dictyostelium discoideum*, *J. Supramol. Struct. Cell. Biochem.* **17**:387–401.

Murray, B. A., Niman, H. L., and Loomis, W. F., 1983, A monoclonal antibody recognizing gp 80, a membrane glycoprotein implicated in intercellular adhesion of *Dictyostelium discoideum*, *Mol. Cell. Biol.* **3**:863–870.

Nakajima, T., and Ballou, C. E., 1974, Structure of the linkage region between the polysaccharide and protein parts of *Saccharomyces cerevisiae* mannan, *J. Biol. Chem.* **249**:7685–7694.

Neufeld, E. F., and Ashwell, G., 1980, Carbohydrate recognition systems for receptor-mediated pinocytosis, in: *The Biochemistry of Glycoproteins and Proteoglycans* (W. J. Lennarz, ed.), Plenum Press, New York.

Newell, P. C., 1977, How cells communicate: the system used by slime molds, *Endeavor* **1**:63–68.

Newell, P. C., Ellingson, J. S., and Sussman, M., 1969, Synchrony of enzyme accumulation in a population of differentiating slime mold cells, *Biochim. Biophys. Acta* **177**:610–614.

Newell, P. C., Franke, J., and Sussman, M., 1972, Regulation of four functionally related enzymes during shifts in the developmental program of *Dictyostelium discoideum*, *J. Mol. Biol.* **63**:373–382.

Ochiai, H., Stadler, J., Westphal, M., Wagle, G., Merkl, R., and Gerisch, G., 1982a, Monoclonal antibodies against contact sites A of *Dictyostelium discoideum*: Detection of modifications of the glycoprotein in tunicamycin-treated cells, *EMBO J.* **1**:1011–1016.

Ochiai, H., Schwarz, H., Merkl, R., Wagle, G., and Gerisch, G., 1982b, Stage-specific antigens reacting with monoclonal antibodies against contact site A, a cell-surface glycoprotein of *Dictyostelium discoideum*, *Cell Differ.* **11**:1–13.

Ono, K.-I., Toda, K., and Ochiai, H., 1981, Drastic changes in accumulation and synthesis of plasma-membrane proteins during aggregation of *Dictyostelium discoideum*, *Eur. J. Biochem.* **119**:133–143.

Oohata, A., and Takeuchi, I., 1977, Separation and biochemical characterization of the two cell types present in the pseudoplasmodium of *Dictyostelium mucoroides*, *J. Cell Sci.* **24**:1–10.

Orlowski, M., and Loomis, W. F., 1979, Plasma membrane proteins of *Dictyostelium discoideum*: The spore coat proteins, *Dev. Biol.* **71**:297–307.

Parish, R. W., 1979, Cyclic AMP induces synthesis of developmentally regulated plasma membrane proteins in *Dictyostelium*, *Biochim. Biophys. Acta* **553**:179–182.

Parish, R. W., and Schmidlin, S., 1979a, Synthesis of plasma membrane proteins during development of *Dictyostelium discoideum*, *FEBS Lett.* **98**:251–256.

Parish, R. W., and Schmidlin, S., 1979b, Resynthesis of developmentally regulated plasma membrane proteins following disaggregation of *Dictyostelium discoideum* pseudoplasmodia, *FEBS Lett.* **99**:270–274.

Parish, R. W., Schmidlin, S., and Weibel, M., 1978a, Effect of cyclic AMP pulses on the synthesis of plasma membrane proteins in aggregateless mutants of *Dictyostelium discoideum*, *FEBS Lett.* **96**:283–286.

Parish, R. W., Schmidlin, S., and Parish, C. R., 1978b, Detection of developmentally controlled plasma membrane antigens of *Dictyostelium discoideum* cells in sodium dodecyl sulfate polyacrylamide gels, *FEBS Lett.* **95**:366–370.

Peacey, M. J., and Gross, J. D., 1981, The effect of proteases on gene expression and cell differentiation in *Dictyostelium discoideum*, *Differentiation* **19**:189.

Rahmsdorf, H. J., Cailla, H. L., Spitz, E., Moran, M. J., and Rickenberg, H. V., 1976, Effect of sugars on early biochemical events in development of *Dictyostelium discoideum*, *Proc. Natl. Acad. Sci. USA* **73**:3183–3187.

Raper, K. B., 1935, *Dictyostelium discoideum*, a new species of slime mold from decaying forest leaves, *J. Agric. Res.* **50**:135–147.

Raper, K. B., 1940, Pseudoplasmodium formation and organization in *Dictyostelium discoideum*, *J. Elisha Mitchell Sci. Soc.* **56**:241–282.

Raper, K. B., 1951, Isolation, cultivation, and conservation of simple slime molds, *Quarterly Rev. Bid.* **26:**169–190.

Raper, K. B., and Fennell, D. I., 1952, Stalk formation in *Dictyostelium, Bull. Torrey Bot. Club* **79:**25–51.

Ray, J., and Lerner, R. A., 1982, A biologically active receptor for the carbohydrate-binding protein(s) of *Dictyostelium discoideum, Cell* **28:**91–98.

Ray, J., Shinnick, T., and Lerner, R., 1979, A mutation altering the function of a carbohydrate binding protein blocks cell–cell cohesion in developing *Dictyostelium discoideum, Nature (London)* **279:**215–221.

Reitherman, R. W., Rosen, S. D., Frazier, W. A., and Barondes, S. H., 1975, Cell surface species-specific high affinity receptors for discoidin: Developmental regulation in *Dictyostelium discoideum, Proc. Natl. Acad. Sci. USA* **72:**3541–3545.

Rickenberg, H. V., Rahmsdorf, H. J., Campbell, A., North, M. J., Kwasniak, J., and Ashworth, J. M., 1975, Inhibition of development in *Dictyostelium discoideum* by sugars, *J. Bacteriol.* **124:**212–219.

Rosen, S. D., Kafka, J. A., Simpson, D. L., and Barondes, S. H., 1973, Developmentally regulated, carbohydrate-binding protein in *Dictyostelium discoideum, Proc. Natl. Acad. Sci. USA* **70:**2554–2557.

Rössler, H., Peuckert, W., and Risse, H.-J., 1978, The biosynthesis of glycolipids during the differentiation of the slime mold *Dictyostelium discoideum, Mol. Cell. Biochem.* **20:**3–15.

Rössler, H. H., Schneider-Seelbach, E., Malati, T., and Risse, H.-J., 1981, The dependence of glycosyltransferases in *Dictyostelium discoideum* on the structure of polyisoprenols, *Mol. Cell. Biochem.* **34:**65–72.

Rössler, H. H., Zimpfer, A., and Risse, H.-J., 1982, A differentiation-dependent polyisoprenol kinase in *Dictyostelium discoideum, Mol. Cell. Biochem.* **48:**183–189.

Ryter, A., and de Chastellier, C., 1977, Morphometric and cytochemical studies of *Dictyostelium discoideum* in vegetative phase digestive system and membrane turnover, *J. Cell Biol.* **75:**200–217.

Ryter, A., and Hellio, R., 1980, Electron microscope study of *Dictyostelium discoideum* plasma membrane and its modifications during and after phagocytosis, *J. Cell Sci.* **41:**75–88.

Saxe, C. L., III, and Sussman, M., 1982, Induction of cell cohesion by a membrane-associated moiety in *Dictyostelium discoideum, Cell* **29:**755–759.

Schmidt, J. A., and Loomis, W. F., 1982, Phosphorylation of the contact site A glycoprotein (gp80) of *Dictyostelium discoideum, Dev. Biol.* **91:**296–304.

Shaffer, B. M., 1965, Cell movement within aggregates of the slime mold *Dictyostelium discoideum* revealed by surface markers, *J. Embryol. Exp. Morphol.* **13:**97–117.

Shinnick, T. M., and Lerner, R. A., 1980, The *cbpA* gene: Role of the 26,000 dalton carbohydrate binding protein in intercellular cohesion of developing *Dictyostelium discoideum* cells, *Proc. Natl. Acad. Sci. USA* **77:**4788–4792.

Simpson, D. L., Rosen, S. D., and Barondes, S. H., 1974, Discoidin, a developmentally regulated carbohydrate binding protein from *Dictyostelium discoideum*: Purification and characterization, *Biochemistry* **13:**3487–3493.

Siu, C.-H., and des Roches, B., 1982, Involvement of a surface glycoprotein in cell sorting in *Dictyostelium discoideum, Proc. Can. Fed. Biol. Soc.* **25:**43a.

Siu, C.-H., Lerner, R. A., Ma, C., Firtel, R. A., and Loomis, W. F., 1976, Developmentally regulated proteins of the plasma membrane of *Dictyostelium discoideum*: The carbohydrate binding protein, *J. Mol. Biol.* **100:**157–178.

Siu, C.-H., Lerner, R. A., and Loomis, W. F., 1977, Rapid accumulation and disappearance of plasma membrane proteins during development of wild type and mutant strains of *Dictyostelium discoideum, J. Mol. Biol.* **116:**469–488.

Smart, J. E., and Hynes, R. O., 1974, Developmentally regulated cell surface alterations in *Dictyostelium discoideum*, *Nature (London)* **251:**319–321.

Smith, E., and Williams, K. L., 1979, Preparation of slime sheath from *Dictyostelium discoideum*, *FEMS Lett.* **6:**119–122.

Springer, W. R., and Barondes, S. H., 1980, Cell adhesion molecules: Detection with univalent seconu antibody, *J. Cell Biol.* **87:**703–707.

Springer, W. R., and Barondes, S. H., 1982, Evidence for another cell-adhesion molecule in *Dictyostelium discoideum*, *Proc. Natl. Acad. Sci. USA* **79:**6561–6565.

Stadler, J., Bordier, C., Lottspeich, F., Henschen, A., and Gerisch, G., 1982, Improved purification and N-terminal amino acid sequence determination of the contact site A glycoprotein of *Dictyostelium discoideum*, *Hoppe-Seyler's Z. Physiol. Chem.* **363:**771–776.

Steinemann, C., and Parish, R. W., 1980, Evidence that a developmentally regulated glycoprotein is the target of adhesion blocking Fab in reaggregation of *Dictyostelium discoideum*, *Nature (London)* **286:**621–623.

Sussman, M., 1982, Morphogenetic signalling, cytodifferentiation, and gene expression, in: *The Development of Dictyostelium discoideum* (W. F. Loomis, ed.), Academic Press, New York.

Sussman, M., and Osborn, M. J., 1964, UDP-galactose polysaccharide transferase in the cellular slime mold, *Dictyostelium discoideum*: Appearance and disappearance of activity during cell differentiation, *Proc. Natl. Acad. Sci. USA* **52:**81–87.

Takeuchi, I., 1963, Immunochemical and immunohistochemical studies on the development of the cellular slime mold *Dictyostelium mucoroides*, *Dev. Biol.* **8:**1–26.

Takeuchi, I., 1972, Differentiation and dedifferentiation in cellular slime molds, in: *Aspects of Cellular and Molecular Physiology* (K. Hamaguchi, ed.), pp. 217–236, University of Tokyo Press, Tokyo.

Tarentino, A. L., and Maley, F., 1975, A comparison of the substrate specificities of endo-β-N-acetyl-glucosaminidases from *Streptomyces griseus and Diplococcus pneumoniae*, *Biochem. Biophys. Res. Commun.* **67:**455–461.

Tasaka, M., and Takeuchi, I., 1979, Sorting out behavior of disaggregated cells in the absence of morphogenesis in *Dictyostelium discoideum*, *J. Embryol. Exp. Morphol.* **49:**89–102.

Tasaka, M., and Taekuchi, I., 1981, Role of cell sorting in pattern formation in *Dictyostelium discoideum*, *Differentiation* **18:**191–196.

Telser, A., and Sussman, M., 1971, Uridine-diphosphate-galactose-4-epimerase, a developmentally regulated enzyme in the cellular slime mold *Dictyostelium discoideum*, *J. Biol. Chem.* **246:**2252–2257.

Thilo, L., and Vogel, G., 1980, Kinetics of membrane internalization and recycling during pinocytosis in *Dictyostelium discoideum*, *Proc. Natl. Acad. Sci. USA* **77:**1015–1019.

Toda, K., Ono, K.-I., and Ochiai, H., 1980, Surface labeling of membrane glycoproteins and their drastic changes during development of *Dictyostelium discoideum*, *Eur. J. Biochem.* **111:**377–388.

Toda, K., Ono, K.-I., and Ochiai, H., 1981, Developmentally regulated glycoprotein alterations in *Dictyostelium discoideum*, *J. Biochem. Tokyo* **90:**1429.

Town, C., and Stanford, E., 1979, An oligosaccharide-containing factor that induces cell differentiation in *Dictyostelium discoideum*, *Proc. Natl. Acad. Sci. USA* **76:**308–312.

Town, C. D., Gross, J. D., and Kay, R. R., 1976, Cell differentiation without morphogenesis in *Dictyostelium discoideum*, *Nature (London)* **262:**717–719.

Tschursin, E., and Henderson, E. J., 1983, Rise in fucosylation of glycoprotein-linked oligosaccharides of bacterially-grown vegetative stage *Dictyostelium discoideum*, *J. Cell Biol.* **97:**74a.

Vogel, G., 1981, Recognition mechanisms in phagocytosis in *Dictyostelium discoideum*, *Monogr. Allergy* **17:**1–11.

Vogel, G., Thilo, L., Schwarz, H., and Steinhart, R., 1980, Mechanism of phagocytosis in *Dictyostelium discoideum*: Phagocytosis is mediated by different recognition sites as disclosed by mutants with altered phagocytotic properties, *J. Cell Biol.* **86:**456–465.

Weeks, G., 1973, Agglutination of growing and differentiating cells of *Dictyostelium discoideum* by concanavalin A, *Exp. Cell Res.* **76:**467–470.

Weeks, C., and Weeks, G., 1975, Cell surface changes during the differentiation of *Dictyostelium discoideum, Exp. Cell Res.* **92:**372–382.

West, C. M., and McMahon, D., 1977, Identification of concanavalin A receptors and galactose binding proteins in purified plasma membranes of *Dictyostelium discoideum, J. Cell Biol.* **74:**264–273.

West, C. M., and McMahon, D., 1979, The axial distribution of plasma membrane molecules in pseudoplasmodia of the cellular slime mold *Dictyostelium discoideum, Exp. Cell Res.* **124:**393–401.

West, C. M., and McMahon, D., 1981, The involvement of a class of cell surface glycoconjugates in pseudoplasmodial morphogenesis in *Dictyostelium discoideum, Differentiation* **20:**61–64.

West, C. M., McMahon, D., and Molday, R. S., 1978, Identification of glycoproteins, using lectins as probes, in plasma membranes from *Dictyostelium discoideum* and human erythrocytes, *J. Biol. Chem.* **253:**1716–1724.

White, G. J., and Sussman, M., 1963a, Polysaccharides involved in slime-mold development. I. Water-soluble polymer(s), *Biochim. Biophys. Acta* **74:**173–178.

White, G. J., and Sussman, M., 1963b, Polysaccharides involved in slime-mold development. II. Water-soluble acid mucopolysaccharide(s), *Biochim. Biophys. Acta* **74:**179–187.

Wilcox, D. K., and Sussman, M., 1981a, Defective cell cohesivity expressed late in the development of a *Dictyostelium discoideum* mutant, *Dev. Biol.* **82:**102–112.

Wilcox, D. K., and Sussman, M., 1981b, Serologically distinguishable alterations in the molecular specificity of cell cohesion during morphogenesis in a *Dictyostelium discoideum* mutant, *Proc. Natl. Acad. Sci. USA* **78:**358–362.

Wilhelms, O.-H., Luderitz, O., Westphal, O., and Gerisch, G., 1974, Glycosphingolipids and glycoproteins in the wild-type and in a nonaggregating mutant of *Dictyostelium discoideum, Eur. J. Biochem.* **48:**89–101.

Williams, G. B., and Sussman, M., 1982, The action of NH_3 on the relay system in *Dictyostelium discoideum*, International Cellular Slime Mold Conference, Hartford, Conn.

Williams, K. L., Kessin, R. H., and Newell, P. C., 1974, Parasexual genetics in *Dictyostelium discoideum*: Mitotic analysis of acriflavine resistance and growth in axenic medium, *J. Gen. Microbiol.* **84:**68–78.

Williams, K. L., Robson, G. E., and Welker, D. L., 1980, Chromosome fragments in *Dictyostelium discoideum* obtained from parasexual crosses between strains of different genetic background, *Genetics* **95:**289–304.

Yabuno, Y. Y., 1971, Changes in cellular adhesiveness during the development of the slime mould *Dictyostelium discoideum, Dev. Growth Differ.* **13:**181.

Index